The Ethics of the Environment

The International Research Library of Philosophy
Series Editor: John Skorupski

Metaphysics and Epistemology

Identity *Harold Noonan*
Personal Identity *Harold Noonan*
Scepticism *Michael Williams*
Infinity *A.W. Moore*

Theories of Truth *Paul Horwich*
Knowledge and Justification, Vols I & II *Ernest Sosa*
Space and Time *Jeremy Butterfield, Mark Hogarth, Gordon Belot*

Future Volumes: Substance and Causation; Necessity; The Existence and Knowability of God; Faith, Reason and Religious Language.

The Philosophy of Mathematics and Science

The Ontology of Science *John Worrall*
Mathematical Objects and Mathematical Knowledge *Michael D. Resnik*
Theory, Evidence and Explanation *Peter Lipton*

Future Volumes: Proof and its Limits; Probability; The Philosophy of the Life Sciences; The Philosophy of History and the Social Sciences; Rational Choice.

The Philosophy of Logic, Language and Mind

Understanding and Sense, Vols I and II *Christopher Peacocke*
Metaphysics of Mind *Peter Smith*
Relativity in Logic *Stewart Shapiro*
Events *Achille C. Varzi, Roberto Casati*

Future Volumes: Truth and Consequence; Modality, Quantification, High-order Logic; Reference and Logical Form; The Nature of Meaning; Functionalism; Interpretation; Intentionality and Representation; Reason, Action and Free Will.

The Philosophy of Value

Consequentialism *Philip Pettit*
Punishment *Antony Duff*
Meta-ethics *Michael Smith*

Duty and Virtue *Onora O'Neill*
The Ethics of the Environment *Andrew Brennan*
Medical Ethics *Robin Downie*

Future Volumes: Aesthetics; The Foundations of the State; Justice; Liberty and Community.

The Ethics of the Environment

Edited by

Andrew Brennan
The University of Western Australia

Dartmouth
Aldershot · Brookfield USA · Singapore · Sydney

© Andrew Brennan 1995. For copyright of individual articles please refer to the Acknowledgements.

All rights reserved. No part of this publication may be reproduced, stored in a retrieval system, or transmitted in any form or by any means, electronic, mechanical, photocopying, recording, or otherwise without the prior permission of Dartmouth Publishing Company Limited.

Published by
Dartmouth Publishing Company Limited
Gower House
Croft Road
Aldershot
Hants GU11 3HR
England

Dartmouth Publishing Company
Old Post Road
Brookfield
Vermont 05036
USA

British Library Cataloguing in Publication Data
Ethics of the Environment. –
(International Research Library of Philosophy)
 I. Brennan, Andrew II. Series
 179.1

Library of Congress Cataloging-in-Publication Data
The ethics of the environment / edited by Andrew Brennan.
  p. cm. — (International research library of philosophy)
  Includes bibliographical references (p.).
  ISBN 1-85521-348-6
  1. Environmental ethics. 2. Environmental sciences—Philosophy.
 I. Brennan, Andrew. II. Series.
 GE42.E845 1995
 179'.1—dc20   95-5686
                                                                                          CIP

ISBN 1 85521 348 6

Printed in Great Britain by Galliard (Printers) Ltd, Great Yarmouth

Contents

Acknowledgements	*ix*
Series Preface	*xiii*
Introduction	*xv*

PART I INTRINSIC VALUE AND MORAL STANDING

1   Paul W. Taylor (1981), 'The Ethics of Respect for Nature', *Environmental Ethics*, **3**, pp. 197–218.   3
2   Gerald H. Paske (1989), 'The Life Principle: A (Metaethical) Rejection', *Journal of Applied Philosophy*, **6**, pp. 219–25.   25
3   Andrew Brennan (1984), 'The Moral Standing of Natural Objects', *Environmental Ethics*, **6**, pp. 35–56.   33
4   John O'Neill (1992), 'The Varieties of Intrinsic Value', *The Monist*, **75**, pp. 119–37.   55

PART II SPECIES, ECOSYSTEMS AND INTERESTS

5   Holmes Rolston III (1985), 'Duties to Endangered Species', *BioScience*, **35**, pp. 718–26.   77
6   Holmes Rolston III (1992), 'Disvalues in Nature', *The Monist*, **75**, pp. 250–78.   87
7   Gary E. Varner (1990), 'Biological Functions and Biological Interests', *The Southern Journal of Philosophy*, **28**, pp. 251–70.   117
8   Harley Cahen (1988), 'Against the Moral Considerability of Ecosystems', *Environmental Ethics*, **10**, pp. 195–216.   137

PART III DEEP ECOLOGY AND RADICAL ENVIRONMENTALISM

9   Arne Naess (1986), 'The Deep Ecological Movement: Some Philosophical Aspects', *Philosophical Inquiry*, **8**, pp. 10–31.   161
10   Freya Mathews (1988), 'Conservation and Self-Realization: A Deep Ecology Perspective', *Environmental Ethics*, **10**, pp. 347–55.   183
11   Peter Reed (1989), 'Man Apart: An Alternative to the Self-Realization Approach', *Environmental Ethics*, **11**, pp. 53–69.   193
12   Arne Naess (1990), '"Man Apart" and Deep Ecology: A Reply to Reed', *Environmental Ethics*, **12**, pp. 185–92.   211

13	Thomas H. Birch (1993), 'Moral Considerability and Universal Consideration', *Environmental Ethics*, **15**, pp. 313–32.	219
14	Ramachandra Guha (1989), 'Radical American Environmentalism and Wilderness Preservation: A Third World Critique', *Environmental Ethics*, **11**, pp. 71–83.	239
15	David M. Johns (1990), 'The Relevance of Deep Ecology to the Third World: Some Preliminary Comments', *Environmental Ethics*, **12**, pp. 233–52.	253

PART IV ECOLOGY AND FEMINISM

16	Val Plumwood (1990), 'Women, Humanity and Nature', in Sean Sayers and Peter Osborne (eds), *Socialism, Feminism and Philosophy*, London: Routledge, pp. 211–34.	275
17	Karen J. Warren and Jim Cheney (1991), 'Ecological Feminism and Ecosystem Ecology', *Hypatia*, **6**, pp. 179–97.	299
18	Karen J. Warren (1990), 'The Power and the Promise of Ecological Feminism', *Environmental Ethics*, **12**, pp. 125–46.	319
19	Ariel Salleh (1992), 'The Ecofeminism/Deep Ecology Debate: A Reply to Patriarchal Reason', *Environmental Ethics*, **14**, pp. 195–216.	341

PART V ARE HUMANS PART OF NATURE OR SEPARATE FROM IT?

20	Holmes Rolston III (1979), 'Can and Ought We to Follow Nature?', *Environmental Ethics*, **1**, pp. 7–30.	365
21	Michael F. Smith (1991), 'Letting in the Jungle', *Journal of Applied Philosophy*, **8**, pp. 145–54.	389
22	Eric Katz (1992), 'The Big Lie: Human Restoration of Nature', *Research in Philosophy and Technology*, **12**, pp. 231–41.	399
23	Alastair S. Gunn (1991), 'The Restoration of Species and Natural Environments', *Environmental Ethics*, **13**, pp. 291–310.	411
24	J. Baird Callicott (1991), 'The Wilderness Idea Revisited: The Sustainable Development Alternative', *The Environmental Professional*, **13**, pp. 235–47.	431
25	Holmes Rolston III (1991), 'The Wilderness Idea Reaffirmed', *The Environmental Professional*, **13**, pp. 370–77.	445
26	J. Baird Callicott (1991), 'That Good Old-Time Wilderness Religion', *The Environmental Professional*, **13**, pp. 378–9.	453

PART VI POLICY, DILEMMAS AND PLURALISM

| 27 | Mark Sagoff (1988), 'Some Problems with Environmental Economics', *Environmental Ethics*, **10**, pp. 55–74. | 457 |

28	Kristin Shrader-Frechette (1991), 'Ethical Dilemmas and Radioactive Waste: A Survey of the Issues', *Environmental Ethics*, **13**, pp. 327–43.	477
29	Bryan Norton (1992), 'Sustainability, Human Welfare, and Ecosystem Health', *Environmental Values*, **1**, pp. 97–111.	495
30	Christopher D. Stone (1988), 'Moral Pluralism and the Course of Environmental Ethics', *Environmental Ethics*, **10**, pp. 139–54.	511
31	J. Baird Callicott (1990), 'The Case Against Moral Pluralism', *Environmental Ethics*, **12**, pp. 99–124.	527
32	Gary E. Varner (1991), 'No Holism Without Pluralism', *Environmental Ethics*, **13**, pp. 175–9.	553
33	Andrew Brennan (1992), 'Moral Pluralism and the Environment', *Environmental Values*, **1**, pp. 15–33.	559

Name Index   579

Acknowledgements

The editor and publishers wish to thank the following for permission to use copyright material.

BioScience for the essay: Holmes Rolston III (1985), 'Duties to Endangered Species', *BioScience*, **35**, pp. 718–26.

Thomas H. Birch (1993), 'Moral Considerability and Universal Consideration', *Environmental Ethics*, **15**, pp. 313–32. Copyright © Thomas H. Birch.

Blackwell Publishers for the essays: Gerald H. Paske (1989), 'The Life Principle: A (Metaethical) Rejection', *Journal of Applied Philosophy*, **6**, pp. 219–25; Michael F. Smith (1991), 'Letting in the Jungle', *Journal of Applied Philosophy*, **8**, pp. 145–54.

Andrew Brennan (1984), 'The Moral Standing of Natural Objects', *Environmental Ethics*, **6**, pp. 35–56. Copyright © Andrew Brennan.

Harley Cahen (1988), 'Against the Moral Considerability of Ecosystems', *Environmental Ethics*, **10**, pp. 195–216. Copyright © Harley Cahen.

J. Baird Callicott (1990), 'The Case Against Moral Pluralism', *Environmental Ethics*, **12**, pp. 99–124. Copyright © J. Baird Callicott.

Ramachandra Guha (1989), 'Radical American Environmentalism and Wilderness Preservation: A Third World Critique', *Environmental Ethics*, **11**, pp. 71–83. Copyright © Ramachandra Guha.

Alastair S. Gunn (1991), 'The Restoration of Species and Natural Environments', *Environmental Ethics*, **13**, pp. 291–310. Copyright © Alastair S. Gunn.

Jai Press Inc. for the essay: Eric Katz (1992), 'The Big Lie: Human Restoration of Nature', *Research in Philosophy and Technology*, **12**, pp. 231–41.

David M. Johns (1990), 'The Relevance of Deep Ecology to the Third World: Some Preliminary Comments', *Environmental Ethics*, **12**, pp. 233–52. Copyright © D.M. Johns.

Freya Mathews (1988), 'Conservation and Self-Realization: A Deep Ecology Perspective', *Environmental Ethics*, **10**, pp. 347–55. Copyright © Freya Mathews.

The Monist for the essays: John O'Neill (1992), 'The Varieties of Intrinsic Value', *The Monist*, **75**, pp. 119–37; Holmes Rolston III (1992), 'Disvalues in Nature', *The Monist*, **75**, pp. 250–78. Copyright © 1992, The Monist, La Salle, Illinois. Reprinted by permission.

Arne Naess (1990), '"Man Apart" and Deep Ecology: A Reply to Reed', *Environmental Ethics*, **12**, pp. 185–92. Copyright © Arne Naess.

National Association of Environmental Professionals for the essays: J. Baird Callicott (1991), 'The Wilderness Idea Revisited: The Sustainable Development Alternative', *The Environmental Professional*, **13**, pp. 235–47; Holmes Rolston III (1991), 'The Wilderness Idea Reaffirmed', *The Environmental Professional*, **13**, pp. 370–77; J. Baird Callicott (1991), 'That Good Old-Time Wilderness Religion', *The Environmental Professional*, **13**, pp. 378–9.

Holmes Rolston III (1979), 'Can and Ought We to Follow Nature?', *Environmental Ethics*, **1**, pp. 7–30. Copyright © Holmes Rolston III.

Routledge for the essay: Val Plumwood (1990), 'Women, Humanity and Nature', in Sean Sayers and Peter Osborne (eds), *Socialism, Feminism and Philosophy*, pp. 211–34.

Mark Sagoff (1988), 'Some Problems with Environmental Economics', *Environmental Ethics*, **10**, pp. 55–74. Copyright © Mark Sagoff.

Ariel Salleh (1992), 'The Ecofeminism/Deep Ecology Debate: A Reply to Patriarchal Reason', *Environmental Ethics*, **14**, pp. 195–216. Copyright © Ariel Salleh.

Kristin Shrader-Frechette (1991), 'Ethical Dilemmas and Radioactive Waste: A Survey of the Issues', *Environmental Ethics*, **13**, pp. 327–43. Copyright © Kristin Shrader-Frechette.

The Southern Journal of Philosophy for the essay: Gary E. Varner (1990), 'Biological Functions and Biological Interests', *The Southern Journal of Philosophy*, **28**, pp. 251–70.

Christopher D. Stone (1988), 'Moral Pluralism and the Course of Environmental Ethics', *Environmental Ethics*, **10**, pp. 139–54. Copyright © Christopher D. Stone.

Paul W. Taylor (1981), 'The Ethics of Respect for Nature', *Environmental Ethics*, **3**, pp. 197–218. Copyright © Paul W. Taylor.

Universite Aristote de Thessalonique for the essay: Arne Naess (1986), 'The Deep Ecological Movement: Some Philosophical Aspects', *Philosophical Inquiry*, **8**, pp. 10–31.

University of South Florida for the essay: Karen J. Warren and Jim Cheney (1991), 'Ecological Feminism and Ecosystem Ecology', *Hypatia*, **6**, pp. 179–97.

Gary E. Varner (1991), 'No Holism Without Pluralism', *Environmental Ethics*, **13**, pp. 175–9. Copyright © Gary Varner.

Karen J. Warren (1990), 'The Power and the Promise of Ecological Feminism', *Environmental Ethics*, **12**, pp. 125–46. Copyright © Karen J. Warren.

The White Horse Press for the essays: Bryan Norton (1992), 'Sustainability, Human Welfare, and Ecosystem Health', *Environmental Values*, **1**, pp. 97–111; Andrew Brennan (1992), 'Moral Pluralism and the Environment', *Environmental Values*, **1**, pp. 15–33.

Every effort has been made to trace all the copyright holders, but if any have been inadvertently overlooked the publishers will be pleased to make the necessary arrangements at the first opportunity.

Series Preface

The International Research Library of Philosophy collects in book form a wide range of important and influential essays in philosophy, drawn predominantly from English-language journals. Each volume in the Library deals with a field of inquiry which has received significant attention in philosophy in the last 25 years, and is edited by a philosopher noted in that field.

No particular philosophical method or approach is favoured or excluded. The Library will constitute a representative sampling of the best work in contemporary English-language philosophy, providing researchers and scholars throughout the world with comprehensive coverage of currently important topics and approaches.

The Library is divided into four series of volumes which reflect the broad divisions of contemporary philosophical inquiry:

- Metaphysics and Epistemology
- The Philosophy of Mathematics and Science
- The Philosophy of Logic, Language and Mind
- The Philosophy of Value

I am most grateful to all the volume editors, who have unstintingly contributed scarce time and effort to this project. The authority and usefulness of the series rests firmly on their hard work and scholarly judgement. I must also express my thanks to John Irwin of the Dartmouth Publishing Company, from whom the idea of the Library originally came, and who brought it to fruition; as also to his colleagues in the Editorial Office, whose care and attention to detail are so vital in ensuring that the Library provides a handsome and reliable aid to philosophical inquirers.

John Skorupski
General Editor
University of St. Andrews
Scotland

Introduction

The last two years have seen the publication of a flurry of anthologies on environmental ethics. Most have been targeted on an undergraduate audience. The present collection is different, providing academic researchers with a comprehensive collection of some of the most interesting recent work in environmental ethics. The focus is strictly on the ethics of the environment. Related but relatively self-contained ethical and political issues – such as duties to future generations, ecotheology and the moral status of non-human animals – have been excluded. In keeping with the policy of the series, papers have been omitted which did not appear in academic journals or which have been anthologized elsewhere.

In the following Introduction, I summarize a number of the central issues facing environmental ethics and mention individual contributions that are widely agreed to be central to the field. The references include some of the key English-language literature in the field since the publication of Aldo Leopold's *A Sand County Almanac* in 1949. Use of them along with the essays in the volume is intended to equip researchers in philosophy, the social sciences and environmental policy with the materials necessary for entering into the current debates in the subject.

Environmental Ethics and Environmental Philosophy

Environmental ethics developed in part as a response to calls from outside philosophy. The authors of these calls ranged from scientists to journalists, all alike concerned at the problems posed by pesticide use, fertilizer dependence, pollution, climate change, increasing human population and – more recently – loss of species and threats to the earth's remaining biodiversity. Often the calls consisted of demands for a new set of values (see Meadows 1972) or, more generally, a suggestion that it is time for humans to try travelling by different paths from those that lead to increasing resource consumption and the ultimate devastation of planetary life-support systems (Carson 1963, McKibben 1990).

By way of response there have been two general strategies available to moral philosophers. One – assimilation – suggests that existing moral categories and conceptions of value are adequate for describing our environmental concerns and theorizing about them. The problem for this strategy is to find means of assimilating new cases to existing concepts. One line of thought is that what appear to be new moral responsibilities facing us in the 20th century are really just old ones presented in a new form. For those who take this line, there are no new values to be found in nature; the number of things which count, morally speaking, are just the same as before. Concerns about the health of ecosystems or our duties towards rivers may have to be redescribed as concerns about human beings whose flourishing depends on ecosystems, or whose children will want to use the river that we are polluting today.

A more radical version of the same strategy tries to find ways of extending the concepts of 'good', 'right', 'duty', 'care' and 'value' to non-human beings. For some theorists, answering this question is a central project of environmental ethics. Hence Aldo Leopold, himself a forester and one of the significant influences on modern American environmentalism, wrote: 'The land ethic simply enlarges the boundaries of the community to include soils, waters, plants and animals, or collectively: the land' (Leopold 1949, p.204).

The second strategy is one of challenge rather than assimilation. Instead of finding values in rocky deserts, montane forests and whole ecosystems akin to those identified in human lives, theorists have tried to challenge the values they perceive in conventional moral theory. Instead of accepting the conventional account of the connection between possessing rights and the possession of interests (as argued, for example, in Feinberg 1974), they may try to argue that natural things which lack interests can still be worthy of respect. If the connections between respect, value and interests are severed in this way, an account is required of what the nature of respect owed to natural things is. After all, if rivers have no interests, then how can it be that we are morally bound to treat them one way rather than another? Other projects for this strategy may involve drawing attention to the multiplicity of forms of value, or to the narrow depiction of human nature at the centre of many theories, or to the prevalence of *humanism* in much of our thinking. For this strategy, the problem is not so much how to extend existing moral categories to new cases, but to find ways of revising the categories themselves so as to provide a richer account of our moral situation.

In the present collection, some of the writers in Parts III and IV on *Deep Ecology and Radical Environmentalism* and on *Ecology and Feminism* can be seen as challenging what they take to be the conventional wisdom about our moral situation. Elsewhere, writers like Paul Taylor in Chapter 1 can be seen to be looking for means to underwrite the extension of existing moral categories to apply to new cases (see also Regan 1981). Several writers, however, seem to be doing a bit of each. This is to be expected as boundaries are moved, redefined and then moved again.

As environmental ethics enters its third decade of study by professional philosophers, there are signs that a wider understanding of environmental philosophy is becoming established. In the last 20 or so years, discussion has focused on the intrinsic value of natural things (if any), the question of whether humans are part of nature or apart from it, the role of interests, feeling and goal-directed behaviour in the assessment of what has moral standing and so on. These questions, however, cannot be fully addressed without plotting their connections with the other components of a philosophy or world-view. Despite the existence of some work in this area (see Routley 1983 and de Groot 1992), much remains to be done.

In one fairly popular sense of the term, a 'philosophy' is a general theory which explains or justifies actions, policies or positions. A company can have a sales philosophy, a political party an electoral philosophy and individuals can discuss their philosophies of life without reference to any academic work in philosophy. In this popular sense, there are many environmental philosophies. For present purposes, an environmental philosophy can be defined as a general theory linking humans, nature and values. More specifically, an environmental philosophy will have four components:

- a theory about what nature is, and what kinds of objects and processes it contains;
- a theory about human beings, providing some overall perspective on human life, the context in which it is lived, and the problems it faces;

- a theory of value and an account of the evaluation of human action with reference to the two points above;
- a theory of method, indicating by what standards the claims made within the overall theory are to be tested, confirmed or rejected.

Much of the last 20 years' work in environmental philosophy has not directly confronted the issues involved in producing an environmental philosophy. Rather, writers and theorists have developed fragmentary approaches to this larger task. For example, a great deal of attention has been focused on the question of values, sometimes without much explicit treatment of the other issues. This is not surprising, given our tendency to hold parts of the background constant while we tackle a particular problem. However, any properly articulated environmental philosophy should present a coherent picture linking all of the above components.

The description just given is broad enough to encompass a very wide range of theories. For example, important figures in the history of thought, such as Rousseau, Scheler and Heidegger, have said enough about all of the above matters to count as having environmental philosophies, even though little attention has been paid to them so far in the environmental ethics literature (but see Foltz 1984). Christianity, to take a different case, can also be thought of as an environmental philosophy. However, its account of nature and its processes is ambiguous, leaving it a matter of debate whether humans should be respectful stewards of God's creation or instead regard nature as a mere resource, to do with as they please. In a classic essay on the medieval roots of the environmental crisis, Lynn White claimed that Christianity has encouraged both the explosion of technology and an exploitative attitude to nature (White 1967). Subsequent writers have failed to reach consensus on this issue (see the essays in Hargrove 1985 and compare Passmore 1980).

Marxism, likewise, is an environmental philosophy, emphasizing the transformative capacities of human agents in the presence of nature, but has little to say in detail about natural objects, systems and processes. What little it does say presents nature as a resource for the transformative power of labour rather than as something of value in its own right. Finally, as well as other religious views, pantheism should not be ignored as an environmental philosophy in the sense here defined. A pantheistic metaphysics, as Michael Levine points out, may well be as good a metaphysical foundation for an environmental ethic and environmental philosophy as we are likely to get (see Levine 1994).

Environmental philosophies in this sense are important structures because they both influence, and depend on, other theories in politics, aesthetics and the sciences. In all these cases, it must be borne in mind that the direction of justification need not match the direction of responsibility. Marxism, Christianity and deep ecology each give a systematic account of what human beings are, where they are and how to think about our relations with each other and with the rest of nature. But there is arguably as much reason to believe that Christianity, for example, serves to rationalize and legitimate human technological intervention in nature as that it is itself responsible for such intervention. It is easy to credit a discourse and system of beliefs with power over our behaviour, since we regularly try to justify our actions by appeals to principles drawn from within such systems. The relation of such systems to powerful factions within society has been analysed by social ecologists such as the anarchist theorist, Murray Bookchin, and by radical environmentalists (see Bookchin 1980 and 1990 and the readings in List 1993). It is also important to consider the role of justifying explanations.

in the light of the very real possibility that human nature is prone to self-deception, mythologizing and prey to weakness of will (see Brennan 1992).

From Moral Extensionism to Ecological Sensibility

In a classic essay on forms of ecological consciousness, John Rodman suggests that there are four increasingly radical ways in which our moral concerns can be extended from an initial focus solely on traditional human concerns all the way out to an ultimate reconsideration of values (see Rodman 1977 and 1983). In terms of the distinction between assimilation and challenge, Rodman's spectrum of views includes several forms of assimilation and also embraces one view which is meant to challenge the conventional moral wisdom.

The first awakenings of ecological consciousness, Rodman suggests, involve a concern to *conserve* resources for the sake of our own and future generations of people. In the history of American conservation thought, Gifford Pinchot provides a good example of this form of environmental awareness (see the discussion by Norton 1991). Even at this level, there are difficult philosophical and moral problems to be faced. The moral status of future generations is not something on which everyone agrees. Indeed, it can be argued that no matter how big a mess is created by present social and economic policies, future generations should still be grateful to us for their very existence. For any choice of social policy involves a choice about *who* will live in the next generation. To wish that our forebears had followed quite different social policies may be to wish that we did not exist. Because the issue of future generations and our moral reponsibilities (if any) towards them are so large, and since they do not always involve environmental issues, the present collection has not focused specifically on this level. For preliminary reading in this area, see Partridge 1990.

The next view in Rodman's spectrum is associated more with wilderness *preservation* than with resource conservation. However, Rodman notes that not all ecologically rich places are aesthetically grand and wild, and some rather grand places are far from wild. Any wilderness ethic, he argues, is liable to collapse into a merely conservationist one, especially if it suggests that the values of wilderness are bound up with human spiritual, recreational and aesthetic needs. John Muir and Aldo Leopold are associated with an American tradition of valuing wilderness, found also in contemporary American writers such as Baird Callicott and Holmes Rolston. In recent work, the whole status of wilderness in American thinking has been subject to reassessment, as can be seen in the exchanges between Rolston and Callicott in Part V, Chapters 24–26, *Are Humans Part of Nature or Separate From It?*

A third form of ecological consciousness proposed by Rodman takes seriously the status of non-human animals and other living things as being of value in their own right and having an independent claim to fulfil their own purposes. This kind of moral extensionism does not cavil at granting intrinsic value to many living creatures, although writers espousing this line often take human beings as occupants of the top layer in the intrinsic value hierarchy. Paul Taylor's essay on 'The Ethics of Respect for Nature' (Chapter 1) is a good exemplar of the kind of view Rodman has in mind here. A problem for this kind of extensionism is its maintenance of a strongly anthropocentric value theory and its attempts to assimilate – implausibly, in the opinion of some people – what is valuable in nature to valued human features.

This third form of consciousness is ambiguous, lying between a form of extensionism on the one hand and the exploration of new values on the other. Because extensionism has been much studied, more will be said about it below. Rodman next proposed a fourth form of moral consciousness, what he called 'a new ecological sensibility'. Holmes Rolston, Baird Callicott and Val Plumwood may – at least in some of their writings – represent just such a new form of ecological sensibility, as do the deep ecologists. However, Rodman's attempts to define this view are not entirely satisfactory. Citing Aldo Leopold as a guide, he suggests that anything which has an end – an Aristotelian *telos* – is something that is objectively valuable, has (in many cases) a perspective on the world and is the subject of a life. Such centred lives are not only characteristic of individual living things, but also, he suggests, of communities and systems. Coming to see one's place in the world in terms of centred communities and a multiplicity of points of view (animal and vegetable) is part, Rodman argues, of a paradigm shift away from simple extensionism to a more radical, transformed ecological sensibility.

It is not clear, however, just how radical this suggestion is. Even if the natural environment has been 'officially' overlooked in much of the last 300 years of moral philosophy, this is not evidence that an ecological sensibility and corresponding values have not been widespread. What we face here is a difficulty that is bound to arise whenever new ideas are being canvassed. Indeed, the strategy of challenge – complete with the possibility of exploring new values, focusing the theory of value on new topics, redefining the distinction between subject and object, and so on – is not easily distinguished from putting a new spin on some very old doctrines. The position of writers who think of themselves as 'deep' ecologists, 'post-modernists' or 'anti-humanists' hovers ambiguously between extensionism on the one hand and the exploration of a new moral path on the other.

Further study shows that those who claim to be pioneers of new moral paths often take their directions from old ones. For example, deep ecologists such as Arne Naess make much of the ecological conception of the self, that is of the idea that we are knots in larger biospherical webs, and that what we are depends in part on our relations with other elements in the web (see, in addition to the selections in the present volume, Naess 1989, Fox 1990 and Mathews 1991). These arguments simply extend to our biological surroundings the claims of theorists who argue that our identity within a society is a product of social relations and for whom society itself can be thought of as a means for producing persons (as opposed to human beings). Moreover, Naess himself operates with decidedly traditional philosophical equipment. For example, in charting the range of 'ecosophies' he uses an entirely traditional fact/value distinction to draw attention to the role of norms and hypotheses in moral system-building.

In spite of this, once the idea of an ecological self is introduced, new problems do emerge. Just as my DNA reflects my life in a larger genetic system, my conceptual scheme reflects a particular social and historical set of relations. But now I – the subject – cannot be understood separately from a surrounding context of many layers – biological, economic, social and so on. Life (which includes the gaining of a personality and an identity) is a community business. Yet, however proper these points, the term 'I' nonetheless designates an individual, with its own unique projects, values and possibilities. Such an individual is atomic in the sense that it can be separated from other such individuals, no matter how much effect they have had in its development. Faced with pulls in different directions here, we seem to require some appropriately contextualized idea of 'self' versus 'others' even to make sense of the fact that others have an effect on our own self-development. These issues are very much to the fore in the selections on *Deep Ecology and Radical Environmentalism*.

It is an important truth that an individual can suffer if transplanted to new social and biological surroundings. But such a move need not be like death. Indeed, a change of scene can, at some times, be the best way to promote someone's flourishing. So when deep ecologists and ecofeminists object to standard moral theory's preoccupation with supposedly isolable individuals, subject only to the demands of rationality, what is at stake? The debate here may possibly involve matters of emphasis rather than distinct conceptions of what it is to be an individual person, animal or plant. This consideration is particularly important in understanding the central points at issue in disagreements between deep ecologists and ecofeminists. For the feminist, our embodiment is something that is either overlooked by mainstream Western philosophers or treated as a matter for regret or even shame. Overemphasis on rationality, abstract principle and on a whole bureaucratic style of conducting philosophical and ethical investigation can, it is argued, be traced to our Cartesian penchant for abstracting from bodily, social, historical and physical contingencies. For reflections on this theme, see Cheney 1989 in addition to the readings in the present collection (and contrast Smith 1993).

Finally, note that the ethical importance of recognizing our ecological situation is best explored in the context of a general environmental philosophy in the sense already defined. Suppose we take literally the dictum that we are what we eat. No doubt there is a great deal of biochemical truth in this remark. But however important that truth is, there would need to be considerable argument and ingenuity expended in showing that it had serious ethical implications for us, or in establishing that eating, in as much as it contributes to our being what we are, is an ethically problematic matter. The connections between what we might crudely call the ecological facts of our situation and the requirements of moral decency are best explored within a larger structure which explains the relationship of facts to values, and of our personal identity to our moral situation. By ignoring this dimension in his recent book on green political theory, Robert Goodin manages to force a gap between green theories of value and green theories of agency (see Goodin 1992 and also Dobson 1990). Ecofeminism and deep ecology are best thought of as attempts at delineating just such larger structures.

Moral Extensionism

Whether moral concern for planetary systems is, properly speaking, an extension of concerns focused initially on human beings and subsequently other animals is a controversial issue. It is tempting to think of four possible extensions. First, there is the case of 'higher' animals. They are so like us in physiology that they might be thought to merit much the same respect as that owed to humans in general. Yet we farm, hunt, eat and experiment on our animal cousins as if they were, morally, in a quite different camp from humans altogether. Under the influence of 'animal rights' and 'animal liberation' theorists, practices with respect to non-human animals are now subject to intense moral scrutiny. As soon as such animals figure on the moral agenda, a host of specific problems emerge relating to animal ethics. For example, in many places deer are a menace to forests unless their numbers are strictly controlled. But how should we undertake necessary culling of deer and other ungulates? Is it proper to make a sport of this, or should it be done by professionals paid to carry out the task for as long as it is made necessary by human environmental interference?

In trying to answer such questions it becomes clear that concerns about animal welfare are not just additional to, and independent of, questions about human welfare and human interests. What hunters desire to shoot, for example, does not always coincide with the general requirements of species conservation. Animal ethics theorists, such as Tom Regan and Stephen Clark, have explored issues about animals with sensitivity. As a result of their work, it looks as if studying the moral dimensions of our relationship to other animals – considered as subjects of lives and bearers of pleasure and pain – can add depth and complexity to our moral deliberations. In fact, the results of these reflections may force us to revise some previously held views about what is right and wrong for humans (see Clark 1984, Regan 1983 and Regan and Singer 1989).

A second extension of moral consideration leads to examining the case of 'lower' animals and plants. Think – for the sake of being specific – of the situation of plants. Although they seem to have no capacities for pain and pleasure, they are regarded by many life-respecting ethics and religions as subjects of a life, defending that life against environmental challenges. Even in societies where there seems to be little by way of a generalized respect for life, there is still often a deep, apparently moral, concern for some plants such as trees. It is not uncontroversial to suggest that these are cases of *moral* concern. Trees can be loved for their instrumental value to us, that is, for the fact that their existence is a necessary condition for the production or maintenance of other values such as recreation, aesthetic enjoyment, the work of the woodcarver and so forth. Theorists have made much of the distinction between instrumental and non-instrumental (often called 'intrinsic') value, sometimes arguing that the existence of intrinsic values in nature is required by any serious environmental ethic. The distinction between these two kinds of value has intersected interestingly with two general approaches to value-theory, one human-centred and the other not.

In terms of these two sets of distinctions, the strategies of assimilation and challenge can be fleshed out in more detail. Those who regard the environment as raising no new questions for ethics are, in the main, committed to a value theory which is anthropocentric and denies anything other than instrumental value to natural things. Theorists of this kind can certainly deplore environmental destruction as strongly as anyone else. Their reasons for doing so are, typically, human-centred. For example, they may regard the loss of forest and associated biodiversity as a loss for us and our children, because of lost opportunities for aesthetic and recreational experiences of value (to us). They may also regret the damage done to groups of people who have a special dependence on their environment (aboriginal peoples, for example) or the distress caused to conservationists, biologists and others with a special fondness for natural things.

Such a human-centred focus is sometimes dismissed as mere 'shallow' ecological thinking by more radical theorists. This latter group includes those who are opposed to the arrogance of human-centred ethics, and who are willing to consider values as antecedently 'there' in nature prior to human encounters with them. Between the extremes are a variety of positions of varying stability. Some thinkers, for example, put restrictions on what can have value in itself, such that some (but not all) animals belong to this category. Interesting questions about value arise for artefacts as well as natural objects. Works of art are specially puzzling here (see Chapter 3). In a famous paper, 'Faking Nature', Robert Elliot posed the problem of whether a restored environment has any greater value, *qua* environment, than a forged work of art has *qua* work of art. Elliot's article is anthologized elsewhere, but two essays inspired by it are reproduced in Part V, *Are Humans Part of Nature or Separate From It?*

Once plants are given a place on the moral agenda, then the third extension, to whole species, seems less problematic. The definition of biological species is controversial; for some, such as David Hull, they are best thought of as scattered, long-lived individuals (Hull 1976). Given Hull's view, it is hard to resist the idea that if a Californian redwood or West Australian karri can make a moral claim on us, then so too can an endangered species. An interesting twist emerges in the work of those who, like Rolston, argue that, other things being equal, species should be valued above individuals. If they are right, then species should not be thought of as coming *after* individuals in considering the possibilities for extending moral consideration. For Rolston, one consideration in support of this view is that the species represents a staging point in the evolving history of a genetic lineage: once a species is extinguished, then a whole set of developmental possibilities is lost along with it.

The fourth extension, to ecosystems themselves, is at once challenging and exciting. If ecosystems are to have a moral claim on us, then – in terms of the way contemporary ethical theory works – a standard problem will be: what grounds this latest extension? It should be noted that *if* ecosystems are allowed to figure on the moral agenda, then this will involve conflicts of striking kinds. First, there will be problems about systems versus people. Second, suppose that we do have duties or at least some form of moral responsibility towards other animals and plants. Then our responsibilities towards ecosystems will not just be *in addition to* these other responsibilities but, in some cases, will be in conflict with them.

Over the last 20 years a significant schism has arisen between those whose ethical focus has been primarily environmental and those whose focus of concern has been the welfare and flourishing of non-human animals. While 'animal liberation' theorists are usually vegetarian, opponents of factory farming, animal experimentation and hunting, environmental ethics theorists are often meat eaters, more concerned with the fate of pig manure than the conditions of the animals which produce it and willing to countenance both subsistence and some sport hunting. While environmental ethicists are fascinated by biological relations such as parasitism and predation, extreme animal welfarists sometimes dream of a world free from parasites and in which hungry lions are fed soya substitute while their natural prey graze in peace.

There is a considerable body of literature on animal welfare, the possibility of animal rights and the tensions between these ideas and those of environmental ethics (see especially Callicott 1980, reprinted in Scherer and Attig 1983 and in Zimmerman *et al.* 1993). Given constraints on space, it has proved necessary to leave most of this material out of the present collection. For readers interested in this area, a good starting point is the essays in VanDeVeer and Pierce 1986, as well as material by Regan, Singer and Clark already mentioned. Note, however, that a significant number of environmental ethics theorists do try to accommodate non-human animals along with other natural objects in their theory of value. This is noticeably true of the work of Holmes Rolston. For him, human beings are top products of a rich evolutionary process. Yet we should not value only the fruit, he argues, while ignoring the rest of the plant: individual animals, species and ecosystems all have their own kind of intrinsic value and merit appropriate respect.

Policy, Politics and Radical Environmentalism

Policy debates often reveal the legacy of past philosophies. For example, John Locke argued that humans could claim as property an amount of land with which they mix their labour. A condition of his theory, however, was that it applies where there is more, equally good property left for others. Nowadays, the theory is still taken seriously although the world's supply of potential property has largely been appropriated.

A different account of property is given by Rousseau, whose work is presently being rediscovered by writers in environmental ethics. According to him, property is the symptom of the transition from a simple, authentic form of social organization to one in which power, corruption, domination and exploitation characterize human relationships. His last work, *Reveries of the Solitary Walker*, is an underused source of insights for environmental philosophers. Whether we believe Locke or Rousseau, the issue is not just a question of what property is. Rather, each of them tells us something about the meaning of property and about the justification of actions in connection with it.

A philosophy, in the sense introduced earlier, provides an account of practices, institutions or activities in terms which offer explanations, meanings and justifications for them. It may do this by means of a narrative which purports to depict our situation. The modern debates on pluralism in ethics are concerned with the scope and range of such narrative exercises. Ethical *monists* look to single structures and ordered sets of principles by which to guide our lives. *Pluralists*, by contrast, depict our situation as too complex to be captured by any single set of principles. Monists can be seen as trying to provide a single master narrative within which to place and locate everything that is of moral significance in our lives. Pluralists, by contrast, think of ethics as constituted by a series of narratives, each with its own capacity for illuminating some aspect of our moral lives. Monists normally object to pluralism in that it permits too much by way of *ad hoc* justification for specific ethical decisions. Pluralists, in turn, object to monists' tendency to oversimplify complex situations and to their demand that all values be strictly comparable.

These debates about the foundations of ethics have immediate relevance to matters of public policy. Whereas monistic theorists have a tendency to regard policy as a matter of reaching the *correct* answer and to focus on methods for pricing (morally, aesthetically or economically) the social costs and benefits of environmental decisions, pluralists focus instead on the *process* of reaching policy decisions. From a pluralistic perspective it is important that the different interests and values involved in a situation are identified and given due recognition in the process of reaching a decision on a matter of policy. Process theorists also emphasize the importance of regular evaluation of policy outcomes so that adjustments can be made in the light of further experience.

There are several monistic environmental philosophies. For example, *economism* puts forward a conception of the world and an account of human nature which seem to appeal to many policy-makers. According to it, we are agents of *consumption*, motivated by consumer *preference* living in a world of *resources*. Economism has a clear account of the method of testing its claims and of ascertaining the right answers. Questionnaires, studies of human consumer behaviour and the use of mathematical models are all integrated into an approach which provides straightforward answers to such questions as, how much is this stand of timber worth? What are the gains and losses associated with draining this wetland and establishing an industrial estate on it? Critiques of economism can be found in a number of sources, especially in Sagoff 1988, as well as in some of the selections in Part VI, especially Chapters 27 and 33.

Those who wish to reject economism – perhaps because it reveals all too clearly that those who know the price of everything may at the same time know the value of nothing – may be attracted to other philosophies such as deep ecology and ecofeminism. Since these latter views develop environmental philosophies in the sense defined here, they stand as counter-challenges to economism as a way of making sense of our predicament and finding a route out of it. Ecofeminism is usually pluralistic (in the sense defined above) both in method and values. A radical response which may be prompted by feminism, bioregionalism or deep ecology is to engage in direct action to protect remnants of wilderness or even basic resources. The Chipko tree-huggers and the Edward Abbey-inspired monkey-wrenching movement represent two very different forms that such direct action can take. Deep ecology as it has developed since 1973 has drawn heavily on the theories of Spinoza and Gandhi. The latter's commitment to non-violent action could have led to a reluctance to engage in more than passive resistance. However, groups like *Sea Shepherd* and *Earth First!* have attracted some supporters with a liking for militaristic forms of organization and confrontational action. Nonetheless, the 'official' line of both organizations is to avoid violence against persons; activities such as tree-spiking, staking roads and crippling machinery are presented as non-violent forms of ecotage. For an introduction to the moral considerations in this area, see List 1993.

The future of radical environmentalism and green politics is unclear. What is strikingly clear, however, is that environmental problems are a constant companion of human development (see Ponting 1991 for a concise, if one-sided, overview). The essays in the present collection represent a variety of attempts to help us think more clearly about the present and to explore ways of grounding future policy and individual responses. This brief Introduction has tried to provide for the researcher new to the area a rationale for the selection that follows, as well as to give an indication of the intellectual and philosophical problems that face us at present.

References

Abbey, Edward (1976), *The Monkey Wrench Gang*, Avon Books.
Armstrong, Susan J. and Richard G. Botzler (eds) (1993), *Environmental Ethics: Convergence and Divergence*, New York: McGraw-Hill.
Bookchin, Murray (1980), *Toward an Ecological Society*, Montreal: Black Rose Books.
Bookchin, Murray (1990), *The Philosophy of Social Ecology*, Toronto: Black Rose Books.
Brennan, A. (1988), *Thinking about Nature*, Routledge.
Brennan, A. (1992), 'Environmental Decision-Making', in R.J. Berry (ed.), *Environmental Dilemmas: Ethics and Decisions*, Chapman Hall.
Callicott, J. Baird (1980), 'Animal Liberation: A Triangular Affair', *Environmental Ethics*, **2**.
Carson, Rachel (1963), *Silent Spring*, Hamish Hamilton (originally published as essays in *The New Yorker*, 1962).
Cheney, J. (1989), 'Postmodern Environmental Ethics: Ethics as Bioregional Narrative', *Environmental Ethics*, **11**.
Clark, Stephen R.L. (1984), *The Moral Status of Animals*, Clarendon Press.
Cooper, David E. and Joy A. Palmer (1992), *The Environment in Question: Ethics and Global Issues*, London: Routledge.

de Groot, W. (1992), *Environmental Science Theory*, Elsevier.
Devall, B. and Sessions, G. (eds) (1985), *Deep Ecology*, Salt Lake City: Peregrine Smith Books.
Dobson, Andrew (1990), *Green Political Thought*, Harper Collins.
Elliot, Robert (1982), 'Faking Nature', *Inquiry*, **25**.
Elliot, Robert and Gare, Arran (eds) (1983), *Environmental Philosophy: A Collection of Readings*, Pennsylvania State University Press.
Feinberg, Joel (1974), 'The Rights of Animals and Unborn Generations', in William T. Blackstone (ed.), *Philosophy and Environmental Crisis*, University of Georgia Press.
Foltz, B. (1984), 'On Heidegger and the Interpretation of Environmental Crisis', *Environmental Ethics*, **6**.
Foreman, D. (1991), *Confessions of an Eco-Warrior*, New York: Harmony Books.
Fox, Warwick (1990), *Toward a Transpersonal Ecology*, Boston: Shambhala.
Goodin, R.E. (1992), *Green Political Theory*, Polity Press.
Goodpaster, K.E. (1978), 'On Being Morally Considerable', *Journal of Philosophy*, **75**.
Gruen, Lori and Dale Jamieson (eds) (1994), *Reflecting on Nature: Readings in Environmental Philosophy*, New York: Oxford University Press.
Hargrove, E. (ed.) (1985), *Religion and Environmental Crisis*, University of Georgia Press.
Hargrove, E. (1989), *Foundations of Environmental Ethics*, Prentice Hall.
Hull, D. (1976), 'Are Species Really Individuals?', *Systematic Zoology*, **25**.
Johnson, Lawrence E. (1991), *A Morally Deep World*, Cambridge University Press.
Leopold, Aldo (1949), *A Sand County Almanac*, Oxford University Press.
Levine, Michael (1994), 'Pantheism, Ethics and Ecology', *Environmental Values*, **3**.
List, Peter C. (1993), *Radical Environmentalism*, Wadsworth.
Mathews, F. (1991), *The Ecological Self*, Routledge.
McKibben, Bill (1990), *The End of Nature*, Viking.
Meadows, Donella H. et al. (1972), *The Limits to Growth*, Signet.
Naess, Arne (1973), 'The Shallow and the Deep, Long-Range Ecology Movement: A Summary', *Inquiry*, **16**.
Naess, Arne (1989), *Ecology, Community and Lifestyle*, trans. by D. Rothenberg, Cambridge University Press.
Norton, Bryan (1988), *Why Preserve Natural Variety?*, Princeton University Press.
Norton, Bryan (1991), *Toward Unity Among Environmentalists*, Oxford University Press.
Partridge, E. (1990), 'On the Rights of Future Generations', in D. Scherer (ed.), *Upstream/Downstream: Issues in Environmental Ethics*, Temple University Press.
Passmore, John (1980), *Man's Responsibility for Nature*, 2nd edn, Duckworth.
Ponting, Clive (1991), *A Green History of the World*, Penguin.
Regan, Tom (1981), 'The Nature and Possibility of an Environmental Ethic', *Environmental Ethics*, **3**.
Regan, Tom (1983), *The Case for Animal Rights*, University of California Press.
Regan, Tom (ed.) (1984), *Earthbound, New Introductory Essays in Environmental Ethics*, Random House. Reprinted in 1990 by Prospect Heights, IL: Waveland Press.
Regan, Tom and Singer, P. (eds) (1989), *Animal Rights and Human Obligations*, 2nd edn, Prentice Hall.
Rodman, John (1977), 'The Liberation of Nature?', *Inquiry*, **20**.
Rodman, John (1983), 'Four Forms of Ecological Consciousness Reconsidered', in Scherer and Attig (1983) below.
Rolston III, Holmes (1975), 'Is There an Ecological Ethic?', *Ethics*, **85**.
Rolston III, Holmes (1981), 'Values in Nature', *Environmental Ethics*, **3**.
Rolston III, Holmes (1988), *Environmental Ethics*, Temple University Press.
Rousseau, J.J. (1782) [1979], *Reveries of the Solitary Walker*, trans. by P. France, Penguin Books.
Routley (Sylvan), Richard (1983), 'Roles and Limits of Paradigms', in R. Elliot and A. Gare (eds), *Environmental Philosophy*, Pennsylvania State University Press.
Sagoff, Mark (1988), *The Economy of the Earth*, Cambridge University Press.
Scherer, D. and Attig, T. (eds) (1983), *Ethics and the Environment*, Prentice Hall.
Smith, M. (1993), 'Cheney and the Myth of Postmodernism', *Environmental Ethics*, **15**.
Taylor, Paul (1986), *Respect for Nature*, Princeton University Press.

VanDeVeer, D. and Pierce, C. (eds) (1986), *People, Penguins and Plastic Trees*, Wadsworth.
White, Lynn (1967), 'The Historical Roots of Our Ecological Crisis', *Science*, **55**; reprinted in I.G. Barbour (ed.), *Western Man and Environmental Ethics*, Addison Wesley, 1973.
Zimmerman, Michael E., Callicott, J. Baird, Sessions, George, Warren, Karen J. and Clark, John (eds) (1993), *Environmental Philosophy: From Animal Rights to Radical Ecology*, Englewood Cliffs, NJ: Prentice-Hall.

Part I
Intrinsic Value and Moral Standing

[1]

The Ethics of Respect for Nature

Paul W. Taylor*

> I present the foundational structure for a life-centered theory of environmental ethics. The structure consists of three interrelated components. First is the adopting of a certain ultimate moral attitude toward nature, which I call "respect for nature." Second is a belief system that constitutes a way of conceiving of the natural world and of our place in it. This belief system underlies and supports the attitude in a way that makes it an appropriate attitude to take toward the Earth's natural ecosystems and their life communities. Third is a system of moral rules and standards for guiding our treatment of those ecosystems and life communities, a set of normative principles which give concrete embodiment or expression to the attitude of respect for nature. The theory set forth and defended here is, I hold, structurally symmetrical with a theory of human ethics based on the principle of respect for persons.

I. HUMAN-CENTERED AND LIFE-CENTERED SYSTEMS OF ENVIRONMENTAL ETHICS

In this paper I show how the taking of a certain ultimate moral attitude toward nature, which I call "respect for nature," has a central place in the foundations of a life-centered system of environmental ethics. I hold that a set of moral norms (both standards of character and rules of conduct) governing human treatment of the natural world is a rationally grounded set if and only if, first, commitment to those norms is a practical entailment of adopting the attitude of respect for nature as an ultimate moral attitude, and second, the adopting of that attitude on the part of all rational agents can itself be justified. When the basic characteristics of the attitude of respect for nature are made clear, it will be seen that a life-centered system of environmental ethics need not be holistic or organicist in its conception of the kinds of entities that are deemed the appropriate objects of moral concern and consideration. Nor does such a system require that the concepts of ecological homeostasis, equilibrium, and integrity provide us with normative principles from which could be derived (with the addition of factual knowledge) our obligations with regard

---

* Department of Philosophy, Brooklyn College of the City University of New York, Bedford Avenue and H, Brooklyn, NY 11210. Taylor's special fields are ethics and theory of value. He is the author of *Normative Discourse* (Englewood Cliffs, N.J.: Prentice-Hall, 1961) and *Principles of Ethics: An Introduction* (Encino, Calif.: Dickenson Publishing Co., 1975), and has also edited two books of readings: *The Moral Judgment: Readings in Contemporary Meta-Ethics* (Englewood Cliffs, N.J.: Prentice-Hall, 1963) and *Problems of Moral Philosophy*, 3rd ed. (Encino, Calif.: Dickenson Publishing Co., 1971).

to natural ecosystems. The "balance of nature" is not itself a moral norm, however important may be the role it plays in our general outlook on the natural world that underlies the attitude of respect for nature. I argue that finally it is the good (well-being, welfare) of individual organisms, considered as entities having inherent worth, that determines our moral relations with the Earth's wild communities of life.

In designating the theory to be set forth as life-centered, I intend to contrast it with all anthropocentric views. According to the latter, human actions affecting the natural environment and its nonhuman inhabitants are right (or wrong) by either of two criteria: they have consequences which are favorable (or unfavorable) to human well-being, or they are consistent (or inconsistent) with the system of norms that protect and implement human rights. From this human-centered standpoint it is to humans and only to humans that all duties are ultimately owed. We may have responsibilities *with regard to* the natural ecosystems and biotic communities of our planet, but these responsibilities are in every case based on the contingent fact that our treatment of those ecosystems and communities of life can further the realization of human values and/or human rights. We have no obligation to promote or protect the good of nonhuman living things, independently of this contingent fact.

A life-centered system of environmental ethics is opposed to human-centered ones precisely on this point. From the perspective of a life-centered theory, we have prima facie moral obligations that are owed to wild plants and animals themselves as members of the Earth's biotic community. We are morally bound (other things being equal) to protect or promote their good for *their* sake. Our duties to respect the integrity of natural ecosystems, to preserve endangered species, and to avoid environmental pollution stem from the fact that these are ways in which we can help make it possible for wild species populations to achieve and maintain a healthy existence in a natural state. Such obligations are due those living things out of recognition of their inherent worth. They are entirely additional to and independent of the obligations we owe to our fellow humans. Although many of the actions that fulfill one set of obligations will also fulfill the other, two different grounds of obligation are involved. Their well-being, as well as human well-being, is something to be realized *as an end in itself.*

If we were to accept a life-centered theory of environmental ethics, a profound reordering of our moral universe would take place. We would begin to look at the whole of the Earth's biosphere in a new light. Our duties with respect to the "world" of nature would be seen as making prima facie claims upon us to be balanced against our duties with respect to the "world" of human civilization. We could no longer simply take the human point of view and consider the effects of our actions exclusively from the perspective of our own good.

II. THE GOOD OF A BEING AND THE CONCEPT OF INHERENT WORTH

What would justify acceptance of a life-centered system of ethical principles? In order to answer this it is first necessary to make clear the fundamental moral attitude that underlies and makes intelligible the commitment to live by such a system. It is then necessary to examine the considerations that would justify any rational agent's adopting that moral attitude.

Two concepts are essential to the taking of a moral attitude of the sort in question. A being which does not "have" these concepts, that is, which is unable to grasp their meaning and conditions of applicability, cannot be said to have the attitude as part of its moral outlook. These concepts are, first, that of the good (well-being, welfare) of a living thing, and second, the idea of an entity possessing inherent worth. I examine each concept in turn.

(1) Every organism, species population, and community of life has a good of its own which moral agents can intentionally further or damage by their actions. To say that an entity has a good of its own is simply to say that, without reference to any *other* entity, it can be benefited or harmed. One can act in its overall interest or contrary to its overall interest, and environmental conditions can be good for it (advantageous to it) or bad for it (disadvantageous to it). What is good for an entity is what "does it good" in the sense of enhancing or preserving its life and well-being. What is bad for an entity is something that is detrimental to its life and well-being.[1]

We can think of the good of an individual nonhuman organism as consisting in the full development of its biological powers. Its good is realized to the extent that it is strong and healthy. It possesses whatever capacities it needs for successfully coping with its environment and so preserving its existence throughout the various stages of the normal life cycle of its species. The good of a population or community of such individuals consists in the population or community maintaining itself from generation to generation as a coherent system of genetically and ecologically related organisms whose average good is at an optimum level for the given environment. (Here *average good* means that the degree of realization of the good of *individual organisms* in the population or community is, on average, greater than would be the case under any other ecologically functioning order of interrelations among those species populations in the given ecosystem.)

The idea of a being having a good of its own, as I understand it, does not entail that the being must have interests or take an interest in what affects its life for better or for worse. We can act in a being's interest or contrary to its

---

[1] The conceptual links between an entity *having* a good, something being good *for* it, and events doing good *to* it are examined by G. H. Von Wright in *The Varieties of Goodness* (New York: Humanities Press, 1963), chaps. 3 and 5.

interest without its being interested in what we are doing to it in the sense of wanting or not wanting us to do it. It may, indeed, be wholly unaware that favorable and unfavorable events are taking place in its life. I take it that trees, for example, have no knowledge or desires or feelings. Yet is is undoubtedly the case that trees can be harmed or benefited by our actions. We can crush their roots by running a bulldozer too close to them. We can see to it that they get adequate nourishment and moisture by fertilizing and watering the soil around them. Thus we can help or hinder them in the realization of their good. It is the good of trees themselves that is thereby affected. We can similarly act so as to further the good of an entire tree population of a certain species (say, all the redwood trees in a California valley) or the good of a whole community of plant life in a given wilderness area, just as we can do harm to such a population or community.

When construed in this way, the concept of a being's good is not coextensive with sentience or the capacity for feeling pain. William Frankena has argued for a general theory of environmental ethics in which the ground of a creature's being worthy of moral consideration is its sentience. I have offered some criticisms of this view elsewhere, but the full refutation of such a position, it seems to me, finally depends on the positive reasons for accepting a life-centered theory of the kind I am defending in this essay.[2]

It should be noted further that I am leaving open the question of whether machines—in particular, those which are not only goal-directed, but also self-regulating—can properly be said to have a good of their own.[3] Since I am concerned only with human treatment of wild organisms, species populations, and communities of life as they occur in our planet's natural ecosystems, it is to those entities alone that the concept "having a good of its own" will here be applied. I am not denying that other living things, whose genetic origin and environmental conditions have been produced, controlled, and manipulated by humans for human ends, do have a good of their own in the same sense as do wild plants and animals. It is not my purpose in this essay, however, to set out or defend the principles that should guide our conduct with regard to their good. It is only insofar as their production and use by humans have good or ill effects upon natural ecosystems and their wild inhabitants that the ethics of respect for nature comes into play.

(2) The second concept essential to the moral attitude of respect for nature is the idea of inherent worth. We take that attitude toward wild living things

---

[2] See W. K. Frankena, "Ethics and the Environment," in K.E. Goodpaster and K.M. Sayre, eds., *Ethics and Problems of the 21st Century* (Notre Dame, University of Notre Dame Press, 1979), pp. 3–20. I critically examine Frankena's views in "Frankena on Environmental Ethics," *Monist*, vol. 64, no. 3 (July, 1981), pp. 313-324.

[3] In the light of considerations set forth in Daniel Dennett's *Brainstorms: Philosophical Essays on Mind and Psychology* (Montgomery, Vermont: Bradford Books, 1978), it is advisable to leave this question unsettled at this time. When machines are developed that function in the way our brains do, we may well come to deem them proper subjects of moral consideration.

(individuals, species populations, or whole biotic communities) when and only when we regard them as entities possessing inherent worth. Indeed, it is only because they are conceived in this way that moral agents can think of themselves as having validly binding duties, obligations, and responsibilities that are *owed* to them as their *due*. I am not at this juncture arguing why they *should* be so regarded; I consider it at length below. But so regarding them is a presupposition of our taking the attitude of respect toward them and accordingly understanding ourselves as bearing certain moral relations to them. This can be shown as follows:

What does it mean to regard an entity that has a good of its own as possessing inherent worth? Two general principles are involved: the principle of moral consideration and the principle of intrinsic value.

According to the principle of moral consideration, wild living things are deserving of the concern and consideration of all moral agents simply in virtue of their being members of the Earth's community of life. From the moral point of view their good must be taken into account whenever it is affected for better or worse by the conduct of rational agents. This holds no matter what species the creature belongs to. The good of each is to be accorded some value and so acknowledged as having some weight in the deliberations of all rational agents. Of course, it may be necessary for such agents to act in ways contrary to the good of this or that particular organism or group of organisms in order to further the good of others, including the good of humans. But the principle of moral consideration prescribes that, with respect to each being an entity having its own good, every individual is deserving of consideration.

The principle of intrinsic value states that, regardless of what kind of entity it is in other respects, if it is a member of the Earth's community of life, the realization of its good is something *intrinsically* valuable. This means that its good is prima facie worthy of being preserved or promoted as an end in itself and for the sake of the entity whose good it is. Insofar as we regard any organism, species population, or life community as an entity having inherent worth, we believe that it must never be treated as if it were a mere object or thing whose entire value lies in being instrumental to the good of some other entity. The well-being of each is judged to have value in and of itself.

Combining these two principles, we can now define what it means for a living thing or group of living things to possess inherent worth. To say that it possesses inherent worth is to say that its good is deserving of the concern and consideration of all moral agents, and that the realization of its good has intrinsic value, to be pursued as an end in itself and for the sake of the entity whose good it is.

The duties owed to wild organisms, species populations, and communities of life in the Earth's natural ecosystems are grounded on their inherent worth. When rational, autonomous agents regard such entities as possessing inherent worth, they place intrinsic value on the realization of their good and so hold themselves responsible for performing actions that will have this effect and for refraining from actions having the contrary effect.

III. THE ATTITUDE OF RESPECT FOR NATURE

Why should moral agents regard wild living things in the natural world as possessing inherent worth? To answer this question we must first take into account the fact that, when rational, autonomous agents subscribe to the principles of moral consideration and intrinsic value and so conceive of wild living things as having that kind of worth, such agents are *adopting a certain ultimate moral attitude toward the natural world.* This is the attitude I call "respect for nature." It parallels the attitude of respect for persons in human ethics. When we adopt the attitude of respect for persons as the proper (fitting, appropriate) attitude to take toward all persons as persons, we consider the fulfillment of the basic interests of each individual to have intrinsic value. We thereby make a moral commitment to live a certain kind of life in relation to other persons. We place ourselves under the direction of a system of standards and rules that we consider validly binding on all moral agents as such.[4]

Similarly, when we adopt the attitude of respect for nature as an ultimate moral attitude we make a commitment to live by certain normative principles. These principles constitute the rules of conduct and standards of character that are to govern our treatment of the natural world. This is, first, an *ultimate* commitment because it is not derived from any higher norm. The attitude of respect for nature is not grounded on some other, more general, or more fundamental attitude. It sets the total framework for our responsibilities toward the natural world. It can be justified, as I show below, but its justification cannot consist in referring to a more general attitude or a more basic normative principle.

Second, the commitment is a *moral* one because it is understood to be a disinterested matter of principle. It is this feature that distinguishes the attitude of respect for nature from the set of feelings and dispositions that comprise the love of nature. The latter stems from one's personal interest in and response to the natural world. Like the affectionate feelings we have toward certain individual human beings, one's love of nature is nothing more than the particular way one feels about the natural environment and its wild inhabitants. And just as our love for an individual person differs from our respect for all persons as such (whether we happen to love them or not), so love of nature differs from respect for nature. Respect for nature is an attitude we

---

[4] I have analyzed the nature of this commitment of human ethics in "On Taking the Moral Point of View," *Midwest Studies in Philosophy*, vol. 3, *Studies in Ethical Theory* (1978), pp. 35–61.

believe all moral agents ought to have simply as moral agents, regardless of whether or not they also love nature. Indeed, we have not truly taken the attitude of respect for nature ourselves unless we believe this. To put it in a Kantian way, to adopt the attitude of respect for nature is to take a stance that one wills it to be a universal law for all rational beings. It is to hold that stance categorically, as being validly applicable to every moral agent without exception, irrespective of whatever personal feelings toward nature such an agent might have or might lack.

Although the attitude of respect for nature is in this sense a disinterested and universalizable attitude, anyone who does adopt it has certain steady, more or less permanent dispositions. These dispositions, which are themselves to be considered disinterested and universalizable, comprise three interlocking sets: dispositions to seek certain ends, dispositions to carry on one's practical reasoning and deliberation in a certain way, and dispositions to have certain feelings. We may accordingly analyze the attitude of respect for nature into the following components. (a) The disposition to aim at, and to take steps to bring about, as final and disinterested ends, the promoting and protecting of the good of organisms, species populations, and life communities in natural ecosystems. (These ends are "final" in not being pursued as means to further ends. They are "disinterested" in being independent of the self-interest of the agent.) (b) The disposition to consider actions that tend to realize those ends to be prima facie obligatory *because* they have that tendency. (c) The disposition to experience positive and negative feelings toward states of affairs in the world *because* they are favorable or unfavorable to the good of organisms, species populations, and life communities in natural ecosystems.

The logical connection between the attitude of respect for nature and the duties of a life-centered system of environmental ethics can now be made clear. Insofar as one sincerely takes that attitude and so has the three sets of dispositions, one will at the same time be disposed to comply with certain rules of duty (such as nonmaleficence and noninterference) and with standards of character (such as fairness and benevolence) that determine the obligations and virtues of moral agents with regard to the Earth's wild living things. We can say that the actions one performs and the character traits one develops in fulfilling these moral requirements are the way one *expresses* or *embodies* the attitude in one's conduct and character. In his famous essay, "Justice as Fairness," John Rawls describes the rules of the duties of human morality (such as fidelity, gratitude, honesty, and justice) as "forms of conduct in which recognition of others as persons is manifested."[5] I hold that the rules of duty governing our treatment of the natural world and its inhabitants are forms of conduct in which the attitude of respect for nature is manifested.

---

[5] John Rawls, "Justice As Fairness," *Philosophical Review* 67 (1958): 183.

IV. THE JUSTIFIABILITY OF THE ATTITUDE OF RESPECT FOR NATURE

I return to the question posed earlier, which has not yet been answered: why *should* moral agents regard wild living things as possessing inherent worth? I now argue that the only way we can answer this question is by showing how adopting the attitude of respect for nature is justified for all moral agents. Let us suppose that we were able to establish that there are good reasons for adopting the attitude, reasons which are intersubjectively valid for every rational agent. If there are such reasons, they would justify anyone's having the three sets of dispositions mentioned above as constituting what it means to have the attitude. Since these include the disposition to promote or protect the good of wild living things as a disinterested and ultimate end, as well as the disposition to perform actions for the reason that they tend to realize that end, we see that such dispositions commit a person to the principles of moral consideration and intrinsic value. To be disposed to further, as an end in itself, the good of any entity in nature just because it is that kind of entity, is to be disposed to give consideration to *every* such entity and to place intrinsic value on the realization of its good. Insofar as we subscribe to these two principles we regard living things as possessing inherent worth. Subscribing to the principles is what it *means* to so regard them. To justify the attitude of respect for nature, then, is to justify commitment to these principles and thereby to justify regarding wild creatures as possessing inherent worth.

We must keep in mind that inherent worth is not some mysterious sort of objective property belonging to living things that can be discovered by empirical observation or scientific investigation. To ascribe inherent worth to an entity is not to describe it by citing some feature discernible by sense perception or inferable by inductive reasoning. Nor is there a logically necessary connection between the concept of a being having a good of its own and the concept of inherent worth. We do not contradict ourselves by asserting that an entity that has a good of its own lacks inherent worth. In order to show that such an entity "has" inherent worth we must give good reasons for ascribing that kind of value to it (placing that kind of value upon it, conceiving of it to be valuable in that way). Although it is humans (persons, valuers) who must do the valuing, for the ethics of respect for nature, the value so ascribed is not a human value. That is to say, it is not a value derived from considerations regarding human well-being or human rights. It is a value that is ascribed to nonhuman animals and plants themselves, independently of their relationship to what humans judge to be conducive to their own good.

Whatever reasons, then, justify our taking the attitude of respect for nature as defined above are also reasons that show why we *should* regard the living things of the natural world as possessing inherent worth. We saw earlier that, since the attitude is an ultimate one, it cannot be derived from a more fundamental attitude nor shown to be a special case of a more general one. On what sort of grounds, then, can it be established?

The attitude we take toward living things in the natural world depends on the way we look at them, on what kind of beings we conceive them to be, and on how we understand the relations we bear to them. Underlying and supporting our attitude is a certain *belief system* that constitutes a particular world view or outlook on nature and the place of human life in it. To give good reasons for adopting the attitude of respect for nature, then, we must first articulate the belief system which underlies and supports that attitude. If it appears that the belief system is internally coherent and well-ordered, and if, as far as we can now tell, it is consistent with all known scientific truths relevant to our knowledge of the object of the attitude (which in this case includes the whole set of the Earth's natural ecosystems and their communities of life), then there remains the task of indicating why scientifically informed and rational thinkers with a developed capacity of reality awareness can find it acceptable as a way of conceiving of the natural world and our place in it. To the extent we can do this we provide at least a reasonable argument for accepting the belief system and the ultimate moral attitude it supports.

I do not hold that such a belief system can be *proven* to be true, either inductively or deductively. As we shall see, not all of its components can be stated in the form of empirically verifiable propositions. Nor is its internal order governed by purely logical relationships. But the system as a whole, I contend, constitutes a coherent, unified, and rationally acceptable "picture" or "map" of a total world. By examining each of its main components and seeing how they fit together, we obtain a scientifically informed and well-ordered conception of nature and the place of humans in it.

This belief system underlying the attitude of respect for nature I call (for want of a better name) "the biocentric outlook on nature." Since it is not wholly analyzable into empirically confirmable assertions, it should not be thought of as simply a compendium of the biological sciences concerning our planet's ecosystems. It might best be described as a philosophical world view, to distinguish it from a scientific theory or explanatory system. However, one of its major tenets is the great lesson we have learned from the science of ecology: the interdependence of all living things in an organically unified order whose balance and stability are necessary conditions for the realization of the good of its constituent biotic communities.

Before turning to an account of the main components of the biocentric outlook, it is convenient here to set forth the overall structure of my theory of environmental ethics as it has now emerged. The ethics of respect for nature is made up of three basic elements: a belief system, an ultimate moral attitude, and a set of rules of duty and standards of character. These elements are connected with each other in the following manner. The belief system provides a certain outlook on nature which supports and makes intelligible an autonomous agent's adopting, as an ultimate moral attitude, the attitude of respect for nature. It supports and makes intelligible the attitude in the sense that, when an autonomous agent understands its moral relations to the natural world in terms of this outlook, it recognizes the attitude of respect to be the only *suitable* or *fitting* attitude to take toward all wild forms of life in the Earth's biosphere. Living things are now viewed as *the appropriate objects of the attitude of respect* and are accordingly regarded as entities possessing inherent worth. One then places intrinsic value on the promotion and protection of their good. As a consequence of this, one makes a moral commitment to abide by a set of rules of duty and to fulfill (as far as one can by one's own efforts) certain standards of good character. Given one's adoption of the attitude of respect, one makes that moral commitment because one considers those rules and standards to be validly binding on all moral agents. They are seen as embodying forms of conduct and character structures in which the attitude of respect for nature is manifested.

This three-part complex which internally orders the ethics of respect for nature is symmetrical with a theory of human ethics grounded on respect for persons. Such a theory includes, first, a conception of oneself and others as persons, that is, as centers of autonomous choice. Second, there is the attitude of respect for persons as persons. When this is adopted as an ultimate moral attitude it involves the disposition to treat every person as having inherent worth or "human dignity." Every human being, just in virtue of her or his humanity, is understood to be worthy of moral consideration, and intrinsic value is placed on the autonomy and well-being of each. This is what Kant meant by conceiving of persons as ends in themselves. Third, there is an ethical system of duties which are acknowledged to be owed by everyone to everyone. These duties are forms of conduct in which public recognition is given to each individual's inherent worth as a person.

This structural framework for a theory of human ethics is meant to leave open the issue of consequentialism (utilitarianism) versus nonconsequentialism (deontology). That issue concerns the particular kind of system of rules defining the duties of moral agents toward persons. Similarly, I am leaving open in this paper the question of what particular kind of system of rules defines our duties with respect to the natural world.

V. THE BIOCENTRIC OUTLOOK ON NATURE

The biocentric outlook on nature has four main components. (1) Humans are thought of as members of the Earth's community of life, holding that membership on the same terms as apply to all the nonhuman members. (2) The Earth's natural ecosystems as a totality are seen as a complex web of interconnected elements, with the sound biological functioning of each being dependent on the sound biological functioning of the others. (This is the component referred to above as the great lesson that the science of ecology has taught us).

(3) Each individual organism is conceived of as a teleological center of life, pursuing its own good in its own way. (4) Whether we are concerned with standards of merit or with the concept of inherent worth, the claim that humans by their very nature are superior to other species is a groundless claim and, in the light of elements (1), (2), and (3) above, must be rejected as nothing more than an irrational bias in our own favor.

The conjunction of these four ideas constitutes the biocentric outlook on nature. In the remainder of this paper I give a brief account of the first three components, followed by a more detailed analysis of the fourth. I then conclude by indicating how this outlook provides a way of justifying the attitude of respect for nature.

VI. HUMANS AS MEMBERS OF THE EARTH'S COMMUNITY OF LIFE

We share with other species a common relationship to the Earth. In accepting the biocentric outlook we take the fact of our being an animal species to be a fundamental feature of our existence. We consider it an essential aspect of "the human condition." We do not deny the differences between ourselves and other species, but we keep in the forefront of our consciousness the fact that in relation to our planet's natural ecosystems we are but one species population among many. Thus we acknowledge our origin in the very same evolutionary process that gave rise to all other species and we recognize ourselves to be confronted with similar environmental challenges to those that confront them. The laws of genetics, of natural selection, and of adaptation apply equally to all of us as biological creatures. In this light we consider ourselves as one with them, not set apart from them. We, as well as they, must face certain basic conditions of existence that impose requirements on us for our survival and well-being. Each animal and plant is like us in having a good of its own. Although our human good (what is of true value in human life, including the exercise of individual autonomy in choosing our own particular value systems) is not like the good of a nonhuman animal or plant, it can no more be realized than their good can without the biological necessities for survival and physical health.

When we look at ourselves from the evolutionary point of view, we see that not only are we very recent arrivals on Earth, but that our emergence as a new species on the planet was originally an event of no particular importance to the entire scheme of things. The Earth was teeming with life long before we appeared. Putting the point metaphorically, we are relative newcomers, entering a home that has been the residence of others for hundreds of millions of years, a home that must now be shared by all of us together.

The comparative brevity of human life on Earth may be vividly depicted by imagining the geological time scale in spatial terms. Suppose we start with algae, which have been around for at least 600 million years. (The earliest protozoa actually predated this by several *billion* years.) If the time that algae have been here were represented by the length of a football field (300 feet), then the period during which sharks have been swimming in the world's oceans and spiders have been spinning their webs would occupy three quarters of the length of the field; reptiles would show up at about the center of the field; mammals would cover the last third of the field; hominids (mammals of the family *Hominidae*) the last two feet; and the species *Homo sapiens* the last six inches.

Whether this newcomer is able to survive as long as other species remains to be seen. But there is surely something presumptuous about the way humans look down on the "lower" animals, especially those that have become extinct. We consider the dinosaurs, for example, to be biological failures, though they existed on our planet for 65 million years. One writer has made the point with beautiful simplicity:

> We sometimes speak of the dinosaurs as failures; there will be time enough for that judgment when we have lasted even for one tenth as long....[6]

The possibility of the extinction of the human species, a possibility which starkly confronts us in the contemporary world, makes us aware of another respect in which we should not consider ourselves privileged beings in relation to other species. This is the fact that the well-being of humans is dependent upon the ecological soundness and health of many plant and animal communities, while their soundness and health does not in the least depend upon human well-being. Indeed, from their standpoint the very existence of humans is quite unnecessary. Every last man, woman, and child could disappear from the face of the Earth without any significant detrimental consequence for the good of wild animals and plants. On the contrary, many of them would be greatly benefited. The destruction of their habitats by human "developments" would cease. The poisoning and polluting of their environment would come to an end. The Earth's land, air, and water would no longer be subject to the degradation they are now undergoing as the result of large-scale technology and uncontrolled population growth. Life communities in natural ecosystems would gradually return to their former healthy state. Tropical forests, for example, would again be able to make their full contribution to a life-sustaining atmosphere for the whole planet. The rivers, lakes, and oceans of the world would (perhaps) eventually become clean again. Spilled oil, plastic trash, and even radioactive waste might finally, after many centuries, cease doing their terrible work. Ecosystems would return to their proper balance, suffering only the disruptions of natural events such as volcanic eruptions and glaciation. From these the community of life could recover, as it has so often done in the past.

---

[6] Stephen R.L. Clark, *The Moral Status of Animals* (Oxford: Clarendon Press, 1977), p. 112.

But the ecological disasters now perpetrated on it by humans—disasters from which it might never recover—these it would no longer have to endure.

If, then, the total, final, absolute extermination of our species (by our own hands?) should take place and if we should not carry all the others with us into oblivion, not only would the Earth's community of life continue to exist, but in all probability its well-being would be enhanced. Our presence, in short, is not needed. If we were to take the standpoint of the community and give voice to its true interest, the ending of our six-inch epoch would most likely be greeted with a hearty "Good riddance!"

VII. THE NATURAL WORLD AS AN ORGANIC SYSTEM

To accept the biocentric outlook and regard ourselves and our place in the world from its perspective is to see the whole natural order of the Earth's biosphere as a complex but unified web of interconnected organisms, objects, and events. The ecological relationships between any community of living things and their environment form an organic whole of functionally interdependent parts. Each ecosystem is a small universe itself in which the interactions of its various species populations comprise an intricately woven network of cause-effect relations. Such dynamic but at the same time relatively stable structures as food chains, predator-prey relations, and plant succession in a forest are self-regulating, energy-recycling mechanisms that preserve the equilibrium of the whole.

As far as the well-being of wild animals and plants is concerned, this ecological equilibrium must not be destroyed. The same holds true of the well-being of humans. When one views the realm of nature from the perspective of the biocentric outlook, one never forgets that in the long run the integrity of the entire biosphere of our planet is essential to the realization of the good of its constituent communities of life, both human and nonhuman.

Although the importance of this idea cannot be overemphasized, it is by now so familiar and so widely acknowledged that I shall not further elaborate on it here. However, I do wish to point out that this "holistic" view of the Earth's ecological systems does not itself constitute a moral norm. It is a factual aspect of biological reality, to be understood as a set of causal connections in ordinary empirical terms. Its significance for humans is the same as its significance for nonhumans, namely, in setting basic conditions for the realization of the good of living things. Its ethical implications for our treatment of the natural environment lie entirely in the fact that our *knowledge* of these causal connections is an essential *means* to fulfilling the aims we set for ourselves in adopting the attitude of respect for nature. In addition, its theoretical implications for the ethics of respect for nature lie in the fact that it (along with the other elements of the biocentric outlook) makes the adopting of that attitude a rational and intelligible thing to do.

VIII. INDIVIDUAL ORGANISMS AS TELEOLOGICAL CENTERS OF LIFE

As our knowledge of living things increases, as we come to a deeper understanding of their life cycles, their interactions with other organisms, and the manifold ways in which they adjust to the environment, we become more fully aware of how each of them is carrying out its biological functions according to the laws of its species-specific nature. But besides this, our increasing knowledge and understanding also develop in us a sharpened awareness of the uniqueness of each individual organism. Scientists who have made careful studies of particular plants and animals, whether in the field or in laboratories, have often acquired a knowledge of their subjects as identifiable individuals. Close observation over extended periods of time has led them to an appreciation of the unique "personalities" of their subjects. Sometimes a scientist may come to take a special interest in a particular animal or plant, all the while remaining strictly objective in the gathering and recording of data. Nonscientists may likewise experience this development of interest when, as amateur naturalists, they make accurate observations over sustained periods of close acquaintance with an individual organism. As one becomes more and more familiar with the organism and its behavior, one becomes fully sensitive to the particular way it is living out its life cycle. One may become fascinated by it and even experience some involvement with its good and bad fortunes (that is, with the occurrence of environmental conditions favorable or unfavorable to the realization of its good). The organism comes to mean something to one as a unique, irreplaceable individual. The final culmination of this process is the achievement of a genuine understanding of its point of view and, with that understanding, an ability to "take" that point of view. *Conceiving of it as a center of life, one is able to look at the world from its perspective.*

This development from objective knowledge to the recognition of individuality, and from the recognition of individuality to full awareness of an organism's standpoint, is a process of heightening our consciousness of what it means to be an individual living thing. We grasp the particularity of the organism as a teleological center of life, striving to preserve itself and to realize its own good in its own unique way.

It is to be noted that we need not be falsely anthropomorphizing when we conceive of individual plants and animals in this manner. Understanding them as teleological centers of life does not necessitate "reading into" them human characteristics. We need not, for example, consider them to have consciousness. Some of them may be aware of the world around them and others may not. Nor need we deny that different kinds and levels of awareness are exemplified when consciousness in some form is present. But conscious or not, all are equally teleological centers of life in the sense that each is a unified system of goal-oriented activities directed toward their preservation and well-being.

When considered from an ethical point of view, a teleological center of life is an entity whose "world" can be viewed from the perspective of *its* life. In looking at the world from that perspective we recognize objects and events occurring in its life as being beneficent, maleficent, or indifferent. The first are occurrences which increase its powers to preserve its existence and realize its good. The second decrease or destroy those powers. The third have neither of these effects on the entity. With regard to our human role as moral agents, we can conceive of a teleological center of life as a being whose standpoint we can take in making judgments about what events in the world are good or evil, desirable or undesirable. In making those judgments it is what promotes or protects the being's own good, not what benefits moral agents themselves, that sets the standard of evaluation. Such judgments can be made about anything that happens to the entity which is favorable or unfavorable in relation to its good. As was pointed out earlier, the entity itself need not have any (conscious) *interest* in what is happening to it for such judgments to be meaningful and true.

It is precisely judgments of this sort that we are disposed to make when we take the attitude of respect for nature. In adopting that attitude those judgments are given weight as reasons for action in our practical deliberation. They become morally relevant facts in the guidance of our conduct.

IX. THE DENIAL OF HUMAN SUPERIORITY

This fourth component of the biocentric outlook on nature is the single most important idea in establishing the justifiability of the attitude of respect for nature. Its central role is due to the special relationship it bears to the first three components of the outlook. This relationship will be brought out after the concept of human superiority is examined and analyzed.[7]

In what sense are humans alleged to be superior to other animals? We are different from them in having certain capacities that they lack. But why should these capacities be a mark of superiority? From what point of view are they judged to be signs of superiority and what sense of superiority is meant? After all, various nonhuman species have capacities that humans lack. There is the speed of a cheetah, the vision of an eagle, the agility of a monkey. Why should not these be taken as signs of *their* superiority over humans?

One answer that comes immediately to mind is that these capacities are not as *valuable* as the human capacities that are claimed to make us superior. Such uniquely human characteristics as rational thought, aesthetic creativity, auton-

---

[7] My criticisms of the dogma of human superiority gain independent support from a carefully reasoned essay by R. and V. Routley showing the many logical weaknesses in arguments for human-centered theories of environmental ethics. R. and V. Routley, "Against the Inevitability of Human Chauvinism," in K. E. Goodpaster and K. M. Sayre, eds., *Ethics and Problems of the 21st Century* (Notre Dame: University of Notre Dame Press, 1979), pp. 36–59.

omy and self-determination, and moral freedom, it might be held, have a higher value than the capacities found in other species. Yet we must ask: valuable to whom, and on what grounds?

The human characteristics mentioned are all valuable to humans. They are essential to the preservation and enrichment of our civilization and culture. Clearly it is from the human standpoint that they are being judged to be desirable and good. It is not difficult here to recognize a begging of the question. Humans are claiming human superiority from a strictly human point of view, that is, from a point of view in which the good of humans is taken as the standard of judgment. All we need to do is to look at the capacities of nonhuman animals (or plants, for that matter) from the standpoint of *their* good to find a contrary judgment of superiority. The speed of the cheetah, for example, is a sign of its superiority to humans when considered from the standpoint of the good of its species. If it were as slow a runner as a human, it would not be able to survive. And so for all the other abilities of nonhumans which further their good but which are lacking in humans. In each case the claim to human superiority would be rejected from a nonhuman standpoint.

When superiority assertions are interpreted in this way, they are based on judgments of *merit*. To judge the merits of a person or an organism one must apply grading or ranking standards to it. (As I show below, this distinguishes judgments of merit from judgments of inherent worth.) Empirical investigation then determines whether it has the "good-making properties" (merits) in virtue of which it fulfills the standards being applied. In the case of humans, merits may be either moral or nonmoral. We can judge one person to be better than (superior to) another from the moral point of view by applying certain standards to their character and conduct. Similarly, we can appeal to nonmoral criteria in judging someone to be an excellent piano player, a fair cook, a poor tennis player, and so on. Different social purposes and roles are implicit in the making of such judgments, providing the frame of reference for the choice of standards by which the nonmoral merits of people are determined. Ultimately such purposes and roles stem from a society's way of life as a whole. Now a society's way of life may be thought of as the cultural form given to the realization of human values. Whether moral or nonmoral standards are being applied, then, all judgments of people's merits finally depend on human values. All are made from an exclusively human standpoint.

The question that naturally arises at this juncture is: why should standards that are based on human values be assumed to be the only valid criteria of merit and hence the only true signs of superiority? This question is especially pressing when humans are being judged superior in merit to nonhumans. It is true that a human being may be a better mathematician than a monkey, but the monkey may be a better tree climber than a human being. If we humans value mathematics more than tree climbing, that is because our conception of civilized life makes the development of mathematical ability more desirable than the ability to climb trees. But is it not unreasonable to judge nonhumans by the values of human civilization, rather than by values connected with what it is for a member of *that* species to live a good life? If all living things have a good of their own, it at least makes sense to judge the merits of nonhumans by standards derived from *their* good. To use only standards based on human values is already to commit oneself to holding that humans are superior to nonhumans, which is the point in question.

A further logical flaw arises in connection with the widely held conviction that humans are *morally* superior beings because they possess, while others lack, the capacities of a moral agent (free will, accountability, deliberation, judgment, practical reason). This view rests on a conceptual confusion. As far as moral standards are concerned, only beings that have the capacities of a moral agent can properly be judged to be *either* moral (morally good) *or* immoral (morally deficient). Moral standards are simply not applicable to beings that lack such capacities. Animals and plants cannot therefore be said to be morally inferior in merit to humans. Since the only beings that can have moral merits *or be deficient in such merits* are moral agents, it is conceptually incoherent to judge humans as superior to nonhumans on the ground that humans have moral capacities while nonhumans don't.

Up to this point I have been interpreting the claim that humans are superior to other living things as a grading or ranking judgment regarding their comparative merits. There is, however, another way of understanding the idea of human superiority. According to this interpretation, humans are superior to nonhumans not as regards their merits but as regards their inherent worth. Thus the claim of human superiority is to be understood as asserting that all humans, simply in virtue of their humanity, have *a greater inherent worth* than other living things.

The inherent worth of an entity does not depend on its merits.[8] To consider something as possessing inherent worth, we have seen, is to place intrinsic value on the realization of its good. This is done regardless of whatever particular merits it might have or might lack, as judged by a set of grading or ranking standards. In human affairs, we are all familiar with the principle that one's worth as a person does not vary with one's merits or lack of merits. The same can hold true of animals and plants. To regard such entities as possessing inherent worth entails disregarding their merits and deficiencies, whether they are being judged from a human standpoint or from the standpoint of their own species.

The idea of one entity having more merit than another, and so being superior to it in merit, makes perfectly good sense. Merit is a grading or ranking concept, and judgments of comparative merit are based on the different degrees to which things satisfy a given standard. But what can it mean to talk about one thing being superior to another in inherent worth? In order to get at what

---

[8] For this way of distinguishing between merit and inherent worth, I am indebted to Gregory Vlastos, "Justice and Equality," in R. Brandt, ed., *Social Justice* (Englewood Cliffs, N. J.: Prentice-Hall, 1962), pp. 31–72.

is being asserted in such a claim it is helpful first to look at the social origin of the concept of degrees of inherent worth.

The idea that humans can possess different degrees of inherent worth originated in societies having rigid class structures. Before the rise of modern democracies with their egalitarian outlook, one's membership in a hereditary class determined one's social status. People in the upper classes were looked up to, while those in the lower classes were looked down upon. In such a society one's social superiors and social inferiors were clearly defined and easily recognized.

Two aspects of these class-structured societies are especially relevant to the idea of degrees of inherent worth. First, those born into the upper classes were deemed more worthy of respect than those born into the lower orders. Second, the superior worth of upper class people had nothing to do with their merits nor did the inferior worth of those in the lower classes rest on their lack of merits. One's superiority or inferiority entirely derived from a social position one was born into. The modern concept of a meritocracy simply did not apply. One could not advance into a higher class by any sort of moral or nonmoral achievement. Similarly, an aristocrat held his title and all the privileges that went with it just because he was the eldest son of a titled nobleman. Unlike the bestowing of knighthood in contemporary Great Britain, one did not earn membership in the nobility by meritorious conduct.

We who live in modern democracies no longer believe in such hereditary social distinctions. Indeed, we would wholeheartedly condemn them on moral grounds as being fundamentally unjust. We have come to think of class systems as a paradigm of social injustice, it being a central principle of the democratic way of life that among humans there are no superiors and no inferiors. Thus we have rejected the whole conceptual framework in which people are judged to have different degrees of inherent worth. That idea is incompatible with our notion of human equality based on the doctrine that all humans, simply in virtue of their humanity, have the same inherent worth. (The belief in universal human rights is one form that this egalitarianism takes.)

The vast majority of people in modern democracies, however, do not maintain an egalitarian outlook when it comes to comparing human beings with other living things. Most people consider our own species to be superior to all other species and this superiority is understood to be a matter of inherent worth, not merit. There may exist thoroughly vicious and depraved humans who lack all merit. Yet because they are human they are thought to belong to a higher class of entities than any plant or animal. That one is born into the species *Homo sapiens* entitles one to have lordship over those who are one's inferiors, namely, those born into other species. The parallel with hereditary social classes is very close. Implicit in this view is a hierarchical conception of nature according to which an organism has a position of superiority or inferiority in the Earth's community of life simply on the basis of its genetic background. The "lower" orders of life are looked down upon and it is considered perfectly proper that they serve the interests of those belonging to the highest order, namely humans. The intrinsic value we place on the well-being of our fellow humans reflects our recognition of their rightful position as our equals. No such intrinsic value is to be placed on the good of other animals, unless we choose to do so out of fondness or affection for them. But their well-being imposes no moral requirement on us. In this respect there is an absolute difference in moral status between ourselves and them.

This is the structure of concepts and beliefs that people are committed to insofar as they regard humans to be superior in inherent worth to all other species. I now wish to argue that this structure of concepts and beliefs is completely groundless. If we accept the first three components of the biocentric outlook and from that perspective look at the major philosophical traditions which have supported that structure, we find it to be at bottom nothing more than the expression of an irrational bias in our own favor. The philosophical traditions themselves rest on very questionable assumptions or else simply beg the question. I briefly consider three of the main traditions to substantiate the point. These are classical Greek humanism, Cartesian dualism, and the Judeo-Christian concept of the Great Chain of Being.

The inherent superiority of humans over other species was implicit in the Greek definition of man as a rational animal. Our animal nature was identified with "brute" desires that need the order and restraint of reason to rule them (just as reason is the special virtue of those who rule in the ideal state). Rationality was then seen to be the key to our superiority over animals. It enables us to live on a higher plane and endows us with a nobility and worth that other creatures lack. This familiar way of comparing humans with other species is deeply ingrained in our Western philosophical outlook. The point to consider here is that this view does not actually provide an argument *for* human superiority but rather makes explicit the framework of thought that is implicitly used by those who think of humans as inherently superior to nonhumans. The Greeks who held that humans, in virtue of their rational capacities, have a kind of worth greater than that of any nonrational being, never looked at rationality as but one capacity of living things among many others. But when we consider rationality from the standpoint of the first three elements of the ecological outlook, we see that its value lies in its importance for *human* life. Other creatures achieve their species-specific good without the need of rationality, although they often make use of capacities that humans lack. So the humanistic outlook of classical Greek thought does not give us a neutral (nonquestion-begging) ground on which to construct a scale of degrees of inherent worth possessed by different species of living things.

The second tradition, centering on the Cartesian dualism of soul and body, also fails to justify the claim to human superiority. That superiority is supposed to derive from the fact that we have souls while animals do not. Animals are mere automata and lack the divine element that makes us spiritual beings. I won't go into the now familiar criticisms of this two-substance view. I only add the point that, even if humans are composed of an immaterial, unextended soul and a material, extended body, this in itself is not a reason to deem them of greater worth than entities that are only bodies. Why is a soul substance a thing that adds value to its possessor? Unless some theological reasoning is offered here (which many, including myself, would find unacceptable on epistemological grounds), no logical connection is evident. An immaterial something which thinks is better than a material something which does not think only if thinking itself has value, either intrinsically or instrumentally. Now it is intrinsically valuable to humans alone, who value it as an end in itself, and it is instrumentally valuable to those who benefit from it, namely humans.

For animals that neither enjoy thinking for its own sake nor need it for living the kind of life for which they are best adapted, it has no value. Even if "thinking" is broadened to include all forms of consciousness, there are still many living things that can do without it and yet live what is for their species a good life. The anthropocentricity underlying the claim to human superiority runs throughout Cartesian dualism.

A third major source of the idea of human superiority is the Judeo-Christian concept of the Great Chain of Being. Humans are superior to animals and plants because their Creator has given them a higher place on the chain. It begins with God at the top, and then moves to the angels, who are lower than God but higher than humans, then to humans, positioned between the angels and the beasts (partaking of the nature of both), and then on down to the lower levels occupied by nonhuman animals, plants, and finally inanimate objects. Humans, being "made in God's image," are inherently superior to animals and plants by virtue of their being closer (in their essential nature) to God.

The metaphysical and epistemological difficulties with this conception of a hierarchy of entities are, in my mind, insuperable. Without entering into this matter here, I only point out that if we are unwilling to accept the metaphysics of traditional Judaism and Christianity, we are again left without good reasons for holding to the claim of inherent human superiority.

The foregoing considerations (and others like them) leave us with but one ground for the assertion that a human being, regardless of merit, is a higher kind of entity than any other living thing. This is the mere fact of the genetic makeup of the species *Homo sapiens*. But this is surely irrational and arbitrary. Why should the arrangement of genes of a certain type be a mark of superior value, especially when this fact about an organism is taken by itself, unrelated to any other aspect of its life? We might just as well refer to any other genetic makeup as a ground of superior value. Clearly we are confronted here with a wholly arbitrary claim that can only be explained as an irrational bias in our own favor.

That the claim is nothing more than a deep-seated prejudice is brought home to us when we look at our relation to other species in the light of the first three elements of the biocentric outlook. Those elements taken conjointly give us a certain overall view of the natural world and of the place of humans in it. When we take this view we come to understand other living things, their environmental conditions, and their ecological relationships in such a way as to awake in us a deep sense of our kinship with them as fellow members of the Earth's community of life. Humans and nonhumans alike are viewed together as integral parts of one unified whole in which all living things are functionally interrelated. Finally, when our awareness focuses on the individual lives of plants and animals, each is seen to share with us the characteristic of being a teleological center of life striving to realize its own good in its own unique way.

As this entire belief system becomes part of the conceptual framework through which we understand and perceive the world, we come to see ourselves as bearing a certain moral relation to nonhuman forms of life. Our ethical role in nature takes on a new significance. We begin to look at other species as we look at ourselves, seeing them as beings which have a good they are striving to realize just as we have a good we are striving to realize. We accordingly develop the disposition to view the world from the standpoint of their good as well as from the standpoint of our own good. Now if the groundlessness of the claim that humans are inherently superior to other species were brought clearly before our minds, we would not remain intellectually neutral toward that claim but would reject it as being fundamentally at variance with our total world outlook. In the absence of any good reasons for holding it, the assertion of human superiority would then appear simply as the expression of an irrational and self-serving prejudice that favors one particular species over several million others.

Rejecting the notion of human superiority entails its positive counterpart: the doctrine of species impartiality. One who accepts that doctrine regards all living things as possessing inherent worth—the *same* inherent worth, since no one species has been shown to be either "higher" or "lower" than any other. Now we saw earlier that, insofar as one thinks of a living thing as possessing inherent worth, one considers it to be the appropriate object of the attitude of respect and believes that attitude to be the only fitting or suitable one for all moral agents to take toward it.

Here, then, is the key to understanding how the attitude of respect is rooted in the biocentric outlook on nature. The basic connection is made through the denial of human superiority. Once we reject the claim that humans are superior either in merit or in worth to other living things, we are ready to adopt the attitude of respect. The denial of human superiority is itself the result of taking the perspective on nature built into the first three elements of the biocentric outlook.

Now the first three elements of the biocentric outlook, it seems clear, would be found acceptable to any rational and scientifically informed thinker who is fully "open" to the reality of the lives of nonhuman organisms. Without denying our distinctively human characteristics, such a thinker can acknowledge the fundamental respects in which we are members of the Earth's community of life and in which the biological conditions necessary for the realization of our human values are inextricably linked with the whole system of nature. In addition, the conception of individual living things as teleological centers of life simply articulates how a scientifically informed thinker comes to understand them as the result of increasingly careful and detailed observations. Thus, the biocentric outlook recommends itself as an acceptable system of concepts and beliefs to anyone who is clear-minded, unbiased, and factually enlightened, and who has a developed capacity of reality awareness with regard to the lives of individual organisms. This, I submit, is as good a reason for making the moral commitment involved in adopting the attitude of respect for nature as any theory of environmental ethics could possibly have.

X. MORAL RIGHTS AND THE MATTER OF COMPETING CLAIMS

I have not asserted anywhere in the foregoing account that animals or plants have moral rights. This omission was deliberate. I do not think that the reference class of the concept, bearer of moral rights, should be extended to include nonhuman living things. My reasons for taking this position, however, go beyond the scope of this paper. I believe I have been able to accomplish many of the same ends which those who ascribe rights to animals or plants wish to accomplish. There is no reason, moreover, why plants and animals, including whole species populations and life communities, cannot be accorded *legal* rights under my theory. To grant them legal protection could be interpreted as giving them legal entitlement to be protected, and this, in fact, would be a means by which a society that subscribed to the ethics of respect for nature could give public recognition to their inherent worth.

There remains the problem of competing claims, even when wild plants and animals are not thought of as bearers of moral rights. If we accept the biocentric outlook and accordingly adopt the attitude of respect for nature as our ultimate moral attitude, how do we resolve conflicts that arise from our respect for persons in the domain of human ethics and our respect for nature in the domain of environmental ethics? This is a question that cannot adequately be dealt with here. My main purpose in this paper has been to try to establish a base point from which we can start working toward a solution to the problem. I have shown why we cannot just begin with an initial presumption in favor of the interests of our own species. It is after all within our power as moral beings to place limits on human population and technology with the deliberate intention of sharing the Earth's bounty with other species. That such sharing is an ideal difficult to realize even in an approximate way does not take away its claim to our deepest moral commitment.

DISCUSSION ARTICLE

The Life Principle: a (metaethical) rejection

GERALD H. PASKE

ABSTRACT *In* Respect for Nature *Paul W. Taylor argues that there is a moral obligation to respect all living things. I argue that there is no such obligation. Taylor presents three basic premises for his position. The first two are shown to be mistaken but not necessary for Taylor's argument. The third, that being a nonsentient teleological centre of life confers moral significance, while necessary, fails to be rationally compelling. I argue: (1) The relevant concept of teleology as readily applies to inanimate objects as it does to nonsentient life forms. (2) The inanimate-nonsentient distinction (at the relevant molecular level) is founded upon a continuum which offers no basis sufficient to justify The Life Principle. (3) The concept of teleology, as used by Taylor, is too unclear and ill-founded to serve as the basis for a rationally compelling argument.*

Introduction

Animal rights and environmental ethics movements have called attention to the frequently foolish and perhaps inexcusable treatment of animals and the environment by human beings. Unfortunately, as is the case with many reform movements, the most vociferous adherents tend to carry the movement to extremes. Such extremism occurs within the animal rights movement when a right to life is attributed either to all mammals or to all animals, and within the environmental ethics movement when it is claimed that there is a moral duty to respect all life forms, sentient and nonsentient alike.

While I reject these extreme claims, I recognise that the best arguments for them are very attractive. I certainly see no reason to quarrel with one who personally chooses to become a vegetarian, or with one who personally chooses to respect all life. These 'ways of life' are attractive and even admirable. But, as I shall argue, they are not morally obligatory. If anything, they are quasi-religious rather than moral [1].

In this paper I shall deal only with the claim that *all* life merits respect [2]. I interpret this claim to mean that all life merits direct moral concern from moral agents [3]. I will call this latter assertion 'The Life Principle'. In discussing The Life Principle, I shall focus upon the work of Paul W. Taylor [4]. I will explicitly reject what Taylor calls the central tenet of the theory of environmental ethics: "that actions are right and character traits are morally good in virtue of their expressing or embodying a certain ultimate moral attitude, which I call respect for nature..." (p. 80). In contrast to this claim I shall argue that in some cases it is morally acceptable to prefer, for their own sake, inanimate objects over nonsentient life forms.

The arguments in favour of The Life Principle are difficult to summarise, in part because there is no canonical language in which the arguments can be stated. However, considerations in favour of The Life Principle tend to take the form of arguments purporting to show that any more restrictive criterion is arbitrary. Such arguments often take the form of claiming that morality is based upon some natural property and that all living things have that property. Roughly stated, proponents of The Life Principle tend to argue that, like us, all living beings have the capacity to be benefited or harmed in some sense and that, since we ought not be harmed, it is arbitrary to discount harm to any other living entity, even if that entity is nonsentient and the harm, therefore, painless. The arguments for The Life Principle tend to focus upon 'proving' that all life merits moral concern. The question of whether inanimate entities merit moral concern, though logically requiring an answer, tends to be treated in a somewhat cavalier manner.

My criticism of Taylor, and of The Life Principle itself, is that there is no non-arbitrary morally relevant way of distinguishing between nonsentient life and inanimate objects. Hence if nonsentient life merits direct moral concern then so many things would merit direct moral concern that everyday life would become morally overwhelming.

Taylor and the Argument from Perspective

Taylor's claim that all life forms deserve respect depends upon being able to non-arbitrarily separate nonsentient living entities from inanimate entities. In attempting to establish this Taylor offers the following three interrelated and interdependent premises:

(1) That humans can adopt the point of view of nonsentient life forms and that humans cannot adopt the point of view of inanimate entities.

(2) That we can "make sense of benefiting or harming" nonsentient life forms and that it is "intelligible to speak of acting benevolently toward" nonsentient life forms, but it does not make sense and is not intelligible to speak of acting similarly towards inanimate objects.

(3) That being a nonsentient *teleological* centre of life confers moral significance upon an entity, and that there is no analogous characteristic of inanimate objects.

I shall challenge Taylor's arguments for all three of these premises.

Premise One

Taylor's *modus operandi* depends upon the notion the 'standpoint of a living thing'. He asserts that "Perhaps the most ethically significant fact about moral subjects is that it is always possible for a moral agent to *take a moral subject's standpoint and make judgments from its standpoint about how it ought to be treated*" (p. 17) [5].

When doing this we discover that all living things, sentient or nonsentient, simple or complex, have a good of their own:

> ... organisms like trees and one-celled protozoa do not have a conscious life.... Yet they have a good of their own around which their behaviour is organized. All organisms, whether conscious or not, are teleological centers of life in the sense that each is a unified, coherently ordered system of goal-oriented activities that has a constant tendency to protect and maintain the organism's existence. (p. 122)

While it is possible to take the standpoint of any living organism and to understand that that living organism has a good of its own, "This mode of understanding a particular individual is not possible with regard to inanimate objects" (p. 123). This is because "... stones do not have points of view. In pure fantasy, of course, we can play at performing the imaginative act of taking a stone's standpoint and looking at the world from its perspective. But we are then moving away from reality, not getting closer to it" (p. 123).

The claim that we are moving away from reality is supported as follows:

> ... What makes our awareness of an individual stone fundamentally different from our awareness of a plant or animal is that the stone is not a teleological center of life, while the plant or animal is. The stone has no good of its own. We cannot benefit it by furthering its well-being or harm it by acting contrary to its well-being, since the concept of well-being simply does not apply to it. (p. 123)

The conclusion which we are to draw from this is that *all* living organisms are moral subjects and no inanimate objects are moral subjects.

I have no quarrel with the point that animate objects are teleological centres while inanimate objects are not. Nor do I doubt the importance of this distinction. But the distinction and its importance are independent of our being able to take the viewpoint of various objects. The relevance of being a teleological centre will be discussed below; at this point I merely want to discuss whether it is more difficult to take the point of view of an inanimate object than it is to take the point of view of a nonsentient living thing, say a protozoan.

One might argue that since protozoa have no viewpoints one cannot take their viewpoint, but I shall not so argue, at least not quite. However, even if one grants that humans can, *in some sense,* take the viewpoint of nonsentient entities, Taylor's conclusion does not follow. First, and somewhat trivially, the fact that rational agents can imagine themselves to be, or to be taking the perspective of, some nonsentient entity shows nothing about that entity. It only shows that human beings have quite good imaginations.

Secondly, Taylor's claim that in taking a stone's standpoint we are engaging in pure fantasy and performing an imaginative act, is equally true with respect to nonsentient living organisms. In such cases it is *imagination* which comes into play because, in any literal sense, nonsentient entities have no perspective at all. To 'take the perspective' of a nonsentient entity is to imagine what it would be like to be such an entity *if that entity were conscious.* If we do not imaginatively attribute consciousness to the nonsentient entity then there is nothing there which is appropriately referred to as a 'standpoint' or a 'point of view'. Such terms are merely metaphorical for the point that living entities can be harmed, destroyed, and/or killed. This generates the following dilemma for Taylor.

On the one hand, if to take the standpoint of a nonsentient organism is imaginatively to attribute consciousness to it, then the same can be done for inanimate objects. I, for one, can as easily take the perspective of—say—a stalactite as I can that of a protozoan. That is, I can equally imagine (or fail to imagine!) what it would be like to be a stalactite or a protozoan if they were conscious. On the other hand, if one does not imaginatively attribute consciousness, then there simply is no 'standpoint' or 'point of view' to be taken. Furthermore, though inanimate objects cannot be killed, they certainly can be harmed or destroyed. At least it does seem reasonable to say that smashing a stalactite to dust in some sense harms it.

Premise Two

Premise two is slightly different from premise one in that premise two introduces the notion of benevolence and connects that notion with the concepts of interests and ends. According to Taylor, "Since piles of sand, stones, puddles of water, and the like do not pursue ends they have no interests. Not having any interests, they cannot be benefited by having their interests furthered, nor harmed by having their interests frustrated" (p. 62).

I will postpone the discussion of Taylor's concept of 'pursuing ends' until the discussion of premise three. What needs to be examined here is the presupposition that having interests—in the sense in which nonsentient life forms can have interests—depends upon being able to pursue ends.

Taylor explains his concept of "interest" as follows:

> To understand more clearly how it is possible for a being to have a good of its own and yet not have interests, it will be useful to distinguish between an entity having an interest in something and something being in an entity's interest. Something can be in a being's interest and so benefit it, but the being itself might have no interest in it. Indeed, it might not even be the kind of entity that can have interests at all. In order to know whether something is (truly) in X's interests, we do not find out whether X has an interest in it. We inquire whether the thing in question will in fact further X's overall well-being... (p. 63)

The distinction that Taylor has in mind seems to be the standard distinction between being aware that something is in one's interest (i.e. promotes one's good) and something's being in one's interest (i.e. promoting one's good even though one isn't aware of it).

It is in this latter sense, and only in this sense, that the concept of interests applies to nonsentient life. Although in one sense, a protozoan has no (conscious) interest in anything, placing it in salt solution is not in its interest since it would be harmed by being killed.

This I find non-controversial; it makes perfectly good sense to speak of certain things being good for a protozoan. The same is true of any living thing. What is controversial, however, is the claim that in this sense of "interest" inanimate things have no interests and no good of their own.

A protozoan has a good of its own because there are certain things that we can do which are good for it. "Good for it" means contributing to its health and/or keeping it alive. We cannot do this for inanimate objects. However, we can do analogous things to inanimate objects which are clearly good (or bad) for them.

A stalactite must have a certain structure. In order to develop that structure there must be a source of slowly dripping, mineral-laden water. Such a source contributes to the development (to the "growth") of the stalactite and is therefore good for it; good in the sense that it will continue to be a stalactite and will even become a larger stalactite. True, it has no (conscious) interest in continuing to exist or in becoming larger, but then neither does a protozoan.

I conclude that at least some inanimate objects can meaningfully—though perhaps metaphorically—be said to have a good of their own and that, hence, some of our actions may be in or against their (non-conscious) interests. I believe this is to be true of all the inanimate objects Taylor mentions. Piles of sand can be blown away, stones can be smashed, and puddles of water can be dried up. In each case they are destroyed, and to be destroyed, *from the 'viewpoint' of that object*, is to be harmed. Of course, whether the notion of the 'viewpoint' of inanimate or nonsentient objects makes sense is another question. But if it makes sense it makes as much sense for inanimate objects as it does for nonsentient ones.

Even if it can be said that inanimate objects have a 'good of their own' and hence have (nonconscious) interests, it does not follow that it is meaningful to think of them in the same way that we think of nonsentient life. Perhaps what is valid in Taylor's argument is that the concept of "acting to benefit" or "being benevolent towards" is restricted to living entities. Taylor seems to suggest this in an earlier work:

> If an entity has a good of its own it at least makes sense to speak of placing constraints on our conduct out of respect for it. It is also intelligible to speak of acting benevolently toward it by intending to further its good *for its sake*.... It is significant that we cannot act benevolently (or for that matter, malevolently) toward inanimate things like rocks. This is a logical "cannot," since rocks have no good of their own. [6]

However Taylor's suggestion that it is *logically* impossible to act benevolently or malevolently toward inanimate things like rocks is surely false. Actions of these kinds are intentional, human actions. As such they depend upon the mental state of the human agent and not upon the object to which they are directed. Acting benevolently or malevolently toward inanimate things may be misdirected or foolish, but it is not logically impossible. Anyone who has kicked a tyre or smashed a golf club knows this.

Although Taylor has established neither premise one nor premise two, The Life Principle could still be salvaged if the interests of nonsentient living things differ in some relevant manner from the interests of inanimate objects. Taylor suggests that such a difference may lie in the fact that living things are teleological centres.

Premise Three

Premise three is that being a nonsentient *teleological* centre confers moral significance upon an entity, and that there is no analogous, morally relevant characteristic of inanimate entities. This is the crucial step in Taylor's argument and it is established as follows:

> The biocentric outlook on nature also includes a certain way of perceiving and understanding each individual organism. Each is seen to be a teleological (goal-oriented) center of life, pursuing its own good in its own unique way... (pp. 44-45)

> To say it is a teleological center of life is to say that its internal functioning as well as its external activities are all goal-oriented, having the constant tendency to maintain the organism's existence through time and to enable it successfully to perform those biological operations whereby it reproduces its kind and continually adapts to changing environmental events and conditions. It is the coherence and unity of these functions of an organism, all directed toward the realization of its good, that make it one teleological center of activity. (pp. 121-122)

These functions are not merely species functions, they are individualised functions as well:

> Under this (teleological) conception of individual living things, each is seen

> to have a single, unique point of view... When observed in detail, its way of existing is seen to be different from that of any other organism, including those of its species. To be aware of it not only as *a* center of life, but as *the* particular center of life that it is, is to be aware of its uniqueness and individuality. The organism is the individual it is precisely in virtue of its having its own idiosyncratic manner of carrying on its existence in the (not necessarily conscious) pursuit of its good... (pp. 122-123)

The importance of recognising the individuality and uniqueness of each living thing is crucial:

> When our consciouness of the life of an individual organism is characterized by both objectivity and wholeness of vision, we have reached the most complete realisation, cognitively and imaginatively, of *what it is to be that particular individual*... This sets a relevant frame of reference for our conduct as moral agents... Shifting out of the usual boundaries of anthropocentricity, the world-horizon of our moral imagination opens up to encompass all living things. Seeing them as we see ourselves, we are ready to place the same value on their existence as we do on our own... (pp. 127-128)

In short, Taylor claims that we ought to view all life as having the same value as our own because all life forms are teleological centres of life. This leads directly to the question of why being a teleological centre is so significant.

Here we find that the entire weight of Taylor's argument rests upon the (metaphorical) use of the concept of being goal-oriented: "To say it is a teleological center of life is to say that its internal functioning as well as its external activities are all *goal-oriented*...." But nonsentient life is not conscious and hence, literally, has no goals.

A basic problem with Taylor's argument is that the concept of teleology is itself so obscure that it is inappropriate to use it as a fundamental concept in what is supposed to be a rational, universal argument for a particular moral stance. If "teleological" merely refers to the fact that living things tend to maintain their existence and to reproduce and adapt to environmental changes, there is no need to anthropromorphise this by calling it "goal-oriented". Rather than depending upon a vague and obscure concept, it is more useful to utilize descriptive rather than metaphorical terms.

The real difference between stalactites and protozoa is that stalactites come about by a physical-chemical process whereas protozoa come about via biophysical and biochemical processes. At bottom, however, "bio" and "nonbio" processes are physical and chemical; the "biological" distinction being non-essential except for human purposes. "Biochemistry", for example, is not a precise term. Within the scientific community it is simply applied to the study of those chemical reactions which are associated with living organisms, most of which, to date, cannot be duplicated independently of such organisms.

Bearing this in mind, what is literally true of nonsentient life is that the biophysical and biochemical processes which occur in and around it normally or frequently result in the continuation of those processes and in the production of new entities in which those processes also occur. But this is analogously true of many inanimate objects. With regard to inanimate existence, what is true is that the physical-chemical processes which occur in and around it normally or frequently result in the continuation of those physical-chemical processes and, in some cases, the production of new entities in which those physical-chemical processes also occur. (Stalactites usually occur in clusters of individual stalactites and some crystals, in supersaturated solution, reproduce.)

Given these descriptive analogies between nonsentient life and at least some inanimate objects, for The Life Principle to have any validity it must be the case that mere biophysical processes have a higher moral standing than do physical-chemical processes. But the difference between these two processes, at least at the nonsentient level, is minor indeed. This difference simply cannot carry the moral weight that would be necessary in order to endow the life of a protozoan with a value equal to that of a human being while assigning no value to the existence of a stalactite. The Life Principle fails as a moral principle because it lacks sensitivity to the fact that the inanimate-animate-sentient distinction is based upon a continuum and not upon a categorical difference.

As for me, given the necessity of choosing between preserving the life of a protozoan or preserving the existence of a stalactite, all things being equal, I will continue to choose the latter. I will do so because I believe that the stalactite, in itself, has more value than does the protozoan. If Taylor and others prefer to choose differently, I honour their choice. But I do not believe it appropriate that they should view my choice as being immoral.

Gerald H. Paske, Department of Philosophy, Wichita State University, Wichita, Kansas 67208, USA.

NOTES

[1] I am here using the terms 'moral' and 'religious' somewhat loosely. Along with many philosophers, I assume that moral claims are rationally compelling (hence universal for all rational beings) while religious claims ultimately appeal to a faith or a commitment which is not rationally compelling.
[2] I have previously discussed the animal rights claim in 'Why animals have no right to life', forthcoming in *The Australasian Journal of Philosophy*.
[3] I here adopt the language of Tom Regan. Direct concerns are validated by reference to the individual lifeform itself. Indirect concerns are validated by reference to something other than the individual lifeform, usually by reference to human interests or human good. See TOM REGAN (1983) *The Case for Animal Rights* (University of California Press), pp. 150-151.
[4] PAUL W. TAYLOR (1986) *Respect for Nature* (Princeton University Press). All references in the body of the text will refer to this book. Taylor recommends that the term "moral rights" not be applied to animals (p. 254). I sympathise with this suggestion, but I think that the linguistic circumlocutions necessary to follow it preclude doing so. At any rate, since Taylor himself speaks of the moral duty to respect nature (p. 80), as well as speaking of the conflict between animal good and human good as a moral conflict (p. 259), I will also use the language of morality.
[5] I have serious doubts about whether taking the standpoint of nonsentient organisms is possible at all, but for the purposes of this paper I will assume that it is. For a critique of this possibility see THOMAS NAGEL (1979) What is it like to be a bat?, *Mortal Questions* (Cambridge University Press).
[6] PAUL W. TAYLOR (1981) Frankena on environmental ethics, *Monist*, 64, pp. 314-315.

[3]

The Moral Standing of Natural Objects

Andrew Brennan*

> Human beings are, as far as we know, the only animals to have moral concerns and to adopt moralities, but it would be a mistake to be misled by this fact into thinking that humans are also the only proper objects of moral consideration. I argue that we ought to allow even nonliving things a significant moral status, thus denying the conclusion of much contemporary moral thinking. First, I consider the possibility of giving moral consideration to nonliving things. Second, I put forward grounds which justify this extension of morality beyond its conventional boundaries. Third, I argue that natural objects have a status different from a special class of artifacts —works of art. Fourth, I discuss the notion of interest, and fifth I look briefly at the status of natural systems and at ways we might link the proposed extension of moral considerability with the rest of our moral thinking.

I. THE SCOPE OF MORALITY

There is considerable agreement among writers that if anything deserves moral consideration, then normal adult humans do, but given, in Waismann's phrase, the "open texture" of our language, we can easily imagine extending the language of rights, duty, respect, and obligation to children, the senile, the deranged, foetuses, the comatose, higher animals, human and animal corpses, and—perhaps less easily—also to other animals, trees, shrubs, vegetables, bacteria, cells, forests, valleys and even minerals. The length, and the ordering, of such a list is obviously a matter of considerable disagreement. Following Warnock and Goodpaster,[1] I take *moral considerability* as the core notion and consider extending it beyond our fellow humans to four progressively larger groups of things: (1) sentient beings whose psychological states are *models* of our own; (2) sentient beings of any kind; (3) living things; and (4) natural objects of any sort.

The notion of *model* in (1) is borrowed from Matthews who argues that there is a certain psychological unity within the animal kingdom.[2] Human

---

* Philosophy Department, University of Stirling, Stirling, Scotland FK9 4LA. The author is grateful to his colleagues at Stirling, especially Murray MacBeath and Antony Duff, for advice and comments on this paper, to various audiences who commented on earlier versions of this paper, and to the referees of *Environmental Ethics* whose comments were useful in shaping the final version of this paper.

[1] G. J. Warnock, *The Object of Morality* (London: Methuen, 1971) and Kenneth Goodpaster, "On Being Morally Considerable," *Journal of Philosophy* 75 (1978): 308–25. W. Murray Hunt, in objecting to Goodpaster's extension of moral considerability to all living things, asks the question: "Are *Mere Things* Morally Considerable?" *Environmental Ethics* 2 (1980): 59–63.

[2] In G. B. Matthews, "Animals and the Unity of Psychology," *Philosophy* 53 (1978): 437–54.

beings are not unique in forming plans and carrying them out, cooperatively if necessary. Lionesses on a hunt do the same. Cows weep when parted from their calves: do they not then feel the pain of separation? The point is so obviously true that it would be irritating for someone to defend it, as Hume noted long ago:

> Next to the ridicule of denying an evident truth is that of taking much pains to defend it; and no truth appears to me more evident, than that beasts are endow'd with thought and reason as well as man. The arguments are in this case so obvious that they never escape the most stupid and ignorant.[3]

Of course, an appeal to our unity with other animals has its limits, for at some stage our psychological relatedness to other species becomes so attenuated as no longer to count for very much. Problems begin to arise when one tries to determine what the moral significance of this psychological unity is supposed to be.

An answer to this question may have something to do with the related question of *rights*. Passmore argues that legal obligations and rights can only be generated among beings who belong to one community of "mutual obligations" and "common interests," while others, like Feinberg, maintain that for something to be a potential rights holder it must have interests, for the holder of rights must be able to be represented (and what has no interests cannot be represented), and must be able to benefit in its own right (which again requires it to possess interests).[4] It looks as if higher animals will certainly be candidates for moral consideration if Feinberg is right. And it can also be urged that humans and other animals do at least sometimes form communities of common interests. A striking example of this is the relationship between Bush people and the honey guide so lovingly described in Laurens van der Post's novel *A Far-Off Place*. The honey guide is in no way a domesticated creature, but benefits from leading humans to beehives by obtaining a generous helping of honeycomb. Post writes:

> ... this partnership of the honey-guide and man differs from all others because it is voluntary, free and equal, formed out of a sense of mutual obligation to a common purpose of life and love of the honey that is the product of the purpose.... It is proof miraculous of what life could become when a sense of common purpose and interdependence of all living and existing things is recognized and wholeheartedly served.[5]

---

[3] David Hume, *A Treatise of Human Nature*, ed. L. A. Selby-Bigge (Oxford: Clarendon Press, 1968), bk. 1, pt. 3, sec. 3.

[4] Joel Feinberg, "The Rights of Animals and Unborn Generations," in W. T. Blackstone, ed., *Philosophy and Environmental Crisis* (Athens: University of Georgia Press, 1974), pp. 43–68.

[5] Laurens van der Post, *A Far-Off Place* (Harmondsworth: Penguin Books, 1976), p. 217.

Closer to home, we can think of the relationship between dolphins and those who study them—or if we include cases in which the participation of those involved is less than equal, the bond between people and pets. Even so, we can again ask: just what does the possession of interests, or the shared interests of a community, amount to?

Perhaps we can make sense of the appeal to interests, psychological models, and communities if we think of moral values as bound up with a framework of primarily human interests, needs, and purposes. The notion that a *moral* code provides a means of maximizing welfare within a society, for example, would be one way of accommodating this perspective, since considerations regarding welfare only seem to arise for beings who have interests. Alternatively, we can consider what set of principles it would be rational to choose for a society given both information and ignorance about one's place in that society.[6] Or again, like Gewirth, we can seek to establish some fundamental moral principle on the ground that since we claim rights for ourselves in virtue of our having certain qualities, then others who possess these same qualities will also be able to claim similar rights.[7]

What lies behind all these notions is a human- (or at least animal-) centered conception of ethics, a kind of ethical egoism once removed. Whereas the egoist is interested in the welfare of one particular individual above all others, these more sophisticated views recognize that what it is *moral* to do may not always result in maximizing welfare or benefit to the agent, although a code based on these insights supposedly maximizes the welfare of a group, society, or community of appropriately characterized individuals. If my own access to certain benefits is through my membership in such groups, then I will be able to "identify" in some sense with the group and may come to associate my good with that of the group.

It is not my intention to undermine this strategy or to deny that something of moral interest can result from the strategy. Indeed, rather like Gewirth, I appeal to qualities common to humans and to some inanimate objects in order to make out a case for the moral considerability of the latter. Unlike Gewirth and most other theorists, however, the shared quality which I suggest is not anything like rationality, purposive agency, linguistic ability, or even sentience. This is not to deny that such qualities may be morally important. Perhaps a creature which is rational and talks has more of a moral claim on me than one which is rational but lacks language (assuming, for the moment, that rationality and linguistic ability can be separated). But if I am right, my argument at least challenges the notion that by the time we have reached sentience we are sure to have exhausted the fund of morally relevant features.

---

[6] For such starting points, see R. B. Brandt, *A Theory of the Good and the Right* (Oxford: Oxford University Press, 1979).

[7] Alan Gewirth, *Reason and Morality* (Chicago: University of Chicago Press, 1978).

Significantly, then, I suggest that what it is moral to do may, on occasion, be something that does not benefit individual humans, communities of humans, or communities of humans and other beings with qualities like those of agency, rationality, or sentience. Does this mean that we, along with a wide variety of natural objects—perhaps all of them—form a community the welfare of which is the proper object of morality? Perhaps, though as I show later this question may be more terminological than real. For the time being, it should be noted that I am not denying that for social life to be possible at all there have to be rules that take account of the often competing interests of those who live together in communities and that some of these will be moral rules. Nor am I dismissing the possibility of putting forward human-regarding arguments for the preservation of wildlife and wild places. On the contrary, such arguments seem to me to be very important and may themselves be— morally or otherwise—already decisive.

To pursue the hypothesis, should we not then think straightway about the extension of considerability to all sentient and living things? Pleasure and pain is something that is a central feature of human (and probably animal) experience, and interests, we might think, can still exist in the absence of sentience anyway.[8] We need food, and it is a good thing for us, even if we lack the warm glow of contentment a full stomach sometimes brings. The defender of sentience may wonder why we should care for the goals of a being, albeit a *living* one, who is incapable of feeling frustration, fear, disappointment, hurt, or satisfaction. Now there is undoubtedly an *explanatory* route linking sentience with moral respect. If an item has feelings, this explains its possession of interests. And, as we have seen, its possession of interests can in turn explain why it is worthy of moral consideration. But what I suggest is another route to just this same destination—moral considerability. My route may not appear to be so obviously explanatory—this may be because it is not so well worn as the first one—but so long as the route I suggest is a possible one, then the dogma that moral respect is conceptually tied to sentience will be wide-open to challenge.

In this connection Callicott has argued that it cannot simply be the pain we inflict on animals in the course of factory farming or experimentation that makes these procedures wrong, for in their natural state animals are exposed to cruel pain, lingering death, and the ravages of disease and predators.[9] What if the total pain endured in the farm or the laboratory is less for a given species than its total in the wild? For Callicott this seems to show that what is immoral

---

[8] For the centrality of pleasure and pain to human experience, see John Rodman, "The Liberation of Nature?" *Inquiry* 20 (1977): 83–145, and for remarks on pleasure and pain as adaptational signals, not goals in themselves, see Goodpaster, "On Being Morally Considerable."

[9] J. Baird Callicott, "Animal Liberation: A Triangular Affair," *Environmental Ethics* 2 (1980): 311–38. Although here, and elsewhere, I am critical of Callicott, I found his paper extremely stimulating, and useful in preparing the final draft of the present paper.

in our treatment of animals is not the infliction of pain, but "the transmogrification of organic to mechanical processes."[10] Callicott is wrong, however, if he thinks that this weighing of natural against unnatural pains in itself reveals that we wrong animals in some way that is independent of the infliction of pain, for, morally, there is all the difference in the world between those pains that occur in the natural course of events and those that are deliberately and knowingly inflicted by intelligent moral agents. Of course, I agree with Callicott that the evils of factory farming involve more than the evil of deliberately inflicting pain, and it is precisely to let us argue to such a conclusion that we need some ground for moral consideration that is itself independent of sentience.

Suppose, then, that at least for the sake of the argument the possibility of moral consideration to our third group is allowed. Is there any good reason why we should pause there instead of sliding all the way down the slippery slope? Both works of art and natural objects like great deserts and mountains seem to command a certain respect for their own sakes. Is this a kind of moral respect? Or is it simply absurd to suggest so? My strategy here is slightly devious. I start by arguing that it is at least not absurd to consider inanimate natural objects as worthy of moral consideration in their own right. Having thus established the possibility of a morality that gives nature its due, I then turn to the problem of what might motivate or justify such a moral stance.[11] In the West, we are not used to taking seriously the idea that natural objects can have a moral claim on us. What seem to be clear moral intuitions, though, are quite often local to a time and a culture. Thus, in a culture strongly influenced by Buddhism, Jainism, or Shinto it would hardly have been necessary to argue as I have been doing for the considerability of living things. By now taking an example from another culture I intend to establish the possibility of extending moral consideration to at least some things in our fourth class.

In a speech delivered in 1854, Chief Seattle, a North American Indian, laments the coming of the whites and the demise of both the Indian and the Indian's environment:

> Our dead never forget this beautiful earth, for it is the mother of the red man. We are part of the earth, and it is part of us. The perfumed flowers are our sisters; the deer, the horse, the great eagle; these are our brothers. The rocky crests, the juices of the meadow, the body heat of the pony, and man—all belong to the same family....

---

[10] Ibid., p. 336.

[11] Some may balk at the notion that any individualist approach could give nature its due, for nature itself is a kind of society of structures in each of which millions of individuals participate. I deal with this issue in the final section of the paper, and until then—and at the acknowledged risk of oversimplifying—I continue the argument at the level of the individual.

> The rivers are our brothers, they quench our thirst. The rivers carry our canoes, and feed our children. If we sell you our land, you must remember, and teach your children, that the rivers are our brothers, and yours, and you must henceforth give the rivers the kindness you would give any brother....
>
> A few more hours, a few more winters, and none of the children of the great tribes that once roamed on this earth or that roam now in small bands in the woods will be left to mourn the graves of a people once as powerful and hopeful as yours. But why should I mourn the passing of my people? Tribes are made of men, nothing more. Men come and go like the waves of the sea.[12]

In these extracts we can see how Chief Seattle runs together references to the living (flower, horses, eagles) with reference to the nonliving (rocky crests, rivers). The third paragraph suggests far less concern with survival than we might expect from a chief who foresees the extinction of his own tribe.

Yet, struck as we may be by Chief Seattle's words, is there any way we can attach the perspective they reveal to those of us who bask in the material wealth of the twentieth century? Toward the end of his speech, he says:

> ... we do not understand why the buffalo are all slaughtered, the wild horses are tamed, the secret corners of the forest heavy with the scent of many men, and the view of the ripe hills blotted by talking wires. Where is the thicket? Gone. Where is the eagle? Gone. And what is it to say goodbye to the swift pony and the hunt? The end of living and the beginning of survival.[13]

The questions are good ones for a people who live by hunting. For us, when we face the next energy crisis, the worry is more likely to be that we must say goodby to the fast car, the dishwasher, the air conditioner, and the television. Yet, we should be wary of dismissing the alien perspective out of hand; parochial matters of our society's values are not the point. And there are many reasons for thinking that acting morally may well involve the abandonment of many of our materialist goals, along with an end to the exploitation of poorer and weaker societies required to support a way of life both comfortable and corrupting.

So from Chief Seattle I take it that it is possible to extend moral considerability to at least some nonliving natural objects. Of course, this is not to claim that the morality of his tribe was a consistent one; indeed, I am not sure that this would matter, since it would be foolish to pretend that all moralities are

---

[12] I found the text of the speech in the journal *New Internationalist* 3, no. 31 (September 1975): 16–17.

[13] A book by Helen Muir, *Many Men and Talking Wives*, published by Duckworth in September 1981, purports to take its title from this section of Seattle's (or Seatlh's) address. This reading of the text is perpetuated by Jill Tweedie (*Guardian*, 26 October 1981), who draws attention to "the odd title" of Muir's book. I have not undertaken the research necessary to discover which text has the misprint.

consistent. But there is no better evidence that a moral posture is possible than the fact that a group managed to live by it, and Chief Seattle's words reveal just such a stance giving a degree of consideration even to inanimate things. In the next section, I show that this stance is possible *for us* by proposing a criterion for such considerability.

II. THINGS AND THEIR FUNCTIONS

Supposing, at least for the sake of the argument, that there are possible principles which can be said to be moral principles and which allow consideration not only to living things but also to some nonliving natural objects, we might wonder if there are any other objects which might have a claim to moral status. Natural objects contrast with artifacts. Both sorts of object have structure of varying complexity, and both can have functions. The function of scissors is to cut, and good scissors cut well; the function of a heart is to pump blood, and good hearts pump well. But the function of an artifact is the result of design, and this design is intended to satisfy some end. By saying this, I mean to count beaver lodges as artifacts, but not coral reefs.

Now, a badly designed object may not fulfill its function well (or function instead as a different artifact), but the structure of complete natural objects is not the result of design, for they have no functions to fulfill. Parts of natural objects—like hearts, kidneys, leaves, and roots—do have functions, and these are determined by the contribution they make to the growth, maintenance, and survival of the complete, living thing of which they are parts. And, by extension, we may even count whole conglomerates of natural things—ecosystems—as containing whole objects within them which function to preserve the system as a whole. But we must be cautious with this extended use, for whole natural individuals, whether microbes or tigers, have no intrinsic functions at all.

It is important to be clear about this matter of intrinsic function. A cotoneaster shrub, let us suppose, functions to screen the compost heap in a garden. This is typical of the countless ways we—and other living things—use natural objects in the fulfilment of our schemes. Yet, it would be silly to try to define cotoneasters in terms of their contributions to gardens, or in terms of any other functions that we, or any other creature, might assign them. By contrast, to describe an item as a root, or a leaf, is to describe it in terms of its functions; an object that looked like a leaf, yet failed to promote growth by photosynthesis, and took no part in transpiration, would hardly be a genuine leaf. Thus, there seems to be an important distinction between whole natural objects and their functional parts.

It is not my intention here to contribute to the already large literature on function and teleology. Any reasonable account of function, nevertheless, will allow for the existence of defective or diseased things of a kind which fail to carry out the functions characteristic of that kind. Thus, we should no doubt want to distinguish false leaves (which do not function as leaves at all) from defective leaves (which would have functioned as normal leaves had not certain disturbing factors intervened). My notion of intrinsic function is rather like Enç's notion of function *simpliciter*. As he puts it:

> What I am asserting here is that when we discover what the function of the heart is, we also discover part of the identity conditions of a heart. Part of what it is to be a heart is to be capable of pumping blood under normal conditions....[14]

It is no part of the identity conditions of cotoneasters that they screen eyesores. It follows that I can know that a certain bush screens my compost heap without knowing what kind of bush it is; and a grasp of what a cotoneaster is involves no reference to such overlaid (non-intrinsic) functions as that of screening other things.

Now for a problem case. Suppose we find an ecosystem in which stability is preserved in part by the appetite of a large predator. To the predator we assign the function within the system of keeping, say, the population of voles at a reasonable level. Moreover, the system which the predator helps to control may itself have determined a place in it for such a predator. In this case, it looks as if the predator is functionally adapted to the system. Now there are two ways in which we can try to take the sting out of this example and argue that the predator in fact has no function. Despite the initial attractiveness of the first route out, I suggest that the second is in fact more plausible.

The first argument might be called an appeal to the "come off it" strategy. To view an ecosystem as a living machine involves a number of assumptions. We view the system as if designed for stability, as if a certain niche had been made with a predator in mind, as if the predator were custom built to carry out the role of vole control. But none of this is, literally, how selection and adaptation work. To use Dennett's terminology, we can distinguish *design-stance* descriptions from *physical-stance* ones.[15] Although the former stance can be rich in metaphor and useful explanation, the "come off it" strategy simply reminds us that the physical facts of the matter involve blind natural forces and chance selection.

Despite its initial appeal, there are drawbacks to this strategy. What about the heart, and its generally accepted function of pumping blood? We could use

---

[14] Berent Enç, "Functional Attributions and Functional Explanations," *Philosophy of Science* 46 (1979): 349. Enç's view arises from a criticism of Wright's influential views: see L. Wright, "Functions," *Philosophical Review* 82 (1973): 139–68, and his book *Teleological Explanations* (Berkeley: University of California, 1976). Wright's view is also criticized by C. Boorse, "Wright on Functions," *Philosophical Review* 84 (1976): 70–86 and by A C Purton, "Biological Function," *Philosophical Quarterly* 29 (1979): 10–24.

[15] See the introduction to D. Dennett, *Brainstorms* (Sussex: Harvester Press, 1978).

the same strategy here, pointing out that hearts just evolved, that the forces at work were no different from those involved in other cases of evolution. Yet, following Wright's account of functions, we may want to say that the heart is there because it pumps the blood, and its pumping the blood is a consequence of its being there.[16] The joint truth of these conditions suffices for the recognition that the function of the heart is to pump the blood. But the "come off it" line threatens our conception of functional parts within organisms. Additionally, it may well make sense to regard a predator in an ecosystem as having a function. At least, it would be better not to rule this possibility out *a priori*. And, luckily, there is a way in which we can allow this without being committed to the claim that this hawk, or that tiger, has any intrinsic functions at all.

Our second way out of the difficulty involves no more than pointing out that the assignment of functions to individual predators, or to any other individual, risks a division fallacy. Suppose our large predator is an eagle, and suppose further that eagles die out in the system under consideration. As they become extinct, their place is taken by some other species—by hawks of some kind. So long as the hawks control the voles to the same extent as the eagles did, our model of the system will not be substantially altered. The claim that the eagle had this function prior to its extinction is not a claim about any individual bird. Rather it was a claim about a kind of animal, about the need for a group fulfilling the role in question to be represented in the ecosystem. But what is true of a group need not be true of any representative of that group. And the truth, if any there be, in talking about the function of a predator in a natural system is, at best, a truth about a group, not about any individuals.[17] One eagle dies, while another is hatched to replace it. Just as the heart cells do not function to pump blood (although hearts do) individual eagles do not function to control voles (although we may regard groups of them as doing so).[18]

My point, however, is more than the claim that nothing has a function *qua* individual. On the contrary, individual things may acquire this or that function in particular circumstances. And—if we broaden the notion of non-intrinsic

---

[16] I agree with Enç that Wright's account gives a sufficient, but not necessary, set of conditions for the possession of an (intrinsic) function.

[17] Obviously, truths about groups and truths about species should also be kept distinct. We may find nonaggressive groups of individuals, for example, belonging to a species generally characterized as aggressive.

[18] For readers who find the analogy with a physical thing of this sort inappropriate, a more convincing example may come from the "Ant Fugue" in D. R. Hofstadter's *Gödel, Escher, Bach* (Sussex: Harvester, 1979). Teams of ants form "signals," characterized by their function of conveying specialized ants to a part of a colony. But only the team, not the individual ants, have this function. The individual ants have the functions of nursing, cleaning, hunting, and so forth. Incidentally, I regard ant colonies as individuals, with ants as functional components: there is no more to the individual ant than its roles in the colony, whereas there is more to an eagle than its roles, say, as one of a nesting pair.

function to include roles—how well an individual discharges a given function (as mother, chairperson, or whatever) may itself be a matter of moral concern. The claim that nothing has a function *qua* individual would, at best, be true only of intrinsic functions.[19] And it is just such functions which individuals that are not physically parts of other objects are lacking. As we have seen, this intrinsic functionlessness is coupled with a capacity to take on multifarious functions in different contexts. But what makes a factory worker more than a machine operator also makes an elm tree more than a windbreak: in each case we have an assigned function coupled with the potential for taking on many other functions—voluntarily or not—overlaid on an individual that is designed specifically neither for this nor for that, since the individual was not designed at all. And if we are to look for a quality by virtue of which all natural things may claim moral considerability, I tentatively suggest that we have come up with a candidate: their lack of intrinsic function.

III. ART AND AUTONOMY

We celebrate the intrinsic lack of function of persons in various ways: they are not merely the means to others' ends, but have the potential for all sorts of different roles; within institutions they can acquire all sorts of functions, but none of this can undermine their fundamental autonomy. The thrust of my argument is that we do have some grounds, albeit slender ones, for recognizing a similar autonomy in other natural things.[20]

Few would deny that scissors, cars, and other products of human and animal invention lack this autonomy shared by natural objects. But there are some products of human labor which have seemed to many to have a value beyond mere functional utility, and have instead appeared to have an intrinsic value for their own sakes. Works of art are perhaps the best examples of such products, and many modern writers have followed Hegel in ascribing a higher value to art than to nature. Thus, Savile writes:

> Of course it is true that natural beauty has value for us too, but we may observe that at least in the highest examples of art we find functions exemplified and

---

[19] See Enç, "Functional Attributes," p. 361.

[20] Murray MacBeath has impressed on me that it can seem odd to hold mountains as autonomous if by this we mean that they do things in a self-regulating way, getting on—as it were—with the business of being mountains. Intuitions may vary here, but much the same oddness can be found, if we work at it, in ascribing a form of agency, however diminished, to some living things and living systems. Interestingly, Wiggins writes of "geographical or geological terms like *river, lake, spring, sea, glacier* or *volcano*" that "it will not be wildly inappropriate to speak of principles of activity" (see David Wiggins, *Sameness and Substance* [Oxford: Basil Blackwell, 1980], p. 86). Consider also my remarks about the slowness of change in mountains in the final section of the paper.

values conveyed which nature and experience are scarcely able to yield. For instance, through art, though rarely through nature, we find ourselves made sharply aware of the splendours and defects of the society in which we live....[21]

Ignoring the human-regarding stance taken by this writer, we can still see an attempt made to compare natural objects with artifacts. And if we are going to compare the claims of these two kinds of objects, it is interesting to know if the objects compete on the same basis. It may be that works of art are examples of the sort of objects whose existence we wondered about at the start of the preceding section: artifacts that have a claim to moral standing.

Some support for the idea that works of art are more than mere instruments of communication between artist, or composer, and audience is rendered by noting the affront people display when an important art work is vandalized, or one of a country's art treasures is threatened with export. To give a complete account of why art is viewed as something transcendent, and of value for its own sake, would take me too far from the central themes of this paper. Yet, I want to maintain that art works and natural objects do not compete for moral consideration on any similar basis. Unlike natural objects, art works do not seem to me to fall within the scope of morality.

Part of the "magic" of art is that masterpieces of music, literature, painting, and the rest are packed with symbolic richness. Not only is there no simple message conveyed by a great painting or sculpture, but rather a number of messages, allusions, and suggestions conveyed on many different levels. The symbolic content of such objects represents a great deal to a great many people. They are messages of great richness. This sheer expressive power that they represent enables us, I think, to account for their potency while staying within an account of art which treats it as a mode of communication between the creative agent and the audience.

If we do stay within such an account, then artworks are functional objects, and hence not on the same footing as complete natural objects. This is not to say that there is no room for talk of respect, duty, and obligation when we are dealing with art; rather, we need to be clear on the Kantian distinction between duties owed directly to objects, and duties we may have regarding an object. We owe no respect, no duties, to any of Leonardo's work, but we do have duties regarding it, duties which are owed to the many people for whom his work is of immense symbolic power. Likewise, a government contemplating a road building program owes no duty to a valley on the grounds that it is sacred to a particular local tribe, but the government does, of course, owe a duty to the tribe regarding the valley. If a road is built through the valley despite the protests of the tribe, it is the tribe, not the valley, that has been wronged.

---

[21] A. Savile, "The Place of Intention in the Concept of Art," in H. Osborne, ed., *Aesthetics* (Oxford; Oxford University Press, 1972).

Yet, the valley is a natural object, and if what I have argued so far has any plausibility, it may be that we owe any valley, consecrated to ancestors or not, a certain moral respect which we do not owe any artifact. Subtract the expressive power and the fitness for its purposes from a painting and you are left with an artifact of no particular value: the canvas, the wood for the frame, even the frame itself and the pigments in the oils, might all have been put to better use. But subtract the functions assigned by people and animals to a valley and its river, take away the ski lifts, the beaver dams, and the scenic views and you are left with an object containing within it hundreds of self-regulating systems living in a kind of natural anarchy, an object that partly determines its own climate, serving no one's purpose, but still worthy of respect purely in its own right.

To argue in this way raises an interesting problem: could there be items of human or animal creation which, nonetheless, have a value purely for and in themselves, not merely one derived from their function? Gardens, parks, canals, and the like might seem to be of this type. And many animals may be thought, at first sight, to be artifacts themselves. As Callicott says, "Domestic animals are creations of man. They are living artifacts, but artifacts nevertheless. . . . It is literally meaningless to suggest that they be liberated".[22] Certainly, intervention by unnatural selection, or by gardening, has allowed the production of living things and systems particularly suited to human ends and needs. But we need to be wary about classifying such items. Suppose an alien biologist is puzzled to find a breed of (domestic) sheep particularly ill adapted to surviving in the wild. The biologist is enlightened by the discovery that the breed is the result of human intervention aimed at maximizing wool production and yielding a high proportion of edible flesh. Does the discovery tell the biologist more about the *nature* of the breed in question? Of course not: the puzzle was about the etiology of the breed, not about what kind of thing it was. It might be objected that one thing the biologist has found out is that the animals are of a certain kind, namely, the domestic kind. Little hangs on this point. The label "domestic" identifies no natural kind, and is dispensible from the taxonomist's point of view. It alerts us to the likelihood that the animal or plant in question is the result of selective breeding. As Darwin so cautiously made the point:

> The key is man's power of accumulative selection: nature gives successive variations; man adds them up in certain directions useful to him. In this sense he may be said to have made himself useful breeds.[23]

---

[22] Callicott, "Animal Liberation," p. 330.
[23] Charles Darwin, *The Origin of Species*. 6th ed., 1872, chap. 1; reproduced in P. Appleman, ed., *Darwin* (New York: Norton, 1979), p. 45.

Selective breeding, then, yields "artifacts" in only an attenuated sense of that term.

Plants and rocks in a garden, however, represent a different case. They are incorporated as parts into a whole which is the result of human design and that design is aimed at satisfying our own ends. Gardens, parks, and game reserves are thus artifacts in a perfectly literal sense. But their components are natural objects and our use of them within the artifact will more or less restrict their autonomy. A garden, then, has a double value. It has the sum of the values of the individual things within it; and additionally it has its value as an artifact —that is, value for those who use it and benefit from it.

With these cases distinguished, and put to one side, there is a more difficult case we must now consider. In a story by Stanislaw Lem a creator-benefactor named Trurl builds a microscopic kingdom for the "entertainment" of a dispossessed tyrant. The kingdom is housed in a portable glass case and is meant to contain simulations of people, armies, villages, and so on. But Trurl does his job too well: the simulations are perfect; the case contains tiny persons living their life under the decrees of the tyrant who rules over them like a god.[24] Here is a situation in which we would say that something rather strange has happened. What Trurl has invented was meant to be no more than a game, so that the wicked ruler could use up his energies and be happy in his tyranny while no one suffered. And the world in the glass case could no doubt be functionally described in just such a way. Lem's story, though, enables us to see how misleading such a functional description is. Trurl is induced in the course of the tale to feel more and more uncomfortable as he comes to realize that what he has created are real persons with real feelings, undergoing real suffering, and fighting real wars at the tyrant's command. So here at least we have a case where something has been produced with a certain function in mind, but where a merely functional description of the thing produced is quite inadequate.

Of course, human beings are autonomous, and we should surely want to maintain this whether or not we hold particular religious (or other) beliefs about how we came into being. The Lem story has allowed us to conclude that not all purposeful creation results in items that can be adequately defined in terms of intrinsic function. It matters little whether we decide to call such items "artifacts" or not. If Trurl's creations are artifacts, then so—according to some are we. This shows that there can be artifacts that lack intrinsic functions, that in some way go beyond what their creator has programmed into them. My own inclination is to use a different term for such entities—but it is clear, I hope, that nothing of any substance hinges upon this decision.

---

[24] "The Seventh Sally," from *The Cyberiad,* trans. M. Kandel (Seabury Press, 1974). The piece occurs also as chap. 18 of Daniel Dennett and D. R. Hofstadter, eds., *The Mind's I* (Sussex: Harvester, 1981).

Doesn't such a move, however, just open the door to the following objection? Some human works of art, my critic suggests, are precisely artifacts that have a value in and for themselves over and above what can be captured by any functional account of the sort sketched earlier. It is interesting to note the result of taking such an objection seriously. If the critic is right, and at least some works of art are lacking in intrinsic function, then we need some explanation of why such items are often valued far more highly than natural objects. It would be fascinating to pursue this issue, and I do not mean to give the impression that I am entirely unsympathetic to such an enterprise, but, for the moment, I am content to rest happily with the functional model of art as a form of multilevel communication by means of a suitable notational system.[25] Perhaps I err, but if I do, then more will need to be said by the critic about identity conditions for works of art and about why we do not pay nature the same respect we so dutifully give to art.

IV. INTEREST

The position I have now reached has both benefits and drawbacks. From such a stance, we can see why the retention of certain areas in national parks, carefully stocked with selected species, seems to some people a poor attempt at giving nature her due, for such parks and game reserves become large artifacts, no different in kind from zoos and gardens. Of course, not all protected wilderness areas are managed in this way, and even if some are, the objects within them are, as I have argued, often pretty well autonomous. On the other side, much work remains to be done to show whether I am right in lumping together many different kinds of natural objects as all sharing the same autonomy. And the details of how to distinguish parts of objects from whole objects have been conveniently skipped. Are clouds whole objects, or just parts of one object, the atmosphere? Are human bodies whole objects, or just temporary swarms of atoms? Ignoring these difficulties, nothing has been said to support the view that wilderness areas should be left alone, rare species protected and industrial pollution reduced—for the acknowledgement that natural objects are worthy of respect for themselves does not require in itself any prohibition on our use of them, although we might expect that taking their autonomy seriously would mean putting limits on our present somewhat selfish exploitation of them.

But before we take seriously talk about limits in this context, we have to consider how we might give weight to claims on behalf of natural objects at all. Do their interests compete with ours in any way? Does it even make sense to think of an inanimate thing having interests at all? Much of the recent work

---

[25] For requirements on notational systems, see Nelson Goodman's *Languages of Art* (Oxford: Oxford University Press, 1969), especially chaps. 4 and 5.

on the question of respect—or rights—for natural objects has been stimulated by the case of the Mineral King Valley described in Stone's book *Should Trees Have Standing?* Stone's article of the same title was in part responsible for the Douglas dissent, in which an American Supreme Court judge argued for the extension of rights to natural objects (the text of the Douglas dissent is printed in Stone's book).[26] The trouble, though, with Stone's original position is that it seems to require us to recognize that such items as forests, rivers, and valleys have interests. Stone's arguments involve an appeal to the fact that we already accord recognition to the interests of such merely legal "persons" as countries, corporations, and so on, even though in the nineteenth-century jurists found such notions almost unintelligible. Thus, he writes:

> Perhaps injury to the Sierra Club was tenuous, . . . but the injury to Mineral King —the park itself—wasn't. If I could get the courts thinking about the park itself as a jural person—the way corporations are 'persons'—the notion of nature having rights would here make a significant operational difference. . . .[27]

As we have already seen, Feinberg argues that "the sorts of beings who *can* have rights are precisely those who have (or can have) interests,"[28] and so it is hardly surprising to find Stone taking the position he does.

Such a view, indeed, seems plausible when we focus on living things, or simple aggregates of living things, like forests. And let us suppose, perhaps implausibly, that the interests of a corporation can be identified by a reductionist strategy: we can appeal to the interests of employees, shareholders, customers, and so on. In a similar way, perhaps the interests of a forest can be identified in terms of the interests of its individual trees, the birds who nest in them, the fungi around their roots, and so forth. Yet, even in this case there is an overwhelming difference. In the case of a natural forest, there was no design, no purpose, no contracts, no statutes—in brief, none of the hallmarks of the artifacts which are corporations. Corporations have ends to serve, at least economically, but forests have none. But I have already argued that living things in general do have an interest in survival, growth, and freedom from disease. So perhaps we should include forests as special cases of living things. Alas, none of this comes close to establishing any sort of interest for mountains, the air, deserts, and rocky crests.

---

[26] Christopher D. Stone, *Should Trees Have Standing?* (Los Angeles: Kaufmann, 1974). The article appears in *Southern California Law Review* 45 (1972): 450. Also on the same topic, it is worth looking at L. H. Tribe's "Ways not to Think about Plastic Trees," *Yale Law Journal* 83 (1974): 1315–48, and the same author's "From Environmental Foundations to Constitutional Structures," *Yale Law Journal* 84 (1975): 545–56.
[27] In his introduction to the book cited in the preceding note.
[28] Feinberg, "The Rights of Animals," p. 51.

Another point about interests is that while we are expert as far as human interests are concerned (so we think), we can never be in a position to say just what the interests of other living or nonliving things might be. Perhaps rivers enjoy being dammed—how could we ever tell? As Sagoff wryly remarks:

> Make no mistake: the policy which turns our remaining wilderness areas into amusement parks, highlights the scenery with *son et lumiere*, and fetes the animals with garbage increases the general satisfaction of man, beast and mountain.[29]

Yet, if, in Tribe's words, we are to give "institutional expression to the perception that nature exists for itself," how could we do so while abandoning talk of interests? Aping Feinberg, we might suggest that a mountain has no interest unless we assign it a function within a scheme of human interests, and so a mountain *per se* could not be represented in any institutional legal process, especially in an adversarial system of law like our own.[30] Seen in this way, Stone's appeal to corporations, municipalities, and the rest is not a very helpful precedent.

An opponent of species-centered morality might point out at this stage that our predicament merely confirms our alienation from nature. Since we are unable to determine the wants, needs, harms, injuries, and benefits of rivers and mountains, we are unable to let them figure in our institutional processes which are designed for handling just such matters. Rodman, for example, suggests that we should abandon the property paradigm and adopt instead a

> principle of propriety, i.e., the principle that action should be appropriate to the nature of all the parties involved in the transaction accompanied by the corollary recognition that non-human species exist "in their own rights" ... and not simply "for us."[31]

For us there will be the additional problem of determining the "nature" of a river, desert, or mountain. Yet, Rodman couples this principle with the suggestion that "we may need to become less moralistic and less legalistic," and maybe this is a solution to our difficulty. If moral theory tells us that any morality will concentrate paradigmatically on social goals, the harmonizing of essentially divergent human interests, the protection of minorities and so on, then perhaps we should follow Rodman and think of our relationship to nature in less moral terms.

The considerations I have advanced in this paper suggest that all natural objects share a certain functionlessness, unlike their parts, and that this may

---

[29] Mark Sagoff, "On Preserving the Natural Environment," *Yale Law Journal* 84 (1974): 244.
[30] See Feinberg, "The Rights of Animals," p. 54.
[31] Rodman, "The Liberation of Nature," p. 109.

provide a basis for a fellow feeling, a respect, and a care for natural objects which is no less important for being outside the scope of social morality. My own view here is that to accept this way out, however attractive, is to restrict the scope of morality unduly. Although in my account we can talk of the harm caused to the fish and flora of a river by excessive fertilizer runoff, we can only talk of the harm to the river itself metaphorically. I admit that the river itself can have no interest, literally,[32] but to take this fact as showing that the river itself commands no moral respect is to fall back on a conception of morality that ties it to the social goals already mentioned. We already have notions that are at odds with such a conception. The pointless destruction of inanimate things is as much vandalism as the destruction in a similar spirit of living things. Passmore, in speaking of our reverence for life, says that we can link this reverence with

> another more explicitly western tradition, that it is wrong unnecessarily to destroy—a principle embodied in the concept embodied in the concept of "vandalism" ... One could at least go this far: the moral onus is on anyone who destroys.[33]

The arguments in this paper might be thought to give some plausibility to an attempt to divorce an account of vandalism from simple reverence for life and to tie it instead to a recognition of the common predicament of all natural uncreated things. But I doubt whether, in the end, vandalism is the notion we should give our attention to. Artifacts are obvious victims of vandalism, and it seems clear that destruction for its own sake is not a necessary condition of a vandalistic act.

V. INDIVIDUALS AND SYSTEMS

So far I have been dealing with easy cases—with individuals, or simple aggregates (like forests) whose parts are individuals. Before inquiring further into the connections, if any, between the viewpoint I am urging and the accepted notions of what is of moral concern, it is necessary to take a look at one very distinctive view of environmental morality—Leopold's land ethic. This view can be neatly captured by the motto: "A thing is right when it tends to preserve the integrity, stability and beauty of the biotic community. It is wrong when it tends otherwise."[34] It could be argued that the autonomous,

---

[32] However, granted the open texture of our language and the associated possibilities of conceptual development and revision, we could—as I admit in the final section—probably work up a concept of interest that would apply to rivers.

[33] John Passmore, *Man's Responsibility for Nature* (London: Duckworth, 1974), p. 124.

[34] Aldo Leopold, *A Sand County Almanac* (New York: Oxford University Press, 1949), pp. 224–25. Leopold's views are concisely summarized in Callicott, "Animal Liberation."

intrinsically functionless individuals with which I have so far been concerned are themselves *parts* of certain larger wholes—ecosystems. Moreover, the good of such systems and the good of the individuals living within them are not wholly disconnected. As Stephen Clark puts it:

> Plants too, and every clod of earth, are animate: not mystically so, but in straightforward biological terms. The earth itself, the biosphere itself is made up of living things and their products in a single interconnected whole. We are all members one of another, and the lowliest organism may be as vital to the whole as any Nobel prize-winner. More so, indeed. For the very fact which can be immediately adduced to mark the difference between man and plants, or men and micro-organisms, reveals that the latter are strictly very much more important than any one of us.[35]

How can I, calling as I do for respect for rocky crests or great oceans, ignore the claims of individual ecosystems let alone those of that "multi-millionfold life-support system that is the terrestrial biosphere"?[36]

One problem here is knowing just what the good of the biosphere is if it is something different from the sum of the goods of all the individuals living within it. Another, more technical, issue is whether we, or anything else are, literally, *parts* of ecosystems. It is certainly not right to think of our relationship to larger systems in the way that our components stand to us. At least, such an analogy is no more helpful than that which sees the relation of parts of a body to a whole body as similar to that of members of a family to the whole family, or citizens to the state.[37] To discuss this technical issue in detail here would not, I think, be rewarding. So let us stay neutral on the matter of parts and wholes, but bearing in mind that when I speak of the relation of members to a group I am in no way intending to suggest that this is at all the same as that of a component to a whole.

Anyone who is wary of facile reductionism would want to question the suggestion implicitly made at the start of the preceding paragraph that the good of the biosphere may be no more than the sum of the goods of its members. Think of an analogy with families. The good of a family is not merely the sum of the goods of its members, at least not if saying this involves ignoring the fact that I may make a sacrifice for the good of the family (to which I belong). As I suggested in section one, I can come to identify with a group to which I belong, and cease to see a conflict between my good and its good. So

---

[35] Stephen R. L. Clark, *The Moral Status of Animals* (Oxford: Clarendon Press, 1977), p. 170.
[36] Ibid., p. 160.
[37] A *part* is not the same thing as a *member*. Part/whole questions raise complex problems about identity. For a glimpse of such problems see my critical study of Eli Hirsch's book *The Concept of Identity* (New York: Oxford University Press, 1982), forthcoming in *Nous*. A critique of the view that citizens are parts of society can be found in David-Hillel Ruben, "Social Wholes and Parts," *Mind* 93 (1983): 219–38.

my good (being identified with the good of the family) can involve sacrifices on my part. On the other hand, the good of the family is not something quite distinct from the good of its members. My sacrifices will benefit the family only because they benefit other members of it, and it is, no doubt, the fact that I care about their good that involves me in making the sacrifices I do. Now is a family an intrinsically functionless thing? It is hardly an *individual* in the philosopher's sense, but it is a unitary entity of a sort. And even if we are troubled about agreeing that it is intrinsically functionless, there will be some intrinsically functionless groups in which I have some degree of membership. To be consistent, perhaps I ought to argue for the moral standing of such items, although such a claim would not be nearly so clear as one made on behalf of individuals, since the former depends on further clarifications of the status of such entities.

It should now be clear why I focused on individual natural things to begin with. This was no mere post-Renaissance individualism on my part, but rather a concern to work from the simpler cases. It is arguable that ecosystems fulfill the conditions for functionless objects and are therefore candidates for moral consideration every bit as worthy as trees, valleys, rivers and stones. If you take such a view, I have no objection to your taking my references to natural objects in this broad way. But I hope I have said enough to show that it is not a necessary corollary of this view that the good of such a large object will be distinct from, or even at odds with, the good of its members. Just as in the case of the family, the adoption of an appropriate environmental ethic may make it (morally) impossible for any opposition to arise between our good as humans and the good of the biosphere or of the planet.

But what is the focus of an appropriate environmental ethic? A question mark now hangs over notions of *welfare* or *interest,* for I am prepared to concede that rivers and deserts have, quite literally, no welfare or interests, and I would concede the point with regard to ecosystems as well. Of course, since the good of a system or group is, as I have admitted, not distinct from the good of its members, and since at least some members of every ecosystem have interests, we might be tempted by some reductionist account of the interests of such systems. Indeed, we could no doubt try to make sense of acting charitably toward a river, or of being benevolent to a desert. The open texture of our language is perfectly hospitable to just such conceptual development, but it is not part of my aim to argue for it here. Nor am I urging any kind of reductionist ploy. So it may be that a utilitarian, for example, will need either to reinterpret what I say so that notions of welfare, benevolence, and interest can have some kind of application in such cases or else to dismiss my arguments as morally irrelevant.

One small hope I entertain is that reflection on the points I have tried to make may undermine the sympathetic reader's allegiance to a morality based exclusively on notions like those of welfare or interest. We are terribly tempted to simplify and systematize, to think that simple rules and formulas will work for much of our moral life. Suggestions like the ones made in this paper threaten the neat systematizations others have made, and raise problems that few have ever taken seriously. It is bad enough trying to count the interests of sentient things in our calculations: but how are we to proceed when we count in things that are not sentient and have no interests? We are inclined, at least if we take views like utilitarianism seriously, to dismiss this latter group of things simply because they cannot be *counted* or given weight in our *calculations* in any obvious way. My response is to reject this kind of systematic approach to the problem. Let us give considerability to all intrinsically functionless natural things. The next step—not one I can take here—is to look at lots of cases, taking extended moral considerability seriously, and see how we can start to give due weight to the moral claims of the diverse items in the cases. I am pleased to see that others likewise distrust the appeal of systems in morality. Clark, for one, inveighs against moral systems which "present a sort of ghastly *reductio ad absurdum* of their own pretensions".[38]

There are a couple of concepts that figure in our everyday moral thinking which do, I think, have application to the kinds of items for which I have been suggesting we take moral considerability seriously. The first is the idea of *freedom* which seems to make little sense when applied to a functional component or to a merely functional artifact. For something to have an intrinsic function, its very existence as an item of the kind that it is depends upon its fulfilling whatever causal roles its function requires. An intrinsically functionless item, by contrast, can change, develop, and organize itself subject to quite different constraints. Of course, to be a flower of a certain sort is to have certain components organized in a certain structure—a structure determined in the end by the flower's microscopic genetic structure. But the same kind of flower will grow differently in different environments, play host to different sorts of insects, and fulfill all sorts of different imposed roles. We too have the freedom to adapt to various circumstances and environments, a freedom of which we ourselves are aware and a freedom that is far greater than that of a primula or a dandelion. But the difference between dandelions, primulas, and us is one of degree, not of kind.

What may come as a shock is not my notion that ecosystems may differ in degree rather than in kind from us, for after all, such systems are governed by internal principles which enable them to survive changes and crises almost as if they too were living things. Rather, the shock may come in the claim that deserts, rocks, and rivers are similar to us, too. The vagaries of human existence, our limited point of view, are no doubt important in explaining the difficulty we have in seeing such objects as self-organizing systems. If millenia were but seconds to us, we would be immediately aware of the logic of the rock

---

[38] Clark, *Moral Status of Animals,* p. 187.

cycle, the growth of mountains, and the ebb and flow of deserts. In such a way we can perhaps start to make sense of Wiggins' remarks about a principle of activity.[39] As it is, our knowledge of these things is only remote and indirect, but the claim that rivers, mountains, and wildernesses can be "tamed" is an implicit acknowledgement that they have, so to speak, their own wild, free way of existing.

If freedom is one notion that has a recognizable role in our moral thinking, another, I would suggest, is that of the *natural.* We can perhaps recognize that we ourselves are natural objects sharing the accident of existence with other autonomous natural objects. This separateness and independence of all natural things gives rise to a kind of dignity noted even by the Romantic poets on the occasions when they were able to transcend their often self-indulgent enjoyment of nature. Iris Murdoch writes:

> A self-directed enjoyment of nature seems to me to be something forced. More naturally, as well as more properly, we take a self-forgetful pleasure in the sheer alien pointless independent existence of animals, birds, stones and trees.[40]

In my account, the propriety of this self-forgetful pleasure is based on a certain common predicament that we and our fellow natural objects find ourselves in. That we, and other animals, can act purposively, set ourselves projects and strive intelligently to fulfill them should not lead us to forget the intrinsic functionlessness of uncreated things. We already have a concept of naturalness which leads us to protect and cherish the natural and mistrust the artificial. At its fashionable worst, respect for the natural is invoked in order to promote the sales of herbal shampoos and ginseng root, but such respect is also linked with our growing concern about factory farming, monoculture, aforestation, global pollution, genetic engineering, and loss of wilderness.

It may seem to some that what we are dealing with here might more properly be regarded as *aesthetic* rather than purely moral matters. But if this paper has set out mainly to challenge the notion that sentience, rationality, or even life itself exhaust the bases for moral concern, then a corollary would be a challenge to the view that respect for the natural can be dismissed as merely aesthetic. How, indeed, are we to separate the aesthetic and the moral? If we claim that morality, unlike aesthetics, deals with interest or welfare, then this would be merely question-begging. Perhaps our care for what is natural is both aesthetic and moral: the burden, I would argue, is on the objector to establish that such care is in some distinctive way nonmoral.

My conclusion, then, is that it is quite unclear that appeals to rationality and the rest give rise to claims that are morally more urgent, or more central, than

---

[39] See note 20.
[40] Iris Murdoch, *The Sovereignty of Good* (London: Routledge and Kegan Paul, 1970), p. 85.

the appeal to intrinsic lack of function. I recognize, of course, that much of my argument here has been sketchy in the extreme, but such an approach suits the hypothetical and exploratory nature of the undertaking. If my suggestions have worth, then a great deal will need to be done by way of filling out the details, and justifying claims which at the moment stand on insecure foundations. But if by writing this I have persuaded at least some readers to look at the scope of morality from a new viewpoint, and to consider the subtleties that would be involved in trying to reconcile our moral respect for different kinds of thing and adjudicating the different claims they make on us, then I will have succeeded in overcoming some of the bias to which our moral reflections are so regularly prone.

[4]

THE VARIETIES OF INTRINSIC VALUE*

To hold an environmental ethic is to hold that non-human beings and states of affairs in the natural world have intrinsic value. This seemingly straightforward claim has been the focus of much recent philosophical discussion of environmental issues. Its clarity is, however, illusory. The term 'intrinsic value' has a variety of senses and many arguments on environmental ethics suffer from a conflation of these different senses: specimen hunters for the fallacy of equivocation will find rich pickings in the area. This paper is largely the work of the underlabourer. I distinguish different senses of the concept of intrinsic value, and, relatedly, of the claim that non-human beings in the natural world have intrinsic value; I exhibit the logical relations between these claims and examine the distinct motivations for holding them. The paper is not however merely an exercise in conceptual underlabouring. It also defends one substantive thesis: that while it is the case that natural entities have intrinsic value in the strongest sense of the term, i.e., in the sense of value that exists independently of human valuations, such value does not as such entail any obligations on the part of human beings. The defender of nature's intrinsic value still needs to show that such value contributes to the well-being of human agents.

I

The term 'intrinsic value' is used in at least three different basic senses:
(1) **Intrinsic value**$_1$ Intrinsic value is used as a synonym for non-instrumental value. An object has instrumental value insofar as it is a means to some other end. An object has intrinsic value if it is an end in itself. Intrinsic goods are goods that other goods are good for the sake of. It is a well rehearsed point that, under pain of an infinite regress, not everything can have only instrumental value. There must be some objects that have intrinsic value. The defender of an environmental ethic argues that among the entities that have such non-instrumental value are non-human beings and states. It is this claim that Naess makes in defending deep ecology:

> The well-being of non-human life on Earth has value in itself. This value is independent of any instrumental usefulness for limited human purposes.[1]

(2) **Intrinsic value**$_2$ Intrinsic value is used to refer to the value an object has solely in virtue of its 'intrinsic properties'. The concept is thus employed by G. E. Moore:

> To say a kind of value is 'intrinsic' means merely that the question whether a thing possesses it, and in what degree it possesses it, depends solely on the intrinsic nature of the thing in question.[2]

This account is in need of some further clarification concerning what is meant by the 'intrinsic nature' of an object or its 'intrinsic properties'. I discuss this further below. However, as a first approximation, I will assume the intrinsic properties of an object to be its non-relational properties, and leave that concept for the moment unanalysed. To hold that non-human beings have intrinsic value given this use is to hold that the value they have depends solely on their non-relational properties.

(3) **Intrinsic value**$_3$ Intrinsic value is used as a synonym for 'objective value' i.e., value that an object possesses independently of the valuations of valuers. As I show below, this sense itself has sub-varieties, depending on the interpretation that is put on the term 'independently'. Here I simply note that if intrinsic value is used in this sense, to claim that non-human beings have intrinsic value is not to make an ethical but a meta-ethical claim. It is to deny the subjectivist view that the source of all value lies in valuers—in their attitudes, preferences and so on.

Which sense of 'intrinsic value' is the proponent of an environmental ethic employing? To hold an environmental ethic is to hold that non-human beings have intrinsic value in the first sense: it is to hold that non-human beings are not simply of value as a means to human ends. However, it might be that to hold a defensible ethical position about the environment, one needs to be committed to the view that they also have intrinsic value in the second or third senses. Whether this is the case is the central concern of this paper.

II

In much of the literature on environmental ethics the different senses of 'intrinsic value' are used interchangeably. In particular senses 1 and 3 are often conflated. Typical is the following passage from Worster's *Nature's Economy*:

> One of the most important ethical issues raised anywhere in the past few decades has been whether nature has an order, a pattern, that we humans are bound to understand and respect and preserve. It is the essential question prompting the

environmentalist movement in many countries. Generally, those who have answered 'yes' to the question have also believed that such an order has an intrinsic value, which is to say that not all value comes from humans, that value can exist independently of us: it is not something we bestow. On the other hand, those who have answered 'no' have tended to be in an instrumentalist camp. They look on nature as a storehouse of 'resources' to be organised and used by people, as having no other value than the value some human gives it.[3]

In describing the 'yes' camp Worster characterises the term in sense 3. However, in characterising the 'no's' he presupposes an understanding of the term in both senses 1 and 3. The passage assumes that to deny that natural patterns have value independently of the evaluations of humans is to grant them only instrumental value: a subjectivist meta-ethics entails that non-humans can have only instrumental value. This assumption is widespread.[4] It also underlies the claims of some critics of an environmental ethic who reject it on meta-ethical grounds thus: To claim that items in the non-human world have intrinsic values commits one to an objectivist view of values; an objectivist view of values is indefensible; hence the non-human world contains nothing of intrinsic value.[5]

The assumption that a subjectivist meta-ethics commits one to the view that non-humans have only instrumental value is false. Its apparent plausibility is founded on a confusion of claims about the source of values with claims about their object.[6] The subjectivist claims that the only sources of value are the evaluative attitudes of humans. But this does not entail that the only ultimate objects of value are the states of human beings. Likewise, to be an objectivist about the source of value, i.e., to claim that whether or not something has value does not depend on the attitudes of valuers, is compatible with a thoroughly anthropocentric view of the object of value—that the only things which do in fact have value are humans and their states, such that a world without humans would have no value whatsoever.

To enlarge, consider the emotivist as a standard example of a subjectivist. Evaluative utterances merely evince the speaker's attitudes with the purpose of changing the attitudes of the hearer. They state no facts. Within the emotivist tradition Stevenson provides an admirably clear account of intrinsic value. Intrinsic value is defined as non-instrumental value: ' "intrinsically good" is roughly synonymous with "good for its own sake, as an end, as distinct from good as a means to something else" '.[7] Stevenson then offers the following account of what it is to say something has intrinsic value:

> 'X is intrinsically *good*' asserts that the speaker approves of X intrinsically, and acts emotively to make the hearer or hearers likewise approve of X intrinsically.[8]

There are no reasons why the emotivist should not fill the X place by entities and states of the non-human world. There is nothing in the emotivist's meta-ethical position that precludes her holding basic attitudes that are biocentric. Thus let the H! operator express hurrah attitudes and B! express boo attitudes.[9] Her ultimate values might for example include the following:

    H! (The existence of natural ecosystems)
    B! (The destruction of natural ecosystems by humans).

There is no reason why the emotivist must assume that either egoism or humanism is true, that is that she must assign non-instrumental value only to her own or other humans' states.[10]

It might be objected, however, that there are other difficulties in holding an emotivist meta-ethics and an environmental ethic. In making humans the source of all value, the emotivist is committed to the view that a world without humans contains nothing of value. Hence, while nothing logically precludes the emotivists assigning non-instrumental value to objects in a world which contains humans, it undermines some of the considerations that have led to the belief in the need to assign such value. For example, the standard last man arguments[11] in defence of an environmental ethic fail: the last man whose last act is to destroy a rain forest could on a subjectivist account of value do no wrong, since a world without humans is without value.

This objection fails for just the same reason as did the original assumption that subjectivism entails non-humans have only instrumental value. It confuses the source and object of value. There is nothing in emotivism that forces the emotivist to confine the objects of her attitudes to those that exist at the time at which she expresses them. Her moral utterances might evince attitudes towards events and states of affairs that might happen after her death, for example,

    H! (My great grand-children live in a world without poverty).

Likewise her basic moral attitudes can range over periods in which humans no longer exist, for example,

    H! (Rain forests exist after the extinction of the human species).

Like the rest of us she can deplore the vandalism of the last man. Her moral utterances might evince attitudes not only to other times but also to other possible worlds. Nothing in her meta-ethics stops her asserting with Leibniz that this world is the best of all possible worlds, or, in her despair at the destructiveness of humans, expressing the attitude that it would have been better had humans never existed:

    H! (the possible world in which humans never came into existence).

That humans are the source of value is not incompatible with their assigning value to a world in which they do not exist. To conclude, nothing in the emotivist's meta-ethics dictates the content of her attitudes.

Finally it needs to be stressed that while subjectivism does not rule out non-humans having non-instrumental value, objectivism does not rule it in. To claim that moral utterances have a truth value is not to specify which utterances are true. The objectivist can hold that the moral facts are such that only the states of humans possess value in themselves: everything else has only instrumental value. Ross, for example, held that only states of conscious beings have intrinsic value:

> Contemplate any imaginary universe from which you suppose mind entirely absent, and you will fail to find anything in it you can call good in itself.[12]

Moore allowed that without humans the world might have some, but only very insignificant, value.[13] It does not follow from the claim that values do not have their source in humans that they do not have humans as their sole ultimate object.

The upshot of this discussion is a very traditional one, that meta-ethical commitments are logically independent of ethical ones. However, in the realm of environmental ethics it is one that needs to be re-affirmed. No meta-ethical position is required by an environmental ethic in its basic sense, i.e., an ethic which holds that non-human entities should not be treated merely as a means to the satisfaction of human wants. In particular, one can hold such an ethic and deny objectivism. However, this is not to say that there might not be other reasons for holding an objectivist account of ethics and that some of these reasons might appear particularly pertinent when considering evaluative statements about non-humans. It has not been my purpose in this section of the paper to defend ethical subjectivism and in section IV I defend a version of objectivism about environmental values. First, however, I discuss briefly intrinsic value in its Moorean sense, intrinsic value$_2$—for this sense of the term is again often confused with intrinsic value$_1$.

III

In its second sense intrinsic value refers to the value an object has solely in virtue of its 'intrinsic properties': it is value that 'depends solely on the intrinsic nature of the thing in question'.[14] I suggested earlier that the intrinsic properties of an object are its non-relational properties. What is meant by 'non-relational properties'? There are two interpretations that might be placed on the phrase:

(i) The non-relational properties of an object are those that persist regardless of the existence or non-existence of other objects (weak interpretation).

(ii) The non-relational properties of an object are those that can be characterised without reference to other objects (strong interpretation).[15]

The distinction between the two senses will not concern me further here, although a similar distinction will take on greater significance in the following section.

If any property is irreducibly relational then rarity is. The rarity of an object depends on the non-existence of other objects, and the property cannot be characterised without reference to other objects. In practical concern about the environment a special status is ascribed to rare entities. The preservation of endangered species of flora and fauna and of unusual habitats and ecological systems is a major practical environmental problem. Rarity appears to confer a special value to an object. This value is related to that of another irreducibly relational property of environmental significance, i.e., diversity. However, it has been argued that such value can have no place in an environmental ethic which places intrinsic value on natural items. The argument runs something as follows:

1. To hold an environmental ethic is to hold that natural objects have intrinsic value.
2. The values objects have in virtue of their relational properties, e.g., their rarity, cannot be intrinsic values.

Hence:

3. The value objects have in virtue of their relational properties have no place in an environmental ethic.[16]

This argument commits a fallacy of equivocation. The term 'intrinsic value' is being used in its Moorean sense, intrinsic value$_2$ in the second premise, but as synonym for non-instrumental value, intrinsic value$_1$, in the first. The senses are distinct. Thus, while it may be true that if an object has only instrumental value it cannot have intrinsic value in the Moorean sense, it is false that an object of non-instrumental value is necessarily also of intrinsic value in the Moorean sense. We might value an object in virtue of its relational properties, for example its rarity, without thereby seeing it as having only instrumental value for human satisfactions.

This point can be stated with greater generality. We need to distinguish:

(1) values objects can have in virtue of their relations to other objects; and

(2) values objects can have in virtue of their relations to human beings.[17]

The second set of values is a proper subset of the first. Moreover, the second set of values is still not co-extensive with
> (3) values objects can have in virtue of being instrumental for human satisfaction.

An object might have value in virtue of its relation with human beings without thereby being of only instrumental value for humans. Thus, for example, one might value wilderness in virtue of its not bearing the imprint of human activity, as when John Muir opposed the damming of the Hetch Hetchy valley on the grounds that wild mountain parks should lack 'all . . . marks of man's work'.[18] To say 'x has value because it is untouched by humans' is to say that it has value in virtue of a relation it has to humans and their activities. Wilderness has such value in virtue of our absence. However, the value is not possessed by wilderness in virtue of its instrumental usefulness for the satisfaction of human desires. The third set of values is a proper subset of both the second and the first. Intrinsic value in the sense of non-instrumental value need not then be intrinsic in the Moorean sense.

What of the relation between Moorean intrinsic value and objective value? Is it the case that if there is value that 'depends solely on the intrinsic nature of the thing in question' then subjectivism about values must be rejected? If an object has value only in virtue of its intrinsic nature, does it follow that it has value independently of human valuations? The answer depends on the interpretation given to the phrases 'depends solely on' and 'only in virtue of'. If these are interpreted to exclude the activity of human evaluation, as I take it Moore intended, then the answer to both questions is immediately 'yes'. However, there is a natural subjectivist reading to the phrases. The subjectivist can talk of the valuing agent *assigning* value to objects solely in virtue of their intrinsic natures. Given a liberal interpretation of the phrases, a subjectivist can hold that some objects have intrinsic value in the Moorean sense.

IV

In section II I argued that the claim that nature has non-instrumental value does not commit one to an objectivist meta-ethics. However, I left open the question as to whether there might be other reasons particularly pertinent in the field of environmental ethics that would lead us to hold an objectivist account of value. I will show in this section that there are.

The ethical objectivist holds that the evaluative properties of objects are real properties of objects, that is, that they are properties that objects possess independently of the valuations of valuers. What is meant by 'in-

dependently of the valuations of valuers'? There are two readings of the phrase which parallel the two senses of 'non-relational property' outlined in the last section:
> (1) The evaluative properties of objects are properties that exist in the absence of evaluating agents. (Weak interpretation)
>
> (2) The evaluative properties of objects can be characterised without reference to evaluating agents. (Strong interpretation)

The distinction is a particular instance of a more general distinction between two senses in which we can talk of a property being a real property of an object:
> (1) A real property is one that exists in the absence of any being experiencing that object. (Weak interpretation)
>
> (2) A real property is one that can be characterised without reference to the experiences of a being who might experience the object. (Strong interpretation)

Is there anything about evaluations of the environment that make the case for objectivism especially compelling? I begin by considering the case for the weak version of objectivism. For the purpose of the rest of the discussion I will assume that only human persons are evaluating agents.

1. Weak Objectivity

A popular move in recent work on environmental ethics has been to establish the objectivity of values by invoking an analogy between secondary qualities and evaluative properties in the following manner:
> (1) The evaluative properties of objects are analogous to secondary qualities. Both sets of properties are observer dependent.
>
> (2) The Copenhagen interpretation of quantum mechanics has shown the distinction between primary qualities and secondary qualities to be untenable. All the properties of objects are observer dependent.

Hence,
> (3) the evaluative properties of objects are as real as their primary qualities.[19]

The argument fails at every stage. In the first place the conclusion itself is too weak to support objectivism about values: it is no argument for an objectivist theory of values to show that all properties of objects are observer dependent. The second premise should in any case be rejected. Not only is it the case that the Copenhagen interpretation of quantum theory is but one amongst many,[20] it is far from clear that the Copenhagen interpretation is committed to the ontological extravagance that all properties are observer dependent. Rather it can be understood as a straightforward instrumentalist interpretation of quantum theory. As such it involves no ontological commitments about the quantum domain.[21]

More pertinent to the present discussion, there are also good grounds for rejecting the first premise. The analogy between secondary qualities and values has often been used to show that values are not real properties of objects. Thus Hume remarks:

> Vice and virtue . . . may be compared to sounds, heat and cold, which, according to modern philosophy, are not qualities in objects, but perceptions in the mind . . . [22]

For the Humean, both secondary qualities and evaluative properties are not real properties of objects, but, rather, illustrate the mind's 'propensity to spread itself on external objects': as Mackie puts it, moral qualities are the 'projection or objectification of moral attitudes'.[23] The first premise of the argument assumes this Humean view of the analogy between secondary qualities and values. However, there are good grounds for inverting the analogy and that inversion promises to provide a more satisfactory argument for objectivism than that outlined above.

On the weak interpretation of the concept of a real property, secondary qualities are real properties of objects. They persist in the absence of observers. Objects do not lose their colours when we no longer perceive them. In the kingdom of the blind the grass is still green. Secondary qualities are dispositional properties of objects to appear in a certain way to ideal observers in ideal conditions. So, for example, an object is green if and only if it would appear green to a perceptually ideal observer in perceptually ideal conditions.[24] It is consistent with this characterisation of secondary qualities that an object possesses that quality even though it may never actually be perceived by an observer. Thus, while in the strong sense of the term secondary qualities are not real properties of objects—one cannot characterise the properties without referring to the experiences of possible obervers—in the weak sense of the term they are.[25]

This point opens up the possibility of an inversion of the Humean analogy between secondary and evaluative qualities which has been recently exploited by McDowell and others.[26] Like the secondary qualities, evaluative qualities are real properties of objects. An object's evaluative properties are similarly dispositional properties that it has to produce certain attitudes and reactions in ideal observers in ideal conditions. Thus, we might tentatively characterise goodness thus: x is good if and only if x would produce feelings of moral approval in an ideal observer in ideal conditions. Likewise, beauty might be characterised thus: x is beautiful if and only if x would produce feelings of aesthetic delight in ideal observers in ideal conditions. Given this characterisation, an object is beautiful or good even if it never actually appears as such to an observer. The evaluative properties of objects are real in just the same sense that secondary qualities are. Both sets of properties are independent of observers in the sense that they persist in the absence of observers. The first premise of the argument outlined above should therefore be rejected. Furthermore, in rejecting this premise, one arrives at a far more convincing case for the reality of evaluative properties than that provided by excursions into quantum mechanics.

However, the promise of this line of argument for environmental ethics is, I believe, limited. There are a variety of particular arguments that might be raised against it. For example, the Humean might respond by suggesting that the analogy between secondary and evaluative properties is imperfect. The arguments for and against the analogy I will not rehearse here.[27] For even if the analogy is a good one, it is not clear to me that any point of substance about the nature of values divides the Humean and his opponent. The debate is one about preferred modes of speech, specifically about how the term 'real property' is to be read. For the Humean such as Mackie, the term 'real property' is understood in its strong sense. It is a property that can be characterised without reference to the experiences of an observer. Hence neither secondary qualities nor values are real properties of objects. The opponent of the Humean in employing the analogy to establish the reality of evaluative properties merely substitutes a weak interpretation of 'real property' for the strong interpretation. There may be good reasons for doing this, but nothing about the nature of values turns on this move.[28] Moreover, there seems to be nothing about evaluative utterances concerning the natural environment which adds anything to this debate. Nothing about specifically environmental values tells for or against this argument for objectivism.

2. Strong Objectivity

A more interesting question is whether there are good reasons for believing that there are objective values in the strong sense: are there evaluative properties that can be characterised without reference to the experiences of human observers? I will now argue that there are and that uses of evaluative utterances about the natural world provide the clearest examples of such values.

Consider the gardener's use of the phrase 'x is good for greenfly'. The term 'good for' can be understood in two distinct ways. It might refer to what is conductive to the destruction of greenfly, as in 'detergent sprays are good for greenfly', or it can be used to describe what causes greenfly to flourish, as in 'mild winters are good for greenfly'. The term 'good for' in the first use describes what is instrumentally good for the gardener: given the ordinary gardener's interest in the flourishing of her rosebushes, detergent sprays satisfy that interest. The second use describes what is instrumentally good for the greenfly, quite independently of the gardener's interests. This instrumental goodness is possible in virtue of the fact that greenflies are the sorts of things that can flourish or be injured. In consequence they have their own goods that are independent of both human interests and any tendency they might have to produce in human observers feelings of approval or disapproval.[29] Such goods I will follow Von Wright in terming the 'goods of X'.[30]

What is the class of entities that can be said to possess such goods? Von Wright in an influential passage offers the following account:

> A being, of whose good it is meaningful to talk, is one who can meaningfully be said to be well or ill, to thrive, to flourish, be happy or miserable ... the attributes, which go along with the meaningful use of the phrase 'the good of X', may be called *biological* in a broad sense. By this I do not mean that they were terms, of which biologists make frequent use. 'Happiness' and 'welfare' cannot be said to belong to the professional vocabulary of biologists. What I mean by calling the terms 'biological' is that they are used as attributes of beings, of whom it is meaningful to say they have a *life*. The question 'What kinds or species of being have a good?' is therefore broadly identical with the question 'What kinds or species of being have a life'.[31]

This biological use of the terms 'good for' and 'good of' is at the centre of Aristotelian ethics. The distinction between 'good for' and 'good of' itself corresponds to the Aristotelian distinction between goods externally instrumental to a being's flourishing and those that are constitutive of a being's flourishing.[32] And the central strategy of Aristotle's ethics is to found ethical argument on the basis of this broadly biological use of the term 'good'. I discuss this further below.

The terms 'good' and 'goods' in this biological context characterise items which are real in the strong interpretation of the term. In order to characterise the conditions which are constitutive of the flourishing of a living thing one need make no reference to the experiences of human observers. The goods of an entity are given rather by the characteristic features of the kind or species of being it is. A living thing can be said to flourish if it develops those characteristics which are normal to the species to which it belongs in the normal conditions for that species. If it fails to realise such characteristics then it will be described by terms such as 'defective', 'stunted', 'abnormal' and the like. Correspondingly, the truth of statements about what is good for a living thing, what is conducive to its flourishing, depend on no essential reference to human observers. The use of the evaluative terms in the biological context does then provide good reasons for holding that some evaluative properties are real properties on the strong interpretation of the phrase. Hence, evaluative utterances about living things do have a particular relevance to the debate about the objectivity of values. Specifically biological values tell for objectivism.

However, while the use of value terms in the specifically biological context provides the clearest examples of the existence of objective goods, the class of entities that can be meaningfully said to have such goods is not confined to the biological context. Von Wright's claim that the question 'What kinds or species of being have a good?' is identical with the question 'What kinds or species of being have a life' should be rejected. The problem case for this identity claim is that of collective entities. Von Wright is willing to entertain the possibility that such entities have their own good but only if they can also be said to have their own life in a non-metaphorical sense.

> But what shall we say of social units such as the family, the nation, the state. Have they got a life 'literally' or 'metaphorically' only? I shall not attempt to answer these questions. I doubt whether there is any other way of answering them except by pointing out existing analogies of language. It is a fact that we speak about the life and also the good (welfare) of the family, the nation and the state. This fact about the use of language we must accept and with it the idea that the social units in question *have* a life and a good. What is arguable, however, is whether the life and *a fortiori* also the good (welfare) of a social unit is not somehow 'logically reducible' to the life and therefore the good of the beings—men or animals—who are its members.[33]

This passage conflates two distinct issues: whether collective entities have a life and whether they have their own goods. It does not appear to me that we can talk of collective entities having a life in anything but a metaphorical sense. They clearly lack those properties typical of living things—reproduction, growth, death and such like. However, it does make sense to talk about the conditions in which collective entities flourish and hence of their goods in a non-metaphorical sense. Correspondingly, we can meaningfully talk of what is damaging to them. Furthermore, the goods of collective entities are not reducible to the goods of their members. Thus for example we can refer to the conditions in which bureaucracy flourishes while believing this to be bad for its constituent members. Or to take another example, what is good for members of a workers' cooperative can be quite at odds with what is good for the cooperative itself: the latter is constituted by its relative competitive position in the market place, and members of cooperatives might find themselves forced to forego the satisfaction of their own interests to realise this.[34] The question 'What class of beings has a good?' is identical with the question 'What class of beings can be said to flourish in a non-metaphorical sense?' The class of living things is a proper subset of this class.

This point is central to environmental questions. It makes sense to talk of the goods of collective biological entities—colonies, ecosystems and so on—in a way that is irreducible to that of its members. The realisation of the good of a colony of ants might in certain circumstances involve the death of most of its members. It is not a condition for the flourishing of an individual animal that it be eaten: it often is a condition for the flourishing of the ecosystem of which it is a part. Relatedly, a point central to Darwin's development of the theory of evolution was that living beings have a capacity to reproduce that outstrips the capacity of the environment to support them. Most members of a species die in early life. This is clearly bad for the individuals involved. But it is again essential to the flourishing of the ecosystems of which they are a part. Collective entities have their own goods. In defending this claim one need not show that they have their own life.[35]

Both individual living things and the collective entities of which they are members can be said, then, to have their own goods. These goods are quite independent of human interests and can be characterised without reference to the experiences of human observers. It is a standard at this juncture of the argument to assume that possession of goods entails moral considerability: 'moral standing or considerability belongs to whatever has a good of its own'.[36] This is mistaken. It is possible to talk in an objective sense of what constitutes the goods of entities, without making any claims that these ought to be realised. We can know what is 'good for X' and relatedly what constitutes 'flourishing for X' and yet believe that X is the sort of thing that ought not to exist and hence that the flourishing of X is just the sort of thing we ought to inhibit. The case of the gardener noted earlier is typical in this regard. The gardener knows what it is for greenfly to flourish, recognises they have their own goods, and has a practical knowledge of what is good for them. No moral injunction follows. She can quite consistently believe they ought to be done harm. Likewise one can state the conditions for the flourishing of dictatorship and bureaucracy. The anarchist can claim that 'war is the health of the state'. One can discover what is good both for rain forests and the AIDS virus. One can recognise that something has its own goods, and quite consistently be morally indifferent to these goods or believe one has a moral duty to inhibit their development.[37] That Y is a good of X does not entail that Y should be realised unless we have a prior reason for believing that X is the sort of thing whose good ought to be promoted. While there is not a logical gap between facts and values, in that some value statements are factual, there is a logical gap between facts and oughts. 'Y is a good' does not entail 'Y ought to be realised'.[38]

This gap clearly raises problems for environmental ethics. The existence of objective goods was promising precisely because it appeared to show that items in the non-human world were objects of proper moral concern. The gap outlined threatens to undermine such concern. Can the gap be bridged? There are two ways one might attempt to construct such a bridge. The first is to invoke some general moral claim that linked objective goods and moral duties. One might for example invoke an objectivist version of utilitarianism: we have a moral duty to maximise the total amount of objective good in the world.[39] There are a number of problems of detail with such an approach: What are the units for comparing objective goods? How are different goods to be weighed? However, it also has a more general problem that it shares with hedonistic utilitarianism. Thus, the hedonistic utilitarian must include within his calculus pleasures that ought not to count at all e.g., those of a sadist who gets pleasure from needless suffering. The hedonistic utilitarian fails to allow that pleasures themselves are the direct objects of ethical appraisal. Similarly, there are some entities whose flourishing simply should not enter into any calculations—the flourishing of dictatorships and viruses for example. It is not the case that the goods of viruses should count, even just a very small amount. There is no reason why these goods should count at all as ends in themselves (although there are of course good *instrumental* reasons why some viruses should flourish, in that many are indispensable to the ecosystems of which they are a part). The flourishing of such entities is itself a direct object of ethical appraisal. The quasi-utilitarian approach is unpromising.

A second possible bridge between objective goods and oughts is an Aristotelian one. Human beings like other entities have goods constitutive of their flourishing, and correspndingly other goods instrumental to that flourishing. The flourishing of many other living things ought to be promoted because they are constitutive of our own flourishing. This approach might seem a depressingly familiar one. It looks as if we have taken a long journey into objective value only to arrive back at a narrowly anthropocentric ethic. This however would be mistaken. It is compatible with an Aristotelian ethic that we value items in the natural world for their own sake, not simply as an external means to our own satisfaction. Consider

Aristotle's account of the relationship of friendship to human flourishing.[40] It is constitutive of friendship of the best kind that we care for friends for their own sake and not merely for the pleasures or profits they might bring. To do good for a friend purely because one thought they might later return the compliment not for their own sake is to have an ill-formed friendship. Friendship in turn is a constitutive component of a flourishing life. Given the kind of beings we are, to lack friends is to lack part of what makes for a flourishing human existence. Thus the egoist who asks 'why have friends?' or 'why should I do good for my friends' has assumed a narrow range of goods—'the biggest share of money, honours and bodily pleasures'[41]—and asked how friends can bring such goods. The appropriate response is to point out that he has simply misidentified what the goods of a human life are.

The best case for an environmental ethic should proceed on similar lines. For a large number of, although not all, individual living things and biological collectives, we should recognise and promote their flourishing as an end in itself.[42] Such care for the natural world is constitutive of a flourishing human life. The best human life is one that includes an awareness of and practical concern with the goods of entities in the non-human world. On this view, the last man's act of vandalism reveals the man to be leading an existence below that which is best for a human being, for it exhibits a failure to recognise the goods of non-humans. To outline such an approach is, however, only to provide a promissory note. The claim that care for the natural world for its own sake is a part of the best life for humans requires detailed defence. The most promising general strategy would be to appeal to the claim that a good human life requires a breadth of goods. Part of the problem with egoism is the very narrowness of the goods it involves. The ethical life is one that incorporates a far richer set of goods and relationships than egoism would allow. This form of argument can be made for a connection of care for the natural world with human flourishing: the recognition and promotion of natural goods as ends in themselves involves just such an enrichment.[43]

John O'Neill

University of Sussex,
Brighton, England

NOTES

*Earlier versions of this paper were read to an Open University summer school and to a philosophy seminar at Sussex University. My special thanks to Roger Crisp, Andrew Mason and Ben Gibbs for their comments on these occasions. Thanks are also due to Robin Attfield, John Benson, Stephen Clark, Terry Diffey, Alan Holland and Geoffrey Hunter for conversations on the issues discussed in this paper.

1. A. Naess, 'A Defence of the Deep Ecology Movement', *Environmental Ethics*, 6 (1984), 266. However, Naess's use of the term is unstable and he sometimes uses the phrase 'intrinsic value' to refer to objective value. See n4, below.

2. G. E. Moore, 'The Conception of Intrinsic Value' in *Philosophical Studies* (London: Routledge and Kegan Paul, 1922), p. 260.

3. D. Worster, *Nature's Economy* (Cambridge: Cambridge University Press, 1985), p. xi.

4. Thus, for example, Naess and Rothenberg in *Ecology, Community and Lifestyle* (Cambridge: Cambridge University Press, 1989) initially define 'intrinsic value' as value which is 'independent of our valuation' (*ibid.*, p. 11) but then in the text characterise it in terms of a contrast with instrumental value (*ibid.*, pp. 74-75). In his own account of deep ecology Naess employs the term in the sense of non-instrumental value (see n2 and A. Naess, 'The Shallow and the Deep: Long Range Ecology Movement' *Inquiry*, 16, 1973). Others are more careful. Thus, while Attfield is committed to both an objectivist meta-ethics and the view that the states of some non-humans have intrinsic value, in *A Theory of Value and Obligation* (London: Croom Helm, 1987) ch. 2, he *defines* intrinsic value as non-instrumental value and distinguishes this from his 'objectivist understanding of it'. Callicott in 'Intrinsic Value, Quantum Theory, and Environmental Ethics', *Environmental Ethics*, 7 (1989), 257-75, distinguishes non-instrumental value from objective value, using the term 'inherent value' for the former and 'intrinsic value' for the latter. However, the use of these terms raises its own problems since there is little agreement in the literature as to how they are to be employed. For example, P. Taylor, *Respect for Nature* (Princeton, NJ: Princeton University Press, 1986) pp. 68-77 makes the same distinction but uses 'inherent value' to describe Callicott's 'intrinsic value' and 'intrinsic value' to describe his 'inherent value', while R. Attfield in *The Ethics of Environmental Concern* (Oxford: Blackwell, 1983) ch. 8, uses the term 'inherent value' to refer to something quite different. Another exceptionally clear discussion of the meta-ethical issues surrounding environmental ethics is R. and V. Routley, 'Human Chauvinism and Environmental Ethics' in D. Mannison, M. McRobbie and R. Routley (eds.), *Environmental Philosophy* (Canberra: Australian National University, 1980).

5. This kind of argument is to be found in particular in the work of McCloskey. See H. J. McCloskey, 'Ecological Ethics and its Justification' in Mannison *et al.*, *op. cit.*, and *Ecological Ethics and Politics* (Totowa, NJ: Rowman and Littlefield, 1983).

6. Cf. D. Gauthier, *Morals by Agreement* (Oxford: Oxford University Press, 1986) pp. 46-49 and J. B. Callicott, 'Intrinsic Value, Quantum Theory and Environmental Ethics', *Environmental Ethics* 7 (1985), 257-75, who make this point quite emphatically.

7. C. L. Stevenson, *Ethics and Language* (New Haven, CT: Yale University Press, 1944).

THE VARIETIES OF INTRINSIC VALUE

8. *Ibid.*, p. 178.
9. I take the operators from S. Blackburn, *Spreading the Word* (Oxford: Clarendon Press, 1984), p. 193ff.
10. Cf. R. and V. Routley, 'Human Chauvinism and Environmental Ethics' in D. Mannison, M. McRobbie and R. Routley (eds.), *Environmental Philosophy* (Canberra: Australian National University, 1980).
11. See *ibid.*, pp. 121-23.
12. W. D. Ross, *The Right and the Good* (Oxford: Clarendon Press, 1930), p. 140. Ross held four things to have intrinsic value—'virtue, pleasure, the allocation of pleasure to the virtuous, and knowledge' (*ibid.*, p. 140).
13. G. E. Moore, *Principia Ethica* (Cambridge: Cambridge University Press, 1903), pp. 28, 83ff. and 188ff.
14. G. E. Moore, 'The Conception of Intrinsic Value', *Philosophical Studies* (London: Routledge and Kegan Paul, 1922), p. 260.
15. I do not follow Moore's own discussion here. Moore's own use of the term is closer to the weaker than the stronger interpretation. Thus, for example, the method of isolation as a test of intrinsic value proceeds by considering if objects keep their value 'if they existed *by themselves*, in absolute isolation': G. E. Moore *Principia Ethica* (Cambridge: Cambridge University Press, 1903), p. 187.
16. A similar argument is to be found in A. Gunn, 'Why Should We Care about Rare Species?', *Environmental Ethics*, 2, 1980, pp. 17-37, especially pp. 29-34.
17. J. Thompson partially defines intrinsic value and hence an environmental ethic in terms of a contrast with such values: 'those who find intrinsic value in nature are claiming . . . that things and states which are of value are valuable for what they are in themselves and not because of their relation to us . . . ' (J. Thompson, 'A Refutation of Environmental Ethics', p. 148, *Environmental Ethics*, 12 (1990), 147-60). This characterisation is inadequate, in that it rules out of an environmental ethic positions such as that of Muir who values certain parts of nature because of the absence of the marks of humans. I take it that Thompson intends a contrast to the third set of values—values objects can have in virtue of being instrumental for human satisfaction.
18. Cited in R. Dubos, *The Wooing of Earth* (London: The Athlone Press, 1980) p. 135.
19. A relatively sophisticated version of the argument is to be found in Holmes Rolston, III, 'Are Values in Nature Subjective or Objective?', pp. 92-95 in *Philosophy Gone Wild* (Prometheus Books, Buffalo, NY: 1989). Cf. J. B. Callicott, 'Intrinsic Value, Quantum Theory and Environmental Ethics', *Environmental Ethics* (1985) 7, pp. 257-75.
20. M. Jammer, *The Philosophy of Quantum Mechanics* (New York: John Wiley, 1974) remains a good survey of the basic different interpretations of quantum theory.
21. It should also be noted that the view, popular among some Green thinkers (see, for example, F. Capra, *The Tao of Physics* [London: Wildwood House, 1975]), that the Copenhagen interpretation entails a radically new world-view that undermines the old classical Newtonian picture of the world is false. The Copenhagen interpretation is conceptually conservative and denies the possibility that we could replace the concepts of classical physics by any others (see N. Bohr, *Atomic Theory and the Description of Nature* [Cambridge: Cambridge University Press, 1934], p.

94. Cf. W. Heisenberg, *Physics and Philosophy* [London: Allen and Unwin, 1959] p. 46. I discuss this conservativism in J. O'Neill, *Worlds Without Content* [London: Routledge, in press], ch. 6

22. D. Hume, *A Treatise of Human Nature* (London: Fontana, 1972), Book III, §1, p. 203.

23. J. Mackie, *Ethics* (Harmondsworth, England: Penguin, 1977) p. 42.

24. Cf. J. McDowell, 'Values and Secondary Qualities', p. 111 in T. Honderich (ed.), *Morality and Objectivity* (London: Routledge, 1985).

25. Cf. *ibid.*, p. 113 and J. Dancy, 'Two Conceptions of Moral Realism', *Proceedings of the Aristotelian Society*, Supp. vol. 60, 1986.

26. See J. McDowell, 'Values and secondary qualities' in T. Honderich (ed.), *Morality and Objectivity* (London: Routledge, 1985) and J. McDowell, 'Aesthetic value, objectivity and the fabric of the world' in E. Schaper (ed.), *Pleasure, Preference and Value* (Cambridge: Cambridge University Press, 1983). Cf. D. Wiggins, *Needs, Values, Truth* (Oxford: Blackwell, 1987), Essays III and IV. For critical discussion of this approach see S. Blackburn, 'Errors and the Phenomenology of Value' in T. Honderich (ed.), *Morality and Objectivity*; J. Dancy, 'Two Conceptions of Moral Realism', *Proceedings of the Aristotelian Society*, Supp. vol. 60, 1986; C. Hookway, 'Two Conceptions of Moral Realism', *Proceedings of the Aristotelian Society* Supp. vol. 60, 1986; C. Wright, 'Moral Values, Projections and Secondary Qualities', *Proceedings of the Aristotelian Society*, Supp. vol. 62, 1988.

27. For such a Humean response see Blackburn, 'Errors and the Phenomenology of Value' in T. Honderich (ed.), *Morality and Objectivity*.

28. Cf. Hookway, 'Two Conceptions of Moral Realism', p. 202.

29. Hence I also reject Feinberg's claim that the goods of plants are reducible to those of humans with an interest in their thriving: 'The Rights of Animals and Unborn Generations' in *Rights, Justice and the Bounds of Liberty* (Princeton, NJ: Princeton University Press, 1980), pp. 169-71. For a similar argument against Feinberg see P. Taylor, *Respect for Nature*, p. 68.

30. G. H. von Wright, *The Varieties of Goodness* (London: Routledge and Kegan Paul, 1963), ch. 3.

31. *Ibid.*, p. 50. Cf. P. Taylor, *Respect for Nature*, pp. 60-71.

32. See J. Cooper, *Reason and Human Good in Aristotle* (Cambridge, MA: Harvard University Press, 1975), p. 19ff.

33. Von Wright, *The Varieties of Goodness*, pp. 50-51.

34. I discuss this example in more detail in J. O'Neill, 'Exploitation and Workers' Councils', *Journal of Applied Philosophy*, 8 (1991), 263-67.

35. Hence, there is no need to invoke scientific hypotheses such as the Gaia hypothesis to defend the existence of such goods, as for example Goodpaster does (K. Goodpaster, 'On Being Morally Considerable' p. 323, *Journal of Philosophy*, 75, 1978 pp. 308-25).

36. R. Attfield, *A Theory of Value and Obligation* (Beckenham: Croom Helm, 1987), p. 21. Cf. Holmes Rolston III, *Environmental Ethics* (Philadelphia: Temple University Press, 1988), K. Goodpaster, 'On Being Morally Considerable' and P. Taylor, *Respect for Nature*.

37. This point undermines a common objection to objectivism, i.e., that objectivists cannot explain why value statements necessarily motivate actions. If values

were objective then 'someone might be indifferent to things which he regards as good or actively hostile to them' (S. Blackburn *Spreading the Word*, p. 188). The proper reply to this is that not all value statements do motivate actions, as the example in the text reveals.

38. Compare Wiggins's point that we need to discriminate between 'the (spurious) fact-value distinction and the (real) is-ought distinction' (D. Wiggins, 'Truth, Invention, and the Meaning of Life' in *Needs, Values, Truth: Essays in the Philosophy of Value* [Oxford: Blackwell, 1987] p. 96). Cf. P. Taylor, *Respect for Nature*, pp. 71-72.

39. See R. Attfield, *op. cit.*, for this kind of position. For a different attempt to bridge the gap between objective goods and moral oughts see P. Taylor, *Respect for Nature*, chs. 2-4.

40. Aristotle, *Nicomachean Ethics*, trans. T. Irwin (Indianapolis, IN: Hackett, 1985), Books viii-ix.

41. *Ibid.*, 1168b.

42. This would clearly involve a rejection of Aristotle's own view that animals are made for the sake of humans. (Aristotle, *Politics*, trans. J. Warrington [London: J. A. Dent and Sons, 1959], 1265b.)

43. This line of argument has the virtue of fitting well with Aristotle's own account of happiness, given an inclusive interpretation of his views. Happiness on this account is inclusive of all goods that are ends in themselves: a happy life is self-sufficient in that nothing is lacking. It is a maximally consistent set of goods. (Aristotle, *Nicomachean Ethics*, 1097b14-20; see J. L. Ackrill, 'Aristotle on *Eudaimonia*' in A. O. Rorty (ed.), *Essays on Aristotle's Ethics* (Berkeley, CA: University of California Press, 1980) for a presentation of this interpretation.)

Part II
Species, Ecosystems and Interests

[5]
Duties to Endangered Species

An adequate ethic for preserving species requires an unprecedented mix of biological science and ethics

Holmes Rolston III

In the Endangered Species Act, Congress has lamented the lack of "adequate concern [for] and conservation [of]" species (US Congress 1973). But neither scientists nor ethicists have fully realized how developing this concern requires an unprecedented mix of biology and ethics. What logic underlies duties involving forms of life? Looking to the past for help, one searches in vain through 3000 years of philosophy (back at least to Noah!) for any serious reference to endangered species. Among present theories of justice, Harvard philosopher John Rawls (1971, p. 512) asserts, "The destruction of a whole species can be a great evil," but also admits that in his theory "no account is given of right conduct in regard to animals and the rest of nature." Meanwhile, there is an urgency to the issue. The *Global 2000 Report* (1980–1981) projects a massive loss of species, up to 20% within a few decades.

Duties to persons concerning species

The usual way to approach a concern for species is to say that there are no duties directly to endangered species,

---

Holmes Rolston III is a professor of philosophy at Colorado State University, Fort Collins, CO 80523. He is the associate editor of the journal *Environmental Ethics* and author of *Philosophy Gone Wild*, a collection of essays in environmental ethics, to be published by Prometheus Books in spring 1986. © 1985 American Institute of Biological Sciences.

> Destroying species is like tearing pages out of an unread book, written in a language humans hardly know how to read

only duties to other persons concerning species. From a utilitarian standpoint (Hampshire 1972, pp. 3–4), the protection of nature and "the preservation of species are to be aimed at and commended only in so far as human beings are, or will be, emotionally and sentimentally interested." In an account based on rights, Feinberg (1974, p. 56) reaches a similar conclusion. "We do have duties to protect threatened species, not duties to the species themselves as such, but rather duties to future human beings." Using traditional ethics to confront the novel threat of extinctions, we can reapply familiar duties to persons and see whether this is convincing. This line of argument can be impressive but seems to leave deeper obligations untouched.

Persons have a strong duty not to harm others and a weaker, though important, duty to help others. Arguing the threat of harm, the Ehrlichs (1981) maintain, in a blunt metaphor, that species are rivets in the Earthship in which humans are flying. Extinctions are maleficent rivet popping. In this model, nonrivet species, if there are any, would have no value; humans desire only the diversity that prevents a crash. The care is not for particular species but, in a variant metaphor, for the sinking ark (Myers 1979a). To worry about a sinking ark seems a strange twist on the Noah story. Noah built the ark to preserve each species. In the Ehrlich/Myers account, the species-rivets are preserved to keep the ark from sinking! The reversed justification is revealing.

On the benefits side, species that are not rivets may prove to be resources. Thomas Eisner testified to Congress that only two percent of the flowering plants have been tested for alkaloids, which often have medical value (US Congress 1982, p. 296). A National Science Foundation report (1977) advocated saving the Devil's Hole pupfish, *Cyprinodon diabolis*, because it thrives in extremes and "can serve as useful biological models for future research on the human kidney—and on survival in a seemingly hostile environment." Myers (1979b) further urges "conserving our global stock." At first, this advice seems wise, yet later somewhat demeaning for humans to regard all other species as *stock*.

Destroying species is like tearing pages out of an unread book, written in a language humans hardly know how to read, about the place where they live. No sensible person would destroy the Rosetta Stone, and no self-respecting persons will destroy the mouse lemur, endangered in Madagascar and thought to be the nearest modern animal to the relatively unspecialized primates from which the

718

Cheetah *(Acinonyx jubatus)* and cub, above, Masai Amboseli, Kenya. Photo: Mark Boulton, courtesy World Wildlife Fund-US. Right, Black rhinoceros *(Diceros bicornis)*. Photo: Norman Myers, courtesy World Wildlife Fund-US.

human line evolved. Still, following this logic, humans do not have duties to the book, the stone, or the species but to ourselves, duties of prudence and education. Humans need insight into the full text of ecosystem evolution. It is not endangered species but an endangered human future that is of concern. Such reasons are pragmatic and impressive. They are also moral, since persons are benefited or hurt. But are they exhaustive?

One problem is that pragmatic reasons get overstated. Peter H. Raven testified before Congress that a dozen dependent species of insects, animals, or other plants typically become extinct with each plant that goes extinct (US Congress 1982, p. 293). But Raven knows that such cascading, disastrous extinction is true only on statistical average, since a plant named for him, Raven's manzanita, *Arctostaphylos hookeri* ssp. *ravenii*, is known from a single wild specimen, and its extinction is unlikely to trigger others. Rare species add some backup resilience. Still, if all 79 plants on the endangered species list disappeared, it is doubtful that the regional ecosystems involved would measurably shift their stability. Few cases can be cited where the removal of a rare species damaged an ecosystem.

Let's be frank. A substantial number of endangered species have no resource value. Beggar's ticks, *Bidens* spp., with their stick-tight seeds, are a common nuisance through much of the United States. One species, tidal shore beggar's tick, *B. bidentoides*, which differs little in appearance from the others, is endangered. It seems unlikely to be a potential resource. As far as humans are concerned, its extinction might be good riddance.

We might say that humans ought to preserve for themselves an environment adequate to match their capacity to wonder. But this is to value the *experience* of wonder, rather than the *objects* of wonder. Valuing merely the experience seems to commit a fallacy of misplaced wonder, for speciation is itself among the wonderful things on Earth. Valuing speciation directly, however, seems to attach value to the evolutionary process, not merely to subjective experiences that arise when humans reflect over it.

We might say that humans of decent character will refrain from needless destruction of all kinds, including destruction of species. Vandals destroying art objects do not so much hurt statues as cheapen their own character. But is the American shame at destroying the passenger pigeon only a matter of self-respect? Or is it shame at our ignorant insensitivity to a form of life that (unlike a statue) had an intrinsic value that placed some claim on us?

The deeper problem with the anthropocentric rationale, beyond overstatement, is that its justifications are submoral and fundamentally exploitive, even if subtly. This is not true intraspecifically among humans, when out of a sense of duty an individual defers to the values of fellows. But it is true interspecifically, since *Homo sapiens* treats all other species as rivets, resources, study materials, or entertainments. Ethics has always been about partners with entwined destinies. But it has never been very convincing when pleaded as enlightened self-interest (that one ought always to do what is one's intelligent self-interest), including class self-interest, even though in practice genuinely altruistic ethics often needs to be reinforced by self-interest. To value all other species only for human inter-

The challenge now is to learn interspecific altruism

ests is like a nation's arguing all its foreign policy in terms of national interest. Neither seems fully moral.

Perhaps an exploiting attitude, and the tendency to justify it ethically, has been naturally selected in *Homo sapiens*, at least in the population that has become dominant in the West. But humans—scientists who have learned to be disinterested and ethicists who have learned to consider the interests of others—ought to be able to see further. Humans have learned some intraspecific altruism. The challenge now is to learn interspecific altruism. Utilitarian reasons for saving species may be good ones, necessary for policy. But can we not also discover the best reasons, the full extent of human duties? Dealing with a problem correctly requires an appropriate way of thinking about it. What is offensive in the impending extinctions is not merely the loss of rivets and resources, but the maelstrom of killing and insensitivity to forms of life and the forces producing them. What is required is not prudence but principled responsibility to the biospheric Earth.

Specific forms of life

There are many barriers to thinking of duties to species, however, and scientific ones precede ethical ones. It is difficult enough to argue from the fact that a species exists to the value judgment that a species ought to exist—what philosophers call an argument from *is* to *ought*. Matters grow worse if the concept of species is rotten to begin with. Perhaps the concept is arbitrary and conventional, a mapping device that is only theoretical. Perhaps it is unsatisfactory theoretically in an evolutionary ecosystem. Perhaps species do not exist. Duties to them would be as imaginary as duties to contour lines or to lines of latitude and longitude. Is there enough factual reality in species to base duty there?

Betula lenta uber, round-leaf birch, is known from only two locations on nearby Virginia creeks and differs from the common *B. lenta* only in having rounded leaf tips. For 30 years it was described as a subspecies or merely a mutation. But M. L. Fernald pronounced it a species, *B. uber*, and for 40 years it has been considered one. High fences have been built around all known specimens. If a greater botanist were to redesignate it a subspecies, would this change in alleged facts affect our alleged duties? Ornithologists recently reassessed an endangered species, the Mexican duck, *Anas diazi*, and lumped it with the common mallard, *A. platyrhynchos*, as subspecies *diazi*. US Fish and Wildlife authorities took it off the endangered species list partly as a result. Did a duty cease? Was there never one at all?

If a species is only a category, or class, boundary lines may be arbitrarily drawn. Darwin (1968 [1859], p. 108) wrote, "I look at the term species, as one arbitrarily given for the sake of convenience to a set of individuals closely resembling each other." Some natural properties are used to delimit species—reproductive structures, bones, teeth. But which properties are selected and where the lines are drawn vary with taxonomists. When A. J. Shaw (1981) recently "discovered" a new species of moss, *Pohlia tundrae*, in the alpine Rocky Mountains, he did not find any hitherto unknown plants; he just regrouped herbarium material that had been known for decades under other names. Indeed, biologists routinely put after a species the name of the "author" who, they say, "erected" the taxon.

Individual organisms exist, but if species are merely classes, they are inventions. A. B. Shaw (1969) claims, "The species concept is entirely subjective," and, concluding a presidential address to paleontologists, even exclaims, "Help stamp out species!" He refers, of course, to the artifacts of taxonomists, not to living organisms. But if species do not exist except embedded in a theory in the minds of classifiers, it is hard to see how there can be duties to save them. No one proposes duties to genera, families, orders, or phyla; everyone concedes that these do not exist in nature.

But a biological species is not just a class. A species is a living historical form (Latin *species*), propagated in individual organisms, that flows dynamically over generations. Simpson (1961, p. 153) concludes, "An evolutionary species is a lineage (an ancestral-descendant sequence of populations) evolving separately from others and with its own unitary evolutionary role and tendencies." Mayr (1969a, p. 26) holds, "Species are groups of interbreeding natural populations that are reproductively isolated from other such groups." He (1969b) can even emphasize, though many biologists today would deny this, that "*species are the real units of evolution,* they are the entities which specialize, which become adapted, or which shift their adaptation." Recently, Mayr (1982) has sympathized with Ghiselin (1974) and Hull (1976), who hold that species are integrated individuals, and species names proper names, with organisms related to their species as part is to whole. Eldredge and Cracraft (1980, p. 92) find that "a species is a diagnosable cluster of individuals within which there is a parental pattern of ancestry and descent, beyond which there is not, and which exhibits a pattern of phylogenetic ancestry and descent among units of like kind." Species, they insist, are *"discrete entities in time as well as space."*

It is admittedly difficult to pinpoint precisely what a species is, and there may be no single, quintessential way to define species; a polythetic or polytypic gestalt of features may be required. All we need for this discussion, however, is that species be

Chimpanzee *(Pan troglodytes)*. Photo: G. Teleki, courtesy World Wildlife Fund-US.

objectively there as living processes in the evolutionary ecosystem; the varied criteria for defining them (descent, reproductive isolation, morphology, gene pool) come together at least in providing evidence that species are really there. In this sense, species are dynamic natural kinds, if not corporate individuals. A species is a coherent, ongoing form of life expressed in organisms, encoded in gene flow, and shaped by the environment.

The claim that there are specific forms of life historically maintained in their environments over time does not seem arbitrary or fictitious at all but, rather, as certain as anything else we believe about the empirical world, even though at times scientists revise the theories and taxa with which they map these forms. Species are not so much like lines of latitude and longitude as like mountains and rivers, phenomena objectively there to be mapped. The edges of all these natural kinds will sometimes be fuzzy, to some extent discretionary. We can expect that one species will slide into another over evolutionary time. But it does not follow from the fact that speciation is sometimes in progress that species are merely made up, instead of found as evolutionary lines articulated into diverse forms, each with its more or less distinct integrity, breeding population, gene pool, and role in its ecosystem.

At this point, we can anticipate how there can be duties to species. What humans ought to respect are dynamic life forms preserved in historical lines, vital informational processes that persist genetically over millions of years, overleaping short-lived individuals. It is not *form* (species) as mere morphology, but the *formative* (speciating) process that humans ought to preserve, although the process cannot be preserved without its products. Neither should humans want to protect the labels they use, but the living process in the environment. Endangered "species" is a convenient and realistic way of tagging this process, but protection can be interpreted (as the Endangered Species Act permits) in terms of subspecies, variety, or other taxa or categories that point out the diverse forms of life.

Duties to species

The easiest conclusion to reach from prevailing theories of justice, which involve tacit or explicit "contracts" between persons, is that duties and rights are reciprocal. But reciprocally claiming, recognizing, exercising, and enjoying rights and duties can only be done by reflective rational agents. Humans have entered no contract with other species; certainly they have not with us. There is no ecological contract parallel to the social contract; all the capacities for deliberate interaction so common in culture vanish in nature. Individual animals and plants, to say nothing of species, cannot be reasoned with, blamed, or educated into the prevailing contract.

But to make rights and duties reciprocal supposes that only moral agents count in the ethical calculus. Duties exist as well to those persons who cannot argue back—to the mute and powerless—and perhaps this principle extends to other forms of life. Morality is needed wherever the vulnerable must be protected from the powerful.

The next easiest conclusion to reach, either from rights-based or utilitarian theories, is that humans have duties wherever there are psychological interests involving the capacity for experience. That moves a minimal criterion for duty past rational moral agency to sentience. The question is not whether animals can reciprocate the contract but whether they can suffer. Singer (1979) thinks that the only reason to be concerned about endangered species is the interests of humans and other sentient animals at stake in their loss. Only they can enjoy benefits or suffer harm, so only they can be treated justly or unjustly.

But species, not sentience, generate

Humans ought to respect the lifelines within species that persist genetically over millions of years

some duties. On San Clemente Island, the US Fish and Wildlife Service and the California Department of Fish and Game asked the Navy to shoot 2000 feral goats to save three endangered plant species, *Malacothamnus clementinus, Castilleja grisea,* and *Delphinium kinkiense.* That would kill several goats for each known surviving plant. (Happily, the Fund for Animals rescued most of the goats; unhappily, they could not trap them all and the issue is unresolved.) The National Park Service did kill hundreds of rabbits on Santa Barbara Island to protect a few plants of *Dudleya traskiae,* once thought extinct and curiously called the Santa Barbara live-forever. Hundreds of elk starve in Yellowstone National Park each year, and the Park Service is not alarmed, but the starving of an equal number of grizzly bears, which would involve about the same suffering in psychological experience, would be of great concern.

A rather difficult claim to make under contemporary ethical theory is that duty can arise toward any living organism. Such duties, if they exist, could be easy to override, but by this account humans would have at least a minimal duty not to disrupt living beings without justification.

Here the question about species, beyond individuals, is both revealing and challenging because it offers a biologically based counterexample to the focus on individuals—typically sentient and usually persons—so characteristic in Western ethics. In an evolutionary ecosystem, it is not mere individuality that counts, but the species is also significant because it is a dynamic life form maintained over time by an informed genetic flow. The individual represents (re-presents) a species in each new generation. It is a token of a type, and the type is more important than the token.

It is as logical to say that the individual is the species' way of propagating itself as to say that the embryo or egg is the individual's way of propagating itself. We can think of the cognitive processing as taking place not merely in the individual but in the gene pool. Genetically, though not neurally, a species over generations "learns" (discovers) pathways previously unknown. A form of life reforms itself, tracks its environment, and sometimes passes over to a new species. There is a specific groping for a valued *ought*-to-be beyond what now *is* in any individual. Though species are not moral agents, a biological identity—a kind of value—is here defended. The dignity resides in the dynamic form; the individual inherits this, instantiates it, and passes it on. To borrow a metaphor from physics, life is both a particle (the individual) and a wave (the specific form).

A species lacks moral agency, reflective self-awareness, sentience, or organic individuality. So we may be tempted to say that specific-level processes cannot count morally. But each ongoing species defends a form of life, and these are on the whole good things, arising in a process out of which humans have evolved. All ethicists say that in *Homo sapiens* one species has appeared that not only exists but ought to exist. But why say this exclusively of a late-coming, highly developed form? Why not extend this duty more broadly to the other species (though perhaps not with equal intensity over them all, in view of varied levels of development)? These kinds defend their forms of life, too. Only the human species contains moral agents, but perhaps conscience *ought not* be used to exempt every other form of life from consideration, with the resulting paradox that the single moral species acts only in its collective self-interest toward all the rest.

Extinction shuts down the generative processes. The wrong that humans are doing, or allowing to happen through carelessness, is stopping the historical flow in which the vitality of life is laid. Every extinction is an incremental decay in stopping life processes—no small thing. Every extinction is a kind of superkilling. It kills forms *(species),* beyond individuals. It kills "essences" beyond "existences," the "soul" as well as the "body." It kills collectively, not just distributively. It is not merely the loss of potential human information that is tragic, but the loss of biological information, present independently of instrumental human uses for it.

"Ought species *x* to exist?" is a single increment in the collective question, "Ought life on Earth to exist?" The answer to the question about one species is not always the same as the answer to the bigger question, but since life on Earth is an aggregate of many species, the two are sufficiently related that the burden of proof lies with those who wish deliberately to extinguish a species and simultaneously to care for life on Earth. To kill a species is to shut down a unique story; and, although all specific stories must eventually end, we seldom want unnatural ends. Humans ought not to play the role of murderers. The duty to species can be overridden, for example with pests or disease organisms. But a prima facie duty stands nevertheless.

One form of life has never endangered so many others. Never before has this level of question—superkilling by a superkiller—been faced. Humans have more understanding than ever of the speciating processes, more predictive power to foresee the intended and unintended results of their actions, and more power to reverse the undesirable consequences. The duties that such power and vision generate no longer attach simply to individuals or persons but are emerging duties to specific forms of life. If, in this world of uncertain moral convictions, it makes any sense to claim that one ought not to kill individuals without justification, it makes more sense to claim that one ought not to superkill the species, without superjustification.

Individuals and species

Many will be uncomfortable with this claim because their ethical theory does not allow duty to a collection. Feinberg (1974, p. 55) writes, "A whole collection, as such, cannot have beliefs, expectations, wants, or desires. . . . Individual elephants can have interests, but the species elephant cannot." Singer (1979, p. 203) asserts, "Species as such are not con-

> The appropriate survival unit is the appropriate unit of moral concern

scious entities and so do not have interests above and beyond the interests of the individual animals that are members of the species." Regan (1983, p. 359) maintains, "The rights view is a view about the moral rights of individuals. Species are not individuals, and the rights view does not recognize the moral rights of species to anything, including survival." Rescher (1980, p. 83) says, "Moral obligation is thus always interest-oriented. But only individuals can be said to have interests; one only has moral obligations to particular individuals or particular groups thereof. Accordingly, the duty to save a species is not a matter of moral duty toward it, because moral duties are only oriented to individuals. A species as such is the wrong sort of target for a moral obligation."

Even those who recognize that organisms, nonsentient as well as sentient, can be benefited or harmed may see the good of a species as the sum of and reducible to the goods of individuals. The species is well off when and because its members are; species wellbeing is just aggregated individual well-being. The "interests of a species" constitute only a convenient device, something like a center of gravity in physics, for speaking of an aggregated focus of many contributing individual member units.

But duties to a species are not duties to a class or category, not to an aggregation of sentient interests, but to a lifeline. An ethic about species needs to see how the species *is* a bigger event than individual interests or sentience. Making this clearer can support the conviction that a species *ought* to continue.

Events can be good for the wellbeing of the species, considered collectively, although they are harmful if considered as distributed to individuals. This is one way to interpret what is often called genetic load (Fraser 1962), genes that somewhat reduce health, efficiency, or fertility in most individuals but introduce enough variation to permit improving the specific form. Less variation and better repetition in reproduction would, on average, benefit more individuals in any one next generation, since individuals would have less "load." But on a longer view, variation can confer stability in a changing world. A greater experimenting with individuals, although this typically makes individuals less fit and is a disadvantage from that perspective, benefits rare, lucky individuals selected in each generation, with a resulting improvement in the species. Most individuals in any particular generation carry some (usually slightly) detrimental genes, but the variation is good for the species. Note that this does not imply species selection; selection perhaps operates only on individuals. But it does mean that we can distinguish between the goods of individuals and the larger good of the species.

Predation on individual elk conserves and improves the species *Cervus canadensis.* A forest fire harms individual aspen trees, but it helps *Populus tremuloides* because fire restarts forest succession, without which the species would go extinct. Even the individuals that escape demise from external sources die of old age; their deaths, always to the disadvantage of those individuals, are a necessity for the species. A finite lifespan makes room for those replacements that enable development to occur, allowing the population to improve in fitness or adapt to a shifting environment. Without the "flawed" reproduction that permits variation, without a surplus of young or predation and death, which all harm individuals, the species would soon go extinct in a changing environment, as all environments eventually are. The individual is a receptacle of the form, and the receptacles are broken while the form survives; but the form cannot otherwise survive.

When a biologist remarks that a breeding population of a rare species is dangerously low, what is the danger to? Individual members? Rather, the remark seems to imply a specific-level, point-of-no-return threat to the continuing of that form of life. No individual crosses the extinction threshold; the species does.

Reproduction is typically assumed to be a need of individuals, but since any particular individual can flourish somatically without reproducing at all, indeed may be put through duress and risk or spend much energy reproducing, by another logic we can interpret reproduction as the species keeping up its own kind by reenacting itself again and again, individual after individual. In this sense a female grizzly does not bear cubs to be healthy herself, any more than a woman needs children to be healthy. Rather, her cubs are *Ursus arctos,* threatened by nonbeing, recreating itself by continuous performance. A species in reproduction defends its own kind from other species, and this seems to be some form of "caring."

Biologists have often and understandably focused on individuals, and some recent trends interpret biological processes from the perspective of genes. A consideration of species reminds us that many events can be interpreted at this level too. An organism runs a directed course through the environment, taking in materials, using them resourcefully, discharging wastes. But this single, directed course is part of a bigger picture in which a species via individuals maintains its course over longer spans of time. Thinking this way, the life the individual has is something passing through the individual as much as something it intrinsically possesses. The individual is subordinate to the species, not the other way around. The genetic set, in which is coded the *telos,* is as evidently a "property" of the species as of the individual.

Biologists and linguists have learned to accept the concept of information in the genetic set without any subject who speaks or understands. Can ethicists learn to accept value in, and duty to, an informed process in which centered individuality or sentience is absent? Here events can be significant at the specific level, an additional consideration to whether they are beneficial to individuals. The species-in-environment is an interactive complex, a selective system where individuals are pawns on a chessboard. When human conduct endangers these specific games of life, duties may appear.

A species has no self. It is not a bounded singular. Each organism has its own centeredness, but there is no specific analogue to the nervous hookups or circulatory flows that characterize the organism. But, like the market in economics, an organized system does not have to have a controlling center to have identity. Having a biological identity reasserted genetically over time is as true of the species as of the individual. Individuals come and go; the marks of the species collectively remain much longer.

A consideration of species strains any ethic focused on individuals, much less on sentience or persons. But the result can be a biologically sounder ethic, though it revises what was formerly thought logically permissible or ethically binding. The species line is fundamental. It is more important to protect this integrity than to protect individuals. Defending a form of life, resisting death, regeneration that maintains a normative identity over time—all this is as true of species as of individuals. So what prevents duties arising at that level? The appropriate survival unit is the appropriate level of moral concern.

Species and ecosystem

A species is what it is inseparably from its environment. The species defends its kind against the world, but at the same time interacts with its environment, functions in the ecosystem, and is supported and shaped by it. The species and the community are complementary processes in synthesis, somewhat parallel to but a level above the way the species and the individual have distinguishable but entwined identities. Neither the individual nor the species stands alone; both are embedded in a system. It is not preservation of *species* but of *species in the system* that we desire. It is not just what they are but where they are that we must value correctly.

The species *can* only be preserved in situ; the species *ought* to be preserved in situ. Zoos and botanical gardens can lock up a collection of individuals, but they cannot begin to simulate the ongoing dynamism of gene flow under the selection pressures in a wild biome. The full integrity of the species must be integrated into the ecosystem. Ex situ preservation, while it may save resources and souvenirs, does not preserve the generative process intact. Again, the appropriate survival unit is the appropriate level of moral concern.

It might seem that ending the history of a species now and again is not far out of line with the routines of the universe. But artificial extinction, caused by human encroachments, is radically different from natural extinction. Relevant differences make the two as morally distinct as death by natural causes is from murder. Though harmful to a species, extinction in nature is no evil in the system; it is rather the key to tomorrow. Such extinction is a normal turnover in ongoing speciation.

Anthropogenic extinction has nothing to do with evolutionary speciation. Hundreds of thousands of species will perish because of culturally altered environments radically differing from the spontaneous environments in which such species were naturally selected and in which they sometimes go extinct. In natural extinctions, nature takes away life when it has become unfit in habitat, or when the habitat alters, and supplies other life in its place. Artificial extinction shuts down tomorrow because it shuts down speciation. Natural extinction typically occurs with transformation, either of the extinct line or related or competing lines. Artificial extinction is without issue. One opens doors; the other closes them. Humans generate and regenerate nothing; they only dead-end these lines.

From this perspective, humans have no duty to preserve rare species from natural extinctions, although they might have a duty to other humans to save such species as resources or museum pieces. Humans cannot and need not save the product without the process.

Through evolutionary time, nature has provided new species at a higher rate than the extinction rate; hence, the accumulated diversity. In one of the best documented studies of the marine fossil record, Raup and Sepkoski (1982) summarize a general increase in standing diversity (Figure 1). Regardless of differing details on land or biases in the fossil record, a graph of the increase of diversity on Earth must look something like this.

There have been four or five catastrophic extinctions, each succeeded by a recovery of previous diversity. These anomalies so deviate from the trends that many paleontologists look for extraterrestrial causes. If due to supernovae, collisions with asteroids, or oscillations of the solar system above and below the plane of the galaxy, such events are accidental to the evolutionary ecosystem. Thousands of species perished at the impingement of otherwise unrelated events. The disasters were irrelevant to the kinds of ecosystems in which such species had been selected. If the causes were more terrestrial—cyclic changes in climates or continental drift—the biological processes are still to be admired for their powers of recovery. Even interrupted by accident, they maintain and increase the numbers of species. Raup and Sepkoski further find that the normal extinction rate declines from 4.6 families per million years in the Early Cambrian to 2.0 families in recent times, even though the number of families (and species) enormously increases. This seems to mean that optimization of fitness increases through evolutionary time.

An ethicist has to be circumspect. An argument might commit what logicians call the genetic fallacy to suppose that present value depended on origins. Species judged today to have intrinsic value might have arisen anciently and anomalously from a valueless context, akin to the way life arose mysteriously from nonliving materials. But in a historical ecosystem, what a thing is differentiates poorly from the generating and sustaining matrix. The individual and the species have what value they have,

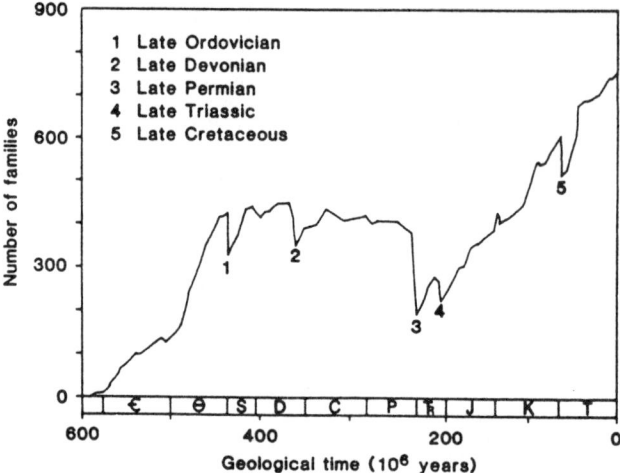

Figure 1. Standing diversity through time for families of marine vertebrates and invertebrates; numbers refer to catastrophic extinctions. Source: Raup and Sepkoski (1982). © 1982 by AAAS; reprinted with permission.

to some extent, in the context of the forces that beget them.

Imagine that Figure 1 is the graph of the performance of a 600-million-year-old business. Is it not a healthy one? But this is the record of the business of life, and the long-term performance deserves ethical respect. There is something awesome about an Earth that begins with zero and runs up toward 5 to 10 million species in several billion years, setbacks notwithstanding.

What is valuable about species is not to be isolated in them for what they are in themselves. Rather, the dynamic account evaluates species as process, product, and instrument in the larger drama toward which humans have duties, reflected in duties to species. Whittaker (1972) finds that on continental scales and for most groups, "increase of species diversity . . . is a self-augmenting evolutionary process without any evident limit." There is a tendency toward "species packing." Nature seems to produce as many species as it can, not merely enough to stabilize an ecosystem or only species that can directly or indirectly serve human needs. Humans ought not to inhibit this exuberant lust for kinds. That process, along with its product, is about as near

to ultimacy as humans can come in their relationship with the natural world.

Several billion years worth of creative toil, several million species of teeming life, have been handed over to the care of this late-coming species in which mind has flowered and morals have emerged. Ought not those of this sole moral species do something less self-interested than count all the produce of an evolutionary ecosystem as rivets in their spaceship, resources in their larder, laboratory materials, recreation for their ride? Such an attitude hardly seems biologically informed, much less ethically adequate. Its logic is too provincial for moral humanity. Or, in a biologist's term, it is ridiculously territorial. If true to their specific epithet, ought not *Homo sapiens* value this host of species as something with a claim to care in its own right?

An endangered ethic?

Contemporary ethical systems seem misfits in the role most recently demanded of them. There is something overspecialized about an ethic, held by the dominant class of *Homo sapiens*, that regards the welfare of only one of several million species as an object of duty. If this requires a paradigm change about the sorts of things to which duty can attach, so much the worse for those ethics no longer functioning in, or suited to, their changing environment. The anthropocentrism associated with them was fiction anyway. There is something Newtonian, not yet Einsteinian, besides something morally naive, about living in a reference frame where one species takes itself as absolute and values everything else relative to its utility.

References cited

Darwin, C. 1968 [1859]. *The Origin of Species*. Penguin Books, Harmondsworth, UK.
Ehrlich, P. R., and A. H. Ehrlich. 1981. *Extinction*. Random House, New York.
Eldredge, N., and J. Cracraft. 1980. *Phylogenetic Patterns and the Evolutionary Process*. Columbia University Press, New York.
Feinberg, J. 1974. The rights of animals and unborn generations. Pages 43–68 in W. T. Blackstone, ed. *Philosophy and Environmental Crisis*. University of Georgia Press, Athens.
Fraser, G. R. 1962. Our genetical "load": a review of some aspects of genetical variation. *Ann. Hum. Genet.* 25: 387–415.
Ghiselin, M. T. 1974. A radical solution to the species problem. *Syst. Zool.* 23: 536–544.
The Global 2000 Report to the President: Entering the Twenty-first Century. 1980–1981. Council on Environmental Quality and the Department of State. US Government Printing Office, Washington, DC.
Hampshire, S. 1972. *Morality and Pessimism*. Cambridge University Press, Cambridge, UK.
Hull, D. L. 1976. Are species really individuals? *Syst. Zool.* 25: 174–191.
Mayr, E. 1969a. *Principles of Systematic Zoology*. McGraw-Hill, New York.
———. 1969b. The biological meaning of species. *Biol. J. Linn. Soc.* 1: 311–320.
———. 1982. *The Growth of Biological Thought*. Harvard University Press, Cambridge, MA.
Myers, N. 1979a. *The Sinking Ark*. Pergamon Press, Oxford, UK.
———. 1979b. Conserving our global stock. *Environment* 21(9): 25–33.
National Science Foundation (NSF). 1977. The biology of aridity. *Mosaic* 8(1): 28–35.
Raup, D. M., and J. J. Sepkoski, Jr. 1982. Mass extinctions in the marine fossil record. *Science* 215: 1501–1503.
Rawls, J. 1971. *A Theory of Justice*. Harvard University Press, Cambridge, MA.
Regan, T. 1983. *The Case for Animal Rights*. University of California Press, Berkeley.
Rescher, N. 1980. *Unpopular Essays on Technological Progress*. University of Pittsburgh, Pittsburgh, PA.
Shaw, A. B. 1969. Adam and Eve, paleontology, and the non-objective arts. *J. Paleontol.* 43: 1085–1098.
Shaw, A. J. 1981. *Pohlia andrewsii* and *P. tundrae*, two new arctic-alpine propaguliferous species from North America. *Bryologist* 84: 65–74.
Simpson, G. G. 1961. *Principles of Animal Taxonomy*. Columbia University Press, New York.
Singer, P. 1979. Not for humans only. Pages 191–206 in K. E. Goodpaster and K. M. Sayre, eds. *Ethics and Problems of the 21st Century*. University of Notre Dame, Notre Dame, IN.
Whittaker, R. H. 1972. Evolution and measurement of species diversity. *Taxon* 21: 213–251.
US Congress, 1973. Sec. 2(a)(1) in *Endangered Species Act*. 87 STAT. 884 (Public Law 93-205).
———. 1982. *Endangered Species Act Oversight*. Hearings before the US Senate, 97th Cong., 1st sess., 8 and 10 December. 97-H34. US Government Printing Office, Washington, DC.

[6]

DISVALUES IN NATURE

1. Disvalues and values

To accentuate the positive one can eliminate the negative. A touchstone for any theory of truth is its theory of error; a theory of value needs to handle disvalue. Philosophers, lately exercised about values in nature, have not yet much asked about disvalues there. They have also cautioned against committing an alleged naturalistic fallacy. The usual version, the *positive naturalistic fallacy*, argues from *is* to *ought*. Nature is described such and such a way; that is a good thing and ought to be so. But, we are warned, "natural" does not imply "right." The unusual version, the *negative naturalistic fallacy*, argues to *ought not*. Nature is such and such a way; this ought not to be so.

This locates disvalue in nature, but on what grounds? Possibly there is a suppressed, implicit premise: I (or other humans) choose for good reasons supporting human welfare to disvalue this matter of natural fact. Now the conclusion does follow, given my premises. But these are only "my" premises. Take away my preferences and reasons, and there is no disvalue in nature. Opportunistic humans operate as a preference sieve through a booming, buzzing confusion, a kaleidoscopic churn where no event is of value or disvalue of itself. But assume our human outlook, and we cannot avoid valuing some and disvaluing other events in nature.

Often there is a stronger claim. A certain natural condition is bad, whether or not any humans are around. When I come on scene, I can evaluate it so. I may also suffer from this preexisting disvalue; I ought to remedy it if I can. So "my" premises are describing the natural case; my subjective preferences are registering an objective disvalue. In this latter case, no fallacy is committed; to the contrary truth is discovered, only, alas, unhappy truth.

Charles Darwin exclaimed that the process is "clumsy, wasteful, blundering, low, and horribly cruel."[1] John Stuart Mill cursed nature as an "odious scene of violence."[2] Thomas Huxley admonished, "Let us understand, once for all, that the ethical process of society depends, not on imitating the cosmic process, still less on running away from it, but in combating it."[3] George Williams insists, "The process and products of evolution are morally unacceptable and act in opposition to the ethical progress of humanity. . . . What is, in the biological world, normally ought not."[4] These seem to be judgments disvaluing objective natural history. If true, there could prove little cause for biological conservation out of respect for nature.

Disvalue might be discovered in natural products oblivious to the systemic processes. If the disvalue comes only by human preferences, origins are irrelevant. But nature is processes as much as products. Possibly, a disvaluable or valueless process has occasionally produced valuable products, though it would be anomalous if there were a longstanding, high statistical correlation between such a disvaluable process and its valuable products. When valuing nature it is difficult to separate the evolutionary past from the ecological present, and we need to connect both with the ethical future.

An intermediate position is that evolution leads sometimes to good, sometimes to bad things, and that on balance this is a zero sum game. The goods and bads cancel each other out, like heads and tails when tossing a coin. Or the dice might be loaded, either toward values or disvalues.

Another position, only seemingly intermediate, holds that the evolutionary processes and products are devoid of value. This does not mean that each result lies still within the category of valuation, their average hitting the scale at zero. Rather, evolution is entirely off the value scale. Hamlet mused, "There is nothing either good or bad, but thinking makes it so."[5] More acurately, preferring makes it so. Disvalue is as added as value; the naked term is non-value, but even that term backs into nature pejoratively, because it defines a thing by finding entities absent value. Natural things—if we can find ourselves up to this thought experiment—just *are*, neither present nor absent value; rather they lie outside the domain of value (and therefore outside nonvalue and disvalue categories), until humans or other conscious valuers arrive and ignite values or disvalues.

We can say that natural processes and products always have the standing possibility of valuation (or disvaluation), but that is to categorize them in terms of possibilities when humans come on scene. Whether such a standing possibility can be retroactively posited as some real potential there absent humans is doubtful. True enough, we cannot formulate the question whether there is value in nature independently of human experience; but can we formulate a question that reaches outside of human experience? Notice that both those who do and those who do not find or assign value, disvalue, or nonvalue, think they can describe what is really there or not there, outside of the perceptions in the eye of the human beholder. Even the null-value view thinks to know what categories are absent in nature.

Sometimes, we view from our human niche, when we value timber and disvalue wildfire that destroys it. But sometimes we wish a view from no niche, that of an ideal Earth-observer. This need not be absolute, but it will be regional or global. This view is made with our human perceptual and cognitive equipment, but we can partially at least distinguish what we see from how we are able to see it.

Notice carefully that the appropriate evaluative category is not nature's *moral* goodness, for there are no moral agents in nonhuman nature. The appropriate category is one or more kinds of nonmoral goodness, better called nature's *value*. Such value is not to be mapped by projection from culture, much less from human moral systems within culture. It may be that we humans, who are moral agents, ought or ought not to imitate this process; such decision and action will be a moral matter. But the evaluation of nature is not yet moral. Nature is a-moral but that does not imply either a-value or neg-value. That may be where the suspected fallacy lies.

One appropriate category on Earth will be its biological goodness, the extent to which the natural system is pro-life, prolific. We must evaluate phenomena such as: the achievement of diversity and complexity out of simplicity, the mixture of order and contingency, the nomothetic and the idiographic, autonomy and interdependence, individualism and community, endurance through struggle. None of these results can be called moral. Rather, these are creative activities. They may have resultant worth.

If we ask whether all natural things are made for human benefit (either by divine design or natural selection), then it is clear that the answer is that they are not. From this perspective, many things in nature will have no value or disvalue. For *Homo sapiens* some things will prove of value, likeliest some found in our evolved niche, if we have one. Or if we have rebuilt our niche culturally, valuable things will be those that we use as natural resources, remaking them into instruments of culture. But if we ask whether natural things are well made in and by themselves, an answer still needs to be sought.

Philosophers have been chary and fastidious about granting objective value in nature, but they have often been quite willing to allege objective disvalue in nature. But by parity of reasoning, those willing to grant objective disvalue must also consider objective value. Indeed, we shall argue that such disvalues point to values more evidently there. The view of nature that results is not value negative, much less value neutral; it embeds values in biotic community. The argument must follow opposites in conflict and resolution, and this will require a series of gestalt switches, finding out what is actually going on in nature, evaluating it *prima facie* and revaluating it at more depth.

Also, take caution against forgetting the pluralism in nature. There are myriad sorts of things and they are differently made.

2. Predation

There is quite a list of candidate disvalues. One is predation. Indisputably, for a prey animal, it is bad to be eaten; death results. If a tiger eats a human, we are horrified. And if a tiger eats a monkey? The monkey screams, and "nature, red in tooth and claw"[6] has seemed quite tragic. Richard Dawkins concludes that Tennyson's famous phrase "sums up our modern understanding of natural selection admirably." All organisms are "like successful Chicago gangsters."[7] Steve Sapontzis finds predation lamentable: "Where we can prevent predation without occasioning as much or more suffering than we would prevent, we are obligated to do so."[8] These judgments, made from a human perspective, are judgments from the perspective of any prey and judgments about processes deeply embedded in ecosystems.

The disvalue to the prey is, however, a value to the predator, and, with a systemic turn, perspectives change. The violent death of the hunted means life to the hunter. There is not value loss so much as value capture; nutrient materials and energy flow from one life stream to another, with selective pressures to be efficient about the transfer. The pains of the prey are matched by the pleasures of the predator. Should we register the amounts of each to compute the net? Or is the hedonistic criterion perhaps not the most relevant one? We need to ask what biological achievements result from predation?

The wolf is not a big, bad wolf; it is one of the most handsome animals on Earth. Many wish to reintroduce wolves to Yellowstone. In Africa, tourists most want to see the big cats. Florida school children chose the lithe, supple panther as their state animal. We admire the muscle and power, the sentience and skills that could only have evolved in predation. Such aesthetic experience is in the eye of the human beholder, but the biological achievements are objective in cat and wolf. Are these good products of a bad process? Or does something about the creative process require predation?

Autotrophs synthesize their own food; heterotrophs eat something else. Could we have had a world with only flora, no fauna? Possibly not, since in a world in which things are assembled something has to disassemble them for recycling. In any case, no one thinks that a mere floral world would be of more value than a world with fauna also. In a floral world, there would be no one to think. Heterotrophs must be built on autotrophs, and no autotrophs are sentient or cerebral.

Could there have been only plant-eating fauna, only grazers, no predators? Possibly, though probably there never was such a world, since predation preceded photosynthesis. Even grazers are predators of a kind, though what they eat does not suffer. Again, an Earth with only herbivores and no omnivores or carnivores would be impoverished. The animal skills demanded would be only a fraction of those that have resulted in actual zoology—no horns, no fleet-footed predators or prey, no fine-tuned eyesight and hearing, no quick neural capacity, no advanced brains.

Nor are all benefits to the predators. The individual prey, eaten, loses all; but the species may gain as the population is regulated, as selection for better skills at avoiding predation takes place, and the prey not less than the predator will gain in sentience, mobility, cognitive and perceptual powers. Being eaten is not always a bad thing, even from the perspective of the prey species. The predator depends on a continuing prey population; they have entwined destinies.

A rattlesnake has fangs—weapons—and people and animals can get hurt by them. That is clearly a disvalue, if bitten. But fangs are of value to the snake, and their disvalue is close-coupled with value. Indeed, in the systemic picture, fangs are teeth, teeth belong in the food chains by which animals eat. Sometimes animals use the same tools that secure food also as weapons to protect themselves against being eaten. There is nothing inherently disvaluable about teeth, including incisors or canines. A human eating an apple is no more, no less ungodly than a snake eating a mouse. A photosynthetic world would be a largely immobile world. Some species must sit around and soak up sunlight; other species will capture this value to fuel mobility. Still other species will rise higher on the trophic pyramid, funded by capturing resources from below for greater achievements in sentience, cognition, and mobility.

We humans stand in this tradition, omnivores over most of our timespan. Our ancestors were hunters until the advent of domestication and agriculture a few thousand years ago; even in domestication meat eating continues. Perhaps we ought to kill food animals humanely, perhaps we even ought now to stop eating them. But meanwhile we face the paradox that without predation in our past we should not be here to deliberate the question at all.

A world without blood would be poorer, but a world without bloodshed would be poorer too. Among other things, it would be a world without humans—not that humans now cannot be vegetarians but that the evolution of humans would never have taken place. We are beginning to see that disvalues are often subtly transmuted into values, especially if we transpose the question from one of human preferences into one about systemic processes.

3. Parasitism

Alexander Skutch concludes, "After long pondering, I believe that I can define good and evil in terms to which even a biologist . . . can hardly take exception. . . . The great evil of life is parasitism."[9] Parasitism is predation minus what we just admired. The lofty power, sentience, cognition, and locomotion are gone; the degenerate parasite just sucks nutrients out of the host and cripples it. Parasites do not even contribute the photosynthetic energy collection supplied in plants. Parasites often lose skills they formerly had—eyes, wings, even their brains. As humans we dislike parasites, but the judgment here is not mere human preference. It is based on objective facts recognizing a bad feature in a generally positive evolutionary system. "Parasitism has taken its tremendous toll of life with scarcely any return that we can see; it has led to retrogression rather than to progress. Hence we may call it the greatest evil of life."[10]

Ichneumon wasps in larval stages are parasites that slowly excavate caterpillars from within. Stephen Jay Gould reels from the grisly business, "I suspect that nothing evokes greater disgust in most of us than slow destruction of a host by an internal parasite—gradual ingestion, bit by bit, from the inside."[11] Gould concludes, rightly, that nature is nonmoral; the wasps cannot be faulted, nor can we morally censure the system they inhabit. But have we a disvalue? Gould finds himself bewildered and is unsure whether he can ask, much less answer, the question. E. O. Wilson remarks that much parasitism is an "evolutionary sink" the end of which is "a state of abject dependence on the host species."[12] Notice, though, that seldom does the system as a whole degenerate. The parasite that loses skills borrows skills that remain in the host.

But contemporary parasitology questions the extent to which parasites are either degenerate or detrimental to their hosts. Their evolution is often doubtfully to be termed "regressive"; the exchange of free-living skills achieves less evident but biochemically sophisticated skills. Arthur W. Jones concludes, "Parasite life cycles are extremely ingenious. . . . Parasites may be, not less, but more evolutionarily progressive than their hosts."[13] Much of this is microscopic; most of it is invisible to the uneducated observer.

Most parasites do not harm their hosts and many are beneficial. We could almost have predicted this, had we listened to evolutionary theory, since the success of a parasite is closely tied to that of its host species. There is much evidence (and some counterevidence) that in the most advanced parasite-host coevolution the parasites show minimal pathogenesis. A. C. Allison concludes, "Long co-evolution of vertebrate hotes and parasites results in infections of low virulence."[14] The ill-adapted parasite species slays its host species and goes extinct; better-adapted parasites co-exist with their hosts, and the best-adapted will benefit their hosts at the species and even at the individual level. Overvirulence to the host decreases the parasite's chances of survival; a parasite can increase its chances by benefiting its host organism. Parasites also regulate animal populations.

There is obvious disvalue in much parasitism; this may be, on average, a glitch in the system. But it is almost as clear that virulence is not intrinsic to parasitism and that the systemic selective pressures are often positive even when the trend is sometimes locally negative. "Parasitism, one of life's great phenomena, possesses a quality of goodness that has largely been overlooked."[15]

Parasitism is a subroutine in a larger value-capture system. It cannot be a bad thing for an organism to depend on another for skills or metabolisms it lacks, else humans (who cannot photosynthesize) eating plants (which can) would be a parasitic evil. All heterotrophs of spectacular evolutionary achievement live in "abject dependence" on plants. Meanwhile, it might be a bad thing if some life forms took degenerating routes. No one wants to say that there are no disvalues in nature; but we do want to keep them in perspective in the fuller picture. The whole idea of parasitism is conceptually parasitic on values elsewhere present and flourishing enough to be parasitized. The disvalue, parasitism, is privative on some value, autonomous life, but all life is interdependent. There could be no world composed entirely of parasites, taking in each other's values, so to speak. However much disvalue parasitism introduces, there must be positive value in excess of it, else the parasite could not parasitize.

4. Selfishness

Selfishness is rightfully lamented in humans but immoral selfishness can be a disvalue in nature only if there is moral agency present. Who or what are such moral agents? Monkeys? Snakes? Trees? Mountains? Lightning bolts? Evolution? Genes? Systemic nature? Every organism is full of "selfish genes," Richard Dawkins tells us, the only kind of genes there are. Why does George Williams disvalue evolutionary nature so? Because it "can honestly be described as a process for maximizing short-sighted selfishness."[16] But if there is no moral agency present, if genes cannot be selfish in any sense that we disvalue in human ethics, that is not an "honest" (=accurate) description after all. Moreover, evolution has no "sight" and cannot be faulted for being short-sighted. What seems to be claimed is that there are processes in evolution that moral agents, when they appear, will disvalue—rectify or repudiate—in their own cultural life. The alleged selfishness is said to be located in one organism's looking out after its own interests or in assisting relatives and offspring, when it shares genes.

Though there is no moral selfishness in nature, there are selves—biological organismic identities to be preserved (not to be confused with psychological egos). Such a self-impulse cannot in itself be a disvalue. Quite to the contrary, this self-impulse just is the life impulse, the principal carrier of biological value. An organismic self is not a bad thing, nor is the defense of it.

Can we describe the natural system more accurately and less pejoratively? The system evolves organisms that attend to their immediate somatic needs (food, shelter, metabolism) and that reproduce themselves in the very next generation. In the birth-death-birth-death system a series of replacements is required. The organism must do this, it has no options; it is "proper" for the organism to do this (Latin *proprium*, one's own proper characteristic). Somatic defense and genetic transmission are the only conservation activities possible to most organisms; they are necessary for all, and they must be efficient about it.

If there is some disvalue, this must lie in an overextension or aberration of the self-impulse. When a subordinate monkey relinquishes a feeding site to a dominant, the dominant may be said to have "selfishly" taken over. Or males may "selfishly" dominate females or defend territories. Such behavior, says Williams, is "not only selfish in some theoretical sense but patently pernicious. Only the morally and intellectually dishonest could label it otherwise."[17] But if we strike out the negative moral overtones and replace this with positive self-preservation, what is going on? The monkey with the superior genes gets fed and bred; the monkey with the inferior genes does not, or at least not first. What is so disvaluable about that? Should it rather be the other way round—that the inferior genes get nourished and propagated, the superior ones not so?

If we cast the event in terms of values defended, what is of value here (the superior genome) gets transmitted, maintained through feeding and breeding, while what is of relative disvalue gets selected against. There is no moral agency at issue; what is at stake is value that is self-actualized. To ask these monkeys to behave as altruistic humans misunderstands the events and misvalues them accordingly. Read out the immorality, and the picture looks different. Take off the dark glasses and put on clearer ones. It is a category mistake to describe (and censure) what goes on in wild nature with terms borrowed from culture and projected onto nature. There is nothing here particularly disvaluable, which moral agents, when they come, will want to deplore and rectify. The alleged selfishness is really the conservation of value intrinsic to the organism in the only manner possible and appropriate to it.

All such contests at feeding and reproduction are endured for "selfish" advantage by males or females only in a problematic sense, since the somatic individual soon dies anyway. A better way of interpreting events is that the contest is to share genes. It is self-defense in one sense, but if males and females spend time, energy, and effort to reproduce, this is self-sacrificing in another sense. By those who resolve to see everything through selfish lenses, this will (rather confusingly) be called selfishness again, seen from the nonmoral genetic level. But we get a much clearer picture of what is going on if we interpret this as values being transmitted over generations.

Although the organism is engaged in a short-range reproduction of its kind, the systemic processes are neither short-range nor do they selfishly maximize only one kind. The system is 3.5 billion years old; it has steadily produced new arrivals, replacements, and elaborations of kinds, going from zero to five (or ten) million species, through five (or ten) billion turnover species in a kaleidoscopic panorama. Any particular organism, in the subroutines of this system, actualizes its own values and transmits these to the next generation (with variations). That is all any one organism has the capacity to do, a capacity of critical value. The result is a quite dramatic story, not just a long, long chain of "patently pernicious" short-sighted selfishness. The value account seems quite descriptively plausible, not at all "morally and intellectually dishonest."

Of course, when humans pass over into culture, they may wish to transcend or revise the now-relaxed processes of natural selection. A good thing in nature may not be a good thing in culture, and vice versa. We certainly do not want to turn to amoral nature for templates with which to judge moral events in culture. That too is a category mistake. Interhuman ethics is nowhere to be educated by watching monkeys, or wildflowers, or bolts of lightning.

5. Randomness

"Nature does nothing in vain. *Natura nihil agit frustra.*"[18] But, contrary to Sir Thomas Browne's famous claim, some biologists find that nature does nothing with any bent toward value. Physical and chemical laws operate there, as do some biological regularities, such as survival of the fittest, but the system is mostly just contingent. Unrelated and probabilistically-related causal lines impinge, resulting in lucky and unlucky events. When an elk, crashing through the forest, accidentally steps on the nest of a thrush, crushing the young, an event of disvalue has occurred, without redeeming construction of value. The elk gains nothing by the accident; had it stepped elsewhere nothing would have been really different for it. The thrush has only lost. Windblown seeds fall, some on rock or unsuitable ground. Some get eaten. Some sprout to get killed by a frost; some die when the rains fail.

Some judge this to be local disvalue in a null-valued system. A random event generator lies outside the domain of value generation, even though its output is accidentally valuable or disvaluable on rare occasions. But when we place local bad luck into the larger system, is this systemically a disvalue? The organism by its genetic programming, instincts, perceptions, and conditioned learning modifies its exposure to luck and acts as a preference sieve, sometimes but not always accumulating the lucky upstrokes and discarding the unlucky downstrokes. Should we value the capacity to catch or discard, while cursing the world of contingency?

But that is to value the retention side as though the supply side were not logically and empirically necessary. It does not seem possible for the world to be otherwise if there is to be autonomy, freedom, adventure, success, achievement, emergents, openness, surprise, and idiographic particularity. The feet of free-ranging elk cannot be bound. We inhabit a world where luck is one of the required ingredients in the arrival, construction, and conservation of value. There is a combination of deterministic and opportunistic processes, a mixture of vitality with necessity and possibility. We are not dealing with probabilities and random walks so much as with developing story lines.

Frances Crick complains that biology has no "elegance." Organisms evolve happenstance structures and wayward functions that have no more overarching logic than the layout of the Manhattan subway system.[19] Stephen Jay Gould insists that the panda's thumb is evolutionary tinkering and that orchids are "jury-rigged."[20] Evolution works with what is at hand, and makes something new out of it.

But what is so disvaluable about that? The achievements of evolution do not have to be optimal to be valuable; and if a reason that they are not optimal is that they had to be reached historically along story lines, it is more valuable to have history plus value as storied achievement than to have "elegant" optimal value solutions without history, autonomy, or adventure. Organismic vitality is better than regimented simplicity. The elegance of the thirty-two crystal classes is not to be confused with the grace of life lived in the midst of perpetual perishing.

How global is this contingency? Viewing natural selection systemically, some hardnosed scientists are reluctant to see any value increase, because their theory, as usually interpreted, does not entitle them to see any. Despite the use of "better" with "adaptation," the theory predicts only survivors and leaves entirely open whether the survivors have progressively more worth. If the environment that species track is drifting, then they do not progress toward anything, they just track drift—buffeted about by aimless geomorphic processes. The only outcome that natural selection can promote is capacity to survive, an independent variable with regard to increasing complexity or diversity. Like a rotating kaleidoscope, there is change without development.

The biological panorama, more or less packed since Cambrian times, is a scene of steady turnover, but later-coming grasses or crustaceans are not any better than earlier, extinct ones; they are just different. Indeed, in climates growing colder or drier fewer species may live there later than did before. There are fewer dinosaurs now than in Cretaceous times, fewer birds than in Pleistocene times. Gould insists, "We are the accidental result of an unplanned process . . . the fragile result of an enormous concatenation of improbabilities, not the predictable product of any definite process."[21] There may result "chance riches,"[22] but the system is without value heading. Note that this claim is intended to describe objective natural history. Both descriptively and axiologically, there is only luck.

But something is increasingly learned across evolutionary history: how to make more kinds and more complex kinds. This may be a truth about natural history, even if neo-Darwinism is incompetent to say much about how this happens. Cold and warm fronts come and go, so do ice ages. There are rock cycles, orogenic uplift, erosion, and uplift again. But there is no natural selection there, nothing is competing, nothing is surviving, nothing has adapted fit, and biology seems different. All those climatological and geomorphological agitations continue in the Pleistocene period more or less like they did in the Precambrian, but the life story is not the same all over again. Where once there were no species, now there are five to ten million. It seems evident that, on average and environmental conditions permitting, the numbers of life forms start low and end high.

J. W. Valentine concludes for marine environments: "A major Phanerozoic trend among the invertebrate biota of the world's shelf and epicontinental seas has been towards more and more numerous units at all levels of the ecological hierarchy. . . . the biosphere has become a splitter's paradise."[23] There is "a gradually rising average complexity."[24] The story of terrestrial life is even more impressive, because the land environment is more challenging. Reptiles can cope in a broader spectrum of humidity conditions than amphibians. Mammals can cope in a broader spectrum of temperature conditions than reptiles. Genetic and enzymatic control is surpassed by neural networks and brains; there are increases in sentient capacity, locomotion, acquired learning, communication, language acquisition, and in manipulation.

Francisco J. Ayala concludes, "Progress has occurred in nontrivial senses in the living world because of the creative character of the process of natural selection."[25] Theodosious Dobzhansky agrees, "Evolution as a whole doubtless had a general direction, from simple to complex, from dependence on to relative independence of the environment, to greater and greater autonomy of individuals, greater and greater development of sense organs and nervous systems conveying and processing information about the state of the organism's surroundings, and finally greater and greater consciousness."[26]

What the random walk omits is the cybernetic, hereditary capacity of organisms to acquire, store, and transmit new information over historical time. Organisms start simple and some of them end up complex; there are trends over longer-range time scales because something is at work additionally to tracking drifting environments. The life process is drifting through an information search, locking onto discoveries. With such a conclusion the value question returns. If the system does produce both diversity and complexity, randomness is not the systemic story, however important it may be in the subplots.

6. Blindness

Williams, who thinks evolutionary nature selfish and pernicious, also thinks it blind. This follows from randomness on the cutting edge. Variations bubble up from the genetic level, as organisms compete for a place in drifting environments. Those few variations that are accidentally useful are selected for; the most, worthless, are discarded; some, to which even natural selection is blind, since they produce no differential survival rates, remain and result in genetic drift. Zig here, zag there, organisms stumble onto a life program. "The evolutionary process is immensely powerful and oppressive, . . . it is abysmally stupid."[27] "Clumsy" and "blundering" were Darwin's words.

Now evolution has no sight, nor is it deliberate. Maybe it is not elegant. Evolution is a problem-solving process. Is this unguided, inept, disvaluable? The genetically originated novelties are formed in a shuffle that, while in some sense blind to the organismic needs, is far from chaotic. Only those variations are tested and selected that are more or less functional. The organism typically only probes the nearby space for possible directions of development. Mutators and animutators increase or trim the mutation rates as a function of population stresses.[28] Repair mechanisms snip out certain genetic errors, and thus eliminate some variation. The genetic program has the capacity to "reject" some of the random recombinants on the basis of information already present in the genetic coding. Individual genetic sets are adept at pumping out their own disorder. But they do not pump out all novelty; that would cease evolutionary development and lead to extinction.

Contemporary geneticists are insisting that we misperceive this process if we think of it as blind. It is not deliberated in the conscious sense; but it is cognitive, somewhat in the way in which computers, likewise without felt experience, can run problem-solving programs. There is a vast array of sophisticated enzymes to cut, splice, digest, rearrange, mutate, reiterate, edit, correct, translocate, invert, and truncate particular gene sequences. There is much redundancy (multiple, variant copies of a gene in multi-gene families) that shields the species from accidental loss of a beneficial gene and provides flexibility on which these enzymes can work.

John H. Campbell concludes, "Cells are richly provided with special enzymes to tamper with DNA structure," enzymes that biologists are extracting for genetic engineering. But this "engineering" is already going on in spontaneous nature. "Gene-processing enzymes also engineer comparable changes in genes in vivo. . . . We have discovered enzymes and enzyme pathways for almost every conceivable change in the structure of genes. The scope for self-engineering of multigene families seems to be limited only by the ingenuity of control systems for regulating these pathways." These pathways may have "governors" that are "extraordinarily sophisticated." "Self-governed genes are 'smart' machines in the current vernacular sense. Smart genes suggests smart cells and smart evolution, . . . the promise of radically new genetic and evolutionary principles."[29]

So far from disparaging the blind groping of genes, computer scientists may deliberately seek to imitate a similar process on their unconscious computers. Some sophisticated programs use what are called "genetic algorithms."[30] Such algorithms involve recombining partial solutions to a problem in order to generate improved solutions. The mode is biological with sexual mating and strings of genes on chromosomes that can be shuffled and selected. The underlying metaphor is natural selection. Scientists may want to program a computer to search for the optimal set of values to solve certain multivalued problems where the values interact with each other, such as solving sets of complex mathematical equations, or detecting patterns against a background of noise, or scheduling the most effective work and meeting times for many dozens of employees in a manufacturing plant.

The computer will generate at random some "bit strings," or "genotypes," analogous to information coded on chromosomes, which are possible values in solution. It will test members of the initial population for effectiveness as a solution, rank them for their "fitness," and select the fittest. The computer will then generate new solutions, stimulating variations on the highest ranking ones, inhibiting the lower ranking ones, evaluate the new possibilities for their fitness, and put them in competition with the previous, partially effective solutions. The computer also "mates" the various solutions, that is, cuts up and splices portions of bit strings that seem to code the most effective values, and then tests these "offspring" for their fitness. It works with coadjusted clusters that probably (but not inevitably) move together during crossover. It may vary the "population size" of the set of solution values that it mates. It will discard solutions with low fitness.

If two or more sets of solutions begin to appear that have little in common (widely separated local optima), the computer will preserve these multiple solution tracks, but try an occasional cross-mixing of segments from the different local optima, some of which will result in offspring that have enough fitness to remain in the working population. Such outbreeding prevents getting trapped in local optima that are not effective solutions globally. The computer will continue with lesser probability (which may be varied during the program) occasionally to explore unlikely solutions. Even in large and complicated search spaces, genetic algorithms tend to converge on solutions that are globally optimal or nearly so. Simple bit strings can encode complicated structures, and reiterated transformations of partial solutions have a striking power to improve them. Computer searches that would take a computer an estimated billion years, if done completely at random, can be accomplished in a few hours.

Genetic problem solving, then, does not seem so tinkering, jury-rigged, and blind. To the contrary, it is remarkably like what some of the smartest scientists are doing. Indeed, Herbert A. Simon finds the cutting edges of science itself like natural selection. "Various paths are tried; some are abandoned, others are pushed further. Before a solution is found, many paths of the maze may be explored. The more difficult and novel the problem, the greater is likely to be the amount of trial and error required to find a solution. At the same time, the trial and error is not completely random or blind; it is, in fact, rather highly selective. . . . Human problem solving, from the most blundering to the most insightful, involves nothing more than varying mixtures of trial and error and slectivity."[31]

In nature, the challenge is to get as much versatility coupled with as much stability as possible, optimizing twin maxima. There is selective advantage in using past knowledge where possible; there is advantage in quickly breaking through to something new where required. The dominant/recessive phenomenon in genetics, with a large number of recessive alleles waiting in a population, is a way of storing variability that is not usually expressed in a stable environment, but which is nevertheless there when the environment shifts.[32]

So what is so disvaluable about genetic systems? It seems rather that there is valuable problem-solving taking place, and that this is taking place whether or not humans are present and doing any evaluating of it. Maybe there is more elegance than we first thought. Certainly there are remarkable success stories.

7. Disaster

Violent forces strike animals, plants, and people; disaster results. When human affairs are not touched, we may be unconcerned. Leopold mused, while sawing wood, "It was a bolt of lightning that put an end to wood-making by this particular oak. We were all awakened, one night in July, by the thunderous crash; we realized that the bolt must have hit near by, but, since it had not hit us, we all went back to sleep. Man brings all things to the test of himself, and this is notably true of lightning."[33] If it didn't help or hurt me, it didn't help or hurt.

But we must try to bring things to some test outside ourselves. If the test is anthropocentric, all and only those bolts that hit humans or destroy what they value have disvalue; after that there is nothing more to be said. But this bolt did end the life of the good oak, and is that not a disvalue? Individually yes, but systemically? Lightning, like storms, fires, and floods, is statistically regular though individually erratic. At this moment 1,800 thunderstorms are raging around the globe, about 44,000 each day. Lightning is striking the Earth 100 times each second, over 8.6 million times each day. Earth is a sometimes turbulent planet. Is this turbulence, evidently bad when it catches individuals, a disvalue systemically? In 1988, lightning triggered Yellowstone fires that burned catastrophically. On July 21, 1987, 175-mile-an-hour tornado winds destroyed 15,000 acres in the Teton Wilderness of Wyoming. Catastrophic floods rip up riparian zones, as did the Big Thompson Flood in Colorado July 31, 1976. Such disruptions escalate the unrelated causal lines that tear things up.

Possibly these violent forces are bad, but there are good ones that overcome them. Possibly the catastrophic, negative forces are integrated with the uniformitarian, positive forces. Floods, windstorms, lightning storms, and such violences would be more or less like wildfire in ecosystems, a bad thing to individuals burned and in short range, but not really all that bad systemically in long range, given nature's restless creativity. Without thunderstorms, Earth would lose to the upper atmosphere, in less than an hour, the negative electrical charge that produces the atmospheric nitrogen upon which most plants depend. Without thunderstorms, playing electric charges over the thin hot soup, life could not have originated. Lightning has been essential to life.[34]

Often such violence comes with enough regularity that life can adapt—the oak with the stalwart trunk, ready for the winds; the lodgepole pines with their serotinous cones; the ecosystem rejuvenated by fires. Most bigscale processes are incremental, like mountain building over millennia; life can track such changing environments, often innovating and respeciating as it follows geomorphic history. Any event too infrequent to be naturally selected for is rare enough to be systemically atypical on most Earthen scales.

Rarely, this violence can come so spasmodically that life is unable to cycle onto it. We will then be hard pressed to say that these things can rejuvenate the ecological succession. Volcanic eruptions come every thousand years, and the ecosystem cannot adjust to their intermittent occurrence. They just disrupt and the ecosystem recovers as best it can. Even then life rejuvenates, value rises from the ashes of disvalue, and the system seems prolific once more. The incremental processes are punctuated by catastrophes, but nature is never too violent for life to continue—at least never yet in the several billion years of Earth history.

In March 1872 John Muir was in Yosemite Valley when it was struck by the Inyo earthquake: "I ran out of my cabin, near the Sentinel Rock, both glad and frightened, shouting, 'A noble earthquake!' . . . a terribly sublime and beautiful spectacle"[35] "It is delightful to be trotted and dumpled on our Mother's mountain knee."[36] Later, Muir concludes that the earthquake was "wild beauty-making business." "On the whole, by what at first sight seemed pure confusion and ruin, the landscapes were enriched; for gradually every talus, however big the boulders composing it, was covered with groves and gardens, and made a finely proportioned and ornamental base for the sheer cliffs."[37] Muir once climbed a Douglas fir to ride out a storm, "to take the wind into my pulses and enjoy the excited forest from my superb outlook."[38] "Many of Nature's finest lessons are to be found in her storms."[39]

Is this Romanticism at its worst? Not really, for behind the poetry there is some botany, life persisting in the midst of its besetting storms. It is objective fact that the adverse, violent forces in dialectic with the prolific, enduring forces yield much of the romance of life. The violent forces of nature are as much to be celebrated for their creativity as for their destruction. The glaciers did carve the most spectacular scenery. Floods cut the valleys. Muir insisted that these are "Nature's modes of working toward beauty and joy." Volcanism is one of the mountain building forces; and, after the violence, life will return. "The cooled lava is forested now. The sun shines lovingly upon it, and all is joyous life."[40] "Storms of every sort, torrents, earthquakes, cataclysms, 'convulsions of nature,' etc., however mysterious and lawless at first sight they may seem, are only harmonious notes in the song of creation, varied expressions of God's love."[41] Muir certainly has an intensive faith in natural systems, but such faith is not without some impressive evidence.

None of this means that, in our culturally rebuilt environments, we should not take shelter. After that, "ecstacy is a monster storm."[42]

8. Indifference

In the California desert, April, May, and early June are usually good months; with rain and before the heat, the fauna and flora flourish. But sometimes the rains fail and the heat comes quickly. Mary Austin, recalls: "The quick increase of suns at the end of spring sometimes overtakes birds in their nesting and effects a reversal of the ordinary manner of incubation. It becomes necessary to keep eggs cool rather than warm. One hot, stifling spring in the Little Antelope I had occasion to pass and repass frequently the nest of a pair of meadowlarks, located unhappily in the shelter of a very slender weed. I never caught them sitting except near night, but at midday they stood, or drooped above it, half fainting with pitifully parted bills, between their treasure and the sun. Sometimes both of them together with wings spread and half lifted continued a spot of shade in temperature that constrained me at last in a fellow feeling to spare them a bit of canvas for permanent shelter."[43]

Barry Lopez recalls, "In the fall of 1973 an October rainstorm created a layer of ground ice that, later, muskoxen could not break through to feed. Nearly 75 percent of the muskox population in the Canadian Archipelago perished that winter."[44] Like the meadowlarks, the muskoxen hit bad luck; the ice layer was of disvalue to the oxen, the blazing heat of disvalue to the larks. This is randomness now with brutal indifference to life. David Hume claimed that nature "has no more regard to good above ill than to heat above cold, or to drought above moisture, or to light above heavy."[45] Or to life above nonlife, he would have added. That indifference can seem true in the short range, sometimes even in the long range. Nature doesn't care.

Is that all there is to be said? Though stressed on these occasions, those individuals are satisfactory fits in their ecosystems, and the nature that doesn't care is also the nature that provides life support. There the meadowlarks are, this pair in trouble, but the species nevertheless flourishes both East and West. So what are we to do? Curse this exception or rejoice in the usual? The muskoxen, decimated this one winter, continue on the tundra, living on for millennia, so well adapted to a polar existence that this is one of the few large animals to have survived the Ice Ages in North America. What are we to do: lament the October rainstorm or celebrate the tundra that has supported them for several million years? Or both?

As a species, organisms get selected for those functions and skills that enable them to do better in their niches, and what is so uncaring about that? Selection for adapted fit is a strange kind of indifference, an odd disvalue. The geomorphic, climatic, and even the biological processes can seem to have no axiological component at all. They just drift around in a mixture of cycles and random walk. Life arises in this geological kaleidoscope but no thanks to the elements. But then a different perspective on this earthen stew strikes us. This churn of materials, perpetually agitated and irradiated with energy, is not the problematic, indifferent resource but the prolific source. Against the indifference, we now must counter that the systemic results have been prolific, five million species flourishing in myriads of diverse ecosystems.

Maybe nature doesn't "care"; but nature does produce increased diversity and complexity. To say that there is nothing but systemic indifference ignores the principal result of natural history. Even if agnostic about the statistical worth of the whole, can anyone deny that repeated movements within it are prolific, values achieved here and there amidst the blooming buzzing confusion of nonvalues and disvalues—and all this objectively so whether or not humans take cognizance of these wild affairs?

There is a kind of "promise" in nature, not only in the sense of potential that is promising but in the reliability in the earthen set-up that is right for life. Perhaps the planetary set-up is an accident, but the ongoing after the set-up seems to be loaded with fertility. In the short-range all lose, death is inevitable; but then again in the long-range life persists, phoenix-like, in the midst of its destruction. How much promise do we need? Isn't several billion years, with a turnover of several billion species, and a cumulation of several million species in almost every nook and cranny of Earth enough? Perhaps to say that nature "has regard" for life is the wrong way to phrase it; we do not want to ascribe purpose to nature. At the same time, nature is a fountain of life. That was the original etymology of "nature," from "natans," giving birth—and on Earth nature as birth is nearer the truth than nature as sheer indifference.

It seems a shame now for humans to break that "promise." We may not bring ourselves to say that nature is "keeping" any promise; that is not the right way to put it. But humans ought not to break what has been so promising in nature—not at least without an argument that the values they make in culture exceed the values they break in nature. It also seems uninformed to turn a philosophically taciturn gaze over all this promising, storied natural history and say, "Because I was not there, nor any other human, the question of value and disvalue cannot be asked. Argument over."

No argument has been won; none has been begun. Rather, we are not really going to be informed about what is going on in nature until we ask how value is achieved there, and we are not going to find an answer until we see that disvalues are regularly transmuted into values. The indifference to be lamented is not that in neg-valued or null-valued nature; the troublesome indifference lies in humans so careless about values outside the human sector.

9. Waste

Here an opposite thought strikes us. The nature indifferent to life is also ridiculously prolific; now the indifference is manifest not in niggard support but in unreasonable fecundity. The teeming kinds are cast forth to die. The nature complained against before because it was indifferent to life is complained against now because it produces it too lavishly. This profusion of creatures can seem inordinate and senseless, vigorous and horrible. Life is a good thing, but must the system waste it so?

The Salt Creek pupfish overwinters in permanent springs in Death Valley, several thousand individuals. With spring runoff, the population expands a hundred fold, spilling into downstream habitat; the waters abound with juvenile pupfish. Summer comes, the streams dry up, and the population is decimated again.[46] The proliferation and die-off is extravagant, indifferent reproductive power. Heavy-seeded trees, such as oaks or hickories, may produce 200 million acorns or nuts to replace themselves once or twice, and light-seeded trees, like cottonwoods produce billions.[47] A pair of robins may produce thirty eggs to replace themselves once. Mother Nature seems spendthrift.

One response to prodigal life is to celebrate it. Thoreau exclaimed: "I love to see that Nature is so rife with life that myriads can be afforded to be sacrificed and suffered to prey on one another; that tender organizations can be so serenely squashed out of existence like pulp,—tadpoles which herons gobble up, and tortoises run over in the road . . . With the liability to accident, we must see how little account is to be made of it. . . . Poison is not poisonous after all, nor are any wounds fatal."[48] This may be exuberance more than waste; the book of Genesis reports that God commanded the earth to bring forth "swarms" of creatures and found the prodigious result to be very good.[49]

Another question to ask is whether all those acorns, hickory nuts, seeds, and dead fishes really go to waste? They do, if nothing is of value unless and until humans value it. But "waste" is not the way that squirrels, blue jays, bears, deer look at acorns and nuts, nor mice the cottonwood seeds, nor the way the insect larvae, fungi, and decomposing bacteria approach the dead fish. One organism's waste is another organism's treasure. A muskox, laid waste by the ice sheet, becomes a carcass that benefits dozens of scavengers and predators. When something dies, something else lives. Nutritious pollens, fruits, and seeds may even evolve to be eaten, if this also facilitates dispersion by mobile animals and birds. Seed predators are often seed dispersers.

This alleged "waste" makes trophic pyramids possible. The lavish primary production of the grasses supports the ungulates; grass seeds support the granivorus rodents, whose fecundity supports the coyotes and owls. Insects pick up detritus and become food for birds. There is episodic surplus—an overkill of muskoxen, a horde of locusts. In odd situations, anomalies result. A mountain lioness got into a fold of sheep in Colorado laid down ninety-nine in one night's kill. But systemically on average, organisms must be eficent. There is capture of valuable nutrients and energy, which are recycled through the ecosystem in myriads of pathways. Animals partition out and use food resources in sequential stages. All living things are marginally pressed for survival; there is little waste. Wherever there is available free energy and biomass, a life form typically evolves to exploit those resources. Nature's exuberance is also nature's economy.

Truth is, there is not much "waste" in natural systems, though there is exuberance and fecundity. What there is, if immediately a disvalue, is systemically transformed into something of value. And all this happens regardless of what humans think about it.

10. Struggle

The relentless struggle to survive can be supposed a disvalue. Adapted fit seems a good thing, but the shadow side is how each organism is doomed to eat or be eaten, to stake out what living it can in competition with others. Perhaps there is more efficiency than waste, more fecundity than indifference, but each organism is ringed about with competitors and limits, forced to do or die. Each is set as much against the world as within it. Physical nature, from which are wrested the materials of life, is brute fact and brutally there, caring naught and always threatening. Organic nature is savage; life preys on life.

Nature as a jungle does not mean that there are no valuers in the wild; it portrays too many claimants contesting scarce worth. Perhaps local achievements of value are wrested out of a disvaluable place? Or does the truth lie deeper? Perhaps the context of creativity logically and empirically requires this context of conflict and resolution. An environment entirely hostile would slay us; life could never have appeared within it. An environment entirely irenic would stagnate us; advanced life, including human life, could never have appeared there either. Oppositional nature is the first half of the truth; the second is that none of life's heroic quality is possible without this dialectical stress. Take away the friction, and would the structures stand? Would they move? Muscles, teeth, eyes, ears, noses, fins, legs, wings, scales, hair, hands, brains—all these and almost everything else comes out of the need to make a way through a world that mixes environmental resistance with environmental conductance. Half the beauty of life comes out of endurance through struggle.

In culture, humans relax these pressures of natural selection, though we cannot and do not eliminate the dimension of struggle. If human children catch pinkeye, physicians prescribe sodium sulfacetamide; but when the bighorns of Yellowstone caught pinkeye, they were left to the ravages of the disease. Being sick is a meaningless disvalue in a medically skilled culture; but the bighorn herd, surviving the epidemic, is stronger now. We count disease in domestic sheep a disvalue, because our resources are threatened; we call the veterinarian to cure it. But park officials let half the bighorn herd perish, letting nature take its course and valuing more the *Chlamydia*-resistant sheep that would survive, wilder and stronger than had they intervened to fix this disvalue in nature.

Indeed, from this wilder perspective the domesticated is the degraded. Muir contrasts wild sheep, which he admires, "elegant and graceful as a deer, every movement manifesting admirable strength and character," with the domestic ones, which he despises, stupid "expressionless, like a dull bundle of something only half alive."[50] If the standard of evaluation is our human subjective preferences, fashions in wool, the domesticated breeds can be better; but objectively, in natural systems, the wild sheep, honed to its strength, alertness, and endurance by the struggle for survival, is the more valuable. From this perspective, struggle is no disvalue (even though many sheep lose); it is the key to value achieved. Indeed, it is difficult to envision any of the properties admired (the horns, the eyesight, the agility, the musculature, the wool) except as created in this arduous environment. After a hunt on Mount Shasta, Muir examines closely the carcasses of a dead ram and ewe, and, repentant and chagrined by his kill, shouts, "Well done for wildness!"[51]

What is this struggle but a history of transvaluing disvalues into values? Disvalues and values are both objectively present in nature (regardless of human evaluators), nor is the struggle a zero sum game, nor null of value; rather, the struggle is prolific creativity.

11. *Suffering*

Over evolutionary history, with the diversity, complexity, and creativity we have celebrated, there emerges the capacity to suffer. Indeed the story could be titled, perversely, "The Evolution of Suffering." Each seeming advance—from plants to animals, from instinct to learning, from ganglia to brains, from sentience to self-awareness, from herbivores to carnivores—steps up the pain. In the planetary drama, struggle deepens through time into suffering. In chemistry, physics, astronomy, geomorphology, meteorology, nothing suffers; in botany life is stressed, but only in zoology does pain emerge. Is not this the evolution of increasing disvalue?

We are not much troubled by seeds that fail, but it is difficult to avoid pity for nestling birds fallen to the ground. In every season, most of the sentient young starve, are eaten, abused, abandoned. Wolves can get at the rump of a deer (avoiding the antlers); they may half-hamstring the deer and eat at it from behind. Sometimes a deer gets away with a hunk of its anus eaten out, to die slowly afterward. The Greek root for suffering is "pathos"; there are pathologies in nature, such as the diseases of parasitism, noted earlier. But pathology is only part of the disvalue; even in health there is suffering. Life is indisputably prolific; it is just as indisputably pathetic, almost as if its *logos* were *pathos*, as if the whole of sentient nature were pathological. "Horribly cruel!" exclaimed Darwin.

We are trying not to be anthropocentric in our evaluation of nature; this means that animal suffering counts. This means also, however, that the human experience of suffering must not be projected indiscriminately onto the animal world. Suffering in some sense seems copresent with neural structures; there are endorphins in earthworms, which indicates both that they suffer and that they are provided with pain buffers.[52] A safe generalization is that pain becomes less intense as we go down the phylogenetic spectrum and is often not as acute in the nonhuman world. It is a mistake to view the sufferings of animals, birds, repiles too anthropopathically, too subjectively. Birds and reptiles typically have fewer nerve endings per surface area of skin, for instance; and the level of consciousness, self-awareness, or experience, or whatever is the proper name for their experiential state, is very different from, more subdued than, less intense and coherent than our own.

Nevertheless pain is objectively present, and is it sheer disvalue? A more adequate answer is that, just as struggle is the dark side of creativity, pain is logically and empirically the shadow side of pleasure; one cannot enjoy a world in which one cannot suffer, any more than one can succeed in a world in which one cannot fail. The logic here is not so much formal or universal as it is dialectical and narrative. In natural history—whatever might be true in other imaginable worlds—the pathway to psychosomatic consciousness, the only kind of experience we know, is through flesh that can feel its way through that world. There is some sentience without much capacity to be pained by it; we do not much suffer through our eyes or ears. But neither would we have those eyes and ears had they not evolved for the protection of the kinesthetic core of an experiential life that can suffer, whether by lack of food for which eyes may search or by predators whom ears may hear. We recouple here with the claims made about predation; levels of achievement and experience are generated in both predator and prey not otherwise possible.

The capacity to suffer is generally accompanied by possibilities of avoiding suffering, some freedom and self-assertion. The capacity to suffer, for instance, drives the capacity for learned behavior; it brings animal life to a central focus in sentient consciousness. This does not and cannot happen in plants. Thought appears in order to prevent pain and to affirm wellbeing, but the thought that cannot feel pain cannot figure out how to escape it. In humans, this evolution of thought seeking comfort drives the transition and exodus from nature to culture.

Pain is eminently useful in survival, and it will be naturally selected, on average, as functional pain. Natural selection requires pain as much as pleasure in the construction of concern and caring; pain is an alarm system in a world where there are helps and hurts through which a sentient organism must move. On the other hand, any population whose members are constantly in counterproductive pain will be selected against and go extinct or develop some capacities to minimize it. In this sense, natural selection, so far from needlessly increasing pain, rather trims it back in the system, so far as the system can remain vital—both conserving past vitalities and developing new ones. Pain is self-eliminating except insofar as it is instrumental of a subsequent, functional good. Intrinsic pain has no logical or empirical place in the system, neither does maladaptive pain. We cannot show this in the detail of every case; perhaps we need not expect it to be true in every case, and there are troublesome anomalies. Nevertheless, the system statistically must select for beneficial pain.

12. Death

Life is the first mystery that comes out of Earthen nature, and death a secondary one. But death comes as surely as life to all higher organisms. Even the lower forms that reproduce by cell fission or plant genets that produce ramets may and do die. So the great value, life, is countered by the great disvalue, death; and have we again a zero sum game? For each organism, the last word is destruction.

But we are trying to see nature systemically, where death is not the last word—at least it has never yet been across three and a half billion years. Death is the key to replacement with new life. If nothing much had ever died, nothing much could have ever lived. Just as the individual overtakes, assimilates to itself, and discards its resource materials, so the evolutionary wave is propagated onward, using and sacrificing particular individuals, which are employed in, but readily abandoned to, the larger currents of life. Thus the prolife evolution both overleaps death and seems impossible without it.

The vast number of creatures sprouted, hatched, or born, are, of necessity, more or less well-endowed genetically and emplaced in a more or less congenial environment, despite or including the fact that in their environment they are spurred to earn their way. Even though most will not live to maturity, their task is a reasonable natural ideal, a *telos* for which they are competently programmed. There are lethal mutants and monstrosities, but these bad ideas, as it were, are aborted immediately without further experiment. Organisms survive in about that proportion in which they are viable, so that life is sustained in any individual in relative proportion to its fitness for it.

A community of life is systemically sustained, and this requires value capture as nutrients, energy, and skills are shuttled round the trophic pyramids. This anastomosing of life threads characterizes an ecosystem. The surplus of young is efficiently used as resource material in alternative life courses and is thus doubly beneficial, permitting both mutational, cybernetic advance and the interdependent synthesis of biotic materials with higher forms at the top of the pyramid. Overlaid on these interconversions natural selection edits life for evolutionary advance. There is the creation and conservation of life. Death *in vivo* is death ultimately; death *in communitate* is death penultimately but life regenerated ultimately.

Individual organisms must die; many of them (as though the genes had accepted this inevitability) are programmed to die. Annual plants die when they make fruit. Most anadromous fishes die after breeding, as do many mollusks, annelids, and insects. In mayflies and in some male salmon, and in some squids, the gut atrophies prior to the onset of breeding and the organism can no longer eat. It only remains to breed, and die. In this built-in senescence, the cycle of births and deaths is not a disvalue, or if it is, it is a disvalue that is overcome by valuable achievement that overleaps it. Reproduction necessitates death as much as does death necessitate reproduction.

Even if death is not preprogrammed, it is inevitable because of aging. Since life ages, a process that is not well understood, it can be perpetuated only by regeneration. This regeneration must be with variation, if there is to be creative advance or even tracking of changing environments. Death is part of the life cycle, not life part of the death cycle. But when nature—hostile from the perspective of the slain—regenerates and regularly elevates life from the perspective of the community, is that a disvalue? When natural selection cuts away what does not fit and leaves what is better adapted, is that a disvalue?

Species do not have to die; their extinction is never programmed. Most, of course, do die. Ninety eight percent of all species that have ever existed did go extinct, so there are high probabilities, but there is no law of nature or inevitability about species extinction. But here a puzzling aspect of the matter strikes us. The death of the organism feeds into the nondeath of the species. Only by replacements can the species track the changing environment; only by replacements can they evolve into something else. Species sometimes do die, go extinct without issue, but they are often transformed into something else, new species; and, on average, there have been more arrivals than extinctions—the increase of both diversity and complexity over evolutionary history. We think of the extinction of species as tragic, if anthropogenic, because there is loss of birth as well as of death. But the loss of species in natural systems has meant more birth than death; perhaps there too it is tragic, but it is not unredeemed tragedy. Death is not the last word; death is transvalued into life.

13. Systemic value

Nature is random, contingent, blind, disastrous, wasteful, indifferent, selfish, cruel, clumsy, ugly, struggling, full of suffering, and, ultimately, death? This sees only the shadows, and there has to be light to cast shadows. Nature is orderly, prolific, efficient, selecting for adapted fit, exuberant, complex, diverse, renews life in the midst of death, struggling through to something higher. There are disvalues as surely as there are values, and the disvalues systemically drive the value achievements. We miss this panoramic creativity when we restrict value to human consciousness; we make value a prisoner of the particular sort of experiential biology and psychology that humans happen to have, or even of the particular sort of culture that humans happen to have chosen. There is every reason to value what we humans have achieved in our particular biology, sponsoring the emergence of culture as this biology does; there is no reason to think that value lies there and there alone.

All those who find nature to be disvaluable are making objective claims, and they are eventually wrong, not about the form of their claims, for they do try to make objective claims, nor altogether about the content of their claims, for they are locally right. Only they are systematically mistaken in evaluating what they describe, because their descriptions are myopic. Both objectively and globally there is both disvalue and value, and the transmuting of disvalue into value. Such nature is of systemic value, and a better description of what is objectively taking place makes this better evaluation possible, an evaluation that is as objective as is the description.

In this evaluation, we have not painted the world as better than it is in the interests of a philosophical metaphysics, nor worse either; rather we have tried to see into the depths of what is taking place in natural history. The view here is not panglossian; it is a sometimes tragic view of life, but one in which tragedy is the shadow of prolific creativity. That *is* the case, and the biological sciences—evolutionary history, ecology, molecular biology—can be brought to support this view, although neither tragedy nor creativity are part of their ordinary vocabulary. Since the world we have, in its general character, is the only world logically and empirically possible under the natural givens on Earth—so far as we can see at these native ranges that we inhabit—such a world *ought* also to be.

Annie Dillard explodes with horror over her Earth story. "I came from the world, I crawled out of a sea of amino acids, and now I must whirl around and shake my fist at that sea and cry Shame."[53] If I were Aphrodite, rising from the sea, I think I would turn back to reflect on that event and rather raise both hands and cheer. And if I came to realize that this rising out of the misty seas involved a long struggle of life renewed in the midst of its perpetual perishing, I might fall to my knees in praise.

Holmes Rolston, III
Colorado State University

NOTES

1. Darwin, in a letter to Joseph Dalton Hooker, quoted in Gavin de Beer, *Reflections of a Darwinian* (London: Thomas Nelson and Sons, 1962), p. 43. In other moods, Darwin can find the process impressive and beautiful.

2. John Stuart Mill, "Nature" [1874], in *Collected Works* (Toronto: University of Toronto Press, 1963-77), vol. 10, pp. 373-402, citation on p. 398.

3. Thomas H. Huxley, "Evolution and Ethics" in T. H. Huxley and Julian Huxley, *Evolution and Ethics* (London: Pilot Press, 1947), p. 82.

4. George C. Williams, "Huxley's Evolution and Ethics in Sociobiological Perspective," *Zygon* 23 (1988), 383-407, citation on p. 383.

5. Shakespeare, *Hamlet*, Act II, Scene 2, Line 259.

6. Alfred, Lord Tennyson, *In Memoriam* [1850], Part LVI, Stanza 4.

7. Richard Dawkins, *The Selfish Gene* (New York: Oxford University Press, 1976), p. 2.

8. Steve F. Sapontzis, "Predation," *Ethics and Animals* 5, no. 2 (June 1984), 27-38, citation on p. 36.

9. Alexander F. Skutch, "Life's Greatest Evil," *The Scientific Monthly* 66 (1948), 514-18, citation on p. 514.

10. Skutch, p. 516.

11. Stephen Jay Gould, *Hen's Teeth and Horses' Toes* (New York: W. W. Norton, 1983), p. 33.

12. Edward O. Wilson, *Sociobiology: The New Synthesis* (Cambridge: Harvard University Press, 1975), p. 371.

13. Arthur W. Jones, *Introduction to Parasitology* (Reading, MA: Addison-Wesley, 1967), pp. 426-27.

14. A. C. Allison, "Co-evolution Between Hosts and Infectious Disease Agents and its Effects on Virulence." See also, with some dissent, J. C. Holmes, "Impact of Infectious Disease Agents on the Population Growth and Distribution of Animals"; and B. R. Levin, et al., "Evolution of Parasites and Hosts," all in R. M. Anderson and R. M. May, eds., *Population Biology of Infectious Diseases* (New York: Springer-Verlag, 1982). In an alternate theoretical outcome, the level of parasite virulence oscillates with host sensitivity. As we have repeatedly learned in ecosystems, relationships are complex, many other factors are involved (intermediate hosts, effects of and on predators, climate changes, geographical contingencies, genetic variability available), and generalizations are hazardous.

15. David Richard Lincicome, "The Goodness of Parasitism: A New Hypothesis," in Thomas C. Cheng, ed., *The Biology of Symbiosis* (Baltimore, MD: University Park Press, 1971), pp. 139-227, citation on p. 139. See also Elmer R. Noble, Glenn A. Noble, Gerhard A. Schad, and Austin J. MacInnes, *Parasitology: The Biology of Animal Parasites*, 6th ed'n., (Philadelphia: Lea and Febiger, 1989); Thomas W. M. Cameron, *Parasites and Parasitism* (New York: John Wiley, 1958); Maurice Caullery, *Parasitism and Symbiosis* (London: Sidgwick and Jackson, 1952).

16. Williams, "Huxley's Evolution and Ethics," p. 385.

17. *Ibid.*, p. 392.

18. Thomas Browne, *Religio Medici* [1643], Part I, Sec. 15, pp. 262-347 in *Harvard Classics* vol. 3 (New York: P. F. Collier and Son, 1909), citation on p. 278.

19. Francis Crick, *What Mad Pursuit: A Personal View of Scientific Discovery* (New York: Basic Books, 1988), p. 6, pp. 137-42.

20. Stephen Jay Gould, *The Panda's Thumb* (New York: W. W. Norton, 1980), p. 20.

21. Stephen Jay Gould, "Extemporaneous Comments on Evolutionary Hope and Reality," in *Darwin's Legacy, Nobel Conference XVIII*, ed. Charles L. Hamrum (San Francisco: Harper and Row, 1983), pp. 95-103, citation on pp. 101-02.

22. Stephen Jay Gould, "Chance Riches," *Natural History* 89, no. 11 (November 1980), 36-44.

23. James W. Valentine, "Patterns of Taxonomic and Ecological Structure of the Shelf Benthos During Phanerozoic Time," *Palaeontology* 12 (1969), 684-709, citation on p. 706.

24. James W. Valentine, *Evolutionary Paleoecology of the Marine Biosphere* (Englewood Cliffs, NJ: Prentice-Hall, 1973), p. 471.

25. Francisco J. Ayala, "The Concept of Biological Progress," in Francisco Jose Ayala and Theodosius Dobzhansky, eds., *Studies in the Philosophy of Biology* (New York: Macmillan, 1974), pp. 339-55, citation on p. 353.

26. Theodosius Dobzhansky, "Chance and Creativity in Evolution," in Ayala and Dobzhansky, eds., *Studies in the Philosophy of Biology*, pp. 307-37, citation on p. 311.

27. Williams, "Huxley's Evolution and Ethics," p. 400.

28. Eldon J. Gardner, *Principles of Genetics*, 5th ed'n., (New York: John Wiley, 1975), pp. 267-303.

29. John H. Campbell, "Evolving Concepts of Multigene Families," in *Isozymes: Current Topics in Biological and Medical Research, Volume 10: Genetics and Evolution*, 1983, pp. 401-17, citations on pp. 408-10, p. 414.

30. John H. Holland, *Adaptation in Natural and Artificial Systems* (Ann Arbor, MI: University of Michigan Press, 1975); John H. Holland, "Adaptive Algorithms for Discovering and Using General Patterns in Growing Knowledge Bases," *International Journal of Policy Analysis and Information Systems* 4 (1980), 245-68; Lawrence Davis, ed., *Genetic Algorithms and Simulated Annealing* (Los Altos, CA; Morgan Kaufman Publishers, 1987); David Goldberg, *Genetic Algorithms in Search, Optimization, and Machine Learning* (Reading, MA: Addison Wesley, 1989); Heinz Muhlenbein, M. Gorges-Schleuter, and O. Kramer, "Evolution Algorithms in Combinatorial Optimization," *Parallel Computing* 7 (1988), 65-85; D. Whitley, T. Starkweather and C. Bogart, "Genetic Algorithms and Neural Networks: Optimizing Connections and Connectivity," *Parallel Computing* 14 (1990), 347-61.

31. Herbert A. Simon, *The Sciences of the Artificial* (Cambridge, MA: MIT Press, 1969), pp. 95, 97.
32. Francisco Ayala, "The Mechanisms of Evolution," *Scientific American* 239, no. 3 (September 1978), 56-69.
33. Aldo Leopold, *Sand County Almanac* (New York: Oxford University Press, 1968), p. 8.
34. John F. Deeks, "Some Electrifying Facts," *Audubon* vol. 83, no. 4 (July 1981), 56.
35. John Muir in *The Wilderness World of John Muir*, ed. Edwin Way Teale (Boston: Houghton Mifflin, 1954), pp. 166-67.
36. John Muir, *To Yosemite and Beyond: Writings from the Years 1863-1875*, ed. Robert Engberg and Donald Wesling (Madison, WI: University of Wisconsin Press, 1980), p. 119.
37. Muir, *Wilderness World*, p. 169.
38. John Muir, *The Mountains of California* (New York: Viking Penguin, 1985), p. 176.
39. John Muir, *Stickeen* [1897] (Berkeley, CA: Heydey Books, 1990), p. 21.
40. John Muir, *John of the Mountains: The Unpublished Journals of John Muir*, ed. Linnie Marsh Wolfe (Boston: Houghton Mifflin, 1938), p. 213.
41. Muir, *Wilderness World*, p. 169.
42. John F. Deeks, "Ecstacy is a Monster Storm," *Audubon* vol. 83, no. 4 (July 1981), 50-55.
43. Mary Austin, *The Land of Little Rain* (Boston, MA: Houghton Mifflin, 1903), pp. 14-15.
44. Barry Lopez, *Arctic Dreams* (New York: Charles Scribner's Sons, 1986), p. 32.
45. David Hume, *Dialogues Concerning Natural Religion*, ed. Henry D. Aiken (New York: Hafner, 1948, 1972), p. 79.
46. David L. Soltz and Robert J. Naiman, *The Natural History of Native Fishes in the Death Valley System* (Los Angeles: Natural History Museum of Los Angeles County, Science Series 30, 1978), pp. 39-42.
47. Herman H. Shugart, Jr., and Darrell C. West, "Long-Term Dynamics of Forest Ecosystems, *American Scientist* 69 (1981), 647-52.
48. Henry David Thoreau, *Walden* (Boston, MA: Houghton Mifflin, 1938), p. 350-51.
49. Genesis 1: 20-23.
50. Muir, *Mountains of California*, pp. 210-11.
51. John Muir, "Wild Wool" (1875) in *Wilderness Essays* (Salt Lake City, UT: Peregrine Smith, 1980), pp. 227-42, citation on p. 229.
52. J. Alumets, R. Hakanson, F. Sundler, and J. Thorell, "Neuronal Localisation of Immunoreactive Enkephalin and Beta-Endorphin in the Earthworm," *Nature* 279 (1979), 805-06.
53. Annie Dillard, *Pilgrim at Tinker Creek* (New York: Bantam Books, 1974), p. 180. Dillard, a poet, knows other moods; sometimes her world is a burning bush, "the tree with the lights in it" (p. 35).

BIOLOGICAL FUNCTIONS AND BIOLOGICAL INTERESTS

Gary E. Varner
Washington University in St. Louis

I. FEINBERG'S DICTUM

In a widely cited article, Kenneth Goodpaster makes the following point about the "intelligibility" of attributing interests to plants:

> There is no absurdity in imagining the representation of the needs of a tree for sun and water in the face of a proposal to cut it down or pave its immediate radius for a parking lot. We might of course, on reflection, decide to go ahead and cut it down or do the paving, but there is hardly an intelligibility problem about representing the tree's interest in our deciding not to.

And immediately thereafter, he writes:

> In the face of their obvious tendencies to maintain and heal themselves, it is very difficult to reject the idea of interests on the part of trees (and plants generally) in remaining alive.[1]

I agree that there is no problem making *intelligible* a proposal to attribute interests to plants because they have such tendencies, but from this it does not follow that there is no *absurdity* in it. For if we are going to attribute interests to plants on the grounds that they exhibit goal-directed behavior, then by the same token we will have to assign an interest in regulating temperature to a home heating system, and to do so would constitute a *reductio ad absurdum* of the proposal.

Goodpaster's remarks dramatize the challenges which face a defender of the claim that plants have interests. Since simple artifacts clearly have needs in certain senses of the word, two things will have to be established before we can understand how the fact that plants have needs suffices to show that they have interests. The first is an empirical claim:

Gary E. Varner is currently visiting assistant professor of philosophy at Washington University in St. Louis. His primary research interests are in environmental ethics, animal rights, and the philosophy of environmental law.

The empirical claim: plants have needs in some sense in which artifacts do not.

The second is a normative claim:

The normative claim: this difference qualifies plants, but not artifacts, for direct moral consideration.[2]

Goodpaster's remarks dramatize these twin challenges. But there is another challenge implicit in establishing the empirical claim, a problem that may have been in the back of Joel Feinberg's mind when he wrote a puzzling section of his seminal essay on "The Rights of Animals and Unborn Generations." Early in the essay Feinberg appears to endorse a disjunctive criterion for the possession of interests. He says that a being can have interests only if it has conations, but under the concept of "conative life" he includes both "*conscious* wishes, desires, and hopes; or urges and impulses" and "*unconscious* drives, aims, and goals; or latent tendencies, direction[s] of growth, and natural fulfillments."[3] In the section on plants, however, Feinberg insists that only *conscious* conations can define interests. Feinberg's insistence on drawing up the wagons around sentience has puzzled several commentators.[4] He gives very little by way of argument for abandoning the second disjunct of his earlier criterion. I suspect that a particular epistemological problem was troubling Feinberg when he wrote the section, and that this, more so than any argument he explicitly gives, explains his insistence on a sentience criterion for the possession of interests.

This interpretation explains what is otherwise a very puzzling remark in the "Vegetables" section of Feinberg's essay. Early in the section, Feinberg admits that

> Plants ... are not "mere things"; they are vital objects with inherited biological propensities determining their natural growth ... They grow and develop according to the laws of their own nature.[5]

Later in the section, however, he insists that "Plants may need things in order to discharge their functions, but their functions are assigned by human interests, not their own."[6] Given that in the same paragraph he defines a thing's needs, in a morally neutral sense, as whatever is "necessary to the achievement of [its] goals, or to the performance of [its] functions,"[7] this is a very puzzling remark to make. If plants are "vital objects with inherited biological propensities," then why can it not be said that they have needs, at least in this morally neutral sense, quite independently of human interests in them?

Feinberg's puzzling remark, which I refer to as "Feinberg's dictum," has drawn a response from every author writing in defense of the claim that plants have interests. Yet I think that most of that ink has been wasted, because most of these responses ignore the epistemological problem. What Feinberg may have meant to say, I am suggesting, is not that plants *have* no functions aside from those which humans assign to them, but rather that plants' functions cannot be *specified* on any other basis, and that therefore any attempt to identify plants' interests with the fulfillment of their functions is doomed. The problem, in sum, is this:

The epistemological problem: even if plants *have* needs in some sense that artifacts do not, is it possible to specify, in a non-arbitrary way, what these needs are?

If we interpret Feinberg's dictum this way, then it becomes evident why certain responses commentators have made to his dictum fall very short of the mark.

In section II, I first look at several of these responses and evaluate them simultaneously as responses to Feinberg's dictum and as responses to the epistemological problem. In the process, I identify and develop a response to Feinberg's dictum that simultaneously solves the epistemological problem and supports the empirical claim. However, this will not by itself establish the claim that plants have interests. After I argue that plants have needs in a sense in which artifacts do not and that a non-arbitrary criterion of what is and is not in a plant's interests can be given by appealing to what does and does not fulfill these needs, I will then show why the "interests" so described are morally significant.

My arguments provide a partial and tentative defense of biocentric individualism, by which I mean the view that all and only living organisms have interests. The defense is partial because, in a few special cases, some living organisms will not have interests on the view I advance. However, I argue, these cases are rare in a way that renders them insignificant. The defense is tentative because I cannot establish the normative claim with deductive certainty. I show only that it would be plausible to respond to a problem with some views that restrict interests to sentient creatures in a way that commits us to recognizing that all (or, in light of the above qualification, *very nearly all*) living organisms have interests. This is, logically, a significant weakness of my argument.

However, my primary goal is to show that an individual who expresses concern for the interests of plants and "lower" animals[8] need not be guilty of maudlin sentimentalism and need not be committing a variant of what Ralph Barton Perry called "the pathetic fallacy."[9] My tentative defense of the normative claim, coupled with my defense of the empirical claim and my response to the epistemological problem, suffices to show this.

Before proceeding, I offer two general clarifying remarks. First, since the burden of this paper is to explain how nonconscious beings such as plants can have interests, and since our paradigm of an interest involves the endeavors of *conscious* beings, it is useful to introduce a terminilogical distinction between those interests that do involve consciousness in some way and those that do not. Tom Regan marks this difference by distinguishing between conscious "preference interests" on the one hand, and "welfare interests" on the other.[10] Although I use Regan's term for the former, I instead call the latter *biological interests*. My reason is that "welfare" suggests too strongly something like the integrated satisfaction of all an individual's interests, and when I say that "A has a biological interest in X" I mean only that X is in *one* of A's interests, that X would be good for A in some respect or other, rather than that X would be best for A, all things considered.

Second, I leave for another paper all questions of how morally significant preference interests may be in comparison to biological interests, or plants' interests in comparison to those of "higher" animals such as human beings. Here I seek only to establish that plants are morally considerable, not that their interests qualify them for any particular level of moral significance.[11]

II. RESPONDING TO FEINBERG'S DICTUM

In a paper with the provocative title, "Animal Chauvinism, Plant-Regarding Ethics and the Torture of Trees," the aptly-named J. L. Arbor says the following in response to Feinberg's dictum:

it is clearly an error to put trees into the class of objects which have their ends determined outside themselves by conscious beings. Trees, like animals and other plants, but unlike machines, have end-states which are not decided by human beings. Given the right conditions and barring interference they will in the course of natural events reach this state. There is nothing mysterious or improper about insisting that whatever helps trees achieve their natural end-state is in their interest.[12]

Arbor agrees with Feinberg that the "end-states" of artifacts are "decided by human beings," while insisting that plants have "natural end-states."

There is an initial air of plausibility to this claim. The view that there is one course of development that is "natural" for each species of organism (including all plants), in the sense that individuals of that species will inevitably develop along that course unless "external" factors "interfere," *is* a feature of popular consciousness. If shown two mature chestnut trees, for instance, a thin, gangly one living in a crowded woodlot, and a massive, spreading specimen in an open field, we have a strong tendency to say that the latter is the "natural" state of a chestnut tree, and we expect a botanist to be able to explain why the gangly one has "failed" to develop "naturally."

However, Arbor's move fails as a response to Feinberg's dictum because it rests on an essentialist view of organic species which, as Elliott Sober has detailed, is thoroughly discredited in modern biology. The essentialist treats variations within a species as deviations from "natural tendencies" caused by "interfering forces." The problem for the biological essentialist is that, in light of modern biological theory, it is impossible to draw a non-arbitrary distinction between "natural tendencies" on the one hand and "interfering forces" on the other.

The natural tendencies of a species cannot be defined in terms of a specific genotype, because the genetic variability found among members of any given species—especially species that reproduce sexually—is staggering. But neither can the distinction be drawn in terms of the phenotypes that develop given a specific genotype. The problem, as Sober puts it, is that

> when one looks to genetic theory for a conception of the relation between genotype and phenotype, one finds no such distinction between natural states and states which are the results of interference. One finds, instead, the *norm of reaction*, which graphs the different phenotypic results that a genotype can have in different environments.... Each of the [phenotypes] indicated in the norm of reaction is as "natural" as any other...[13]

That is why a botanist's explanation of the chestnut trees' different "forms," that is their different phenotypes, would (perhaps to our disappointment) advert to something like the effect root constriction has on the size and shape of these trees' canopies, rather than to "interference" with a "natural" course of development.

An importantly different response to Feinberg's dictum is offered by Paul Taylor in his recent book on environmental ethics. Taylor writes:

> Though [many] machines are understandable as teleological systems . . . [t]he ends they are programmed to accomplish are not purposes of their own, independent of the human purposes for which they were made . . . [and] it is precisely this fact that separates them from living things. . . . The ends and purposes of machines are built into them by their human creators. It is the original purposes of humans that determine the structures and hence the teleological functions of those machines. . . . [A living thing] seeks its own ends in a way that is not true of any teleologically structured mechanism. It is in terms of *its* goals that we can give teleological explanations of why it does what it does. We cannot do the same for machines, since any such explanation must ultimately refer to the goals their human producers had in mind when they made the machines.[14]

Taylor claims that while both plants and artifacts exhibit goal-directed behavior, the goals of artifacts cannot be identified without reference to the intentions of their human designers, whereas plants' goals can be identified without reference to any human purpose, and that in this sense plants' goals are their own in a way that artifacts' are not.

While not biologically naive like Arbor's response to Feinberg, Taylor's response is no more convincing. For on any viable analysis of what goal-directed behavior is, Taylor's claim about the goals of artifacts is going to turn out to be false. Certain artifacts are clearly goal-directed, and their goals can be objectively specified quite independently of reference to any human purpose. Consider, for instance, Ernest Nagel's paradigmatic analysis in *The Structure of Science*. Nagel argues that

> [the] characteristic feature of such systems is that they continue to manifest a certain state or property G (or that they exhibit a persistence of development "in the direction" of attaining G) in the face of a relatively extensive class of changes in their external environments or in some of their internal parts—changes which, if not compensated for by internal modification in the system, would result in the disappearance of G (or in an altered direction of development of the systems).[15]

Although Nagel does not argue the point specifically, on his analysis, not only is it true that "the distinctive features of goal-directed systems can be formulated without invoking purposes and goals as dynamic agents,"[16] it is also true that the goals of goal-directed artifacts can be identified without having ultimately to refer to the goals of their human producers or users.

For example, suppose that a team of alien scientists reach earth after a nuclear holocaust and discover a supply of functional Stinger missiles in a Middle Eastern cave. With a little experimental ingenuity, they will soon discover that the missiles seek heat. Not wanting to put a fine point on it, the scientist assigned to investigate the missiles will tell the head scientist that "The missiles follow any available heat source," or "The missiles turn in order to follow any available heat source." Such explanations of the missiles' flight paths will be teleological, and the relevant goal will have been accurately identified, but without the scientists ever understanding a thing about contemporary aerial warfare.

Moreover, it is not always the case that "The ends and purposes of machines are built into them by their human creators," as Taylor claims. If the Pentagon revealed that the inventor of the Stinger missile intended to build a ballistic missile for gathering weather data, that would not affect the accuracy of the foregoing explanation of its flight path, so long as the finished product in fact follows any available heat source.

Goodpaster's own response to Feinberg's dictum is similar to Taylor's, but, significantly, he speaks of "tasks" rather than "goals":

As if it were human interests that assigned to trees the tasks of growth or maintenance! The interests at stake are clearly those of the living things themselves, not simply those of the owners or users or other human persons involved.[17]

The significance is this: insofar as Goodpaster is talking about biological functions rather than goals or end-states, it *is* possible to draw a sharp distinction between all artifacts, on the one hand, and all living organisms on the other. None of Feinberg's critics carefully distinguishes between ends, or goals, on the one hand, and *functions* on the other. The reason may be simply that Feinberg fails to draw the distinction in his essay. Yet philosophers of biology have emphasized that a distinction must be drawn, and once the distinction is made, a more promising approach to the empirical claim becomes obvious.

Although there is general agreement that functional claims in biology are grounded somehow in reproductive success, it is clear that the functions of a given organism's organs or subsystems[18] cannot be unpacked in terms of behavior directed to this goal. For if they were, then statements like "The function of a mule's eyes is to enable it to see" would all be false. Mules are sterile. So if functional claims about individual organisms' organs were based on the reproductive fitness of the individual in question, mules' organs would have no functions.

Larry Wright has proposed a plausible analysis which avoids this problem. Wright claims that attributions of biological function are best understood by looking at attributions of functions to artifacts. On Wright's view, for both biological and artificial functions, "The function of S is X" is true if and only if

a) X is a consequence of S's being there, and

b) S is there because it results in X.

In short, on Wright's analysis, the function of a subsystem is "that particular consequence of its being where it is which explains why it is there."[19]

For artifacts, in filling in the "because" in clause (b) we advert to conscious selection. The consequence of a floor-mounted headlight dimmer switch being where it is and having the form it has (being connected in certain ways to the car's electrical system), which explains why it is there, is that it allows the headlights to be switched between high-beam and low-beam. The designer consciously chose to put it there because its being there would have this consequence.[20]

Wright's claim is that attributions of biological function began as useful, but purely metaphorical attributions of analogous intentions to organisms, the metaphorical overtones of which were dropped once the theory of natural selection provided us with an alternative and non-metaphorical way of filling in the "because" in clause (b). The consequence of woodpeckers' toes being where they are (in opposing pairs, rather than in a 3/1 opposition as in perching birds), which explains why they are where they are, is that their being so arranged helps woodpeckers cling to the trunks of trees. Although no designer consciously chose to arrange them this way, this arrangement evolved because it was adaptive for creatures filling the woodpeckers' ecological niche.

On Wright's analysis, functional explanations concern the etiology of the system in question. Thus they are backward-looking, and this is why his analysis neatly overcomes the problem involving sterile organisms. Sight has been selected for in horses and in donkeys, and this is what explains the presence of eyeballs in the mule whether or not they contribute to the reproductive fitness of the mule.

For present purposes, what is most important about Wright's analysis is that it provides us with a clear response to

Feinberg's dictum *and* a clear answer to the epistemological problem. I propose the following criterion for the identification of biological interests:

An organism A has a welfare interest in X if and only if X is the biological function of some organ or subsystem S of A, where X is the biological function of S in A if and only if

a) X is a consequence of A's having S, and

b) A has S because achieving X was adaptive for A's ancestors.

This view provides us with a clear answer to the epistemological problem because on it, what interests a given plant has depends entirely on its species' etiology. Admittedly a great deal of evolutionary research would have to go into giving a detailed and accurate answer to the question, "Exactly what is and is not in the interests of this plant?" But on the view defended here, the answer to this question is objective, non-arbitrary, and—at least in principle—fairly precise and specific. The proposed view also provides us with an answer to Feinberg's dictum and support for the empirical claim, because all and only living organisms are subject to natural selection. We can say that plants have needs in a sense in which artifacts do not, because plants' subsystems have biological functions, but artifacts' subsystems do not.

Selective breeding of domesticated organisms might appear to constitute a difficult case for the view I am defending. For here, it might seem, what once were biological functions, determined by natural selection, are replaced by, or transformed into, artificial functions determined by human interests. This sort of worry begins to fade, however, as soon as we look at concrete examples.

In most cases, selective breeding does not alter the biological functions of any subsystem of the organism. Consider, for example, what selective breeding has done to the dairy cow. Today's heifers give much more milk on much less feed, but the selective breeding has not made it false that the (or at least a) biological function of the cows' mammary glands is to nourish their calves. The etiology of the species is still the same at the relevant point. Cows do not have mammary glands because milk fetches a profit for farmers; they have mammary glands because mammary glands produce the milk that sustains their calves.

In most cases selective breeding operates in this way: it alters some norm of reaction without thereby altering or eliminating any biological function. But now suppose that a strain of domestic turkey is produced with breast muscles so large that they cannot fly, like a powerlifter so muscle bound that he can no longer comb his hair. In such a case, I admit, these turkeys' breast muscles have lost their original biological function due to selective breeding. The capacity for flight is no longer a consequence of the turkeys' having breast muscles, and therefore condition (a) in Wright's criterion is no longer met.[21] So selective breeding can affect the biological functions of an organism's subsystems. Moreover, in such a case it is true that the breast muscles have in the process acquired an artificial function. Farmers' getting more profit out of each turkey is a consequence of the turkeys' having the larger breast muscles, and the larger breast muscles are there because farmers wanted more profitable turkeys.

In such a case, I admit, selective breeding has quite literally replaced the biological function of the breast muscles with an artificial function. But this is a very limited sort of case, and even here the organisms still have many subsystems with biological functions. It is still true, for instance, that they have gizzards, stomachs, and intestines, because these organs result in their being nourished. Suppose, however, that in producing breast muscles so large that the turkeys can no longer fly, we have also made it impossible for them to breed except via artificial insemination. In such a case, it might seem plausible to argue, "natural" selection is no longer operative at all, and, since all further evolution of the birds is contingent upon the conscious choices of human beings, such thoroughly domesticated animals are, as Baird Callicott has claimed, "living artifacts." In a contentious passage of his essay, "Animal Liberation: A Triangular Affair," Callicott writes:

> Domestic animals are creations of man. They are living artifacts.... There is thus something profoundly incoherent... in the complaint of some animal liberationists that the "natural behavior" of chickens and bobby calves is cruelly frustrated on factory farms. It would make almost as much sense to speak of the natural behavior of tables and chairs.[22]

The example of the turkeys who cannot breed is more dramatic than, but not relevantly different from, the case of some wild animals like the California condor, and once we recognize and account for this similarity, the turkey example loses its force. The entire population of California condors is now in a captive breeding program designed to maintain maximum genetic diversity within the species. These birds breed when and only when the conservation biologists in charge of the program decide that they should. But do we want to say that here "natural selection" is no longer occurring? I think not, and the reason, apparently, is that the "natural" in "natural selection" does not mean "unaffected by *human* organisms." "Natural selection" refers to the way biological organisms evolve via random genetic mutation under selective pressure from their environment, whether the environment is under human control or not. (This explains, by the way, the propriety of referring to the functions identified by Wright's criterion as *"biological* functions" rather than *"natural* functions.")

It is therefore a mistake to conclude, in the case of the turkeys, that selective breeding has *replaced* all biological functions with artificial ones. Rather, what has happened is that some artificial functions have been added onto the original biological functions of the birds' various organs and subsystems. The turkey example, therefore, in no way threatens my response to Feinberg's dictum. So long as all and only living organisms evolve via random genetic mutation under reproductive pressure from their environments, all and only living organisms will have subsystems with biological functions.

When the claim is restated in this way it becomes apparent that only in one very limited kind of case will a living organism fail to have *any* biological functions. On Wright's criterion first mutations are accidents, which acquire biological functions only via subsequent selective pressure.[23] This leads Christopher Boorse to object that if a species "simply sprang into existence by an unparalleled saltation" then, on Wright's view, its organs would have no functions.[24] However, for this to be literally true of an organism, it would have to have no ancestors, or at least share no non-vestigial organs or subsystems with its ancestors. Although it is *possible* that researchers will one day create a complete complement of DNA *ex nihilo*, all currently foreseeable DNA research either modifies one small portion of a given species' DNA or "splices" in genetic material from another organism, and in either case many biological functions are left unaffected.

Neither selective breeding nor currently foreseeable genetic research constitutes a significant challenge to the claim that all and only living organisms have biological interests, where these interests are identified with the fulfillment of the biological functions of their component subsystems, and where these functions are in turn identified using Wright's criterion. Using Wright's criterion, then, I have (a) developed a response to Feinberg's dictum that (b) supports the empirical claim while (c) solving the epistemological problem. I have (a) explained a sense in which plants have functions that are not assigned to them by human interests, I have (b) shown that plants have "needs" in a sense in which artifacts do not, namely what they need to fulfill the biological functions of their component subsystems, and I have (c) shown that a non-arbitrary account of what is and is not in a plant's interests can be given when a plant is said to have biological interests in the fulfillment of the biological functions of its subsystems. It remains to be shown, however, that these "biological interests" are interests in a morally relevant sense. Why should we think that the biological functions of a plant's subsystems define interests?

III. SUPPORTING THE NORMATIVE CLAIM

After responding to Feinberg's dictum in the first section of "The Good of Trees," Robin Attfield asks us to imagine that the last sentient organism on Earth is a man who "hew[s] down with an axe the last tree of its kind, a hitherto healthy elm . . . which could propagate its kind if left unassaulted." He concludes that

> Most people who face the question would . . . conclude that the world would be the poorer for this act of the 'last man' and that it would be wrong . . . And if, without being swayed by the interests of sentient creatures, we share in these conclusions and reactions, [then] we must also conclude that the interests of trees are of moral significance.[25]

A specific problem with Attfield's thought experiment is that it is not clear why we must attribute *interests* to the tree in order to explain this intuition. The flourishing of trees can be of moral significance without the trees having interests. Lilly-Marlene Russow has convincingly argued that specimens of some endangered species have high aesthetic value.[26] Would "most people" still think it was wrong of the last man to chop down the tree if it were not "the last . . . of its kind"? To the extent that peoples' intuitions change when the question is rephrased in this way, the explanation may be that they think the tree in question is a thing of beauty, and that this gives it some moral significance, albeit of a different (and probably weaker) kind than it would have if it had interests. (Notice that Attfield himself says that "the world would be the poorer for this act of the 'last man'," not that the tree itself would be harmed.)

Apart from the specifics of Attfield's thought experiment, however, I have general misgivings about using appeals to intuitions about such cases. I admit that widely shared intuitions about "normal" cases—cases involving human beings—can serve as fixed points against which to check our moral theories, but I think that the more "marginal" the case in question, the more a theorist should feel called upon to follow theory, rather than intuition. Since the progression from non-human animals to plants provides an example of what I mean by increasingly "marginal" subjects of moral judgments, I maintain that a theorist can always feel justified in following a theory in the face of contrary intuitions about the interests of plants and "lower" animals.

This problem can be avoided, however, by arguing that the biological functions of a *human being*'s body define morally significant interests of that individual, quite independently of his or her (or anyone else's) conscious mental states. J. L. Arbor appears to be taking such a tack when she or he asks us to imagine a society in which certain children are systematically mutilated in ways that normally would be quite painful, but only after brain surgery has left them incapable of feeling the pain and unhappiness which would otherwise attend these operations and mutilations. Arbor concludes that to unpack the "wrong" done to these children as a violation of an indirect duty to other, normal children and adults, would be "an artificial and awkward way of responding to a straightforward ethical intuition."[27]

While Arbor's approach is more promising than Attfield's, it does not provide the strongest possible defense of the normative claim. The reason is that perfectly good sense can be made of the claim that direct duties to the children are being violated in this case, without abandoning a sentience criterion of moral considerability. The sentientist need only appeal to the *loss* of potential, positive experiences of pleasure and/or desire satisfaction in order to claim that the children have been harmed. The sentientist can say that these children have been made drastically less well off than they would otherwise have been, and that they have in that sense been seriously harmed. The strongest possible defense of the normative claim would be one that questions the adequacy of a mental state theory of benefit and harm in relation to human beings.

I therefore believe that Kenneth Goodpaster is on the right track when he suggests that allegiance to a hedonistic theory of individual welfare is responsible for Western philosophers' general reluctance to recognize the moral considerability of non-sentient organisms. Since only creatures capable of being benefitted and harmed can be the objects of direct duties of

263

beneficence and non-maleficence, Western authors' general allegiance to a hedonistic conception of individual welfare would explain their reluctance to recognize the moral considerability of non-sentient creatures.[28] However, there are two problems with Goodpaster's argument, problems that seriously weaken it as a defense of the normative claim.

The first problem is that Goodpaster offers no specific argument *against* the adequacy of a hedonistic conception of welfare and in favor of a non-sentientist conception.[29] Without such an argument, the normative claim is very weakly supported, because no argument has been offered for thinking that the alternative, non-sentientist conception of welfare is superior to the hedonistic conception.

The second problem is that in characterizing as "hedonistic" the dominant theory of individual welfare, Goodpaster has cast his net too narrowly. A common conception of individual welfare runs through Henry Sidgwick's *The Methods of Ethics*, John Rawls' *A Theory of Justice*, and Richard Brandt's *A Theory of the Good and the Right*, a conception that is sentientist without being narrowly hedonistic. On it, to say that "X is in A's interest" or that "X would be good for A" can mean either

1) A now desires (or prefers) X,

or

2) A would desire (or prefer) X, if A were

a) adequately informed, and

b) sufficiently impartial across phases of his or her life.

Where clause (1) is satisfied, X is said to be good for A in a qualified sense, good in some respect or other. Where clause (2) is satisfied, X is said to be best for A, or good for A, all things considered. Harms are in turn conceived as whatever prevents the fulfillment of some actual desire (this is harm in a qualified, conditional sense) or whatever prevents the fulfillment of one's enlightened (clause (2)) desires (this is harm in an unqualified, unconditional sense, and since, presumably, one's enlightened desires form an integrated set, harms of this sort vary greatly in degree).[30] I call this *the mental state theory of welfare*.

Since it would be more plausible to say that this, rather than a narrowly hedonistic conception of welfare, is the dominant conception in contemporary Western philosophy, the strongest defense of the normative claim would be one

264

that called it into question, rather than the hedonistic conception, and that (in light of my earlier remarks) did so without appealing to intuitions about plants. In the remainder of this section, I advance such an argument. My argument consists in showing how this mental state theory of welfare is inferior in two ways to an account of welfare that incorporates the view proffered in the preceding section. I argue that the best account of individual welfare would be one that recognizes the existence of *two* kinds of interests, *preference* interests on the one hand and *biological* interests on the other, where the latter are defined in terms of the account proffered in the preceding section and the former are defined in terms of one's actual and/or enlightened desires (for present purposes it does not matter which).

On this alternative conception of welfare, which I call *the biological theory of welfare*, to say that "X is in A's interest" or that "X would be good for A" can mean either

1) A has a biological interest in X, meaning that X would fulfill a biological function of one or more of A's organs or subsystems,

or

2) A has a preference interest in X, meaning that either

a) A now desires (or prefers) X, or

b) A would desire (or prefer) X, if A were

i) adequately informed, and

ii) sufficiently impartial across phases of his or her or its life.

Here I ignore the issues of how morally significant biological interests are *vis* a vis preference interests and of what constitutes an individual's good on the whole. My criticism of the mental state theory of welfare is independent of these issues, discussion of which I leave to another occasion.[31]

Consider the case of Maude, an unusually intelligent, acutely rational, and generally far-sighted young adult, with a strong desire to smoke. Concerned for her welfare, we bring to her attention the fact that the best available evidence indicates that this smoking will shorten her life by a certain number of years. Suppose that Maude really takes this fact to heart, that (in Sidgwick's words) "the consequences [of her conduct are] accurately foreseen and adequately realised in [her] imagination at the present point in time,"[32] but that she nevertheless goes right on smoking.

On the mental state theory of harm, no sense whatsoever can be made of the claim that Maude's smoking is bad for her. For she does not now desire to stop smoking, and on the mental state theory of harm this implies that continuing to smoke is only bad for her if her enlightened preference would be to stop smoking. But by hypothesis, Maude is both adequately informed and sufficiently impartial across phases of her life. Therefore, her actual preference *is* her enlightened preference, and therefore, on the mental state theory of harm, Maude's smoking is in no way bad for her.

This criticism of the mental state theory of welfare does not imply that the biological theory of welfare is true. All it implies is that we must accept *some* alternative to the mental state theory of welfare, and a conjunction of the proffered account of biological interests with an account of preference interests is just one candidate. But such a view would provide a clear and concise account of the sense in which Maude's smoking is bad for her: regardless of her preferences, smoking is always bad for an individual insofar as it impairs effective oxygenation of one's blood by one's lungs, and this is at least one biological function of one's lungs.

Another virtue of the biological theory of welfare is that it illuminates an important distinction among our interests which the mental state theory of welfare does not capture. It would seem that, in talking about some of our interests—particularly those we label "needs"—substitution of coreferential expressions preserves truth value.[33] For instance, if it is in my interest to ingest at least 10 milligrams of vitamin C each day (the amount needed to prevent scurvy) then it is also in my interest to ingest at least 10 milligrams of ascorbic acid each day. But desire contexts are referentially opaque. Where A desires X and $X = Y$, it can still be false that A desires Y. Nineteenth-century mariners desired to avoid scurvy, and they desired citrus fruit to that end. Ingesting ascorbic acid in crystaline form would have accomplished the same end, but they had no desire for ascorbic acid. So what sense can be made of the claim that 19th-century mariners needed to ingest 10 milligrams of ascorbic acid each day? On the mental state theory of welfare, it would have to be claimed that *if they had known* that it was the ascorbic acid in citrus fruits that prevented scurvy, then the sailors would have formed a desire for it specifically. But would they? When offered ascorbic acid in crystaline form, the most rational of sailors is unlikely to form a desire for it, unless and until citrus fruit becomes unavailable. And this means, on the mental state theory of welfare, that a sailor would first come to need ascorbic acid when the fruit runs out. By contrast, on the biological theory of welfare we can say that getting 10 milligrams of ascorbic acid each day is a biological interest of every sailor, whether or not he desires it, because of the role the vitamin plays in the functioning of his various organs and subsystems. The biological theory of welfare allows us to draw a clear distinction between a class of interests which are referentially opaque (preference interests) and a class of those which are not (biological interests).[34]

My defense of the proffered view is indeed tentative. I have argued only that some alternative to the mental state theory of welfare must be accepted, and that my account of welfare interests has some points to recommend it in this regard. I have argued that the conjunction of a desire-based account of *preference* interests with the proffered account of *biological* interests provides a theory of individual welfare which is in certain ways superior to that provided by the mental state theory of welfare, but I have not argued that this conjunction is the *only* way to remedy the identified faults of the mental state theory. However, most of the "tinkering" which has been done with the mental state theory of welfare has focussed on the second clause, rather than on the first (that is, on the notion of what is *best*, all things considered), whereas my criticism of the mental state theory of welfare focussed on a shortcoming of its first clause. I am therefore skeptical that any alternative will overcome the specific problem I have been focussing on, and it is in relation to this specific problem that I have recommended the biological theory.

IV. CONCLUDING REMARKS

In this paper I have defended an account of biological interests which both is consistent with contemporary views of teleological and functional explanations and allows us to attribute interests to a living organism as such—i.e., independently of its being a *conscious* living organism—without at the same time attributing interests to simple artifacts. I conclude with some observations on the larger significance of this inquiry.

Although the view defended here is what Tom Regan would recognize as a truly *environmental* ethic, because it extends moral consideration well beyond the sentient realm,[35] I doubt that an *adequate* environmental ethic[36] can be teased out of an interest-based approach to ethics. The reason is that sound environmental policy seems always to be focussed on systems or wholes, rather than on individuals, whereas only individual organisms can plausibly be said to have interests.[37] It has been proposed, of course, that an individual ecosystem and/or the earth's ecosphere in its entirety is a living organism,[38] but no sober ecologist today believes that ecosystems are in any literal sense organisms.[39] Even if this were so, however, an ecosystem could not have any biological interests, inasmuch as its component subsystems could not have acquired any biological functions via natural selection. There is general agreement among evolutionary biologists that group selection occurs, but only among "species made of many very discrete, socially cohesive groups in direct competition with each other,"[40] like colonies of social insects. Thus the various "organs" of an ant or bee colony (the *individual* queen, the worker *class*, and so on) may have biological functions which define biological interests of the colony, but unless or until evolutionary biology abandons the individualistic stance it inherited from Darwin, the view defended here will not imply that any more inclusive entity has biological interests.[41]

However, as I said earlier, my primary goal in this paper was to show that an individual who expresses concern for the interests of plants and "lower" animals need not be guilty of maudlin sentimentalism and need not be committing a variant of "the pathetic fallacy." And on the view defended here, we *can* say, quite literally and without anthropomorphizing, that a forest fire sets back the interests of the trees it burns and that a child who tears the wings off of a fly is setting back the interests of that fly.[42]

NOTES

[1] Kenneth Goodpaster, "On Being Morally Considerable," *Journal of Philosophy* 75 (1978), p. 319.

[2] I take it that to say that an entity has interests is to say that it warrants direct moral consideration because its needs, desires, etc. (whatever we take to define its interests) ought, other things being equal, to be fulfilled or satisfied, and that for their own sake. To say that a thing has interests is to say that it has a good of its own, a good which moral agents have direct, *prima facie* duties to protect (a duty of non-maleficence) and further (a duty of beneficence).

[3] Joel Feinberg, "The Rights of Animals and Unborn Generations," in William T. Blackstone, ed., *Philosophy and Environmental Crisis* (Athens: University of Georgia Press, 1974), pp. 49-50, emphasis added.

[4] See, for instance, Goodpaster, "On Being Morally Considerable," p. 320; Robin Attfield, *The Ethics of Environmental Concern* (New York: Columbia University Press, 1983), pp. 144-45, and Attfield, "The Good of Trees," *Journal of Value Inquiry* 15 (1981), pp. 39-40.

[5] Feinberg, pp. 51-52.

[6] *Ibid.*, p. 54.

⁷ *Ibid.*, p. 53.

⁸ Although, for simplicity's sake, I will usually speak simply of "plants," the arguments I advance are intended to hold for all non-conscious organisms.

⁹ Ralph Barton Perry, *General Theory of Value* (New York: Longman's, Green and Company, 1926), p. 56.

¹⁰ Tom Regan, *The Case for Animal Rights* (Berkeley: University of California Press, 1983), pp. 87-88.

¹¹ Goodpaster develops this distinction in "On Being Morally Considerable," pp. 311-12. See also note #31 below.

¹² J. L. Arbor, "Animal Chauvinism, Plant-Regarding Ethics and the Torture of Trees," *Australasian Journal of Philosophy* 64 (1986), p. 337.

¹³ Elliott Sober, "Evolution, Population Thinking, and Essentialism," *Philosophy of Science* 47 (1980), p. 374.

¹⁴ Paul Taylor, *Respect for Nature: A Theory of Environmental Ethics* (Princeton: Princeton University Press, 1986), pp. 123-24.

¹⁵ Ernest Nagel, *The Structure of Science: Problems in the Logic of Scientific Explanation* (London: Routledge & Kegan Paul, 1961), p. 411.

¹⁶ *Ibid.*, p. 418.

¹⁷ Goodpaster, "On Being Morally Considerable," p. 319.

¹⁸ Biological functions do not always (or even usually) attach to organs like the heart or lungs, but rather to the *systems* (circulatory or respiratory) of which these organs form a part. In single celled organisms, biological functions attach to *organelles* or subsystems.

¹⁹ Larry Wright, *Teleological Explanations* (Berkeley: University of California Press, 1976), p. 81.

²⁰ The example is Wright's. See *Teleological Explanations*, pp. 77 and 79-80 (notes 3 and 4).

²¹ This is how Wright treats all vestigal organs. See *Teleological Functions*, pp. 89 and 91.

²² J. Baird Callicott, "Animal Liberation: A Triangular Affair," *Environmental Ethics* 2 (1980), p. 330.

²³ Wright, *Teleological Explanations*, p. 114.

²⁴ Christopher Boorse, "Wright on Functions," reprinted in Elliott Sober, ed., *Conceptual Issues in Evolutionary Biology* (Cambridge: MIT Press, 1984), p. 373.

²⁵ Attfield, "The Good of Trees," p. 51.

²⁶ Lilly-Marlene Russow, "Why Do Species Matter?" *Environmental Ethics* 3 (1981), pp. 101-12.

²⁷ Arbor, "Animal Chauvinism," p. 338.

²⁸ Goodpaster, "On Being Morally Considerable," section five, pp. 320-22.

²⁹ Goodpaster appears to presume that by offering support for the empirical claim he is simultaneously offering support for the normative claim, but, as we have seen, a complete defense of the thesis that plants have morally significant interests requires logically separate defenses of each of these claims.

³⁰ For examples of this view in action, see Henry Sidgwick, *The Methods of Ethics*, 7th edition (London: Macmillan, 1907), pp. 111-12; John Rawls, *A Theory of Justice* (Cambridge: Harvard University Press, 1971), chapter 7, "Goodness as Rationality," especially pp. 407-16; and Richard Brandt, *A Theory of the Good and the Right* (Oxford: Clarendon Press, 1979), chapter 3, "The Cognitive Theory of Action."

³¹ My own view is that a variant of Ralph Barton Perry's "Principle of Inclusiveness" (*General Theory of Value* [New York: Longman's, Green & Co., 1926], p. 648) can be used to show both (1) that while any human desire, even the most trivial or malicious, trumps any interest of a plant, we also have a general duty to rid ourselves of such desires, and (2) that the fulfillment

of a person's "ground projects" or "categorical desires" (terminology adopted from Bernard Williams, "Persons, Character, and Morality," in his *Moral Luck* [Cambridge: Cambridge University Press, 1981], pp. 1-19) are the primary components of a person's good on the whole. See my "Perry's Principle of Inclusiveness and Establishing Priorities Among Interests," in preparation.

[32] Sidgwick, *loc. cit.*

[33] Garrett Thomson makes a similar claim in *Needs* (London: Routledge and Kegan Paul, 1987), p. 101.

[34] Thomson argues that a person's "fundamental needs" are defined in part by his or her biology, but he offers no criteria for determining how and when this is so.

[35] Tom Regan, "The Nature and Possibility of an Environmental Ethic," *Environmental Ethics* 3 (1981), pp. 19-20.

[36] On the notion of an "adequate" environmental ethic, see J. Baird Callicott, "The Search for an Environmental Ethics," in Tom Regan, ed., *Matters of Life and Death*, 2nd ed. (New York: Random House, 1986), p. 383.

[37] This oft-repeated claim was first defended in Bryan Norton's essay, "Environmental Ethics and Nonhuman Rights," *Environmental Ethics* 4 (1982), pp. 17-36.

[38] On the former claim, see Donald Worster's account of the Chicago Organicists in chapter 15 of his *Nature's Economy* (Cambridge: Cambridge University Press, 1985); on the latter claim, see James Lovelock and Sidney Epton, "The Quest for Gaia," *New Scientist* 65 (1975), pp. 304-06.

[39] See R. V. O'Neill et al., *A Hierarchical Concept of Ecosystems* (Princeton: Princeton University Press, 1986), chapters one and two, for a brief overview of the relevant theoretical issues; and see Worster, *loc. cit.*, for a popular account of the meteoric rise (and fall) of organicism among professional ecologists.

[40] Stephen Jay Gould, "Caring Groups and Selfish Genes," reprinted in Elliott Sober, ed., *Conceptual Issues in Evolutionary Biology* (Cambridge: MIT Press, 1984), p. 122.

[41] In a recent article, Harley Cahen argues that what appears to be goal directed behavior in ecosystems is actually a byproduct of the goal directed behavior of individual organisms. Harley Cahen, "Against the Moral Considerability of Ecosystems," *Environmental Ethics* 10 (1988), pp. 195-216. Although Cahen offers no explicit defense of either the empirical claim or the normative claim, and although he does not notice the special significance of function attributions which I emphasize in section two of this paper, his is the most lucid article to date in the literature of biocentric individualism.

[42] I am increasingly convinced that the intuitions of the environmental movement are best unpacked in terms of an ethical commitment to the preservation of aesthetic value. See Eugene C. Hargrove, *Foundations of Environmental Ethics* (Englewood Cliffs, New Jersey: Prentice Hall, 1989). As an individual who is intuitively and politically committed to the more idealistic goals of the environmental movement, however, I worry that these goals cannot be harmonized with the equally intuitive principle that the protection of interests takes precedence over the preservation of beauty. But that is a topic for another paper.

[8]

Against the Moral Considerability of Ecosystems

Harley Cahen*

> Are ecosystems morally considerable—that is, do we owe it to them to protect their "interests"? Many environmental ethicists, impressed by the way that individual nonsentient organisms such as plants tenaciously pursue their own biological goals, have concluded that we should extend moral considerability far enough to include such organisms. There is a pitfall in the ecosystem-to-organism analogy, however. We must distinguish a system's genuine goals from the incidental effects, or byproducts, of the behavior of that system's parts. Goals seem capable of giving rise to interests; byproducts do not. It is hard to see how whole ecosystems can be genuinely goal-directed unless group selection occurs at the community level. Currently, mainstream ecological and evolutionary theory is individualistic. From such a theory it follows that the apparent goals of ecosystems are mere byproducts and, as such, cannot ground moral considerability.

I

If natural areas had no value at all for human beings, would we still have a duty to preserve them? Some preservationists think that we would. Aldo Leopold, for instance, argues brilliantly for the cultural and psychological value of wilderness; yet he insists that even "enlightened" self-interest is not enough.[1] According to Leopold, an "ecological conscience" recognizes "obligations to land."[2] The ecological conscience sees that preservation is a good thing in itself—something we have a prima facie duty to promote—apart from any contribution it makes to human welfare. For convenience, let us call this conviction the *preservationist intuition*.[3]

---

* Department of Natural Resources, Cornell University, Ithaca NY 14853. Cahen is writing a dissertation on obligations to future generations and is teaching a course on ethical issues in national park policy. He thanks Elizabeth Meer, Barry Ingber, John Herring, Diane Wray-Cahen, the anonymous referees from *Environmental Ethics*, and Holmes Rolston, III (who, though disagreeing with the thrust of this manuscript, commented on it with his characteristic generosity). Research for this paper was partially supported by a fellowship from the National Science Foundation.

[1] Aldo Leopold, *A Sand County Almanac: With Essays on Conservation from Round River* (New York: Ballantine, 1966), p. 244. He characterizes economic arguments for preservation as "subterfuges," invented to justify what we know we should do on other grounds (p. 247). He also describes "despoliation of land" as "not only inexpedient but wrong" (p. 239).

[2] Ibid., p. 245.

[3] Eric Katz expresses the preservationist intuition clearly in "Utilitarianism and Preservation," *Environmental Ethics* 1 (1979): 357–64. The "danger" posed by an ethic based exclusively on human interests, Katz says, is that it "can support a policy of preservation only on a contingent basis" (p. 362).

I share this intuition. Can we justify it? I see at least four plausible strategies. We might, first, appeal to the intrinsic value of natural ecosystems.[4] A second strategy relies on the interests of the individual creatures that are inevitably harmed when we disturb an ecosystem.[5] A third possibility is a virtue-based approach. Perhaps what offends us—as preservationists—is that anyone who would damage an ecosystem for inadequate reasons falls short of our "ideals of human excellence."[6] Each of these three strategies has something to recommend it. But none captures the element of the preservationist intuition that involves a feeling of obligation *to* "land." This suggests a fourth strategy, the appeal to what Kenneth Goodpaster calls *moral considerability*. This strategy represents an ecosystem as something that has interests of its own, and thus can directly be victimized or benefited by our actions.[7] If ecosystems do have interests of their own, perhaps we owe it to them to consider those interests in our moral deliberations. This fourth strategy is the one that I wish to call into question.

There is a fifth strategy—an appeal to the moral right of a natural ecosystem to

---

[4] See, e.g., Holmes Rolston, III, "Are Values in Nature Subjective or Objective?" *Environmental Ethics* 4 (1982): 125–51 and Peter Miller, "Value as Richness: Toward a Value Theory for the Expanded Naturalism in Environmental Ethics," *Environmental Ethics* 4 (1982): 101–14. Bryan Norton worries, properly, that there are "questionable ontological commitments involved in attributing intrinsic value to nature." "Environmental Ethics and Weak Anthropocentrism," *Environmental Ethics* 6 (1984): 147–48. Some who argue for the "intrinsic value" of nature might be happy with something less worrisome, ontologically—namely, what C. I. Lewis calls "inherent value." See Lewis, *An Analysis of Knowledge and Valuation* (La Salle, Ill.: Open Court, 1946), pp. 382–92. The problem with this strategy is that aesthetic and other kinds of inherent value, though objective, are fundamentally anthropocentric. See Rolston, "Are Values in Nature Subjective or Objective?" p. 151, and Robin Attfield, *The Ethics of Environmental Concern* (New York: Columbia University Press, 1983), pp. 151–52. These waters are very muddy, and J. Baird Callicott has lately stirred them up some more by defining inherent value in a way that is explicitly at odds with Lewis's conception. "Intrinsic Value, Quantum Theory, and Environmental Ethics," *Environmental Ethics* 7 (1985): 262.

[5] See Paul W. Taylor, *Respect for Nature: A Theory of Environmental Ethics* (Princeton: Princeton University Press, 1986), pp. 70–71. Taylor makes it clear that when he speaks of the "good of a whole biotic community," he does not imagine that the community—as a whole—has a good of its own. The community's good is, he says, a "statistical concept" compounded out of the interests of the individual creatures that comprise the community.

[6] See Thomas E. Hill, Jr., "Ideals of Human Excellence and Preserving Natural Environments," *Environmental Ethics* 5 (1983): 211–24. See also Bryan G. Norton, "Environmental Ethics and Weak Anthropocentrism," *Environmental Ethics* 6 (1984): 131–48. If the "excellence" in question is the ability to recognize intrinsic (or even inherent) value when one encounters it, then this strategy turns out to be a variant of the first one.

[7] Goodpaster coined the term *moral considerability* in "On Being Morally Considerable," *Journal of Philosophy* 75 (1978): 308–25. When I speak of an organism's interests I do not mean to imply anything about its state of mind—or even that it has a mind. Throughout this paper I use the terms *interests* and *good of one's own* interchangeably. I recognize that there are good reasons not to equate these terms (see Taylor, *Respect for Nature*, pp. 62–68), but even Taylor concedes that it is "convenient" (pp. 270–71) to speak of whatever furthers a being's good as also promoting its interests. I find it convenient too.

be left alone. This strategy is similar to the fourth one but may be distinct. Rights, some would say, automatically "trump" other kinds of moral claim.[8] If so, an appeal to ecosystem rights would be much stronger than an appeal to moral considerability. (Too strong, I suspect: I find it best to regard talk of the rights of nonhumans as an enthusiastic way of asserting moral considerability.[9]) We can leave this question open, though, for if they are trumps, moral rights have at least this much in common with moral considerability: they both presuppose interests.

I contend that ecosystems cannot be morally considerable because they do not have interests—not even in the broad sense in which we commonly say that plants and other nonsentient organisms "have interests." The best we can do on behalf of plant interests, I believe, is the argument from *goal-directedness*. Nonsentient organisms—those not capable of consciously taking an interest in anything—have interests (and thus are candidates for moral considerability) in achieving their biological goals. Should ecosystems, too, turn out to be goal-directed, they would be candidates for moral considerability.[10]

Although the argument from goal-directedness fails, we should not dismiss the argument too hastily. Some ecosystems are strikingly stable and resilient. They definitely have a goal-directed look. Yet there are reasons to doubt whether this apparent goal-directedness is genuine. The key is to distinguish the goals of a system's behavior from other outcomes that are merely behavioral *byproducts*. Armed with this distinction, we can see that the conditions for genuine goal-directedness are tougher than environmental ethicists typically realize. Ecosystems seem unlikely to qualify.

In sections two and three of this paper I define *moral considerability* and distinguish it from other ways that something can matter morally. In section four I establish that goal-directedness plays a key role in arguments for the considerability of plants and other nonsentient organisms. In sections five and six I argue that this appeal to goal-directedness is plausible as long as we keep the goal/byproduct distinction in mind. In sections seven through nine, I argue that ecology and evolutionary biology cast serious doubt on the possibility that ecosystems are genuinely goal-directed.

---

[8] See, for example, Ronald Dworkin, *Taking Rights Seriously* (Cambridge, Mass.: Harvard University Press, 1978).

[9] According to Joel Feinberg, rights talk is merely a way of referring to *valid claims*. On this view, having rights is equivalent to being morally considerable. See Feinberg, "The Nature and Value of Rights," *Journal of Value Inquiry* 4 (1971): 263–77 and "The Rights of Animals and Unborn Generations," in *Philosophy and Environmental Crisis*, ed. William T. Blackstone (Athens: University of Georgia Press, 1974).

[10] I classify all beings that have interests as "candidates" for moral considerability. If my analysis of organismic interests is correct, then having interests is only the first step to moral considerability.

II

The literature of environmental ethics is full of appeals to the interests of ecosystems. Consider Aldo Leopold's famous remark: "A thing is right when it tends to preserve the integrity, stability, and beauty of the biotic community. It is wrong when it tends otherwise."[11] Is Leopold suggesting that the biotic community has an interest in its own integrity and stability? Some commentators interpret his remark this way. James Heffernan, for instance, defends Leopold by insisting that "even ecosystems . . . are things that have interests and hence, may be benefited or harmed."[12] Holmes Rolston, III likewise would found an "ecological ethic" upon the obligation to promote "ecosystemic interests."[13]

More often the appeal to ecosystem interests is implicit. Consider John Rodman, criticizing animal liberationists such as Peter Singer for drawing the moral considerability boundary to include only sentient beings. Rodman complains: "The moral atomism that focuses on individual animals . . . does not seem well adapted to coping with ecological systems."[14] Why is "atomism" inadequate? Because, Rodman explains, an ecological community as a whole has a good of its own, a "welfare":

> I need only to stand in the midst of clear-cut forest, a strip-mined hillside, a defoliated jungle, or a dammed canyon to feel uneasy with assumptions that could yield the conclusion that no human action can make any difference to the welfare of anything but sentient animals.[15]

Of course, Rodman believes that individual plants and nonsentient animals are morally considerable, too. That is reason enough for him to feel uneasy with

---

[11] Leopold, *Sand County Almanac*, p. 262.

[12] James D. Heffernan, "The Land Ethic: A Critical Appraisal," *Environmental Ethics* 4 (1982): 242. On balance I think Heffernan is right: Leopold is asserting moral considerability for ecosystems. In that case, Heffernan's point is well-taken: an ecosystem must be seen as the sort of entity that we can not only damage (i.e., put in a state of diminished value or impaired usefulness), but also *harm* (i.e., make worse off from the standpoint of its own interests).

[13] Holmes Rolston, III, "Is There an Ecological Ethic?" *Ethics* 85 (1975): 106. Rolston no longer speaks in these terms, though his talk of "projects" suggests that something like an appeal to goal-directedness is still at work. See "Are Values in Nature Subjective or Objective?" pp. 146–47, where Rolston speaks of achievements that "do not have wills or interests," but do have "headings, trajectories, traits, successions, which give them a tectonic integrity." More recently still, even while conceding that "ecologists . . . have doubted whether ecosystems exist as anything over their component parts," and agreeing that in ecosystems there are "no policy makers, no social wills, no goals," Rolston is still drawn to speak, tentatively, of "the good of the system" and to claim that "a spontaneous ecosystem is typically healthy." "Valuing Wildlands," *Environmental Ethics* 7 (1985): 26, 30.

[14] John Rodman, "The Liberation of Nature?" *Inquiry* 20 (1977): 89.

[15] Ibid. See also Rodman's description of a river as a "victim" of a dam (pp. 114–15). A hillside (or even a river) is not likely to be a complete ecosystem, of course. My arguments apply equally, I think, to ecosystems and to ecological communities.

Singer's assumptions. It cannot be his only reason, however, for it would leave him as guilty of moral atomism as Singer. Whose *welfare* could Rodman have in mind? The welfare, I take it, of the communities themselves.[16]

III

Moral considerability is a potentially confusing term. Let me clarify and defend my use of it. I take moral considerability to be the moral status x has if, and only if (a) x has interests (a good of its own), (b) it would be prima facie wrong to frustrate x's interests (to harm x), and (c) the wrongness of frustrating x's interests is direct—that is, does not depend on how the interests of any other being are affected. It is the concern with interests that distinguishes moral considerability from the other varieties of moral status upon which the preservationist intuition might possibly be based.

Goodpaster plainly means to restrict moral considerability to beings with *interests*. In his first paper on moral considerability he explains that life is the "key" to moral considerability because living things have interests; this, he points out, is what makes them "capable of being beneficiaries."[17] Goodpaster makes a point of agreeing with Joel Feinberg about what Feinberg calls "mere things." "Mere things," Goodpaster says, are not candidates for moral considerability because they are "incapable of being benefited or harmed—they have no 'well-being' to be sought or acknowledged."[18] That is why he insists that "x's being a living being" is not only sufficient for moral considerability but is also *necessary*.[19]

In Goodpaster's subsequent work, he characterizes the entire biosphere as a "bearer of value."[20] Yet he does not appear to have changed his understanding of the requirements for moral considerability. "The biosystem as a whole" is considerable, he says. Why? Because it is, in effect, an "organism"—"an integrated, self-sustaining unity which puts solar energy to work in the service of

---

[16] Here are two more examples of the sort of language that raises my suspicions. (1) "Deep ecologists" Bill Devall and George Sessions, in "The Development of Natural Resources and the Integrity of Nature," assert that when "humans have distressed an ecosystem," we are obliged to make "reparations" (*Environmental Ethics* 6 [1984]: 305, 312). (2) J. Baird Callicott denies that he wishes to extend "moral considerability" to "inanimate entities such as oceans, lakes, mountains, forests, and wetlands"; yet he refers to the "well-being of the biotic community, the biosphere as a whole," and employs "the good of the community as a whole" as a standard for the assessment of the relative value . . . of its constitutive parts" ("Animal Liberation: A Triangular Affair," *Environmental Ethics* 2 [1980]: 337, 324–25).

[17] Goodpaster, "On Being Morally Considerable," pp. 323, 319.

[18] Ibid., p. 318. Feinberg's remarks on "mere things" are in "The Rights of Animals and Unborn Generations," pp. 49–50.

[19] Goodpaster, "On Being Morally Considerable," p. 313.

[20] Kenneth E. Goodpaster, "From Egoism to Environmentalism," in *Ethics and Problems of the 21st Century*, ed. Goodpaster and Sayre (Notre Dame: Notre Dame University Press, 1979), p. 30.

growth and maintenance."[21] Goodpaster's focus remains on interests and he expresses his confidence that the "biosystem as a whole" has them.[22]

Some philosophers speak of moral considerability but do not associate it with interests at all. Andrew Brennan, for instance, asserts that natural objects such as ecosystems, mountains, deserts, the air, rocky crests, and rivers may have this moral status though they have no interests and thus can be harmed only "metaphorically." This is no longer moral considerability as I understand it.[23]

Other philosophers equate moral considerability with intrinsic value, holding that both equally presuppose interests. Robin Attfield, for instance, writes, "I follow Goodpaster in holding that things which lack a good of their own cannot be morally considerable . . . or have intrinsic value.[24] J. Baird Callicott attributes to Goodpaster the view that because "life is intrinsically valuable . . . all living beings should be granted moral considerability."[25] As Callicott sums up his own view:

> If the self is intrinsically valuable, then nature is intrinsically valuable. If it is rational for me to act in my own best interest, and I and nature are one, then it is rational for me to act in the best interests of nature.[26]

The association of intrinsic value with interests seems odd to me. Many readers will suppose that "mere things"—things which have no interests, no good of their own—might conceivably be intrinsically valuable. As Eric Katz puts it, "many natural entities worth preserving [i.e., valuable in their own right] are not clearly the possessors of interests."[27]

---

[21] Ibid., pp. 32, 35, n. 25. Here he picks up a suggestion that he had already made in "On Being Morally Considerable," pp. 310, 323.

[22] Goodpaster now rejects "generalizations of egoism" that extend moral concern to a class of beneficiaries that includes "forests, lakes, rivers, air and land" ("From Egoism to Environmentalsim," p. 28). Moral considerability for ecosystems is in the same "egoistic" spirit, I suppose. Goodpaster prefers to speak of "bearers of value," a term that de-emphasizes the possession of interests. Yet this talk is misleading. It can only obscure Goodpaster's assumptions about what makes the biosphere considerable—including its capacity for "successful self-protection" (p. 32).

[23] Andrew Brennan, "The Moral Standing of Natural Objects," *Environmental Ethics* 6 (1984): 53, 49, 51. Brennan uses the term *moral standing*, which he introduces as a synonym for moral considerability (p. 37). I am not attacking the substance of Brennan's view here; I am objecting to his claim to be explicating Goodpaster's concept (p. 35). Brennan has severed moral considerability completely from the notion that Goodpaster finds crucial—that of being a potential beneficiary. I believe that Goodpaster would not recognize the concept after this surgery.

[24] Attfield, *The Ethics of Environmental Concern*, p. 149. See also p. 159.

[25] Callicott, "Intrinsic Value, Quantum Theory," p. 258.

[26] Ibid., p. 275.

[27] Eric Katz, "Organism, Community, and the 'Substitution Problem,'" *Environmental Ethics* 7 (1985): 243. Oddly, Katz goes on to associate intrinsic value with "autonomy," which he in turn locates (p. 246) in the fact that "natural individuals . . . pursue their own *interests* while serving roles in the community."

Is this just a quibble about words? I think not. We have more than one paradigm of moral relevance, and it makes a difference which one we adopt as the model for our ethical thinking about ecosystems. If we aim to justify preservation by appeal to the intrinsic value of natural ecosystems, our arguments must build on the way ecosystems resemble other things that we preserve for their intrinsic *value*. Moral considerability is another matter. To ground the preservationist intuition upon the *interests* of ecosystems, we have to look for an analogy between ecosystems and beings that clearly have interests. Given that ecosystems are not sentient, the most promising models are plants and other nonsentient organisms.[28]

IV

Some ethicists would object that we cannot even get this argument for ecosystems off the ground—it is absurd, they would say, to think that plants could be morally considerable. Such a dismissal of plants, however, is too quick, for it ignores goal-directedness. Peter Singer, for instance, regards *rocks* as representative of all nonsentient beings. "A stone," he says, "does not have interests because it cannot suffer. Nothing that we can do to it could possibly make any difference to its welfare." He therefore boldly concludes: "If a being is not capable of suffering, or of experiencing enjoyment or happiness, there is nothing to be taken into account."[29]

Although sentience may turn out, after all, to be necessary for moral considerability, this just cannot be as obvious as Singer assumes. There is a world of difference between plants and rocks. Surely there might be something to "take into account" even in the absence of sentience. All we need, as Bryan Norton observes, is something appropriately analogous to sentience. Norton rejects the possibility of ecosystem "rights" because "collectives such as mountain ranges, species, and ecosystems have no significant analogues to human sentience on which to base assignments of interests." Since collectives lack any analogue to sentience, he reasons, "the whole enterprise of assigning interests [to them] becomes virtually arbitrary."[30] Norton reaches this conclusion too quickly, as I argue below, but he makes two crucial points. First, we can plausibly attribute

---

[28] Leaving aside Callicott's argument for nature as one's "extended self."

[29] Peter Singer, *Animal Liberation* (New York: Avon Books, 1975), p. 8. Singer repeats this section verbatim in *Practical Ethics* (Cambridge: Cambridge University Press, 1979), p. 50. The leap is surprisingly common. See, e.g., William Frankena, "Ethics and the Environment," in *Ethics and Problems of the 21st Century*, p. 11, and G. J. Warnock, *The Object of Morality* (London: Methuen & Co., 1971), p. 151. Scott Lehmann judges this to be "the standard view among moral philosophers" and rests his case against the "rights" of wilderness areas on the premise that "only subjects of experience can be harmed or benefited" ("Do Wildernesses Have Rights?" *Environmental Ethics* 3 [1981]: 136–38).

[30] Norton, "Environmental Ethics and Nonhuman Rights," p. 35.

moral considerability to x only when we have a nonarbitrary way of identifying x's interests. Second, this project does not require actual sentience. It is plain enough that plants, for instance, have interests in a straightforward sense, though they feel nothing.[31] Paul Taylor puts it this way:

> Trees have no knowledge or feelings. Yet it is undoubtedly the case that trees can be harmed or benefited by our actions. We can crush their roots by running a bulldozer too close to them. We can see to it that they get adequate nourishment and moisture. . . . It is the good of trees themselves that is thereby affected.[32]

In general, Taylor explains, "the good of an individual nonhuman organism [consists in] the full development of its biological powers." Every organism is "a being whose standpoint we can take in making judgments about what events in the world are good or evil."[33]

Let us grant, in spite of Singer and his allies, that there is something about trees that we might intelligibly "take into account" for moral purposes. Can we be more specific? What is it that plants have and rocks do not? The obvious, but unilluminating answer is "life." Just what is it about being alive that makes plants candidates for moral considerability?

Goal-directedness is the key. Taylor, for instance, describes organisms as "teleological centers of life."[34] Goodpaster points to plants' "tendencies [to] maintain and heal themselves" and locates the "core of moral concern" in "respect for self-sustaining organization and integration."[35] Attfield writes of a tree's "latent tendencies, direction of growth and natural fulfillment."[36] Jay Kantor bases his defense of plant interests on their "self-regulating and homeostatic functions."[37] Rodman condemns actions that impose our will upon "natural entities that have their own internal structures, needs, and potentialities," potentialities that are actively "striving to actualize themselves."[38] Finally, James K. Mish'alani points to each living thing's *self-ameliorative competence:* "that is, a power for coordinated movement towards favorable states, a capacity

---

[31] Devastating critiques of claims to have demonstrated plant sentience include Arthur Galston "The Unscientific Method," *Natural History,* March 1974, and Arthur W. Galston and Clifford L. Slayman, "The Not-So-Secret Life of Plants," *American Scientist* 67 (1979): 337–44.

[32] Paul Taylor, "The Ethics of Respect for Nature," *Environmental Ethics* 3 (1981): 200. Taylor has reiterated this in *Respect for Nature.*

[33] Ibid., p. 199.

[34] Ibid., pp. 210–11.

[35] Goodpaster, "On Being Morally Considerable," pp. 319, 323.

[36] Robin Attfield, "The Good of Trees," *Journal of Value Inquiry* 15 (1981): 37. See also Attfield, *The Ethics of Environmental Concern,* pp. 140–65.

[37] Jay Kantor, "The 'Interests' of Natural Objects," *Environmental Ethics* 2 (1980): 169.

[38] Rodman, "The Liberation of Nature?" pp. 100, 117.

to adjust to its circumstances in a manner to enhance its survival and natural growth."[39]

The goal-directedness of living things gives us a plausible and nonarbitrary standard upon which to "base assignments of interests." If ecosystems, though not sentient, are goal-directed, then we may (without absurdity) attribute interests to them, too. Goodpaster is right: there is no *a priori* reason to think that "the universe of moral considerability [must] map neatly onto our medium-sized framework of organisms."[40] Of course, we must not get carried away with this line of thinking. Goal-directedness is certainly not sufficient for moral considerability. One problem is that some machines are goal-directed—e.g., guided missiles, thermostatic heating systems, chess-playing computers, and "The Terminator."[41] The defender of moral considerability for plants must distinguish plants, morally, from goal-directed but inanimate objects.[42] Still, the possession of goals is what makes the notion of a plant's "standpoint" intelligible. Can we locate an ecosystem's standpoint by understanding its goals? Not if it doesn't have any goals.

V

We often know goal-directedness when we see it. The analysis of goal-directedness is, however, a terribly unsettled subject in the philosophy of science.[43] In light of this unsettledness, one must be cautious. Here are three claims. First, the attribution of goal-directedness to organisms can be scientifically and philosophically respectable—even when the organisms in question are nonsentient. Teleology talk need not be vitalistic, anthropomorphic, or rooted in obsolete Aristotelian biology or physics. It does not imply "backward causation." Nor need it run afoul of the "missing goal-object" problem.[44]

---

[39] James K. Mish'alani, "The Limits of Moral Community and the Limits of Moral Thought," *Journal of Value Inquiry* 16 (1982): 138.

[40] Goodpaster, "On Being Morally Considerable," p. 323.

[41] Goodpaster correctly describes the idea that a house's porch has interests as "simply incoherent" ("On Stopping at Everything: A Reply to W. M. Hunt," *Environmental Ethics* 2 [1980]: 282). We cannot dismiss the machines in my list so easily.

[42] Mish'alani and Taylor recognize this, but I find their solutions unpersuasive. They seem to imply (implausibly) that if we humans should turn out to be the artifacts of a Creator, we should cease to regard ourselves as morally considerable. Taylor, *Respect for Nature*, p. 124 and Mish'alani, "Limits of Moral Community," p. 139.

[43] Even so, it is surprising that environmental ethicists have not looked more often to the philosophy of science literature. Brennan is an exception. See "Moral Standing of Natural Objects," pp. 41–44. (Taylor mentions, in passing, that a philosophical literature about goal-directedness exists, but he does not make much use of it. [*Respect for Nature*, p. 122, n. 8].)

[44] The problem of the "missing goal-object" is that behavior may be directed at a goal that happens to be unattainable or even nonexistent. This fact seems fatal for accounts of goal-directedness in

Second, some of these respectable accounts of goal-directedness are useful for the environmental ethicist. They enable us to resist crude versions of the common slippery-slope argument against the moral considerability of plants and other nonsentient living things. Once we admit nonsentient beings into the moral considerability club, how can we bar the door to ordinary inanimate objects? Porches, paintings, automobiles, garbage dumps, buildings, and other ordinary objects are allegedly lurking just outside, waiting for us to admit plants.[45] Goal-directedness can keep them out.

Third, we ought to recognize a distinction between goals and behavioral byproducts. A defensible conception of goal-directedness must distinguish true goals from outcomes that a system achieves incidentally. Ecosystem resilience and stability look like goals, but this appearance may deceive us. An ecosystem property such as stability might turn out to be just a byproduct, the incidental result of individual activities aimed exclusively at the individuals' own goals.

I shall discuss two of the main approaches to understanding goal-directedness. The approaches differ in important ways. I favor the second, but either will do for my purposes. The first approach is propounded by Ernest Nagel (among many others). Nagel holds that a system is goal-directed when it can reach (or remain in) some particular state by means of behavior that is sufficiently *persistent* and *plastic*.[46] Persistence refers to the system's ability to "compensate" for interfering factors that would otherwise take the system away from its goal.[47]

---

terms of feedback. See Israel Scheffler, *The Anatomy of Inquiry* (New York: Alfred A. Knopf, 1963), pp. 112–16. These worries, nevertheless, still trouble many. It is for this reason that William C. Wimsatt insists, defensively, that "teleology, properly so-called, does have a respectable role in the scientific characterization of non-cognitive systems" ("Teleology and the Logical Structure of Function Statements," *Studies in the History and Philosophy of Science* 3 [1972]: 80). Wimsatt maintains that he is innocent of anthropomorphism (p. 65). See also Michael Ruse, *The Philosophy of Biology* (London: Hutchinson, 1973), pp. 174–76.

[45] W. Murray Hunt claims that a porch's "needs" (e.g., to be painted) are as evident as a lawn's need to be watered ("Are Mere *Things* Morally Considerable?" *Environmental Ethics* 2 [1980]: 59–66). R. G. Frey asserts that a Rembrandt painting has interests in every sense in which a dog has them, in *Interests and Rights: The Case Against Animals* (Oxford: Clarendon Press, 1980), p. 79. Elliott Sober claims that he cannot see how the needs of plants can be plausibly distinguished from the needs of "automobiles, garbage dumps, and buildings." "Philosophical Problems of Environmentalism," in *The Preservation of Species*, ed. Bryan G. Norton (Princeton, New Jersey: Princeton University Press, 1986), p. 184.

[46] Ernest Nagel, *The Structure of Science* (Indianapolis: Hackett, 1961), pp. 398–421 and "Teleology Revisited," in *Teleology Revisited and Other Essays in the Philosophy and History of Science* (New York: Columbia University Press, 1979). See also Richard Braithwaite, *Scientific Explanation* (Cambridge: Cambridge University Press, 1953), pp. 319–41 and Ruse, *Philosophy of Biology*, pp. 174–96.

[47] Nagel says that persistence is "the continued maintenance of the system in its goal-directed behavior, by changes occurring in the system that compensate for any disturbances . . . which, were there no compensating changes . . . would prevent the realization of the goal" ("Teleology Revisited," p. 286).

Plasticity refers to the system's ability to reach the same outcome in a variety of ways.[48]

Nagel assumes that this approach will count all living things as goal-directed. It seems to.[49] There are problems, to be sure. Chief among these is the danger that it will include some behavior that plainly is not goal-directed—the movement of a pendulum, for instance, or the behavior of a buffered chemical solution.[50] Nagel, however, shows that with some plausible tinkering—mainly, by adding a third condition that he calls "orthogonality"—we can deal with these counterexamples.[51]

The second approach, pioneered by Charles Taylor, insists that goal-directed behavior "[really does] occur 'for the sake of' the state of affairs which follows."[52] Subsequent philosophers have developed this basic insight in various ways.

An especially influential exponent of Taylor's approach is Larry Wright. Taylor's considered formulation of his insight requires that the behavior in question be both necessary and sufficient for the goal. Wright finds this unsatisfactory—too generous in some ways and too strict in others.[53] He suggests what he calls an "etiological" account, one that focuses on the causal background of the behavior in question. A system is goal-directed, Wright contends, only if it behaves as it does just because that is the type of behavior that tends to bring about that type of goal. Formally, behavior B occurs for the sake of goal-state G if "(i) B tends to bring about G," and "(ii) B occurs because (i.e. is brought about by the fact that) it tends to bring about G."[54] The key condition is (ii). Some machines, say guided missiles, meet it, for a machine may B because it is designed to B, and it may be designed to B, in turn, because B tends to bring

---

[48] In Nagel's words: "the goal . . . can generally be reached by the system following alternate paths or starting from different initial positions" (Ibid). See also Braithwaite, *Scientific Explanation*, pp. 329–34. Braithwaite conducts his analysis entirely in terms of plasticity.

[49] Nagel also seems to believe, though, that mechanistic accounts of behavior, when they become available, automatically drive out teleological accounts ("Teleology Revisited," pp. 289–90). He doesn't say whether he thinks we have successful mechanistic accounts of plant behavior yet.

[50] For other problems, see Wimsatt, "Teleology," p. 26, David Hull, *Philosophy of Biological Science* (Englewood Cliffs, N.J.: Prentice-Hall, 1974), pp. 107–09, and Larry Wright, "The Case Against Teleological Reductionism," *British Journal for the Philosophy of Science* 19 (1968): 211–23.

[51] Nagel, *Structure of Science*, pp. 418–21, and "Teleology Revisited," pp. 287–90.

[52] Charles Taylor, *The Explanation of Behavior* (London: Routledge and Kegan Paul, 1964), p. 5.

[53] Larry Wright, "Explanation and Teleology," *Philosophy of Science* 29 (1972): 204–18. One problem especially relevant to my thesis here is that, as Wright points out, Taylor's formulation admits "all sorts of bizarre accidents into the category of goal-directed activity" (p. 209).

[54] Ibid., p. 211. This formulation obviously fails in the case of intentional but misguided action unless we understand "tends to bring about G" as "tends *under normal circumstances* to bring about G." If I submit a paper to a defunct journal, that may not tend to bring about the goal of having my paper published, but my behavior is clearly goal-directed. I have submitted the paper in order to have it published.

about some *G* desired by the designer. Organisms meet it, too, because of the way that natural selection operates. The fitness of an organism usually depends on how appropriate its behavior is—that is, the extent to which it does the sort of thing (say, *B*) that tends to help that kind of organism survive and reproduce. If the disposition to *B* is heritable, organisms whose tendency to *B* helps make them fit will leave descendants that tend to *B*. Those descendants are disposed to *B*, then, in part because *B* is an appropriate type of behavior.[55]

Some people emphatically do not find Wright's approach respectable. He has, for example, recently been accused of "misrepresenting" natural selection as a teleological process in the old-fashioned (and discredited) sense according to which nature selects with certain outcomes in mind.[56] This charge, however, misses the mark, for there is nothing wrong with Wright's understanding of natural selection.[57] In addition, Wright has also dealt effectively with other, better-founded criticisms that need not be discussed here.[58]

Wright's development of Taylor's insight is the best approach for my purposes because alternative versions of Taylor's approach are not as good for sustaining attributions of goal-directedness to plants and lower animals.[59] With regard specifically to the slippery slope and the alleged "needs" of paintings and porches, Nagel's approach seems good enough, for these objects do not act persistently or plastically toward any result that we could seriously be tempted to call a goal. With Wright's criteria, however, we sidestep questions of degree that can plague Nagel. Consider my car, which responds to the upstate New York environment by rusting. The car rusts in spite of my efforts to stop it, and it would rust even if I tried much harder. Eventually it will fall apart. Does this unpleasantly persistent behavior count as goal-directed? A dedicated slippery-sloper might suggest that the car has the goal of rusting, a "need" to rust. Both Nagel and Wright can resist this suggestion, but Nagel would have a tougher time due to the vagueness of his persistence and plasticity conditions. Wright

---

[55] This "because" applies, as Wright acknowledges, in a rather "involuted" way. It applies, nevertheless. See "Explanation and Teleology," pp. 216–17.

[56] Kristen Shrader-Frechette, "Organismic Biology and Ecosystems Ecology: Description or Explanation?" in *Current Issues in Teleology*, ed. Nicholas Rescher (Lanham, Md.: University Press of America, 1986), pp. 84–85, n. 28.

[57] See Larry Wright, "Functions," *Philosophical Review* 82 (1973): 139–68, esp. 159–64.

[58] See, for example, Andrew Woodfield, *Teleology* (Cambridge: Cambridge University Press, 1976), pp. 83–88, and Arthur Minton, "Wright and Taylor: Empiricist Teleology," *Philosophy of Science* 42 (1975): 299. Wright defends himself in "The Ins and Outs of Teleology: A Critical Examination of Woodfield," *Inquiry* 21 (1978): 233–45.

[59] Jonathan Bennett, for instance, in *Linguistic Behavior* (Cambridge: Cambridge University Press, 1976), pp. 36–81. Bennett introduces the concept of "registration" and says that a system (*S*) is goal-directed toward *B* when it does *B* because it registers that it is in a situation where *B* will bring about *G*. This analysis does not immediately exclude plants. The question, as Bennett sees it (p. 79), is whether the behavior of plants has a "unitary" mechanistic explanation. Because phototropism in green plants is "controlled by one unitary mechanism," he refuses to count it as a goal-directed behavior.

would simply check the behavior's etiology. My car, we may safely say, does not rust because rusting tends to cause cars to fall apart. It rusts because rust is just what happens when steel meets moisture and road salt. The car's behavior fails Wright's condition (ii).

We can imagine an etiology that would make my car's rusting genuinely goal-directed. Assume that car designers know how to make sturdy rust-free cars. Suppose, however, that they greedily conspire to build cars that are susceptible to rust in order to force people to buy new cars more frequently. We would then be unable fully to understand my car's rusting as a purely chemical process, for—on the conspiracy theory of rust—my car would be rusting (in part) because rusting tends to cause cars to fall apart.

Now, what about ecosystems? I concede that the heralded stability and resilience of some ecological systems make them prima facie goal-directed. When such an ecosystem is perturbed in any one of various ways, it bounces back. The members of the ecosystem do just what is necessary (within limits) to restore the system to equilibrium.[60] But are they cooperating in order to restore equilibrium? That is surely imaginable. On the other hand, each creature might instead be "doing its own thing," with the fortunate but incidental result that the ecosystem remains stable. If this is correct, then we are dealing with a behavioral byproduct, not a systemic goal.

The goal/byproduct distinction is well entrenched in the literature on natural selection and biological adaptation. Let me illustrate this distinction with an example from George Williams. Williams asks us to consider the behavior of a panic-stricken crowd rushing from a burning theater. A biologist newly arrived from Mars, he suggests, might be impressed by

> [the group's] rapid 'response' to the stimulus of fire. It went rapidly from a widely dispersed distribution to the formation of dense aggregates that very effectively sealed off the exits.[61]

If the crowd clogs the exits in spite of strenuous crowd-control efforts, would our Martian be entitled to report that he had observed a crowd that was goal-directed toward self-destruction via the sealing off of the exits? Of course not. We know that the clogging of the exits is just incidental. The people are trying to get out. The crowd clogs the exits in spite of the dreadful consequences.

Any theory of goal-directedness ought to be able to avoid the Martian's conclusion. Wright's theory does that easily via condition (ii): G can be a goal of

---

[60] For a sound discussion of ecosystem stability see John Lemons, "Cooperation and Stability as a Basis for Environmental Ethics," *Environmental Ethics* 3 (1981): 219–30.

[61] George Williams, *Adaptation and Natural Selection: A Critique of Some Current Evolutionary Thought* (Princeton: Princeton University Press, 1966), pp. 210–11. We can, by the way, imagine a system that is goal-directed toward self-destruction. Wimsatt describes a "suicide machine" in "Teleology," pp. 20–22.

behavior *B* only if *B* occurs *because* it tends to bring about *G*. If *G* plays no explanatory role it cannot be a genuine goal.[62]

Nagel's account also permits us to distinguish goal from byproduct. The persistence condition does the work here. There is no reason to think that the theater crowd's behavior is truly persistent toward clogging the exits. If there were more exits, or larger exits, the people would have escaped smoothly. We may be sure that the crowd would not compensate for greater ease of exit by modifying its behavior in order to achieve clogging.

VI

If the idea that organisms have morally considerable "interests" seems plausible, it must, I think, be because organisms are genuinely goal-directed. When Taylor, for instance, characterizes a tree's good as "the full realization of its biological powers," we know what he means. We naturally assume that *powers* does not refer to everything that can happen to a tree—disease, say, or stunting from lack of nutrients. The tree's powers are the capabilities that the tree exercises in the service of its goals of growth, survival, and reproduction. We certify that those are the tree's goals, in turn, by employing criteria such as Wright's or Nagel's.

Should we find moral significance in an organism's goals? Perhaps not. We may coherently admit that plants have goals, yet deny that we have duties to them. Still, there is a tempting analogy between the goal-directed behavior of organisms and the intentional behavior of humans. Recall the rhetorical role that the notion of natural "striving" plays in Paul Taylor's argument for an ethic of respect for nature.[63] Recall Katz's choice of the term *autonomy* to characterize an organism's capacity for independent pursuit of its own interests.[64] Indeed the word *interests* itself conveys the flavor of intention.[65] This flavor lends persuasiveness to arguments such as Taylor's.

Let us, in any event, grant that to have natural goals is to have morally considerable interests. Where does this leave behavioral byproducts? It leaves

---

[62] Wright is keen on distinguishing goals from byproducts, though not in precisely those terms. "Teleological behavior is not simply appropriate behavior," he insists, "it is appropriate behavior with a certain etiology" ("Explanation and Teleology," p. 215). Byproducts result, after all, from behavior that is appropriate in a trivial sense—appropriate for producing those byproducts. Bennett, too, has a good discussion of "fraudulent" attributions of goal-directedness (*Linguistic Behavior*, pp. 75–77).

[63] Taylor, "Respect for Nature," p. 210.

[64] See note 27 above. See also Heffernan, "The Land Ethic," p. 242.

[65] General treatments of teleology often point this out. In *Linguistic Behavior*, Jonathan Bennett treats human intention as a special case of goal-directedness. Andrew Woodfield reverses the analysis, claiming that attributions of goal-directedness to nonsentient things such as plants and machines are extensions of the "core concept" of having an "intentional" object of desire (*Teleology*, pp. 164–66, 201–02).

them where they were—morally irrelevant. We need a nonarbitrary standard for deciding which states of affairs are good ones from the organism's own "standpoint." Sentience gives us such a standard by way of the organism's own preferences (which we are capable of discovering in various ways). By analogy, a nonsentient organism's biological goals—its "preferred" states—can do the same thing. But is there any reason at all for supposing that either mere natural tendencies or behavioral byproducts give rise to interests? I think not. Why, from a given system's "standpoint," should it matter whether some natural tendency, unconnected (except incidentally) to the system's goals, plays itself out?

Consider John Rodman's account of why it is wrong to dam a wild river. Rodman emphasizes that the river "struggles" against the dam "like an instinct struggles against inhibition."[66] One might be tempted to say that this way of talking is unnecessary, that every natural tendency is morally privileged. Such a claim, however, is implausible. What leads Rodman to talk of instinct and struggle is, I take it, the notion that the river actually has goals and would be frustrated, by the dam, in its pursuit of them.

I do not expect this example to be convincing. To see clearly that mere tendencies are in themselves morally irrelevant, we should consider something really drastic—like *death*. Usually, death is something that just happens—by accident, by disease, or simply when the body wears out. Organisms tend to die, but they do not ordinarily aim to die. As Jonathan Bennett puts it: "Every animal is tremendously plastic in respect of becoming dead: throw up what obstacles you may, and death will still be achieved. Yet animals seldom have their deaths as a goal."[67]

Consider a salmon of a species whose members routinely die after spawning. Even here death seems unlikely to be the organism's goal. The salmon dies because the arduous upstream journey has worn it out. If it could spawn without dying, it would do so. Once in a while that actually happens. When it does, do we say (without further evidence) that the salmon has been frustrated in its efforts to die after spawning? No. We would say that the salmon has managed to spawn without having had the misfortune to die.

Behavioral byproducts, like mere tendencies, seem not to generate anything we can comfortably call "interests." The salmon example illustrates this, if we interpret the death of the adult as a byproduct of its spawning. Williams' theater example illustrates it, too. It would be truly bizarre to suggest that the panicky crowd has an interest in being trapped and incinerated.

Although there is much more that needs to be said about whether the argument from goal-directedness can establish the moral considerability of plants, let us go ahead and accept plant moral considerability. But does ecosystem moral considerability follow? No, an obstacle remains: the goal/byproduct distinction. We

---

[66] Rodman, "Liberation of Nature?" p. 115.
[67] Bennett, *Linguistic Behavior*, p. 45.

still need to determine whether stability (or any other property) of an ecosystem is a genuine goal of the whole system rather than merely a byproduct of self-serving individual behavior.

VII

Donald Worster has written in his history of ecological ideas that "More often than not, the ecological text [holistic environmentalists] know and cite is either of their own writing or a pastiche from older, superseded models. Few appreciate that the science they are eagerly pursuing took another fork back yonder up the road."[68] Orthodox ecology, Worster says, has abandoned the "organismic" view of ecosystems and adopted a fundamentally individualistic one.[69] Robert M. May represents this individualistic orthodoxy. Of course, says May, there are "patterns at the level of ecological systems." He insists that these patterns do not represent goals. They are entirely explicable in terms of "the interplay of biological relations that act to confer specific advantages or disadvantages on individual organisms."[70]

What then are we to make of ecosystem stability and resilience? If May is right, the tendency of an ecosystem to bounce back after a disturbance is merely the net result of self-serving responses by individual organisms. We need not view stability as a system "goal." We may not even be entitled to do so. As Robert Ricklefs explains:

> The ability of the community to resist change [is] the sum of the individual properties of component populations. . . . Relationships between predators and prey, and between competitors, can affect the inherent stability of the community, but trophic structure does not evolve to enhance community stability.[71]

Certain forms of trophic structure typically enhance community stability, Rick-

---

[68] Donald Worster, *Nature's Economy: A History of Ecological Ideas* (Cambridge: Cambridge University Press, 1985), pp. 332–33.

[69] Here is an "organismic" characterization of ecosystems from an ecology text popular throughout the 1950s: "The community maintains a certain balance, establishes a biotic border, and has a certain unity paralleling the dynamic equilibrium and organization of other living systems. Natural selection operates upon the whole interspecies system, resulting in a slow evolution of adaptive integration and balance. Division of labor, integration and homeostasis characterize the organism. . . . The interspecies system has also evolved these characteristics of the organism and may thus be called an ecological supraorganism." W. A. Allee et al., *Principles of Animal Ecology* (Philadelphia: W. B. Saunders, 1949), p. 728.

[70] Robert M. May, "The Evolution of Ecological Systems," *Scientific American*, September 1987, p. 161. This sort of individualism by no means excludes altruism. Many individual organisms aim to some extent at the survival of their kin.

[71] Robert E. Ricklefs, *The Economy of Nature* (Portland, Oreg.: Chiron Press, 1976), p. 355.

lefs agrees, but trophic structure does not take on particular form because that form enhances stability.[72]

Someone might be tempted to conclude that my own argument undermines the moral considerability of organisms. Organisms, after all, consist of cells. The cells have goals of their own. Does my individualism require us to regard the behavior of organisms as merely a byproduct of the selfish behavior of cells? It does not. Cells do have their own goals, but these goals are largely subordinated to the organism's goals, because natural selection selects *bodies*, not cells. If the cells do not cooperate for the body's sake, the body dies and the cells die, too. That, very roughly, is how natural selection coordinates the body's activities.[73] Selection tends to eliminate individuals that are not good at the survival "game" (taking kin selection into account). Eventually this process leaves us with organisms that are good at it, and these organisms are goal-directed toward those states of affairs that have in the past made them winners.

So much for organisms. A familiar process—ordinary, individualistic natural selection—ensures that they are goal-directed. Is there a process that could account for goal-directedness in ecosystems? The only candidate I know of for this job is group selection operating at the community level.[74]

VIII

Does group selection have a part to play in the full explanation of the behavior of species populations or ecosystems? I hold that the answer is no. Now this may seem hard to believe. "Ecosystem behavior," you might counter, "is just too well coordinated for stability to be an accident." To undermine this intuition, let us consider a description of a simple situation in which there is a result that we could construe as "good for the group," but which is strictly speaking a byproduct of self-serving individual action, and then a second situation, a more complicated one, in which an extremely stable group property is, again, a byproduct.[75]

Consider any single-species population. Suppose that some individuals (call

---

[72] See also J. Engelberg and L. L. Boyarsky, "The Noncybernetic Nature of Ecosystems," *The American Naturalist* 114 (1979): 317–24. "That a system is stable, that it can resist perturbations, is not a sign that it is cybernetic." Engelberg and Boyarsky's main point is that ecosystems lack the "global information networks" that integrate the parts of goal-directed systems such as organisms.

[73] See, for logic similar to mine, Elliott Sober's treatment of "selfish DNA" in *The Nature of Selection: Evolutionary Theory in Philosophical Focus* (Cambridge, Mass.: MIT Press, 1984), pp. 305–14.

[74] I agree with William Wimsatt, who suggests that "*all* teleological phenomena are ultimately to be explained in terms of selection processes" (Teleology," p. 15). David Hull criticizes Wimsatt (unsuccessfully, I think) in *Philosophy of Biological Science*, p. 113.

[75] Both of my examples concern single-species populations. I picked them for their simplicity. They make a point that seems to hold, a fortiori, for ecosystems and communities as well.

them the A-individuals) run into a stretch of bad luck and consequently fail to reproduce. Their failure to reproduce reduces the intensity of competition. This (other things being equal) permits other members of the population (the B-individuals) to reproduce more effectively than they otherwise would have. Should we regard this population as a goal-directed whole, answering a threat to its survival by redirecting its reproductive effort? Of course not. Williams explains the general difficulty in this way:

> Certainly species survival is one result of reproduction. This fact, however, does not constitute evidence that species survival is a function of reproduction. If reproduction is entirely explainable on the basis of adaptation for individual genetic survival, species survival would have to be considered merely an incidental effect.[76]

There is no reason to think of the B-individuals' increased reproductive success as "compensating" for the failure of the A-individuals. If fact, each of the B-individuals has simply taken advantage of the A-individuals' failure. The net result is survival of the group, to be sure, but a postulated goal of group survival has no explanatory role to play.

Let us now consider a more difficult and controversial example, the clutch size in birds, long a bone of contention between group selectionists and Neo-Darwinians. Clutch size in some species of birds is remarkably constant; certain species of plover, for instance, almost always lay four eggs. If an egg is removed from the plover's nest, the bird lays a replacement, bringing the number back up to four. That is not so strange, in itself; yet it shows that the plover is physiologically able to lay more than four eggs. Why should it lay only four to start with?

Perhaps this is a sign of group selection at work, favoring a population of birds in which individual birds restrain themselves for the good of their group. V. C. Wynne-Edwards, the dean of group selectionism, would say so. Consider what Wynne-Edwards says about "reproductive rate":

> If intraspecific selection was all in favor of the individual, there would be an overwhelming premium on higher and ever higher individual fecundity, provided it resulted in a greater posterity than one's fellows. Manifestly this does not happen in practice; in fact, the reproductive rate in many species . . . is varied according to the current needs of the population.[77]

---

[76] Williams, *Adaptation and Natural Selection*, p. 160. See also pp. 107–08. Jan Narveson also makes this point, observing that as long as some people have children for selfish reasons, then the race is "perpetuated willy-nilly" whether or not anyone has children *in order* to perpetuate the race. Narveson, "On the Survival of Humankind," in *Environmental Philosophy*, ed. Robert Elliot and Arran Gare (St. Lucia: University of Queensland Press, 1983), pp. 51–52.

[77] V. C. Wynne-Edwards, *Animal Dispersion in Relation to Social Behavior* (New York: Hafner, 1962), p. 19. See pp. 484–90.

If this group-selectionist account is correct, then the plover population's behavior is goal-directed, even by Wright's criteria, for the individual birds are laying exactly four fertilized eggs just because of the consequences this activity has—that is, just because their self-restraint meets the "current needs of the population"—and we are entitled to speak of the group's goal of maintaining a certain specified average clutch size.

There is, however, an alternative account, an individualistic Neo-Darwinian explanation. Each individual bird seeks to maximize its own inclusive fitness. If laying more than four eggs were a sound strategy for the individual, then that is the strategy it would pursue. Chances are, however, that if a pair of plovers divide their parental energy and attention among five offspring instead of four, fewer of the offspring will survive than if the parents had been conservative. "Exactly four eggs" is a sound strategy from the standpoint of each individual. Seen in this way, it does not represent individual self-restraint for the good of the group. There is no group goal.[78]

Evolutionary biologists are by and large skeptical about group selection. For one thing, the argument for group selection in nature is essentially negative: as Wynne-Edwards puts it, group selection simply *must* occur, since normal natural selection would not be "at all effective" in generating "the kind of social adaptations . . . in which the interests of the individual are actually submerged or subordinated to those of the community as a whole."[79] This negative argument for group selection is undermined when we discover plausible individualistic explanations—when, as in the clutch size case, we find that the interests of the individual are not "submerged" at all. Williams and others, including Richard Dawkins, have shown that we do not need group selection to explain any of the phenomena upon which Wynne-Edwards builds his case.[80]

Worster is correct about which fork ecology has taken. To be sure, a number of theorists have shown how something they label "group selection" could occur

---

[78] David Lack offers evidence that (other things being equal) larger-than-normal clutches typically reduce the number of surviving offspring per nest. David Lack, *The Natural Regulation of Animal Numbers* (Oxford: Clarendon Press, 1954), pp. 21–32, and *Population Studies of Birds* (Oxford: Clarendon Press, 1966), pp. 3–7 and throughout. Reducing the number of eggs does not sufficiently improve each offspring's chances to make that a worthwhile strategy, either. Michael Ruse claims (*The Philosophy of Biology*, pp. 179–81) that the tendency of a bird population toward laying clutches of a particular size *is* goal-directed behavior even though (he assumes) it isn't a result of group selection. He thinks he is using Nagel's conception of goal-directedness. The problem, I suspect, is that he does not subject the population's behavior to a robust version of the persistence test. It is not clear that Ruse can get the right answer even in the theater-crowd case.

[79] Wynne-Edwards, *Animal Dispersion*, p. 18.

[80] Richard Dawkins, *The Selfish Gene* (Oxford: Oxford University Press, 1976), esp. chaps. 5–7. See, for example, Dawkins' explanation of how sterile castes have evolved in the social insects. Compare Wynne-Edwards' view that it is *inconceivable* that sterile castes could have evolved except where "selection had promoted the interests of the social group, as an evolutionary unit in its own right" (*Animal Dispersion*, p. 19).

under the right circumstances.⁸¹ These particular theories, however, insofar as they are extensions of kin selection, are fundamentally "individualistic," and are not much like the theories that earlier advocates of group selection had hoped for.⁸² We have little or no reason to believe that evolution by group selection, as traditionally conceived, is significant in nature.⁸³

IX

When we turn from group selection operating on single-species populations to community selection, the result is much the same. According to Robert May, for instance:

> Natural selection acts almost invariably on individuals or on groups of related individuals. Populations, much less communities of interacting populations, cannot be regarded as units subject to Darwinian evolution.⁸⁴

This view has been seconded by Elliott Sober. "Darwinism," Sober asserts, "is a profoundly individualistic doctrine":

> [It] rejects the idea that species, communities, and ecosystems have adaptations that exist for their own benefit. These higher-level entities are not conceptualized as goal-directed systems; what properties of organization they possess are viewed as artifacts of processes operating at lower levels of organization.⁸⁵

To be fair, I should report Robert McIntosh's recent lament that "organismic ecology is alive and well." McIntosh worries that parts of the ecosystem-as-organism view survive in "systems" ecology.⁸⁶ John L. Harper shares this worry and he warns against "one of the dangers of the systems approach to community

---

[81] See, e.g., David Sloan Wilson, *The Natural Selection of Populations and Communities* (Menlo Park, Calif.: Benjamin/Cummings, 1980) and Michael Gilpin, *Group Selection in Predator-Prey Communities* (Princeton: Princeton University Press, 1975).

[82] John Cassidy, "Philosophical Aspects of Group Selection," *Philosophy of Science* 45 (1978): 575–94. See also Sober, *The Nature of Selection*, pp. 255–66, 314–68.

[83] Some parasites may be an exception. See Peter W. Price, *Evolutionary Biology of Parasites* (Princeton: Princeton University Press, 1980). I am not sure, either, exactly what to say about cases of exceedingly close symbiosis, as in lichens.

[84] Robert M. May, "The Evolution of Ecological Systems," *Scientific American*, September 1978, p. 161.

[85] Sober, "Philosophical Problems for Environmentalism," p. 185. The upshot? "An environmentalism based on the idea that the ecosystem is directed toward stability and diversity," he says, "must find its foundation elsewhere." Thus Sober has anticipated the central theme of my argument, though in my opinion he is much too quick to draw this conclusion.

[86] Robert McIntosh, "The Background and Some Current Problems of Theoretical Ecology," in *Conceptual Issues in Ecology*, ed. Esa Saarinen (Dordrecht, Holland: D. Reidel, 1982), p. 10. The context makes it clear that McIntosh finds this survival lamentable.

productivity"—namely, the temptation to "treat the behavior that [one] discovers as something that can be interpreted *as if* community function is organized." Harper insists that we must resist this temptation: "What we see as the organized behavior of systems is the result of the fate of individuals. Evolution is about individuals and their descendants."[87]

Some systems ecologists contend that ecosystems have some "organismic" features while conceding that "natural selection operates only on a community's constituent populations, not on the community as a whole."[88] These sources, as I read them, hold small comfort for the advocate of ecosystem interests. They support at best an analogy that is too weakly organismic to generate ecosystem goals.[89]

Obviously there is room for rebuttal here. Still, this testimony suggests the scorn with which ecologists and evolutionary biologists typically regard group selection.[90] Could anything else cause individuals to cooperate for the sake of ecosystem goals? I know of no plausible candidates. If the verdict against group selection stands up, I see no way to justify ecosystem moral considerability with the argument from goal-directedness.

X

Earlier I mentioned several distinct strategies for justifying what I call the "preservationist intuition"—intrinsic value, the good of individual plants and animals, and ideals of human excellence. Any of these might be enough. Still, we may find ourselves tempted to believe that whole ecosystems have interests and are therefore morally considerable. This avenue, however, is not promising. Genuine goal-directedness is a step—an essential step—toward moral considerability. It makes sense (as I have argued) to claim that plants and other nonsentient organisms are morally considerable—but only because those beings'

---

[87] John L. Harper, "Terrestrial Ecology," in *Changing Scenes in the Natural Sciences, 1776–1976* ed. Clyde E. Goulden (Philadelphia: Academy of Natural Sciences, 1977), pp. 148–49 (emphasis added).

[88] See J. L. Richardson, "The Organismic Community: Resilience of an Embattled Ecological Concept," *Bioscience*, July 1980, pp. 465–71. See also R. V. O'Neill et al., *A Hierarchical Concept of Ecosystems* (Princeton: Princeton University Press, 1986), pp. 37–54.

[89] See, e.g., Richardson's discussion of "keystone" species.

[90] Could my dismissal of group selection be too hasty? A referee points out that group selection is "not a scientific joke like 'Creation Science'." True enough. One respected ecologist who assumes group selection at the community level is Eugene Odum. See *Fundamentals of Ecology*, 3rd ed. (Philadelphia: W. B. Saunders, 1971), pp. 251–75. See also M. J. Dunbar, "The Evolution of Stability in Marine Environments: Natural Selection at the Level of the Ecosystem," *American Naturalist* 94 (1960): 129–36. Nevertheless, Wynne-Edwards (whom the referee mentions favorably) has something in common with Creation scientists I have read—he carefully ignores the explanations that his opponents offer. See, for example, Lack's annoyed reply to Wynne-Edwards in *Population Studies*, pp. 299–312.

own biological goals provide a nonarbitrary standard for our judgments about their welfare. Were ecosystems genuinely goal-directed, we could try for the next step.[91]

Some ecosystems do indeed appear to have goals—stability, for example. There is a complication, however. Mere behavioral byproducts, which are outcomes of no moral significance, can look deceptively like goals. Moreover, on what I take to be our best current ecological and evolutionary understanding, the goal-directed appearance of ecosystems is in fact deceptive. Stability and other ecosystem properties are byproducts, not goals. Ecosystem interests are, I conclude, a shaky foundation for the preservationist intuition.

---

[91] The next step—to ecosystem moral considerability—might not be as tempting as some have thought. It tends to devalue the individual, perhaps too much. See Katz's criticism of Callicott's ethical holism in "Organism, Community, and the 'Substitution Problem'." See also H. J. McCloskey's criticism of holistic political philosophies in "The State as an Organism, as a Person, and as an End in Itself," *Philosophical Review* 72 (1963): 306–26.

Part III
Deep Ecology and Radical Environmentalism

THE DEEP ECOLOGICAL MOVEMENT: SOME PHILOSOPHICAL ASPECTS

Arne Naess*
Institutt for Ecophilosophy, Oslo

1. Deep Ecology on the Defensive

Increasing pressures for growth have placed the vast majority of ecologists and other environmental professionals into a defensive position. Let me illustrate.

The field ecologist K, who both professionally and personally vigorously advocated deep ecological principles in the late sixties, encountered considerable resistance. Colleagues in the university said he should keep to his science and not meddle in philosophical and political matters. He should resist the temptation to become a prominent "popularizer" through exposure in the mass media. Nevertheless, he continued and influenced thousands (including myself). He became a recognized professional 'expert' in assessing the damage done when bears killed or maimed sheep or other domestic animals in Norway. According to the law, their owners are to be paid damages. Licensed hunters can get permission to shoot a bear if its misdeeds become considerable.[1] Growth pressures required consolidating the sheep industry, and sheepowners became fewer, richer, and more prone to live in towns. Due to wage increases, they could not afford to hire shepherds to watch the flock, so the sheep were left alone even more than before. And growth now required placing sheep on what were traditionally "bear territories." In spite of this invasion, bear populations grew, and troubles multiplied.

What was K's reaction? Setting limits to human encroachments on bear territory? Direct application of his deep ecology perspective? Quite the contrary. He adopted a shallow wildlife management perspective which defended the sheepowners: more money in compensation for losses, quicker compensation, and immediate hiring of hunters to reduce the

---

* The MS has been edited by Professor George Sessions in 1984.

bear population. Other deep ecologists noted with concern the change of his public "image;" had K really abandoned his former value priorities? Privately he insisted: No. But, in public, he was silent.

The reason for K's unexpected actions was not difficult to find: the force of economic growth was so strong that the laws protecting bears would be changed in a direction highly unfavorable to the bears if the sheepowners were not soon pacified by accepting some of their demands. And some of their demands seemed reasonable. After all, it did cost a lot of money to hire and equip rescuers to locate a flock of sheep which had been harassed by a bear and, further, to prove the bear's guilt. And the bureaucratic procedures involved were time consuming. In short, K had not changed his basic value priorities at all. Rather, he had adopted a purely defensive compromise. He stopped promoting his deep ecological philosophy publicly in order to retain credibility and standing among opponents of his principles.

And what is true of K is true of thousands more. These people often hold responsible positions in society, where they might strengthen responsible environmental policy, but, given the exponential forces of growth, their publications are limited to narrowly professional and specialized concerns. Their writings are surely competent, but lack a deeper and more comprehensive perspective (although I admit that there are some brilliant exceptions). If professional ecologists persist in voicing their value priorities, their jobs are often in peril, or they tend to lose influence and status among those who are in charge of general policies. Privately, they may admit the necessity for deep and far-reaching changes, but they remain silent in public. As a result, their positive impact on the public has largely vanished. Deeply concerned people feel abandoned by the 'experts'.

In ecological debate many participants know a lot about particular conservation policies in particular places, and many others have strong opinions regarding fundamental philosophical questions of environmental ethics, but only a few have both qualities. When they are silent, the loss is formidable.

Let me illustrate again. A family of four decides to acquire four chairs for a small room, newly added to the home. They buy the chairs and all have peace of mind. But then one of them gets an urge to put ten more chairs into the room. Two of the family who are technically talented and eager to satisfy any "need," use their time to solve the sophisticated physical and mathematical problems involved. They ask the fourth member to work overtime to get the money to purchase the ten chairs. But she answers that the chairs are unnecessary for a life rich in intrinsic values and simple in means. She begins to argue for her view, but the two technocrats insist that first she should work through all the alternative solutions to the 10-Chair problem. At last she wonderfully simplifies the argument. If the ten chairs are not a desired end, it is pointless to discuss the means by which this might be achieved. The technically talented find other outlets for their surplus energy, for there are always enough legitimate problems to work on.

The complicated question of how industrial societies can increase energy production with the least undesirable consequences is of the same kind: a waste of time if the increase is pointless in relation to ultimate ends. When thousands of experts hired by government and other big institutions devote their time to this complicated problem, it is difficult for the public to learn that many of them judge the problem pointless and irrelevant. What is relevant, according to them, are the problems of how to stabilize and eventually decrease consumption without loss of life quality.

2. A Call to Speak Out

What I advocate and argue for is this: even those who completely subsume ecological policies under the narrow ends of human health and well-being cannot attain their more modest aims, at least not fully and easily, without being joined by supporters of deep ecology. They need what these people have to contribute, as this will work for them more often than it works against them. Those in charge of environmental policies, even if they are resource-oriented (and growth tolerating?) decision makers, will increasingly welcome if only for tactical and not fundamental reasons, what deep ecologists have to say. Even though the more radical ethic may seem nonsensical or untenable to them, they know that its advocates are doing in practice conservation work that sooner or later must be done. They concur with the practice, although they operate from diverging theories. If I am right, the time is ripe for professional deep ecologists to break their silence and freely express their deepest concerns. A bolder advocacy of deep ecology by those who are working within the shallow, resource-oriented 'environmental' sphere is the best strategy for regaining some of the strength of this movement among the general public, and thereby to contribute, however modestly, toward a turning of the tide.

What do I mean by saying that even the more modest aims of

shallow environmentalism have a need for deep ecology? We can see this by considering the World Conservation Strategy prepared by the International Union for Conservation of Nature and Natural Resources (IUCN) with the advice, cooperation and financial assistance of the United Nations Environmental Programme (UNEP) and the World Wildlife Fund (WWF). The argument in this important publication is through and through homocentric in the sense that all its recommendations are justified in terms of their effects upon human health and well-being. Even the recomended environmental ethic, with its attendant environmental education campaign, has humans in harmony with nature for human good. "A new ethic, embracing plants and animals as well as people, is required for human societies to live in harmony with the natural world on which they depend for survival and well-being."[2] Such an ethic would surely be more effective if it were acted upon by people who believe in its validity, rather than by those who merely believe in its usefulness. This, I think, will come to be understood more and more by those in charge of educational policies. Quite simply, it is indecent for a teacher to proclaim an ethic only for tactical reasons. Further, this point applies to all aspects of world conservation strategy. Conservation strategy will be more eagerly implemented by persons who love what they are conserving, and who are convinced that what they love is intrinsically lovable. Such lovers will not want to hide their attitudes and values, but rather will increasingly give voice to them in public. They have a genuine ethics of conservation, not merely a tactically useful instrument for social and political ends.

In short, environmental education campaigns can fortunately combine anthropocentric arguments with a practical land and sea ethic based either on a deeper and more fundamental naturalistic philosophical or religious perspective, and on a set of norms resting on intrinsic values. But the inherent strength of this overall position will be lost if those who work professionally on environmental problems do not give public testimony to these fundamental norms.

This article is hortatory, in the positive etymological sense of that word. I seek "to urge, incite, instigate, encourage, cheer" (Latin: hortari). This may seem unacademic in a philosophical journal, but I consider it justifiable because of an intimate relationship between hortatory sentences and basic philosophical views which I will formulate in Section 8 below.

3. What is Deep Ecology?

The term 'deep ecological movement' has so far been used without trying to define it. One should not expect much from definitions of movements. Think of terms like 'conservatism,' 'liberalism,' or 'feminist movement.' And there is no need that supporters should adhere to exactly the same definition. In what follows, a set of principles, or key terms and phrases, agreed upon by George Sessions and myself, are tenatively proposed as basic to deep ecology.[3]

(1) The well-being and flourishing of human and non-human Life on Earth have value in themselves (synonyms: intrinsic value, inherent value). These values are independent of the usefulness of the non-human world for human purposes.

(2) Richness and diversity of life forms contribute to the realization of these values and are also values in themselves.

(3) Humans have no right to reduce this richness and diversity except to satisfy vital needs.

(4) The flourishing of human life and cultures is compatible with a substantial decrease of the human population. The flourishing of non-human life requires such a decrease.

(5) Present human interference with the non-human world is excessive, and the situation is rapidly worsening.

(6) Policies must therefore be changed. These policies affect basic economic, technological, and ideological structures. The resulting state of affairs will be deeply different from the present.

(7) The ideological change is mainly that of appreciating life quality (dwelling in situations of inherent value) rather than adhering to an increasingly higher standard of living. There will be a profound awareness of the difference between big and great.

(8) Those who subscribe to the foregoing points have an obligation directly or indirectly to try to implement the necessary changes.

Comments on the Basic Principles:
RE (1):
> This formulation refers to the biosphere, or more accurately to the ecosphere as a whole. This includes individuals, species, populations, habitat, as well as human and nonhuman cultures. From our current knowledge of all-pervasive intimate relationships, this implies a fundamental deep concern and respect. Ecological processes on the planet should, on the whole, remain intact. "The world environment should remain 'natural' " (Gary Snyder).
>
> The term "life" is used here in a more comprehensive non-technical way to refer also to what biologists classify as "non-living": rivers

(watersheds), landscapes, ecosystems. For supporters of deep ecology, slogans such as "let the river live" illustrate this broader usage so common in most cultures.

Inherent value, as used in (1), is common in deep ecology literature ("The presence of inherent value in a natural object is independent of any awareness, interest, or appreciation of it by any conscious being.")[4]

RE (2):

More technically, this is a formulation concerning diversity and complexity. From an ecological standpoint, complexity and symbiosis are conditions for maximizing diversity. So-called simple, lower, or primitive species of plants and animals contribute essentially to richness and diversity of life. They have value in themselves and are not merely steps toward the so-called higher or rational life forms. The second principle presupposes that life itself, as a process over evolutionary time, implies an increase of diversity and richness. The refusal to acknowledge that some life forms have greater or lesser intrinsic value than others (see points 1 and 2) runs counter to the formulations of some ecological philosophers and New Age writers.

Complexity, as referred to here, is different from complication. Urban life may be more complicated than life in a natural setting without being more complex in the sense of multi-faceted quality.

RE (3):

The term "vital need" is left deliberately vague to allow for considerable latitude in judgment. Differences in climate and related factors, together with differences in the structures of societies as they now exist, need to be considered. (For some Eskimos, snowmobiles are necessary today to satisfy vital needs, not to tourists).

RE (4):

People in the materially richest countries cannot be expected to reduce their excessive interference with the non-human world to a moderate level overnight. The stabilization and reduction of the human population will take time. Interim strategies need to be developed. But this in no way excuses the present complacency. The extreme seriousness of our current situation must first be realized. But the longer we wait the more drastic will be the measures needed. Until deep changes are made, substantial decreases in richness and diversity are liable to occur: the rate of extinction of species will be ten to one hundred times greater than any other period of earth history.

RE (5):

This formulation is mild. For a realistic assessment of the situation, see the unabbreviated version of the I.U.C.N.'s World Conservation Strategy. There are other works to be highly recommended, such as Gerald Barney's **Global 2000 Report to the President of the United States.**

The slogan of "noninterference" does not imply that humans should not modify some ecosystems as do other species. Humans have modified the earth and will probably continue to do so. At issue is the nature and extent of such interference.

The fight to preserve and extend areas of wilderness or near-wilderness should continue and should focus on the general ecological functions of these areas (one such function: large wilderness areas are required in the biosphere to allow for continued evolutionary speciation of animals and plants). Most present designated wilderness areas and game preserves are not large enough to allow for such speciation.

RE (6):

Economic growth as conceived and implemented today by the industrial states is incompatible with (1) - (5). There is only a faint resemblance between ideal sustainable forms of economic growth and present policies of the industrial societies. And "sustainable" still means "sustainable in relation to humans."

Present ideology tends to value things because they are scarce and because they have a commodity value. There is prestige in vast consumption and waste (to mention only several relevant factors).

Whereas "self-determination," "local community," and "think globally, act locally," will remain key terms in the ecology of human societies, nevertheless the implementation of deep changes requires increasingly global action, action across borders.

Governments in Third World countries are mostly uninterested in deep ecological issues. When the governments of industrial societies try to promote ecological measures through Third World governments, then practically nothing is accomplished (e.g., with problems of desertification). Given this situation, support for global action through non-governmental international organizations becomes increasingly important. Many of these organizations are able to act globally "from grassroots to grassroots" thus avoiding negative governmental interference.

Cultural diversity today requires advanced technology, that is, techniques that advance the basic goals of each culture. So-called soft, intermediate, and alternative technologies are steps in this direction.

RE (7):

Some economists criticize the term 'quality of life' because it is supposed to be vague. But on closer inspection, what they consider to be vague is actually the non-quantitative nature of the term. One cannot quantify adequately what is important for the quality of life as discussed here, and there is no need to do so.

RE (8):

There is ample room for different opinions about priorities: what should be done first, what next? What is most urgent? What is clearly necessary as opposed to what is highly desirable but not absolutely pressing?

The above formulations may be useful for many supporters of the deep ecology movement. But others will certainly feel they are imperfect, even misleading. If they need to formulate in a few words what is basic in deep ecology, they will propose an alternative set of sentences. I shall of course be glad to refer to them as alternatives. There ought to be a measure of diversity in what is considered basic and common.

Should we call the movement 'the deep ecological movement'?[5] There are at least six other designations which cover most of the same issues: "Ecological Resistance", used by John Rodman in important discussions; "The New Natural Philosophy" coined by Joseph Meeker; "Eco-philosophy," used by Sigmund Kvaloy and others to emphasize (1) a highly critical assessment of industrial growth societies from a general ecological point of view and (2) the ecology of the human species; "Green Philosophy and Politics," while the term "green" is often used in Europe, in the United States "green" has a misleading association with the rather "blue" Green Revolution; "Sustainable Earth Ethics," as used by G. Tyler Miller; and "Ecosophy", eco-wisdom, which is my own favorite term. Others could also be mentioned.

Why use the adjective "deep"? This question will be easier to answer after the contrast is made between shallow and deep ecological concerns.

What I am talking about is not a philosophy in any academic sense, nor is it institutionalized as a religion or an ideology. Various persons come together in campaigns and direct actions. They form a circle of friends supporting the same kind of lifestyle, which others term "simple," but they themselves think is rich and many sided. They agree on a vast array of political issues, although they may otherwise support different political parties. As in all social movements, slogans and rhetoric are indispensible for ingroup coherence. They react together against the same threats in a predominately nonviolent way. Perhaps the most influential participants are artists and writers who do not articulate their insights in terms of professional philosophy, but do express themselves in art or poetry. For these reasons, I use the term 'movement' rather than 'philosophy.'

4. Deep versus Shallow Ecology

A number of key terms and slogans from the environmental debate will clarify the contrast between the shallow and deep ecology movements.

A. Pollution

Shallow approach: Technology seeks to purify the air and water and to spread pollution more evenly. Laws limit permissible pollution. Polluting industries are preferably exported to developing countries.

Deep approach: Pollution is evaluated from a biospheric point of view,[6] not centering on its effects on human health, but on life as a whole, including life conditions of every species and system. The shallow reaction to acid rain is to avoid action by demands of more research, demands to find species of trees tolerating high acidity etc., whereas the deep approach concentrates on what is going on in the total ecosystem and asks for high priority fight against the economy and technology responsible for acid rain.

The priority is to fight deep causes of pollution, not merely the superficial, short range effects. The third and fourth worlds cannot afford to pay the total cost of the war against pollution in their regions, and consequently they require the assistance of the first and second worlds. Exporting pollution is not only a crime against humanity, but also against life.

B. Resources

Shallow approach: The emphasis is upon resources for humans, especially for the present generation in affluent societies. On this view, the resources of the earth belong to those who have the technology to exploit them. There is confidence that resources will not be depleted because, as they get rarer, a high market price will conserve them, and substitutes will be found through technological progress. Further, animals, plants, and natural objects are valuable only as resources for humans. If no human use is known, they can be destroyed with indifference.

Deep approach: The concern here is with resources and habitat for all life forms for their own sake. No natural object is conceived of solely as a resource. This then leads to a critical evaluation of human modes of production and consumption. It is asked: to what extent does an increase here favor ultimate values in human life? To what extent does it satisfy vital needs, locally and globally? How can economic, legal, and educational institutions be changed to counteract destructive increases? How can resource use serve the quality of life rather than the economic standard of living as generally promoted in consumerism? There is an emphasis here on an ecosystem approach rather than just the consideration of isolated life forms or local situations. There is a long-range maximal perspective of time and place.

C. Population

Shallow approach: The threat of (human) overpopulation is seen mainly as a problem for developing countries. One condones or even cheers population increases in one's own country for shortsighted economic, military, or other reasons; an increase in the number of humans is considered a value in itself or as economically profitable. The issue of "optimum population for humans" is discussed without reference to the question of the "optimum population" of other life forms. The destruction of wild habitats caused by an increasing human population is accepted as an inevitable evil. Drastic decreases of wild life forms tend to be accepted as long as species are not driven to extinction. Animal social relations are ignored. The long term substantial reduction of the global human population is not seen as a desired goal. One has a right to defend one's own borders against "illegal aliens," no matter what the population pressures elsewhere.

Deep approach: It is recognized that excessive pressures on planetary life conditions stem from the human population explosion. The pressure stemming from industrial societies is a major factor, and population reduction must have a high priority in those societies, as well as in developing countries. Estimates of an optimal human population vary. Some quantitative estimates are 100 million, 500 million, and 1000 million, but it is recognized that there must be a long range, humane reduction through mild but tenacious political and economic measures. This will make possible, as a result of increased habitat, population growth for thousands of species which are now constrained by human pressures.

D. Cultural diversity and appropriate technology

Shallow approach: Industrialization of the kind manifested in the West is held to be the goal for developing countries. The universal adoption of Western technology is compatible with mild cultural diversity and the conservation of good (from the Western point of view) elements in present nonindustrial societies. There is a low estimate of deep cultural differences which deviate significantly from Western standards.

Deep approach: Cultural diversity is an analogue on the human level to the biological richness and diversity of life forms. We should give high priority to cultural anthropology in education in industrial societies. We should limit the impact of Western technology upon presently existing nonindustrial countries and defend the fourth world against foreign domination. Political and economic policies should favor subcultures within industrialized societies. Local, soft technologies will allow a basic cultural assessment of any technical innovations, freely criticizing socalled advanced technology and concepts of "progress."

E. Land and sea ethics

Shallow approach: Landscapes, ecosystems, rivers, and other wholes of nature are cut into fragments, disregarding larger units and gestalts. These fragments are regarded as the properties and resources of individuals, organizations, or states. Conservation is argued in terms of "multiple use" and "cost/benefit analysis." Social costs and long term ecological costs are not included. Wildlife management conserves nature for "future generations of humans." The erosion of soils or of ground water quality is noted as a human loss, but a strong belief in future technological progress makes deep changes seem unnecessary.

Deep approach: Earth does not belong to humans. The Norwegian landscapes, rivers, fauna and flora, and the surrounding sea are not the property of Norwegians. Humans only inhabit the lands, using resources to satisfy vital needs. If their nonvital needs conflict with the vital needs of nonhumans, humans might yield. The destruction now going on will not be cured by a technological fix. Current arrogant notions in industrial (and other) societies must be resisted.

F. Education and scientific enterprise

Shallow approach: The degradation of the environment and resource depletion necessitates the further training of experts who can advise on how to combine economic growth with maintaining a healthy environment. We are likely to need highly manipulative technology when global economic growth makes further degradation inevitable. The scientific enterprise must continue giving priority to the "hard" sciences. This necessitates high educational standards with intense competition in relevant "tough" areas of learning.

Deep approach: Education should concentrate on increased sensitivity to nonconsumptive goods and on such consumables as we have enough of for all, provided sane ecological policies are adopted. Education will therefore counteract the excessive valuation of things with a price tag. There should be a shift from concentration upon "hard" to "soft" sciences, stressing local and the global culture. The eductional objective of the World Conservation Strategy, "building support for conservation," should be accorded priority within the deeper framework of respect for the biosphere.

In the future, there will be no shallow movement, if shallow policies are increasingly adopted by governments and, thus, need no support by a special social movement.

5. But why a "Deep" Ecology?

The decisive difference concerns willingness to question and to appreciate the importance of questioning every economic and political policy in public. The questioning is "deep" and public. It asks "why" more insistently and consistently, taking nothing for granted. Deep ecology can readily admit the practical effectiveness of homocentric arguments. "It is essential for conservation to be seen as central to human interests and aspirations. At the same time, people -from heads of state to the members of rural communities- will most readily be brought to demand conservation if they themselves recognize the contribution of conservation to the achievement of their needs, as perceived by them, and the solution of their problems, as perceived by them."[7] Since most policies serving the biosphere also serve humanity in the long run, they may, at least initially, be accepted on the basis of narrow "homocentric" arguments.

But such a tactical approach has significant limitations. There are three dangers: some policies based on successful homocentric arguments turn out to violate or compromise unduly the objectives of deeper argumentation; the strong motivation to fight for decisive change and the willingness to serve a great cause is weakened; and the complicated arguments in human-centered conservation documents such as the World Conservation Strategy go beyond the time and ability of many people to assimilate and understand and also tend to provoke interminable technical disagreements among experts. Special interest groups with narrow short-term exploitative objectives which run counter to saner ecopolicies often exploit these disagreements and thereby stall the debate and steps toward effective action. When arguing from deep ecological premises, most of the complicated proposed technological fixes need not be discussed at all. The relative merits of alternative technology proposals in industrial societies concerned with how to increase energy production are pointless if our vital needs have already been met. The focus on vital issues activates mental energy and strengthens motivation. The shallow environmental approach, on the other hand, tends to make the human population more passive and disinterested in environmental issues.

The deep ecological movement tries to clarify the fundamental presuppositions underlying our economic approach in terms of value priorities, philosophy, and religion. In the shallow movement, argument comes to a halt long before this. The deep ecology movement is therefore "the ecology movement which questions deeper."

The terms 'egalitarianism,' 'homocentrism,' 'anthropocentrism,' and 'human chauvinism' are often used to characterize points of view on the shallow-deep ecology spectrum. But these terms usually function as slogans which are open to misinterpretation. They can imply that man is in some respects only a "plain citizen" (Aldo Leopold) of the planet on a par with all other species, but they are sometimes interpreted as denying that humans have any "extraordinary" traits, or that in situations involving vital interests, humans have no overriding obligations towards their own kind. They have!

In any social movement, rhetoric has an essential function in keeping members fighting together under the same banner. Rhetorical formulations also serve to provoke interest among the outsiders. Of the better known slogans, one might mention "Nature knows best," "Small is beautiful," and "All things hang together." But clearly all things in the universe do not hang together at the levels of quantum physics or relativity theory. The slogan only expresses a doctrine of global, not cosmic, relevance.

Only a minority of deep ecologists are academic philosophers, such as myself. While deep ecology need not be a finished philosophical system, this does not mean that its philosophers should not try to be as clear as possible. So a discussion of deep ecology as a derivational system may be of value.

6. Deep Ecology illustrated as a Derivational System

Underlying the eight tenets or principles presented in section three, there are still more basic positions and norms which reside in philosophical systems and various world religions. Schematically we may represent the total views implied in the movement by streams of derivation from the most fundamental norms and descriptive assumptions to particular decisions in actual life situations. See illustration in page 24.

This pyramidal model has some features in common with hypothetico-deductive systems. The main difference, however, is that some sentences at the top (=deepest) level are normative, and are preferably expressed by imperatives. This makes it possible to arrive at imperatives at the lowest derivational level, the crucial level in terms of decisions. Thus, there are oughts in our premises, as well as in our conclusions. We do not move from an is to an ought.

Just as in a hypothetico-deductive system in physics, where only the two upper levels of the pyramid are thought of as forming physics as a system, so also in normative systems, only the upper levels are considered to be part of the total system. The sentences in the lowest part are changing from day to day as life situations change.

The above derivational structure of a total view must not be taken too seriously. It is not meant in any restrictive way to characterize creative thinking within the deep ecological movement. That thinking moves freely in any direction. But some of us with a professional background in science and analytical philosophy find it helpful.[8]

Answers to ultimate questions, i.e., the highest normative principles and basic assumptions about the world, occur in the upper part of the derivational pyramid. The first three basic principles of deep ecology (see section 3 above) belong to the upper level of the pyramid because they assert, in a general way, life in its diversity as a value in itself, thus forming a norm against undue human interference. The next four (4-7) tenets belong to the middle region because they are more local, they view what is going on at the present. This involves factual claims and projections about the consequences of present policies in industrial and nonindustrial countries. An application of the last tenet (8) is at the lowest derivational level because it imposes an obligation to take part in actions to change policies. Such an obligation must be derivable from principles higher up in the pyramid.

24

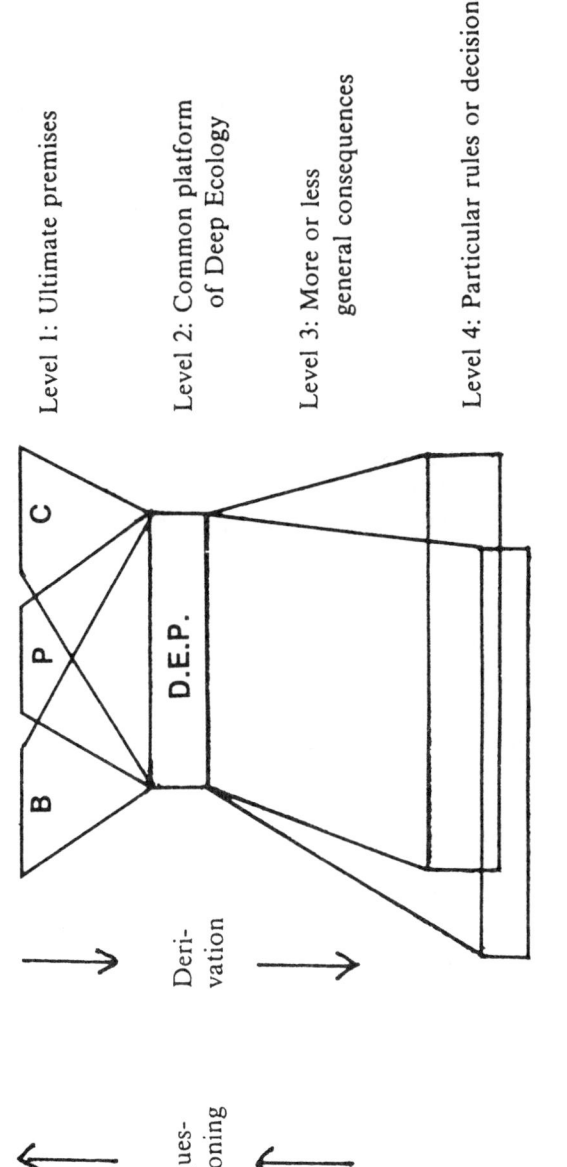

There are few propositions at the top of the pyramid, a great variety at the middle level, and innumerable recommendations at the bottom.

7. Multiple Roots of the Deep Ecology Principles

The deep ecological movement seriously questions the presuppositions of shallow argumentation. Even what counts as a rational decision is challenged, because "rational" is always defined in relation to specific aims and goals. If a decision is rational in relation to the lower level aims and goals of our pyramid but not in relation to the highest level, then the decision should not be judged to be rational. If an environmentally oriented policy decision is not linked to intrinsic values, its rationality is yet undetermined. The deep movement connects rationality with a set of philosophical and religious foundations. One cannot expect the ultimate premises to constitute rational conclusions. There are no "deeper" premises available.

The deep ecological questioning reveals the fundamental normative orientations. Shallow argumentation stops before reaching fundamentals or jumps from the ultimate to the particular, that is, from level 1 to level 4.

It is not only normative claims that are at stake. Most (perhaps all) norms presuppose ideas about how the world functions. Typically the vast majority of propositions needed in normative systems are descriptive. This holds of all levels.

Notice, however, that it does not follow that supporters of deep ecology must have, on ultimate issues, identical beliefs. They do have common attitudes about intrinsic values in nature, but these can, in turn (at a still deeper level), be derived from different, mutually incompatible sets of ultimate beliefs.

Thus, while a specific decision may be judged as rational from within the derivational system (if there is such) of shallow ecology, it might be judged irrational from within the derivational system of deep ecology. What is rational within the deep ecology derivational pyramid does not require unanimity in ontology and fundamental ethics. Deep ecology as a conviction, with its subsequently derived practical recommendations, can follow from several more comprehensive worldviews.

Those engaged in the deep movement have so far revealed their philosophical or religious homes mainly to be in Christianity, Buddhism, Taoism, or philosophy. The top level of the derivational pyramid can therefore be made up of normative and descriptive principles which belong to forms of Christianity, Buddhism, Taoism, and various philosophical creeds.

Since the late seventies, numerous Christians in Europe and America, some of them teachers of theology, have taken part actively in the deep ecological movement. Their interpretations of the Bible and their theological positions in general have been reformed from what was, until recently, a dominating crudely anthropocentric emphasis within Christianity.

There is an intimate relation between some forms of Buddhism and the deep ecological movement. The history of Buddhist thought and practice, especially the principles of non-violence, non-injury and reverence for life, sometimes makes it easier for Buddhists to understand and appreciate that movement than it is for Christians, despite a (sometimes overlooked) blessedness which Jesus recommended in peace-making. I mention taoism chiefly because there is some basis for calling John Muir a Taoist.[9]

Ecosophies are not religions in the classical sense, but are general philosophies inspired by ecology. In the next section I will further introduce Ecosophy T.

The adherents of different religions and philosophies disagree and may not even untimately understand each other at the foundational levels of conviction and experience. But they can have important derived views in common, and these, though themselves derived, are nevertheless deep enough to form what I wish to call the upper level of the deep ecology derivational pyramid.

Some have worried that the mixture of religion and environmentalism could prove a source of dogmatism, intolerance, and "mysticism" (in the sense of obscurantism). So far, there is no evidence that this is happening. Nature mysticism has little to do with obscurantism.[10]

8. Ecosophy T

The main theoretical complaint against shallow ecology is not that it is based on a well-articulated but incorrect philosophical or religious foundation. It is, rather, that there is a lack of depth -or complete absence- of guiding philosophical or religious foundations.

In his excellent book on how to "live in the environment," G. Tyler Miller writes:

> The American attitude (and presumably that of most industrialized nations) toward nature can be expressed as eight basic beliefs [four of which are reproduced here].

1. Humans are the source of all value.
2. Nature exists only for our use.
3. Our primary purpose is to produce and consume. Success is based on material wealth.
4. Production and consumption must rise endlessly because we have a right to an ever increasing material level of living.

But he adds an important reservation:

Although most of us probably would not accept all of these statements, we act individually, corporately, and governmentally as if we did—and this is what counts.[11]

When they are so baldly exposed, we might find that very few persons would actually subscribe to what Miller characterizes as "the American attitude." Nevertheless, as Miller notices, most modern people (and not only Americans!) behave as if they believed such a creed. There is no articulated philosophical or religious view from which "the American attitude" is carefully justified.

Referring back to the illustration, the shallow movement has not offered examples of total views comprising the four levels. I am tempted to say that there will be no examples. Serious attempts to find a deep justification of the way life on the planet is treated today (including the threats of using nuclear "weapons") are doomed to failure. What I say is meant as a challenge: is there a philosopher somewhere who would like to try?

My main purpose in announcing that I feel at home in "Ecosophy T" is didactic and dialectic. I hope to get others to announce their philosophy. If they say they have none, I maintain that they have, but perhaps don't know their own views, or are too modest or inhibited to proclaim what they believe. Following Socrates I want to provoke questioning until others know where they stand on basic matters of life and death. This is done using ecological issues, and also by using Ecosophy T as a foil. But Socrates pretended in debate that he knew nothing. My posture seems to be the opposite. I may seem to know everything and to derive it magically from a small set of hypotheses about the world. But both interpretations are misleading! Socrates did not consistently claim to know nothing, nor do I in my Ecosophy T pretend to have all that comprehensive a knowledge. He claimed to know, for instance, about the fallibility of humans claims to know.

So, here is Ecosophy T!

Its fundamental norm is 'Self-realization!' But I do not use this expression in any narrow, individualistic sense. I want to give it an expanded meaning based on the distinction between Self and self conceived in certain Eastern traditions of **âtman**, comprising all the life forms, and selves (jivas) as usually interpreted in social and personal life. Using five words: maximum (long range, universal) Self realization![12] If I had to give up the term fearing its inevitable misunderstanding, I would use the term 'symbiosis'. "Maximize Self-realization!" could be interpreted in the direction of colossal egotrips. But "Maximize symbiosis!" could be interpreted in the opposite direction, that of the elimination of individuality in favor of collectivity.

Viewed systematically, not individually, maximum Self-realization implies maximizing the manifestations of life. So I next derive the second term, "Maximize (long range, universal) diversity!" A corollary is that the higher the levels of Self-realization which are attained by a person, the more any further increase depends upon the Self-realization of others. Increased self-identification is increased identification with others. "Altruism" is a natural consequence of this identification.

This leads to an hypothesis about an inescapable increase of identification with other beings when one's own self-realization increases. We increasingly see ourselves in others, and others in ourselves. The self is extended and deepened as a natural process of the realization of its potentialities in others.

Universalizing, we can derive the norm, "Self-realization for every being!" From "Diversity!" and a hypothesis that maximum diversity implies a maximum of symbiosis, is derived the norm "Maximum symbiosis!" Further, we work for life conditions such that there is a minimum of coercion in the life of others. And so on![13]

A philosophy as a world view inevitably has implications in practical situations. Ecosophy T therefore without apology moves on to concrete questions of life style. These will obviously show great variation because of differences in hypotheses about the world in which each of us living, and in the 'factual' statements about the concrete situation in which we make a decision. I shall limit myself to a couple of areas where my "style" of thinking and behaving seem somewhat strange to friends and others who know a little about my philosophy. Firstly, a somewhat extreme appreciation of diversity; positive appreciation of the existence of styles and behaviors which I personally detest or find nonsensical (but not clearly incompatible with symbiosis); enthusiasm for "the mere" diversity of species or varieties within a genus of plants or animals; support, as the head of a department of philosophy, of doctrinal theses completely at odds with my own inclinations, with only the requirement that the authors are able to understand fairly adequately some basic features of the kind of philosophy I myself feel at home with; combination of **seemingly** incompatible interests and behaviors, which makes for an increase of subcultures within industial states and might to some extent help future cultural diversity. So much about "diversity!"

Secondly, I have a somewhat extreme appreciation of what Kant calls beautiful actions (good actions based on inclination), in contrast to dutiful ones. The choice of the formulation 'Self-realization!' is in part motivated by the belief that maturity in humans can be measured along a scale from selfishness to Selfishness, that is, broadening and deepening the self, rather than measures of dutiful altruism. I see joyful sharing and caring as a natural process (which, I regret, is somewhat retarded in myself).

Thirdly, I believe that many-sided, high level Self-realization is more easily reached through a "spartan" life-style than through the material standard of average citizens of industrial states.

The simple formulations of the deep ecology platform and Ecosophy T are not meant primarily to be used among philosophers, but in dialogues with "the experts." When I wrote to them personally, asking whether they accept the 8 points of the platform, many answered positively in relation to most or all points. And this includes top people in ministries of oil and energy! But it is still an open question to what extent they are willing to let their written answers be widely published. It is also an open question to what extent they try to influence their colleagues who use only shallow argumentation. The main conclusion is moderately encouraging: there is a philosophy of the man/nature relationship widely accepted among established experts responsible for environmental decisions which requires a pervasive, substantial change of present policies -in favor of our "living" planet, and not only for shortsighted human interests.

FOOTNOTES

1. For more about interspecific community relationships, see Arne Naess, "Self-realization in Mixed communities of Humans, Bears, Sheep, and Wolves," **Inquiry** 22 (1979): 321-341.
2. Quotation from Section 13, "building support for conservation."
3. I cannot here do justice to the many authors who have contributed to the understanding of the emerging deep ecology movement. Only two will be mentioned. The newsletters written by George Sessions, Dept. of Philosophy, Sierra College, Rocklin, CA, are indispensable. There are six letters, April 76, May 79, April 81, May 82, May 83, May 84, about 140 pages in all. The significant contributions by poets and artists are fully recognized. Most of the materials are summarized by Sessions in "Shallow and Deep Ecology: A Review of the Philosophical Literature," in an excellent collection of articles, **Ecological Consciousness,** eds. R.C. Schultz and J.D. Hughes (Washington: University Press of America, 1981). Bill Devall provides a short survey, in part historical, in his potent article, "The Deep Ecology Movement," **Natural Resources Journal** 20 (1980). See also Devall and Sessions, **Deep Ecology: Living As If Nature Mattered.** (Layton, Utah: Peregrine Smith Press, 1984).
4. Tom Regan, "The Nature and Possibility of an Environmental Ethics," **Environmental Ethics** 3 (1981): 19-34, citation on p. 30.
5. I proposed the name 'Deep, Long-Range Ecology Movement' in a lecture at the 3rd World Future Research conference, Bucharest, September 1972. A summary of that lecture: "The Shallow and the Deep, Long-Range Ecology Movement" was published in **Inquiry** 16 (1973): 95-100. Within the deep ecology movement it is fairly common to use the term 'deep ecologist', whereas 'shallow ecologist', I am glad to say, is rather uncommon. Both terms may be considered arrogant and slightly misleading. I prefer to use the awkward, but more egalitarian expression 'supporter of the deep (or shallow) ecology movement', avoiding personification. Also, it is common to call deep ecology consistently anti-anthropocentric. This has led to misconceptions, see my "A Defence of the Deep Ecology Movement," **Environmental Ethics** 5 (1983).
6. The technical term 'biospheric' should perhaps be avoided because it favors the scientifically fruitful distinction between biosphere and ecosphere. I use the term 'life' in a broad sense common in every day speech, and may therefore speak of landscapes and larger systems of the ecosphere as "living" -ultimately speaking of the life of the planet. The biospheric point of view referred to in the text is not a narrower point of view than the ecospheric because bios is used in a broad sense.

7. **World Conservation Strategy,** section 13, concluding paragraph.

8. Many authors take some steps towards derivational structures, offering mild systematizations. The chapter on "environmental ethics and hope" in G. Tyler Miller, **Living in the Environment,** 3rd ed. (Belmont: Wadsworth, 1983) is a valuable start, but the derivational relations are unclear. The logic and semantics of simple models of normative systems is briefly discussed in my "Notes on the Methodology of Normative Systems," **Methodology and Science** 10 (1977): 64-79. For defense of the thesis that as soon as persons assert anything at all we assume a total view, implicit with ontology, methodology, epistemology and ethics, see my "Reflections about Total Views," **Philosophy and Phenomenological Research** 25 (1964-65): 16-29. The best and wittiest warning against taking systematizations too seriously is to be found in Soren Kierkegaard, **Concluding Unscientific Postscript.**

9. Trusting Bill Devall, one may say that "Muir is now understood as the first Taoist of American ecology." Devall, "John Muir as Deep Ecologist," **Environmental Review** 6 (1982); see also Michael Cohen, **The Pathless Way: John Muir and American Wilderness,** (Madison: Univ. of Wisconsin Press, 1984).

10. For empirical studies of attitudes of "Wilderness-users," see the survey by Chris. R. Kent (16438 Clymer St., Granada Hills, CA 91344) in his thesis "The Experiential Process of Nature Mysticism...," Humboldt State Univ., 1981.

11. **Living in the Environment,** 489.

12. The term **âtman** is not taken in its absolutistic senses, not as a permanent indestructible "soul." This makes it consistent with those Buddhist denials (the **avâtman doctrine**) that the **âtman** is to be taken in absolutist senses. Within the Christian tradition some theologians distinguish "ego" and "true self" in ways similar to these distinctions in Eastern religion. See the ecophilosophical interpretation of the gospel of Luke in Stephen Verney's **Onto the New Age,** (Glasgow: Collins 1976) 33-41.

13. For criticism and defence of this fundamental norm, and my answer, see **In Sceptical Wonder, Essays in Honor of Arne Naess,** (Oslo: University Press, 1982). My main exposition of Ecosophy T was originally offered in the Norwegian work **Okologi, samfunn og livsstil,** (Oslo: University Press, 5th ed. 1976). Even there, the exposition is sketchy.

[10]

Conservation and Self-Realization: A Deep Ecology Perspective

Freya Mathews*

> Nature in its wider cosmic sense is not at risk from human exploitation and predation. To see life on Earth as but a local manifestation of this wider, indestructable and inexhaustible nature is to shield ourselves from despair over the fate of our Earth. But to take this wide view also appears to make interventionist political action on behalf of nature—which is to say, conservation—superfluous. If we identify with nature in its widest sense, as deep ecology prescribes, then the "self-defence" argument usually advanced by deep ecologists in support of conservation appears not to work. I argue that the need for eco-activism can be reconciled with a rejection of despair within the framework of deep ecology, and that in the process of this reconciliation the meaning of the term *conservation* acquires a new, spiritual dimension.

If we adopt the more "spiritual" approach to nature that deep ecology prescribes, will we in fact find that the kind of interventionist political action involved in conservation is no longer appropriate? Will we come to see that nature, in its widest sense, stands in no need of our "protection"? Does the same chest-beating self-importance lie behind the urge to conserve, guard, look after things, as lies behind the urge to blow everything sky-high? Is it the same human hubris, the same conviction that we are somehow outside of nature, that leads us to think that we can either destroy or "save" it? Does not destructiveness fall under the "law" of nature, just as much as creativeness does? Does not humanity the exploder and wrecker mirror nature just as faithfully as humanity the worshipful creator and conserver? Does deep ecology in fact require of us this kind of Zen-like surrender to the reality of our own aggressive tendencies, an acceptance of ourselves as a natural phenomenon—albeit a natural disaster perhaps—on a par with ice ages and interplanetary collisions, events which our Earth, Gaia, has incidentally succeeded in turning to the long-term advantage of life?

To see nature in this way, as able to take care of itself, as ultimately out of our destructive reach, poses a dilemma for conservationists: if we identify deeply enough with such an indestructable nature, seeing our Earth as a single man-

---

* Department of Philosophy, University of Melbourne, Parkville, Victoria, Australia 3052. Mathews is currently completing a book titled *Self in a Seamless World*.

ifestation of an infinite, inexhaustible principle, a cosmic principle of life, then this alleviates our angst and despair at the prospect of ecocatastrophe, because it means that ecocatastrophe does not spell the "death of nature" in its widest sense. On the other hand, such a deeper understanding of nature, which enables us to be life-affirming in the face of ecocatastrophe, seems to obviate the need for conservation, for if Gaia is but a single manifestation of a deeper and inexhaustible principle of life, then it no longer matters whether or not Gaia is preserved.

I raise this dilemma because I think that it is vital that we do not feed that soul-destroying angst and despair to which we as conservationists tend to be prey—and I am speaking of the despair that we feel not merely at the prospect of our own extinction, but at the awesome prospect of the end of nature, the end of the very Wheel of Birth and Death. I think that it is vital for our own spiritual health, and for the health of the environmental movement, that we reassure ourselves that we cannot bring about such a death of nature. Life-affirming people cannot be drawn to a movement which lives forever in the shadow of the belief that "the end is nigh"—and life-affirming people are, I try to show, the kind of friends that Gaia needs right now.

Indeed, I think that we cannot hope to change Western civilization, in the ways that we as conservationists dream of, if we constantly threaten it with the wholesale ecocidal consequences of its aggressive activities. From a psychotherapeutic point of view such threats are likely to bring about denial and repression rather than growth and enlightenment. Just as the young child needs to be reassured that its angry tantrums cannot really destroy the needed—and beloved—mother, so I think that Western civilization, at this stage, needs a degree of reassurance rather than terrorization. The child who is frightened of its anger represses and denies it, thereby giving it a life of its own and a license to act out its impulses in a literal and dissociated fashion. If Western society is too frightened of the consequences of its exploitation of nature, it will simply refuse to claim those consequences; it will banish its aggressive tendencies to the night, to unconsciousness, where they will be free to take their most literal, dissociated, and barbaric form. As psychotherapists always insist, you cannot break through to the next stage of consciousness unless you embrace your devil, and you cannot embrace your devil if you are too terrified of its destructiveness. Yet if you do not embrace it, but resist it, not only can you not break through, but you feed its power. The attitude that we may need to assume, as people seeking to shift the values of Western civilization, is thus not one of point-blank rejection of the evils embodied in this civilization, but a recognition of their origins in nature itself, and an acceptance and assimilation of them. This evil will then find its expression in integrated, perhaps symbolic, forms, and will not be left to act itself out in autonomous fashion.

The disadvantage of arguing that nature cannot be destroyed by human agency is, however, as I have said, that it appears to obviate the need for conservation. I think that deep ecology does indeed provide the reassurance that we need vis-à-vis the viability of nature, and protects us from despair, while at the same time providing an ongoing impetus for conservation—though we may find that it brings about a certain "deepening" of the meaning of conservation.

To begin at the beginning, then, let us approach that inevitable question, "What *is* deep ecology?" Saying what deep ecology is is a bit like trying to say what Christianity is, or science, or Marxism: the basic principles that were initially set forth as the axioms of deep ecology[1] are so extraordinarily rich in meaning, and figure in so many traditions of thought, that they inevitably accrete different layers of meaning with each interpreter. I focus on that principle of deep ecology which is most relevant to the issue that I have just outlined—and which, in my view, constitutes the very heart of deep ecology philosophy—self-realization.

Deep ecology is concerned with the metaphysics of nature, and of the relation of self to nature. It sets up ecology as a model for the basic metaphysical structure of the world, seeing the identities of all things—whether at the level of elementary particles, organisms, or galaxies—as logically interconnected: all things are constituted by their relations with other things. How exactly this view is supposed to cash out within the framework of theoretical physics is still uncertain, though hints and clues abound. In the framework of biological ecology, however, the idea of interconnectedness is, of course, completely at home: our ecological definition, or identification, of a blue whale, for instance, will involve reference to krill, and our definition of krill may involve reference to the blue whale. From the viewpoint of the interconnectedness thesis, organisms are not logically constituted by their physical structure only, but by their relations to the elements of their environment: organisms are seen as essentially interactive beings, logically incapable of existing independently of other beings.

Deep ecology takes this kind of interconnectedness as absolutely fundamental to the identity of things—as nonreducible to any kind of mechanistic base. This has the further consequence of rendering the relation between part and whole holistic as opposed to merely aggregative. Let me explain this. If the interconnected elements are seen as constituting a greater whole, then each element, being logically constituted by its relations with the other elements, is conditioned by the whole. The elements are given simultaneously with the whole, since they cannot be given independently. In this sense, the relation between part and whole is genuinely holistic, in that the parts constitute the whole, but are not given independently of it, i.e., the nature of the whole

---

[1] The axioms of deep ecology were initially set forth by Arne Naess in his now historic paper, "The Shallow and the Deep, Long-Range Ecology Movement," *Inquiry* 16 (1973): 95–100. They are elaborated by Bill Devall and George Sessions in *Deep Ecology: Living as if Nature Mattered* (Salt Lake City: Peregrine Smith, 1985).

conditions the parts. Contrast this with the ordinary—atomistic—conception of the relation between part and whole. In the ordinary way of thinking, the parts exist independently of the whole which they aggregatively constitute: while the parts determine the nature of the whole, the whole in no way determines the nature of the parts.

It should be pointed out that interconnectedness does not imply that organisms do not possess a genuine individuality: their functional unity confers on them an essential ontological distinctness and integrity, but this individuality is strictly relative—it is itself a function of the particular environment which is capable of sustaining such a self-realizing, self-maintaining system. A relative ontological individuality, on the one hand, and interconnectedness, on the other, are thus not in this framework mutually exclusive; on the contrary, they entail each other.

Applying this principle of interconnectedness to the human case, it becomes apparent that the individual denoted by "I" is not constituted merely by a body or a personal ego or consciousness. I am, of course, partially constituted by these immediate physical and mental structures, but I am also constituted by my ecological relations with the elements of my environment—relations in the image of which the structures of my body and consciousness are built. I am a holistic element of my native ecosystem, and of any wider wholes under which that ecosystem is subsumed. It is accordingly part of my essence that I stand in certain relations to the relevant elements of my environment. Since this is part of my essence, I cannot be said to flourish, to actualize the potentialities of my nature—in a word, to be fully self-realized—unless I do stand in these relations. Moreover, in order for me to stand in these relations, and thus achieve self-realization, the elements in question must exist. Thus, it is in *my* interests to ensure that *those* elements exist.

Through such reasoning we may be led to a tenet of deep ecology I call the *identification thesis*, according to which I identify with the wider systems of nature: if my identity is logically interconnected with the identity of other beings, then, as I have explained, my chances of self-realization depend on the existence of those beings.[2] It is therefore in my interest (viz., my interest in self-realization) to ensure their existence, and in this sense our interests converge. I identify with the wider systems of nature in the sense that I recognize their interests as my own.

There is a very important assumption involved in this line of reasoning, which is not universally accepted or acknowledged in the literature of deep ecology, namely, that the wider systems of nature do possess interests. That is to say, interconnectedness alone is a necessary but not a sufficient condition for the identification of self with nature. It is logically possible for a world of rocks to

---

[2] See for example, Arne Naess, "Identification as a Source of Deep Ecological Attitudes," in Michael Tobias, ed., *Deep Ecology* (San Diego: Avant Books, 1985).

exhibit interconnectedness—perhaps matter, in this world, is just a particular type of topological disturbance in a substantival spatial continuum. In this case, a topological disturbance, or rock, in one region of space is "interconnected" with topological disturbances, or rocks, in other regions, in the sense that a configuration in one region of this elastic space helps to determine the configurations in the surrounding regions. But this "interconnectedness" does not entail the identity of a given rock with the rock world as a whole; nor does it make sense to say that one rock can or should "identify" with others, since the rock lacks interests, as does the rock world as a whole, and to "identify with" another is essentially a matter of it assuming the interests of the other.

This point raises the question of what it means to say that a being has interests. A being may be said to have interests if it has needs, and it may be said to have needs if it is seeking to maintain or realize its own existence—it *needs* those things which contribute to its self-maintenance or self-realization. A rock cannot be said to have interests because it has no investment in its own existence—it in no way resists causal inroads into its physical integrity. The rock possesses a contingent unity only—it waits passively to be ground to dust, or worn away by wind and rain. An organism, in contrast, is essentially a system for realizing and perpetuating itself, and the more complex it is the more efficiently it resists the forces of disintegration. To have interests, then, is to be ruled by the principle of *self-realization*, a principle to which medieval philosophers referred as the "conatus." The conatus was the impulse, not only for self-preservation or self-maintenance, but also for self-increase or self-perfection—an impulse that is present in all living beings. Spinoza defined it as the "effort by which each thing endeavours to persist in its own being."[3] Clearly the conatus includes the impulse to realize in full the potentialities of one's nature or logical essence as well as the impulse merely to survive on a physical level, i.e., to maintain a certain physical structure. It is in this fuller sense that it is in our interests to promote the interests of the wider systems of nature—and clearly the identification of self with nature is only possible when nature is itself possessed of conatus.

What justification can be offered for this apparently counterscientific idea of nature as animated by a principle of telos, or a will-to-exist? In many of the expositions of deep ecology this idea is present merely as an implicit axiom, accepted on intuitive rather than rationalistic grounds. I think that some justification is called for, and I have attempted to construct one elsewhere.[4] As I do not wish to enter into that argument here, however, I shall just accept, for the purposes of our present discussion, that nature in its widest cosmic sense is

---

[3] Spinoza, *The Ethics*, trans. R. H. M. Elwes (New York: Dover Publications, 1951), part 3, prop. 7.

[4] This justification occurs in a book on ecological metaphysics which I am presently completing.

indeed a being endowed with conatus, a self-realizing entity. I use the term *ecocosm* to characterize such a self-realizing, internally interconnected cosmos.

What then is the normative significance of the identification thesis? Does our identification with the widest possible whole—with the ecocosm—entail that we should practice conservation? This is certainly the implication that has been assumed to hold within the literature of deep ecology. It has generally been argued that my identification with the wider wholes of nature entails that I will defend them when they are under attack. Conservation is in this way seen to be purely a matter of self-defence.

Before we can evaluate this argument, we need to distinguish between different levels of nature, which are also, in the light of the identification thesis, levels of my self. Or rather, identification may be pictured in terms of a series of concentric circles: the center point is my immediate bodily self, the next circle my ecological self, and the final circle is my cosmic self. Looking at our question in the light of these widening circles of self, we ask: does achieving identification with the cosmic self cancel out my need to defend the more immediate circles of self—the bodily and the ecological. From the viewpoint of the ecocosm, does it matter whether local individuals and systems are conserved or not? If it does not matter to the ecocosm whether local individuals or systems are conserved, does my identification with the ecocosm imply that the fate of individuals and systems—including myself and my own system—does not matter to me? In other words, if I become sufficiently spiritual in my approach to nature, and see nature as a whole as my true Self, does this cancel out my natural impulse to defend and maintain my individual self—and hence too the ecosystem on which this self depends?

In order to answer this question, we have first to settle the question whether the fate of particular individuals affects the self-realization of the ecocosm. On the face of it, the answer appears to be no. Even when the cosmos is viewed—as, I have argued, deep ecology must view it—as a self-realizing being, which has to work at keeping itself in existence, it cannot be denied that the process of cosmic self-realization involves a continuous arising and passing away at the level of particular forms. Indeed the death and decay of particulars is obviously as intrinsic to this process as their reemergence in new forms: new life arises from the often violent destruction and deterioration of the old. Planets are born out of the shattered remains of blown-out stars. Elementary particles are constantly created and annihilated in the endless dance of energy at the subatomic level. If this is the "law" of the ecocosm, then when we adopt the ecocosmic viewpoint, the arising and passing away of individual life-forms will not bother us, since it is consistent with the continued existence of our true—cosmic—self.

To argue in this way, however, is, I think, to misconstrue the identification thesis. Certainly there is a psychological sense in which I can "identify with" anything, or with any being with interests, so that the interests of that being displace my own. For instance, I could, in a psychological sense, identify with someone who wanted to kill me, and then it could be said that it would be "in my interest" for me to be killed, in the sense that I have taken on the other person's interests as my own. But such psychological identifications may be arbitrary or irrational—as the tendency to make bizarre identifications in psychosis shows. The identification thesis put forward in deep ecology is rationally justified, and it supervenes on interconnectedness, as I explained earlier: I am only required to identify with a system, S, if my identity, and hence my self-realization, are bound up with S. In other words, the motivation for identifying with S is that S's interests complement mine. If it turns out that S's interests in fact conflict with mine, then this justification for the identification dissolves, for interconnectedness is not, as we saw, equivalent to complete identity: interconnectedness is just interconnectedness, no more.[5] Identification, in the context of deep ecology, is premised on a convergence of interests.

Hence, my essence, and therefore my chances of self-realization, are bound up with the fact that the ecocosm permits me to identify with it. The fact that my individual fate and the fate of my local ecosystem are not necessarily of great moment to the ecocosm does not mean that I should, on the strength of my cosmic identification, give up responsibility for self and ecosystem, for it was the interests of this immediate self which prompted the identification in the first place. My identification with the more immediate circles of self, viz., ecosystem and biosphere, thus requires that I defend and maintain them, and this self-defence argument is *not* nullified by identification with a cosmic whole that is tolerant of destruction at the level of particulars.

Although I have said that the ecocosm appears to be unscathed by the passing away of particulars, this is not to say that it is invulnerable, indestructible. It is, as a self-realizing being, like us, sustained by will, not cause, and will, unlike cause, is capable of faltering, failing. If the will falters, then the fabric of nature may indeed begin to unravel. However, the only way in which the cosmic will could be undermined by us would be if we experienced a radical failure in our own conatus, our own will-to-exist, for it is perhaps by way of our conatus, rather than through our physical fate, that we contribute to the cosmic flourishing, that we truly participate in the cosmic essence. If we developed a will-to-oblivion in place of conatus, as in the Arthurian myth of the wasteland,[6] then indeed might the universe begin to "sicken." (Note that will-to-oblivion is not the

---

[5] Even a Leibnizian monad is not strictly *identical with* the manifold of all other monads, even though the individual monad is wholly and logically constituted by its relations with the other monads. A monad may properly "identify with" the manifold of monads, inasmuch as the existence of the other monads is a condition for its self-realization.

[6] The Arthurian myth of the wasteland is one example of the recurring archetype of the land which has lost the will-to-live, in which fertility declines to zero, and all beings turn away from life.

same as will-to-destruction. The latter is a form of will-to-power, which is in turn a form of will-to-exist.) It is clear, however, that whatever it is that is at present wrong with our culture, it is not our conatus which is at fault: our impulse toward self-maintenance is still robust. From the point of view of deep ecology, what is wrong with our culture is that it offers us an inaccurate conception of the self. It depicts the personal self as existing in competition with and in opposition to nature. Our conatus is accordingly unduly limited, and therefore cannot succeed either in its long-term aim of maintaining us in existence or in its immediate aim of self-realization. If we destroy our environment, we are destroying what is in fact our larger self. We commit this mistake because we are suffering from a maladaptation in the form of a faulty belief system that misrepresents our identity to us. Like many another species healthily eager to live and flourish, yet handicapped by a functional flaw, we are innocently selecting ourselves out of existence. In doing so we are not, as I have said, letting nature down, any more than deer with outsize antlers are. Nor are we even letting down our ecosystem, since in light of our interconnectedness, our dysfunction is really its, and it is through us that it is selecting *itself* out of existence. Since our maladaptation is correctable, however, in a way that outsize antlers are not, our conatus calls on us to rectify it, to change our belief system to serve our ultimate interests.

Although I have said that our conatus is clearly still functioning, it is important to recognize that conatus is a matter of degree. When our sense of self is diminished, then our sense that the self is worth preserving must also diminish. A strong conatus requires a strong, rich sense of self, which means a self encompassing the widest possible circles of being. The stronger our conatus, the greater our contribution to the flourishing of the ecocosm. Like any self-realizing being, the universe may achieve varying degrees of self-realization. At present the sense of self entailed by the world view of Western culture is clearly relatively cramped and poor. By correcting this world view, and expanding our conatus, we might be able to help to realize the ecocosm more fully.

Many traditional or primal peoples have shared this sense that our *attitude* to nature, to life, is important to the welfare of the cosmos, not merely in the sense that a sympathetic attitude brings about ecologically desirable consequences, but in the more spiritual sense that our very affirmation helps to sustain the universal process. John Collier, in his book on American Indian cultures, *On the Gleaming Way*, quotes the Hopi belief that to be effective in sustaining the natural order,

> man must participate not merely by performing certain rites at prescribed intervals in certain ways; but he must also participate with his emotions and thoughts, by prayer and willing. . . . Hopi traditional philosophy therefore ascribes to man a purposive, creative role in the development of his will. The universe is not conceived as a sort of machine at the mercy of the mechanical law. Nor is it viewed as a sum total of hostile, competitive forces struggling for existence. It is by nature

a harmonious, integrated system operating rhythmically according to the principles of justice, and in it the key role is played by man.[7]

Collier comments on the "intense romanticism" of the religion of the southwest Indians, which teaches that "the universe is a living being, and that the universe requires of man an inner concentration and a sustained action of desire and will, to the end that the universe itself may 'carry on'."[8]

I do not think that deep ecology sees human will as *central* to cosmic self-realization. To do so would be anthropocentrism in another form, to which deep ecology would have to be opposed. Nevertheless, I do think that deep ecology permits us to believe that we can enhance nature by helping to increase the cosmic conatus. This is not a matter of conservation at the physical level, though such conservation is, as we have seen, a commitment of the identification thesis. It is rather a matter of attitude, a spiritual matter, calling for an outright affirmation of nature, that may be expressed in an infinite number of possible ways—from the private revering of a flower in hand, through the devotional study of the natural sciences, to full-blown ritual celebrations in the style of the American Indians. Indeed, the affirmation of nature in its widest sense may be present in every thought and gesture, may animate every detail of our lives— leading to a life which is lived in the image of the cosmic law, viz., the conatus, the will-to-exist. Such a life may indeed strengthen the pulse of the cosmos, and this may be the deeper meaning of conservation: to cultivate the affirmation of life and to sustain it in one's heart for a lifetime. This is why I said, at the start, that the friends that nature needs—now and always—are life-affirming people.

My conclusion, then, is that deep ecology does enable us to deepen our sense of self, to an almost unimaginable degree, and this identification with the underlying, cosmic principle of life does enable us to confront the possibility of the extinction of life on Earth with equanimity, and in a life-affirming way. Cultivating this equanimity is perfectly consistent with our taking all reasonable steps to preserve life on Earth. Such equanimity in the face of ecocatastrophe no more obviates the need for conservation than the fact that one has the capacity to face one's own personal death in a positive, life-affirming way obviates the need for one to take all reasonable steps to preserve one's life. And the celebration of existence inspired by deep identification is not only important for personal sanity, but is, I have suggested, the deeper dimension of conservation itself.

---

[7] John Collier, *On the Gleaming Way* (Denver: Sage Books, 1949), pp. 101–02.
[8] Ibid., p. 29.

[11]

Man Apart: An Alternative to the Self-Realization Approach

Peter Reed*

> Seeing nature as ultimately separate from us rather than as a part of us is the source of a powerful environmental ethic. The work of Martin Buber, Rudolf Otto, and Peter Wessel Zapffe forms the conceptual framework for a view of nature as a *Thou* or a "Wholly Other," a view which inspires awe for the nonhuman intrinsic value in nature. In contrast to the Self-realization approach of Naess and others, intrinsic value is here independent of the notion of a self. This approach suggests an ethic of humility and respect for nonhuman nature—to the degree that the continued existence of humans should be considered an open question.

I

One night in times long since vanished, man awoke and saw himself. He saw that he was naked under the cosmos, homeless in his own body. Everything opened up before his searching thoughts, wonder upon wonder, terror upon terror, all blossomed in his mind.

Then woman awoke, too, and told him that it was time to go out and kill something. And man took up his bow, fruit of the union between the soul and the hand, and went out under the stars. But when the animals came to their waterhole, where he waited for them as was his wont, he knew no longer the spring of the tiger in his blood; but heard instead a great psalm to the brotherhood of suffering shared by all that lives.

That day he came home with empty hands, and when they found him later by the rising of the new moon, he sat dead by the waterhole.[1]

---

*Reed worked at the Council of Environmental Studies in Oslo, Norway from fall 1986 until his death in March 1987, when he was killed in an avalanche in the Jotunheimen mountain area. After finishing research on Norwegian ecophilosophy and ecopolitics as a Fulbright fellow (1985–86), he received funding from the Norwegian Ministry of Environment to examine environmental attitudes of bureaucrats at the county level of Norwegian environmental management, a project left unfinished by his death. He also edited an anthology with David Rothenberg, *Wisdom and the Open Air: Selections from Norwegian Ecophilosophy*, which has not yet been published (photocopies of the manuscript are available from the Council of Environmental Studies, PB 1116, Blindern, 0317 Oslo 3, Norway).

[1] Peter Wessel Zapffe, "Den sidste Messias," in *Peter Wessel Zapffe*, ed. Guttorm Fløistad (Oslo: Pax Forlag A/S, 1969), p. 105 (translated by Peter Reed).

Peter Wessel Zapffe's retelling of the Fall of Man confronts us with one of the central problems in environmental philosophy: what is the human relationship with nature? In its normative voice, the question is: what *should* that relationship be? The question is an urgent one: we are a powerful species, and we hold the world hostage to whatever it is we think we are destined for. It must be hard to live with such a terrorist, and there is every indication that the Earth is getting fed up with us.

Environmental philosophers have assumed a role as a sort of counselor, analyzing ways of improving our relationship with *Ecos*. They do this in a number of ways, but two approaches crop up frequently in their analyses. Briefly, one approach sees humans as part of nature; the other sees nature as part of humans.

An example of the first approach is E. O. Wilson's sociobiology. "The urge to affiliate with other forms of life," writes Wilson, "is to some degree innate," and is part of the "program" of the brain.[2] Iltis, Loucks, and Andrews also see the ties that bind us to the Earth as biopsychological. "It is likely," they write,

> that we are as genetically programmed to a natural habitat of clean air and a varied green landscape as any other mammal. To be relaxed and to feel healthy usually means simply allowing our bodies to react as evolution equipped them to for 100 million years.[3]

In the strongest form of this approach, moral thinking itself is a kind of evolutionary outgrowth.[4] Here the supposed gap between facts and values disappears, and a value becomes as much a fact as a tree, a rock, or a cloud.

The basic idea is simple: we need to think of ourselves as biological beings, in close kinship with the beasts of the field. "Men are not *like* animals," writes Joseph Meeker, "they *are* animals. . . . [We] need to find ethical principles and practices which will agree better with the biological requirements of the human species."[5]

The second approach, which I am most concerned with, goes at the problem from the other direction. The problem is still that humans and nature have

---

[2] E. O. Wilson, *Biophilia: The Human Bond with Other Species* (Cambridge: Harvard University Press, 1984), p. 85.

[3] Hugh H. Iltis, Orie L. Loucks, and Peter Andrews, "Criteria for Optimum Human Environments," *Bulletin of the Atomic Scientists* 26, no. 1 (January 1970): 2.

[4] See May Leavenworth, "On Bridging the Gap between Fact and Value," in *Human Values and Natural Science*, ed. E. Laszlo and J. B. Wilbur (New York: Gordon and Breach Science Publishers, 1970), pp. 133–43.

[5] Joseph Meeker, *The Comedy of Survival: Studies in Literary Ecology* (New York: Charles Scribner's Sons, 1972), p. 152.

become estranged, but the solution is to expand our concept of self to include nature. Arne Naess, for example, argues that the mature human being achieves a "Self-realization" by identifying with all of nature, something Warwick Fox calls "a state of being which sustains the widest possible identification."[6] For convenience I call this the "Self-realization approach."

What does Self-realization mean? What is the relation between the Self and the individual human being? For Arne Naess, *Self* (with a capital *S*) is the Hindu Atman, which "corresponds to the notion of the Absolute in Western philosophy—something completely beyond ordinary description but somehow basic to both God and the World."[7] The human individual is a self (with a small *s*), and of course part of the Self. The problem is that the small self does not see that it is a part of the *Atman* because the small self is too busy with its narrow self-interests—egotism. To get around this problem and "realize" itself, the small self is not expanded to encompass everything, but rather *diminished*. When it has reduced its egotism to zero (though this is unlikely to happen), it is realized, that is, it experiences a "oneness in diversity."[8] Self-realization has so little to do with the individual ego, in fact, that the true Self-realization of one individual is impossible without the Self-realization of all individuals.

Naess' formulation of Self-realization differs in particulars from other wider identification approaches. Nevertheless, the general idea seems to be that Self-realization is not egoism, a psychological projection of ourselves onto the world. Instead it is really the losing of the individual self in an "experientially incorporated" world.[9] We substitute a narrow sense of the individual with a broader sense of self which includes the whole universe.

Distinguishing the two approaches may seem like splitting hairs, because the end result is the same. That is, we try to break down the distinction between the human (subject) and the natural (object). We try to bridge the gap between ourselves and the world, emphasizing how we are the same instead of how we are different.

No matter which side of the gap we start from, the moral assumption behind this bridge building is also the same: that once we understand how closely we are related to nature, we will lose interest in destroying it. In many ways both

---

[6] Warwick Fox, *Approaching Deep Ecology: A Response to Richard Sylvan's Critique of Deep Ecology*, Environmental Studies Occasional Paper no. 20 (Hobart: Centre for Environmental Studies, University of Tasmania, 1986), p. 67.

[7] Arne Naess, *Gandhi and Group Conflict: An Exploration of Satyagraha Theoretical Background* (Oslo: Universitetsforlaget, 1974), p. 41.

[8] Arne Naess, "Identification as a Source of Deep Ecological Attitudes," in Michael Tobias, ed., *Deep Ecology* (San Diego, Avant Books), 1985, p. 261.

[9] Fox, *Approaching Deep Ecology*, p. 67.

approaches are arguments from self-interest, even though the meaning of *self* is revised: "If it is rational for me to act in my own best interest, and I and nature are one, then it is rational for me to act in the best interests of nature."[10]

I sympathize with this assumption, but I think the two approaches sketched above ignore another possibility: it is our very *separateness* from the Earth, the gulf between the human and the natural, that makes us want to do right by the Earth. According to this approach, nature is a stranger. It seems to me that recognizing the moral distance between ourselves and the world helps us recognize the values in nature that are totally independent of what we humans think is beautiful, right, and good. Moreover, recognizing these values, some strong conclusions for how we ought to behave suggest themselves. If we turn our attention from our petty hubris, insignificance, and appalling ignorance to a universe vast beyond our ability to comprehend, we might treat the Earth a little less arrogantly.

II

In drawing a picture of nature as something separate from us, as a stranger, I draw on the work of Martin Buber and Rudolf Otto. As theologians, their work is primarily about God, not nature. Still, they give us some very suggestive analyses of the way we relate ourselves to other beings. These analyses are useful when we are thinking about our relationship with the environment.

Buber's seminal *I and Thou* builds on the intuition that we are apart from something before we are related to something. Before we become part of a socioecological organism, we are a single cell—alone. This isolation, however, leads neither to loneliness nor to solipsism. There is separation, but not a desolate or aggressive alienation. We can think, "I am not lonely in the Universe, but comforted. For when I look at nature, I know that *I am not it*. There is something else out there, with me." Buber's work is a call back to a commonsense idea that there *is* something that is not just part of us, a solid something we can kick in refutation of Berkeley's *esse es percipi*. It is a reminder of the power and being that exists independently of the human mind.

"To man," writes Buber, "the world is twofold, in accordance with his twofold attitude."[11] Two "primary words" characterize these attitudes and express the two ways we relate ourselves to our surroundings. "The one primary word is the combination '*I-Thou*'. The other primary word is the combination '*I-It*'."[12] Most commonly we relate to the world as I-It: the world is filled with tools or hin-

---

[10] J. Baird Callicott, "Intrinsic Value, Quantum Theory, and Environmental Ethics," *Environmental Ethics* 7 (1985): 275.

[11] Martin Buber, *I and Thou*, trans. Ronald Gregor Smith (Edinburgh: T. & T. Clark, 1944), p. 3.

[12] Ibid.

drances, dependable, even beautiful objects—*for us*. But, says Buber, when we do so, we are really relating to our own selves, without regard to the integrity of what it is we experience.

"Experience," in Buber's usage, is very self-centered. "The man who experiences has no part in the world," he writes:

> For it is 'in him' and not between him and the world that the experience arises. The world has no part in the experience. It permits itself to be experienced, but has no concern in the matter.[13]

When you "experience" the world, it becomes "your object, remains it as long as you wish, and remains a total stranger, within and without."[14]

Nonetheless, our relation to nature need not be a relation to ourselves. It can happen that we meet nature itself. Here we confront it as a *Thou*, a self-sufficient being of whom we have an inkling, a relationship with something *else* which we cannot pass off as a figment of our imagination. Nature as a *Thou* is

> no impression, no play of my imagination, no value depending upon my mood; but it is bodied over and against me, and has to do with me, and I with it. . . . [15]

We participate in a relation with something outside of ourselves, which commands our attention as long as the relationship lasts.

Two questions come to mind here: how can we tell if we actually are in a genuine relation with a *Thou* and how can we enter into such a relationship? Both of these questions are questions of a mind examining itself, and doubting what it sees. The essence of the *I-Thou* relationship is that *there is no room, no time, for reflection*. We are seized by the relationship; we cannot think about it as we would an object. It is here, now, and while it lasts, there is only now. Since we have no time to ourselves to think about the relationship, there is never any question of doubting its reality. This is akin to the clear and distinct ideas of Cartesian introspection, but with an important difference—what is clearly apprehended is *outside* the thinker, not inside her or his own consciousness. This perception "has more certitude for you than the perceptions of your senses"; "No deception penetrates here; here is the cradle of Real Life."[16] Asking "Is this a real *I-Thou* relationship?" makes about as much sense as asking "Am I having fun yet?"

As for how we can have this relationship with a *Thou*, Buber reminds us that it is not something that we "have." It is something that, in part, is given to us. "The

---

[13] Ibid., p. 6.
[14] Ibid., p. 32.
[15] Ibid., p. 8.
[16] Ibid., pp. 110, 9.

Thou meets me through grace," Buber writes, "it is not found by seeking."[17] There is an undeniable mutuality to the *I-Thou* relation—while we step toward it, it must also step toward us. Buber believes there are no sure-fire techniques for taking these steps:

> [E]verything that has ever been devised and contrived . . . as precept, alleged preparation, practice, or meditation, has nothing whatsoever to do with the primal, simple fact of the meeting. . . . Going out into the relation cannot be taught in the sense of precepts being given.[18]

The *Thou* comes like a gift, or a thief in the night.

RUDOLF OTTO AND THE IDEA OF THE HOLY

Rudolf Otto's idea of "the Holy" is parallel in many respects to Buber's *Thou*. The "Holy" is, like the *Thou*, something that stands radically apart from us. It is the "Wholly Other," the "mysterium tremendum et fascinans."[19] When we meet the "Other," we are not devoured by it, as powerful as that meeting can be. We know that it is *we* who are faced with the Holy, the *Numen*, precisely because it is the Holy who confronts us.

Otto most neatly complements Buber when he tries to describe "by hint and suggestion" what it is we meet when we meet the Holy Thou. He uses words like *majestic, fascinating, aweful,* but stresses that in the end none of these descriptions are really adequate. We lack the adjectives to describe the Holy in its manifest glory; we stand dumb, overcome by an encounter with something that is so obviously beyond our ability to capture in words. Not even value terms like *good* or *evil* will do: "this 'extra' in the meaning of 'holy' [is] beyond the meaning of goodness."[20] It is only in the wake of such a meeting that we can creep trembling back into the familiar structure of our language and try to describe what it was we saw.

The adjectives Otto sometimes uses give the impression that the Numen is something overwhelmingly large and mighty:

> whatever among natural occurrences or events in the human, animal, or vegetable kingdoms has set him [humans] a-stare in wonder and astonishment—such things have ever aroused in man, and become endued with . . . numinous feeling.[21]

---

[17] Ibid., p. 11.
[18] Ibid., p. 77.
[19] Rudolf Otto, *The Idea of the Holy* (Middlesex: Penguin Books, 1959), p. 39.
[20] Ibid., p. 20.
[21] Ibid., p. 79.

But we should not let our scanty language bewitch us even so much, for Otto also writes that the numinous is represented in things as small as paintings or songs. Likewise, Buber makes it clear that we can enter into an *I-Thou* relation with any part of physical nature. In all natural objects there is the potential to fascinate us, to turn our minds away from ourselves in wondrous contemplation of natural complexity: in a tree, a leaf, a gram of soil, even Lewis Thomas' "lives of a cell."

NATURE AS THE THOU OR THE NUMEN

My point here is that we can think of nature as a sort of *Thou* or Numen, something wholly other and awe inspiring. Put this way, the point sounds rather weak. Of course, someone might object, nature is something other than us—on the surface—but as proponents of the new physics, Buddhist thought, and the central intuitions of deep ecology have shown, this sense of separateness stems from a mistaken ontology. The truth is actually somewhere in between: we are interwoven into the fabric of nature; it is a part of us and we are a part of it.

What I am arguing here is that the view that "everything hangs together" is only *one* truth of the matter, one way of seeing the relation between humans and the cosmos. Otto and Buber see it in another way. The physical connections, the interchange of atoms and molecules between creatures and through time is an insubstantial link, merely physical. Certainly we are parts of food chains; certainly we have fellow feelings for nature; nevertheless, there is an existential gulf of awesome depth between ourselves and the Other, a gulf which no amount of "identification in otherness" can span. The Other is really other.

It is easy to see why we slide so readily into the "we are the world" mentality. We have lost, in our daily lives at least, a precious sense of our own insignificance. There are few enclaves of nature which have not been smudged obviously and unerasably with human fingerprints. We are mesmerized by our own cleverness, whether it be the building of physical bridges across a canyon or philosophical bridges across the canyon between ourselves and nature.

Nature, we know, sometimes takes it upon herself to remind us of our insignificance. Human-caused "natural" disasters and the boomerang effects of our ecological malpractice are creeping into our public agendas. We are learning, slowly, a healthy contempt for our stinking cities, shrieking aircraft, and petty works of art, which are, of course, of more interest to *Homo sapiens* than to anyone else.

III

Who else is there? Are there things which matter apart from humans wanting them to matter? Or is value a gift from man to nature, which can be granted or rescinded on our whim? I want to argue that there is value apart from man, value that is intrinsic to nature, wholly independent of whether we perceive it or not.

What sort of value is this? In "Values in Nature," Holmes Rolston, III has listed ten kinds of values nature has. These range from the evident economic values of natural resources to the intangible "sacramental values" that "generate poetry, philosophy, and religion, no less than science."[22] Some are values that are only "good for" something else; others are also inherently good experiences: the awe we feel observing the thundering surf is, so to speak, its own reward. All of these values are, nevertheless, values for humans. Realizing that such a list is incomplete, Rolston has supplemented it in a later article by adding "intrinsic values in nature."[23] Here, like some utilitarians, Rolston argues that an animal or plant has needs and desires, interests which make the creature a locus of value, an end in itself. Although this addition is quite a leap, it falls short of the idea of intrinsic value suggested by Buber and Otto. For Rolston, intrinsic value is very closely tied to a subject having interests, specifically, self-interests. Without a subject, there are no interests, and no loci of intrinsic value. True, Rolston (and others) stretch the notion of the interest-having subject very far to include even a mineral crystal,[24] but the feat of imagination required to see a crystal as having interests produces an overstrained metaphor. A better approach, in my view, is to uncouple "having interests" from "being of intrinsic value."

INTRINSIC VALUE AS THE OTHER

Without the self-interest argument, what criteria are left for deciding what things have an intrinsic value? For Otto, the answer is none—at least none that can be described unambiguously. Recall Otto's argument that our ethical vocabulary is inadequate to meet the demands the Holy puts on it. Faced by the power of nature, its potential *Thou*-ness, there is very little that we can say about it. We end up trying to describe the indescribable. We try to speak of what we encountered, and then what was valuable about it, but we cannot really distinguish between the thing and its value, because they are too tightly entangled. The value is essential in, not an attribute of, the thing. When we try to express what we mean, we wind up saying with Rodman that "all natural entities have intrinsic worth simply by virtue of being what they are,"[25] or with McHarg, "That which is, is justified by being; it is unique, it needs no other justification."[26]

---

[22] Holmes Rolston, III, "Values in Nature," *Environmental Ethics* 3 (1981): 127.
[23] Holmes Rolston, III, "Beyond Recreational Value: The Greater Outdoors," in *A Literature Review: President's Commission on American Outdoors* (Washington: U.S. Government Printing Office, 1986), pp. 103–13.
[24] Holmes Rolston, III, "Are Values in Nature Subjective or Objective?" *Environmental Ethics* 4 (1982): 146.
[25] John Rodman, "The Liberation of Nature?" *Inquiry* 20 (1977): 108.
[26] Ian McHarg, *Design With Nature* (Garden City: Doubleday/Natural History Press, 1971), p. 125.

Of course, the experience of the Holy can be very enlightening, and this enlightenment has an inherent value—for humans—but the enlightening experience is not the whole value:

> There will, in fact, be two values to distinguish in the numen; its 'fascination' (fascinans) will be that element in it whereby it is of *subjective* value (= beatitude) to man; but it is 'august' (augustum) in so far as it is recognized as possessing in itself *objective* value that claims our homage.[27]

It is this sense of objective value that I call *intrinsic value*.

So defined intrinsic value is a value for no one, and there are some quirks to this kind of value that have to be admitted. First, there is no such thing as "greater" or "lesser" intrinsic value, at least not for humans to judge. If we made such gradations, we would be back at the idea that intrinsic value is the value that an experiencing subject compares with other values. What I am suggesting is that an "objective" intrinsic value is independent of being valued by something or a value for something.

The second quirk is that intrinsic value is not additive. That is, we cannot say, "such-and-such has various instrumental values, and in addition to these it has an intrinsic value, which brings its total value up to so-and-so much." It may be tempting for environmentalists to say this, but we are really talking about two distinct kinds of value here. We can certainly use intrinsic value in an argument for environmental preservation, but intrinsic values cannot be an entry in utilitarian ledgers.

SO WHAT?

At this point there are some of objections that this theory of intrinsic value has to address. I am claiming that anything which we can relate to as a *Thou*, anything that can suggest the Numen, has an objective value. Because I have argued that we can relate to anything as a *Thou*, this means that *everything* has this intrinsic value. With this admission, a number of questions arise. Is everything equally good? If we are to care for good things, how are we supposed to care for everything simultaneously? Is there anything that has intrinsic *dis*value? If not, isn't this an empty ethical theory? How could it possibly be a guide for action?

A few answers can be proposed. First, I am dealing here with only one special kind of value, and I admit that there are others. Even if all objects in the world were in the present sense good objects, that would not mean the end of ethics: there would still be questions about rights and duties, how to treat objects with both instrumental and intrinsic value, and so on.

---

[27] Otto, *The Idea of the Holy*, p. 67.

More importantly, though, I want to argue that the intrinsic values that we are able to recognize not only can, but *do* suggest right actions to us. Again, the crux is the meeting with the *Thou* or the revelation of the Holy, as Buber describes it. Buber claims, echoing Otto, that there is very little we can say about what we get out of a meeting with a *Thou*. "As no prescriptions lead us to the meeting, so none leads from it."[28] Nonetheless:

> Man receives, and he receives not a specific 'content' but a Presence, a Presence as Power . . . [and] the inexpressible confirmation of meaning. . . . What does the revealed and the concealed meaning purpose with us, desire from us? It does not wish to be explained (nor are we able to do that) but only to be done by us. . . . [T]his meaning is not that of 'another life', but that of this life of ours, not of a world 'yonder', but that of this world of ours, and it desires its confirmation in this life and in relation to this world. . . . But just as the meaning itself does not permit itself to be transmitted and made into knowledge generally current and admissible, so confirmation of it cannot be transmitted as a valid Ought: it is not prescribed, it is not specified on any tablet to be raised above all mens' heads. The meaning that has been received can only be proved true by each man only in the singleness of his life.[29]

What Buber is trying to say is that having met a *Thou*, we cannot be indifferent. We have to go out and do something about it. What we do will vary from person to person, but a meeting with the intrinsically valuable is not prescriptively empty.

Both Buber and Otto insist that the *Thou*, the Holy, is beyond our moral categories. Thus our subjective experience of the Numen is not like our experiences of other moral urges: "We cannot approach others," writes Buber,

> and say, 'You must know this, you must do this.' We can only go, and confirm its truth. And this, too, is no 'ought', but we can, we *must*.[30]

Nonetheless, it is clear that the feeling of moral obligation is something we carry *away* from the meeting, and not something that is valid only so long as we are *in* the meeting. The encounter with the *Thou* is not mystical in the sense of having nothing to do with our everyday life. On this reading, the problem of what moral prescriptions we get from meeting a *Thou* is not really a problem—though to describe the voice of the *Thou* in conventional ethical language is.

Focusing more on the subjective side of the relationship with the Holy, Otto is a little more specific about what the content of the prescription is likely to be.

---

[28] Buber, *I and Thou*, p. 111.
[29] Ibid., pp. 110–11. Buber is speaking here of the "ultimate" *Thou*, God. Without wanting to identify God with nature, I think we can use these ideas as models for our relationship with nature.
[30] Ibid., p. 111.

When we meet the *mysterium tremendum*, he says, our natural reaction is to feel impressed, and small:

> there is a feeling of one's own submergence, of being but 'dust and ashes' and nothingness. And this forms the numinous raw material for the feeling of religious humility.[31]

According to Otto, the Wholly Other contains *in itself* the "further element" of majesty. This majesty inspires respect, and a corresponding sense of "moral delinquency,"[32] a fear of affronting the Numen.

Here, then, is the backdrop for acting morally. "Inhuman nature in its towering reality"[33] inevitably arouses a feeling of obligation to it. As we play out this obligation in the ways each of us thinks is most appropriate, we use "things in accordance with the priorities established by the things themselves."[34] Although the outlines of this play are sketchy, I hope to have established that there is value we can recognize, that it is "there" whether we recognize it or not, and that recognition of it being there brings with it an indubitable obligation to respect that value. In the next section I flesh out what an appropriate response to the recognition of independent intrinsic value might be.

IV

I have argued that nature's intrinsic value is beyond humanity, that nature matters even if it doesn't matter to us. But do we matter to it? Can we destroy this distant, nonhuman good? Experientially, I think the answer has to be yes, that nature's values are *not* impervious to us. As humans we are different from nature, but we are related to it. When we create a work of culture, we usually destroy a thing in nature, and what is lost is often greater than what we gain. The continued existence of things as they were before the advent of man is not a matter of indifference; it is a matter of value. I want to suggest below how this is so.

When we modify or destroy natural systems or objects, we destroy something that did not take shape under our hands or spring from our brow. We destroy something that is essentially other. What we lose here are the "subjective values" (in Otto's sense) of nature, the value *to us* of a relationship with them.

This does not mean that we cannot lay hands on nature without killing it. It is a

---

[31] Otto, *The Idea of the Holy*, p. 34.
[32] Ibid., p. 67.
[33] Robinson Jeffers, from "The Beauty of Things," in *Robinson Jeffers: Selected Poems* (New York: Vintage Books, 1963), p. 94.
[34] William Leiss, "A Value Basis for Conservation Policy," *Policy Studies Journal* 9 (1980): 613–22.

question of dominance or appropriateness. Although what is appropriate depends partly on culturally conditioned perceptions, we are entitled to criticize our culture's way of seeing nature on the basis of our encounter with it as a *Thou*. When the human elements of a landscape dominate, its wonder can die for us. We then know the landscape all too well, and we slip into thinking of the landscape as an *It* instead of a *Thou*.

The danger is not only in physical interferences with nature. Nature's wonder also recoils under the onslaught of our mental models and our sciences. The awe of the Numen cannot coexist with the drabness of the known:

> No one says, strictly and in earnest, of a piece of clockwork that is beyond his grasp, or of a science he cannot understand: 'That is "mysterious" to me.'[35]

There is peril in becoming too knowledgeable. When we can conquer a wilderness either by main strength or by ingenuity, its dying face appears as a reflection of our own. The myth of progress is the myth of Narcissus: it merely brings us back to ourselves.

As it happens, models are invented and discarded, and research programs usually raise more questions than they answer. Although the danger that we will "really" come to know the cosmos is negligible, the danger that we will come to think we *do* know it may not be. What is at stake for us is the potential for richness of experience that would be lost if we destroy the "holy" in the world.

Clearly, the reason we as humans want to preserve nature has a lot to do with nature's value *for us*, the potential for rich relationships with the nature/*Thou*. But if we can meet all things as *Thous*, what is so special about the intrinsic value of the nature/*Thou* as opposed to the intrinsic value of the human/*Thou*? Suppose our manifest obligations to one conflict with those to the other. Is there a hierarchy of more or less "valuable" *Thous*? Buber clearly thinks there is such a hierarchy. The *Thou* of true "fulfillment" is the "*Thou* that, by its nature, cannot become an *It*," in other words, God.[36] Ignoring for the moment this special case, what are we to do with all the other *Thous*?

While we are in a relation with one *Thou*, of course, there is no question of comparing it with another. Comparisons are only a problem afterwards, when we have transformed the *Thou* into an *It* and can shuffle it around as an object of consciousness. Nevertheless, the problem of comparisons between these objects may be more theoretical than practical. While it is possible to see the whole world and all its parts as *Thou*, we are unlikely to be able to keep up an *I-Thou* relationship with all of them. We will seek an *I-Thou* relation to some things, and

---

[35] Otto, *The Idea of the Holy*, p. 42.
[36] Buber, *I and Thou*, p. 75.

depending on whether bonds are established (remembering that it is not only we as individuals who decide this), those things will give us a "call and a sending": attentiveness to them will take a natural priority over attentiveness to others.

The problem of ranking *Its* is not thereby solved, though, any more than moral choice can be reduced to rote rules by any ethical system. There will always be hard cases which challenge our intuitions as free actors. Recognizing the intrinsic value of a *Thou* is an *incomplete* guide for action. Nevertheless, because nonhuman nature is in some sense "more Other" than people, it is in our human interest to place a high priority on preserving the potential for meeting it.

In short, I argue that we can only experience the mysterious otherness of nature through meetings with a dominant nature. Although this is a familiar argument, it is nonetheless one that goes beyond human interests, for the crux of my argument is that there are values in nature that matter even if they do nothing for us. It is in this sense that the *Numen* demands our respect even when we do not experience the inherent benefits of meeting it. In other words, there are more important things than humanity, things that awe and stupefy us with their longevity, their imperturbability, and their indifference to us. Encountering them as *Thous*, we recognize that they have vast, nonhuman greatness and value, value that we do wrong to threaten. The gossamer Milky Way and the unmoving stars are still beyond us and above us, and they teach us of the Other that is not man. This Other, however, is also the curve of a petal, the delicate lattice of a diatom's shell, things we can easily crush. Faced with such values, the appropriate human virtues are self-restraint, hesitation, a respect for the mystery of the world, and a willingness to leave it at that.

But in practice we aren't willing to: our populations swell, as does our desire for an abundance of *Its* (instead of an abundance of *Thous*). *Terra incognito* shrinks; wonder flees. Do we really love ourselves enough that we are willing to lose the Other? Has our humanism puffed itself up until we cannot imagine that we are neither necessary nor sufficient for value to exist in the world? Who do we think we are?

What I suggest here is that we take self-restraint seriously, even if it is in our power to do without nature. Technological advances may make us independent of our natural life support system; we may learn to live on a barren planet or even to pour our essence into the durable shell of a robot. Faced with such a clash between human interests and the survival of the planet's nature, I think it is humans who must give way.

While this is certainly an inhuman suggestion, it is important to make clear just how inhuman it is. Consider Garret Hardin's well-known lifeboat ethic, i.e., that resourceful nations should work to ensure their own survival in the coming ecoholocaust.[37] If poorer or stupider nations drown under a tide of their own

---

[37] Garrett Hardin, "Living on a Lifeboat," *Bioscience* 24 (1974): 561–68.

humanity, that is their own fault. Rich "lifeboat" nations should not jeopardize their own survival by helping the suffering. Compassion for the victims means catastrophe for the race.

As inhuman as this sounds, Hardin insists that he is arguing for the sake of future humanity. To serve that end, questions of distributional justice are irrelevant. In a way, my modest proposal is even more ghastly than Hardin's: I am suggesting that for the sake of the austere mystery of nature as *Thou*, we have to ask ourselves whether the preservation of humanity is of overriding importance.

The misanthropy of my argument runs parallel with the misanthropy in the writings of Peter Wessel Zapffe. Zapffe sees humanity as a sort of evolutionary monster, one that is outfitted with an overgrown brain that tries to figure out the meaning of the universe in human terms. The attempt is tragically doomed to failure because nature in itself is quite beyond our attempts to find meaning in it.

According to Zapffe, nature gives us special inspirations and joys; it makes possible human values that cannot be found except in relation to that which is not human. We find these values, and hints of values beyond them, in situations in which nature is terribly dominant. Encounters with these values are vital, providing challenges for the development of our peculiar talents as a species. It is for this reason, Zapffe says, that we are reluctant to lose the chance to meet free nature. Still there is no reason for us to think of ourselves as being the alpha and the omega of the universe. The universe can get on quite well without us, diminished, perhaps, but not greatly so. "For me," writes Zapffe, "a desert island is no tragedy, and neither is a deserted planet."[38] Or, as one of Zapffe's alter egos admonishes the throng, "The life on many planets is like a rushing river, but life on this world is like a stagnant puddle and a backwater. . . . Know thyselves—be unfruitful and let there be peace on earth after thy passing."[39]

Zapffe is *not* arguing for collective suicide. What Zapffe (and I) want to do is put a question mark—only a question mark—on the whole of human existence. When we question the right of other species to exist on the planet, it is only fair that we pose the same question with regard to ourselves: do we need, or deserve, to survive? When it is clear that there is value and beauty and wonder and greatness that is wholly independent of us, we cannot conclude that the universe would be a whole lot worse off without us.

V

What I have offered here is a picture of the world and our part in it that departs from the identification or Self-realization approach in ecophilosophical dis-

---

[38] Peter Wessel Zapffe, *Jeg velger sannheten* (a dialogue between Peter Wessel Zapffe and Herman Tønnesen) (Oslo: Universitetsforlaget, 1983), p. 60 (trans. Peter Reed).
[39] Peter Wessel Zapffe, "Den sidste Messias," p. 117.

cussions. There is room for a plurality of approaches in environmental philosophy, and my proposal is only an alternative. In the main, I think its emphasis on human restraint is in line with other approaches even though the reasoning is a little different.

The present approach, though, may be more useful than Self-realization in some respects. As a practical matter, it seems to me that when we try to operationalize Self-realization we are put in an uncertain position. We are supposed to retain a sense of our individuality as we work to save the big Self from destruction—but at the same time we are supposed to *lose* interest in our individuality as we cultivate our identification with the big Self. True, Self-realization is not absolute holism, in which the individual is identical with the big Self. Nor is it absolute separatism, in which the individual is completely apart from the Self. However, those practicing Self-realization seem to want it both ways: we are, somehow, both the big and the small self. How do we set to work with this ambiguous notion of self?

On the face of it, Buber's *I-Thou* model can seem a lot like the Self-realization approach. Both emphasize the virtues of humility, and both report that very little can be said about the *Thou* or the Self. But the two approaches are not the same. Fox points out that Self-realization depends on the "relative autonomy" of beings in nature.[40] For Buber, however, the autonomy is absolute: the *I* and the *Thou* do not depend on one another. They meet, but only as two ships which pass in the night. There is contact, perhaps a yearning to merge into one identity, but *Thou* remains separate from *I*. The *Thou* is not the *Atman*, because we are not a part of the *Thou*.

The "apartness" of my proposal is able to skirt a few questions that Self-realization raises: what is the position of humans in the Self? How important is it that *Homo sapiens* continue? Can a person who has "realized him or herself" agree to the extinction of the human race, believing that it would still exist in the *Atman*-Self? What about the sense of awe, on which Otto bases religious ethics? We, as small selves, can certainly feel awe toward the enormity of the Self—but at what price? How far must we depart from the path of Self-realization in order to retain this picture? How much must a "realized" person *avoid* "experientially incorporating" the world in order to feel awe for it? Can we feel an awe for nature when what we think we are looking at is a part of ourselves?

It is possible that Self-realization does not rule out an awe for nature. Yet the ethical argument of the Self-realization approach seems to be based not on awe but rather a kind of self-interest: "We ought to care about all entities/beings/things in the world because they are a part of our *Self*; their diminishment is our diminishment."[41] The idea of self has become so important that Callicott writes

---

[40] Fox, *Approaching Deep Ecology*, p. 84.
[41] Ibid., p. 76.

that environmentalists have "hopelessly supposed" that "objective intrinsic value could be persuasively established independently of self."[42]

It is clear that proponents of Self-realization are not talking about egoistic self-interests. Still, I hope we have not been so mesmerized by the concept of self that we think it is the only possible basis for an ethic. What I have tried to do here is suggest how values can exist in nature *independently* of self.

The ethic I have argued for here is intuitionist; it rests on a kind of revelation of value to the individual. This does not mean that other people, or groups of people, cannot share relations with a *Thou* that seem (in retrospect) very similar. What it does is base the ethic on the reality of an *I-Thou* relation.

The weakness of an intuitionist ethic, of course, is that there is no guarantee that everyone will have the same intuition. Nor can we be sure that everyone will act on their intuitions in compatible ways. Encountering the Other might be a revelation of terror and alienation for some, which would lead them to attack the Other in fear.

Despite these weaknesses, I believe an environmental ethic must at some level be connected with an intuition or a feeling for nature. While moral philosophy may involve detached reasoning, moral action must involve caring, committed, nonrational attitudes. As one philosopher puts it, "one cannot prove the beauty of Mozart's composition in a way that separates it from personal experience."[43] The likelihood that an encounter with the nature *Thou* would lead to aggression toward nature seems small: Buber tells us that an *I-Thou* relationship is one of awe and compassion, not terror and alienation.

Whatever the verdict on ethical intuitionism, we should remember that Self-realization is also a kind of intuitionism. Some argue that the intuition it describes is superior because it "gets much closer to what many naturalists and environmentalists feel and *want* to say."[44] It may in some cases be true that the Self-realization approach meshes with the world view of ecoactivists; nevertheless, some environmentalists seem to feel more *responsibility for* than *identification with* their natural surroundings.[45] Some surveys of visitors to wilderness areas, moreover, suggest that people set a high value on the ways that wilderness

---

[42] Callicott, "Intrinsic Value, Quantum Theory, and Environmental Ethics," p. 275.
[43] Richard Cartwright Austin, "Beauty: A Foundation for Environmental Ethics," *Environmental Ethics* 7 (1985): 201.
[44] Fox, *Approaching Deep Ecology*, p. 78.
[45] Peter Reed, "Verdier i miljøpolitikken" (Values in Environmental Decision Making), unpublished manuscript. Because this survey is narrow in scope and unfinished, this conclusion is tentative. It *is* the case that many writers on environmental issues, especially those with a philosophical bent, make a point of identifying themselves with nature. Whether this holds true for a wider environmentalist public is an empirical question. I am unaware of comprehensive surveys that have collected information on precisely this attitude of identification with nature.

seizes their attention and on a sense of nature awe at least as much as on a sense of identification.[46]

IV

In any case, the idea that ecophilosophy *should* mesh with the intuitions of environmentalists is absolutely essential. If environmental philosophy is going to be useful in the environmental movement, it *has to* make sense to activists; it must give them conceptual tools and arguments with which to fight ecological degradation. It has to be common-sensical enough that it is easily graspable, and at the same time revolutionary enough that it points to reasons for changing our behavior.

What I have been trying to do is make awe for nature sensible to the enquiring intellect of environmental philosophers. In fact, there is nothing especially new about the concept of nature awe, and I think it fits quite nicely with what many environmentalists think about nature. Still, conceptual consistency and a handful of supporters is only part of an effective environmental ethic. So far (as often happens in environmental ethics) I have said something like, "If all of us had this feeling about ecological problems, we could solve them." This may be true, but it is not much help. There are steps missing. What about those who do not (basically) have the feeling? In the face of power struggles and inert political institutions, how do we implement an ethic based on awe for nature?

Even if (as is doubtful) people do wrong only because they don't understand nature's values, how is everyone going to be persuaded to understand those values the way environmentalists do? Academic philosophy can and should be a start at persuasion, but it cannot do the job without help. One place to look for help might be art and poetry, which can speak directly to our intuitions of nature's value:

> . . . when the whole human race
> Has been like me rubbed out, they will still be here:
> storms, moon, and ocean,
> Dawn and the birds. And I say this: their beauty has
> more meaning
> Than the whole human race and the race of birds.[47]

---

[46] For a treatment of the "awe-values" in nature, see David Douglas, "The Spirit of Wilderness and the Religious Community", *Sierra* 68, no. 3 (1983): 56–57; Linda H. Graber, *Wilderness as Sacred Space* (Washington, D.C.: Association of American Geographers, 1976); W. E. Hammit, "Cognitive Dimensions of Wilderness Solitude," *Environment and Behavior* 14, no. 4 (1982): 478–93.

[47] Robinson Jeffers, "Their Beauty Has More Meaning," in *Selected Poems* (New York: Vintage Books, 1963), p. 77.

[12]

"MAN APART" AND DEEP ECOLOGY: A REPLY TO REED

Peter Reed has defended the basis for an environmental ethic based upon feelings of awe for nature together with an existentialist absolute gulf between humans and nature. In so doing, he has claimed that there are serious difficulties with Ecosophy T and the terms, *Self-realization* and *identification with nature*. I distinguish between discussions of ultimate norms and the penultimate deep ecology platform. I also clarify and defend a technical use of *identification* and attempt to show that awe and identification may be compatible concepts.

I. MAN APART: A TRAGIC BEING

Among the strong, life-long critics of man's domination and exploitation of the Earth, we find philosophers who would rather use the slogan "man apart" than "not man apart." Two exponents of the former trend are the outstanding Scandinavian existentialist philosopher Peter Wessel Zapffe and the young deep ecology theorist Peter Reed.[1]

Reed begins his paper with a quote from Zapffe in which he beautifully sumarizes some of the main points of his early writings concerning the essential suffering of all beings capable of joy or suffering, and the terror upon terror in nature. Zapffe seems to conclude with a "no!" to *human* life in this universe of death and annihilation. At the same time, however, Zapffe expresses his awe in contemplating the serenity and majesty of wild nature: the mountains, the ocean, the stars. . . . Since the 1930s, Zapffe has written passionate and witty criticisms of man's destruction of nature. Nevertheless, his philosophy has another side.

According to Zapffe, humans, when not distracted by all sorts of ultimately meaningless activities, understand through their metaphysical attitudes (e.g., the requirement of justice and the rejection of death and utter annihilation) that they are utterly different from the other living beings that evolution has so far produced. The human brain makes us capable of seeing our own death and our essentially tragic position in nature. Thus, humans are truly and radically apart, man is homeless in nature, and since his body is a part of nature, man is even "homeless in his own body."

Later in life, Zapffe at times did not directly advise humanity to stop begetting children, but he expected that with increasing maturity people would refrain from producing children in such a world. Peter Reed also takes this more moderate

---

[1] See Peter Reed, "Man Apart: An Alternative to the Self-Realization Approach," *Environmental Ethics* 11 (1989): 53–69.

position (p. 66). The significance of Reed's article is, therefore, largely independent of the question of whether humans should disappear from the face of the Earth or not.

II. ALTERNATE SETS OF PREMISES

On the metalevel, Peter Reed admirably emphasizes the positive aspect of a plurality of ultimate philosophical commitments among those who wish to contribute to the lessening of human destructiveness on Earth. He proposes an alternative to my version of philosophical deep ecology, which I call Ecosophy T (Ecosophy T has as its ultimate norm the one word sentence "Self-Realization!" together with the hypothesis that identification functions "as a source of deep ecological attitudes").[2]

Reed is of the opinion, and I am sure that he is right, that some supporters of the deep ecology movement feel more at home with an overall world view more in the direction of his alternative than with Ecosophy T. They might, for example, agree with Reed that "it is our very *separateness* from the Earth, the gulf between the human and the natural, that makes us want to do right by the Earth . . . [where we see] nature as a stranger . . . [and where we respect] values in nature that are totally independent of what we humans think is beautiful, right and good" (p. 56). Reed holds that an appreciation of wilderness and wildness, overwhelming complexity, the enormity of the mountains and the ocean, and the efficiency and ferocity of the splendid, great white shark, are values of this kind.

According to Reed, an acceptance of this kind of absolute separateness from nature makes it unwarranted to use Self-Realization as an ultimate philosophical norm. As Reed puts it, "I hope we have not been so mesmerized by the concept of self that we think it is the only possible basis for an ethics" (p. 68). I am sure "we" have not been so mesmerized! Self-realization is an ultimate norm in only one kind of ecosophy exemplified by Ecosophy T. Ecosophy T is not to be identified as "*the* philosophy of deep ecology." There is no one single philosophy if, as I use the term, a verbally articulated philosophy contains the articulation of *ultimate* premises. The "eight points" of the "basic principles of deep ecology" (or better, the "deep ecology platform"), are, at best, penultimate.[3]

Why should we even wish to have conformity at the ultimate philosophical or religious level? We are on the way from we *know* not where to we *know* not

---

[2] Arne Naess, "Identification as a Source of Deep Ecological Attitudes," Michael Tobias, ed., *Deep Ecology: An Anthology* (San Diego: Avant Books, 1985) pp. 256–70.

[3] The deep ecology platform appears in Arne Naess, "The Deep Ecology Movement: Some Philosophical Aspects," *Philosophical Inquiry* 8 (1986): 10–31; Bill Devall and George Sessions, *Deep Ecology: Living as if Nature Mattered* (Salt Lake City: Gibbs Smith, 1985) pp. 69–73.

where. Ecosophy T is only one of several ecosophies inspired by ecology and by the deep ecology movement. It need not be accepted by everyone. Those around the world who feel that there must be deep changes in human life styles and policies can be members of the deep ecology movement without having to accept ecosophical positions which they find confusing, don't understand, don't feel at home with, or simply dislike. I know I do not feel at home with Reed's ultimate views, but I am glad that they exist and that their proponents are strong supporters of the deep ecology movement.[4]

III. INTERPRETATIONS OF THE TERM IDENTIFICATION

Reed's primary concern in his paper is not the philosophy of *Self-realization*, but rather the notion of *identification*. The term *identification* as used in Ecosophy T "is rather technical, but there is scarcely any alternative."[5] There are several roots for this kind of use of the term.

The Social Science Context. In social science terminology one may say that "A identifies with his corporation more strongly than with his nation or with his family." Clearly, to make sense of this example one must rid oneself of the association of identification with resemblence. Furthermore, if A identifies with B, this does not weaken A's state of individuality. A does not fuse with B. Consider another example: "We must restore that high level of solidarity different groups of workers exhibited thirty years ago. But what can we do when workers today identify more closely with the middle class than with the working class?" The personal identities of these workers are not lost whether they identify with the working class or the middle class.

Nevertheless, acts of solidarity presuppose the process of identification and the same holds true for acts motivated by responsibility. In order to feel responsible for wilderness and for the penguins of Antarctica, a process of identification is a necessary but not a sufficient requirement. In this sense, "a process of identification" can be defined as "a process through which the supposed interests of another being are *spontaneously* reacted to as our own interests." This definition is of course intended to cover only beings which meaningfully, but not necessarily factually, can be said to have interests. The term *spontaneously* is stressed for reasons that might be clear from another example. If A strongly identifies with his organization (or his wife), the question as to how and where to

---

[4] All of this underlines the usefulness of distinguishing the four separate levels or components of ecosophies: the ultimate principles or norms level, the common platform (as a penultimate "philosophy of nature" position), the consequences of more or less general kinds, and the level of decisions in concrete situations (the "choice points," in the terminology of E. C. Tolman in his "Behavior at the Choice Point"). For a diagram of the four levels, see Devall and Sessions, *Deep Ecology*, pp. 225–26.

[5] Naess, "Identification as a Source of Deep Ecological Attitudes," p. 261.

spend his vacation will be answered in a way that spontaneously and implicitly takes care of the interests of his organization (or his wife), for he is not aware of his own interests as being potentially different from that of his organization (or his wife).

Insofar as my terminology parallels that of the social sciences, it is conceptually odd for anyone to hold, as Reed claims, that "some environmentalists seem to feel more *responsibility for* than *identification* WITH their natural surroundings" (p. 68). Reed is right, I think, as long as identification implies some kind of likeness or resemblance as it tends perhaps to do when used in ordinary conversation.

Alienation. The term *alienation* is extensively used in the social sciences and in philosophy. For example, if it is said that A suffers from alienation in relation to some group of people (B), or to a certain environment, a negative feeling or attitude on the part of A may, or may not, be implied. As I use the term *alienation* in Ecosophy T, negative feelings or attitudes are not implied. Rather, I am referring to indifference and "distance," together, implicitly, with some premises concerning what is a normal relationship between A and B. The alienation under discussion is usually partial however, in the sense that the indifference does not cover every aspect of A's relation to B. For example, A may be alienated from his neighbors in every respect except his interest in the amount of their income. A young person may be alienated from his society except for several very narrow personal concerns. As far as he is concerned, society may "go to hell" as long as these narrow personal interests are served.

The term *alienation* as used tentatively in Ecosophy T is an antonym for *identification*. This terminology is not very far from that of Spinoza (and others) when using *in se* (in self) and *in alio* (in other) as an ultimate ontological distinction that is more basic and deeper than, for instance, the ontological distinction between substance/mode or God/mode.

Social Self and Ecological Self. Just as *social self* is an important term in William James' great work, *The Principles of Psychology*, *ecological self* is destined to be an important term in ecosophy,[6] for the process of identification, as interpreted above, is closely connected with the process of the development of the ecological self. It remains to be ascertained, however, just *how* closely identification and ecological self are connected. The small child who identifies strongly with its mother, father, and other humans also identifies just as early with nonpersons (animals and other parts or aspects of its environment). Gestalts are formed which bridge the alleged fundamental gulf between humans and nonhumans.

---

[6] The concept of *ecological self* is introduced in Arne Naess, "Self Realization: An Ecological Approach to Being in the World," Roby Memorial Lecture, Murdoch University, Western Australia, March, 1986 (reprinted in *The Trumpeter* 4, no. 3 [1987]: 35–42).

Up to this point, the concept of a "process of identification" has been defined only for beings who meaningfully can be said to have interests. If there is a need to bring an essentially equivalent process under the same heading for those beings whom we normally would not say have interests, what can we do to satisfy this need? Two ways suggest themselves. First, we might introduce an *or* saying that the term covers beings with *interests or x*, introducing the appropriate term for *x*. Alternatively, we could extend the denotation of *interests* to include *x*. I choose the latter alternative. If we were to see ugly buildings or installations on the very summit of Mount Fujiyama, we would be spontaneously repelled by them, finding that the dignity, majesty, aloofness, etc., of the mountain had been violated. In this context, it would be in the interest of the mountain, as spontaneously experienced by us, to retain those characteristics on which its dignity, etc., rest. We would experience the same reaction if we saw people training an old polar bear to dance or saw the Pope advertising Coca-Cola.

So much for the *term* "process of identification." An empirical study to determine whether people become environmental activists largely through a sense of awe *or* through the process of identification could scarcely, as Reed suggests, rely on direct questions in terms of identification (pp. 68–69), for they would inevitably think in terms of resemblence or likeness, etc., rather than in terms of identification in the technical senses discussed above.

One of the things I most deplore is the predominance of the utilitarian attitude toward nature. This attitude encourages people to identify with nature narrowly in terms of its usefulness to humans and it promotes alienation from the nonhuman—alienation not only from those aspects of nature that elicit awe, but also those that elicit compassion for and decent treatment of nonhuman nature. A pig in a factory farm is not treated as well as a dog who is "a member of our family." The difference is immense. The fulfillment of potentials (of self-realization, I would say) is taken care of to a large extent in the latter case, but not in the former. The dignity of pigs, great waterfalls, or mountains is not considered when narrow economic considerations (bacon, electricity, tourism) are dominant.

IV. AWE AND SEPARATENESS

If A identifies with B, the nontechnical use of *identification* may lead to the assumption that A is familiar to B and, at least sometimes, familiarity may breed contempt. Reed points strongly to the intrinsic value of "things that awe and stupefy us," things that are dramatically different, *solidly apart*, from us:

> . . . it is our very *separateness* from the Earth, the gulf between the human and the natural, that makes us want to do right by the Earth. It seems to me that recognizing the moral distance between ourselves and the world helps us recognize the values in

nature that are totally independent of what we humans think is beautiful, right, and good. (p. 56)

Reed proposes an environmental ethic based on "awe for nature." Although this is a possibility that certainly deserves further study, here I limit myself to the relation of such an ethical possibility to conceptualizations such as *identification* and *self-realization*.

In the traditional Sherpa culture of the Himalayas, the holy mountain they call Tseringma (commonly known as Gaurishankar) is viewed with holy awe. This feeling well suits the descriptions by Rudolph Otto and Martin Buber of this kind of attitude. Nevertheless, the process of identification is also strong in the Sherpa tradition: the very name of the mountain means "the mother of the long (good) life." Although vast snowstorms and deadly avalanches proceed from its flanks, there is still a positive, intimate relationship between the people and the mountain. Traditionally, people throw a little beer in the direction of the "inhuman" mountain before drinking any themselves. The recent "conquest" of the "mother of the long good life" by climbers, the Sherpas believe, was disliked by her and resulted in a bad harvest. Using *is* in a way we are not acquainted with in industrial societies, Tseringma *is* also a beautiful princess and, of course, a mother. A community of about one hundred fifty people live just under the wild immense precipices of Tseringma. For these people, or at least the older generation, it is sacrilegious to "attack" and "conquer" Tseringma. Thus, when the community leader asked the heads of the forty-seven families whether they should try to protect the mountain against Western (and Japanese) climbing expeditions, they unanimously voted for protection, despite the vast amounts of money and goods (measured in their terms) to be gained from such expeditions.[7] In short, for the Sherpas, there was a relationship *both of awe and identification* with the mountain—humans *identifying* with the inhuman. It sounds strange and points to the limitations of the fruitful use of that technical term in ordinary communication. Nevertheless, in the articulation of ecosophies like Ecosophy T, it may sometimes be of help.

Here is another example of the combination of awe and identification. A boy about ten years old, after some years of acquaintance from afar, starts to conceive of a big mountain H as a kind of higher being, perhaps something like a godfather. Some years later, he begins to conceive of H as being serene, equinanimous, unruffled, strong, and great (not only big) in contrast to his own character, which he thinks of as being temperamental, shifting, listless, unbalanced, weak, and small. Clearly, the boy sees the mountain as having these characteristics in sharp contrast with himself. These characterizations of a thing

---

[7] The petition to save Tseringma from assault was not even answered by the Nepalese bureaucracy and the Alpine Clubs of the world showed little interest.

that is not "really" a person are usually called personifications or anthropomorphisms. They are said to be "projections" from the inner life of the boy and attached by him (like a stamp?) to the object. Yet, in spite of its unreachableness and lack of human or animal features, the boy conceives of the mountain as friendly and as somehow communicating this. He would certainly, if he had the strength, defend the mountain against any kind of destruction by fellow humans. Along the alienation/identification axis, the boy may be said to have identified so strongly with the mountain that the term *identification* in this case requires a broader definition than the one used in previous examples above.

What *surpasses* human levels may elicit awe. It need not, however, be big. Even a leaf, an atomic structure, the smallest part of the human body, may elicit awe. (Curiously enough, the Japanese tradition of gardening, which concentrates on small natural "miracles," may actually work against the efforts of a minority to save large areas of "free nature.") If something is vast, inhuman, and utterly indifferent from anything familiar, this does not *in itself* elicit awe. Nor do I see that we are led to protect it, or even to feel an obligation to protect it, contrary to Reed's claims. Rather, it tends to be meaningless. For example, "friends of big numbers," such as octillions, complain that most of their human friends find big numbers to be meaningless. They do not elicit awe.

The feeling of one's own nothingness and insignificance may occur consistently with feelings of nature awe, but in this case a real difference of attitude may be important. To me, the feeling of insignificance is transitory. The starry heavens may elicit this combination of insignificance and awe, but during long meditations, or several nights under the stars, an experience of expansion of the self may be had. To block out the sight of the heavens through artificial lights is felt by millions of people to be one of the saddest things that has happened to industrial societies. One of the sources of integration, of concentration on essentials, and of living at deeper levels, is lost.

The reduction of alienation and the increase of identification are obviously related to the process of increasing meaningfulness. If we look for an alternative to the technical term *identification*, the word *meaning* is worth considering.

V. A GULF BETWEEN NATURE AND ITS EVOLUTIONARY MONSTROSITY: MAN?

The separateness Reed experienced is unbridgeable: ". . . there is an existential gulf of awesome depth between ourselves and the Other, a gulf which no amount of 'identification in otherness' can span" (p. 59). Zapffe experienced this separateness in a terrifying way: he felt even his own body to be a genuine part of nature and as such, unbridgeably separate from him-*self*. He felt himself to be an evolutionary monstrosity with metaphysical needs which nature could not possibly fulfill. In nature, the individual *human* life is meaningless.

Nevertheless, Zapffe has always been a fierce opponent of the destruction of free nature. Whatever his *ultimate* views, his identification with the animal, plant, and mineral worlds has been unusually strong and varied.[8] The essential and highly original existentialist point of view of Zapffe may be summed up as follows: evolution misfired when it created the human brain. The brain allows us to contemplate life as a whole, and we necessarily ask: to what purpose? What purpose does life serve? The valid answer "Nothing" is not good enough. We cannot accept the meaninglessness of life as a whole. "But what if there were no tragedies?" Not decisive. "What if there were no excruciating pain?" Not decisive. "What if we all loved each other?" Not decisive. The hypertrophy of our brain has made us outgrow life. Leave the Earth to beings who cannot contemplate life as a whole!

Although the idea that there is an absolute gulf between humans and nature has not been a clearly recognizable force in the deep ecology movement, there could be a place for it. There is, for example, nothing in the eight points of the deep ecology platform that implies a denial of this absolute gulf as formulated by Reed or Zapffe. If it were included, it would probably belong to "level one" (the very ultimate norms and hypotheses) from which the eight points may possibly follow. Nevertheless, the particular ecosophy which has Self-realization as the ultimate norm is incompatible with the absolute unbridgable gulf hypothesis (on level one). If this hypothesis, together with a norm supporting "awe for nature," is taken to imply a subordinate norm of "extinction of mankind," then there is a clash with point one of the eight points: the intrinsic value of the diversity of human cultures.

The idea that an absolute existential difference and apartness is, as such, a source of awe and of norms of radical protection of nature is, to me and to many other supporters of the deep ecology movement, largely incomprehensible, but so are many of the ideas on the ultimate level. Differences of opinion at this level does not mean that we cannot continue to work together, as we have worked for a long time, at the other levels. What has the practical level to do with environmental *philosophy?* In philosophy conceived of as a total view, the verbal articulations of derived decisions in concrete situations are no less characteristic of that philosophy than articulations of a more abstract and general character.

Arne Naess

Council for Environmental Studies, University of Oslo, P.O. Box 1116, Blindern, 0317 Oslo 3, Norway

---

[8] In his great work *On the Tragic (om Det Tragiske)* Zapffe discusses animals and other natural beings in a way that exhibits a high level of empathy and positive identification in the technical sense.

[13]

Moral Considerability and Universal Consideration

Thomas H. Birch*

> One of the central, abiding, and unresolved questions in environmental ethics has focused on the criterion for moral considerability or practical respect. In this essay, I call that question itself into question and argue that the search for this criterion should be abandoned because (1) it presupposes the ethical legitimacy of the Western project of planetary domination, (2) the philosophical methods that are and should be used to address the question properly involve giving consideration in a root sense to everything, (3) the history of the question suggests that it must be kept open, and (4) our deontic experience, the original source of ethical obligations, requires approaching all others, of all sorts, with a mindfulness that is clean of any a priori criterion of respect and positive value. The good work that has been done on the question should be reconceived as having established rules for the normal, daily consideration of various kinds of others. Giving consideration in the root sense should be separated from giving high regard or positive value to what is considered. Overall, in this essay I argue that universal consideration—giving attention to others of all sorts, with the goal of ascertaining what, if any, direct ethical obligations arise from relating with them—should be adopted as one of the central constitutive principles of practical reasonableness.

We had the experience but missed the meaning.
—T. S. Eliot[1]

Other things enter the same rivers, and other waters pour in upon them.
—Heraclitus[2]

THE CONSIDERABILITY QUESTION AND ITS PRESUPPOSITIONS

It has widely been assumed that if we are to establish "an ethic *of* the environment," in Tom Regan's words, as opposed to "an ethic *for the use* of the environment," or an ethic "about" the environment, we must justify changing our criterion of moral considerability so that nonhuman beings of many sorts can

---

* Department of Philosophy, University of Montana, Missoula, MT 59812. Birch's professional interests include ethics, environmental ethics, and philosophy of ecology. He is the book review editor of *Environmental Ethics*. Portions of this paper were read at the Boston meeting of the International Society for Environmental Ethics in December 1990. He wishes to thank especially Jim Cheney, Anthony Weston, Andrew Brennan, John Lawry, Eric Katz, Holmes Rolston, III, and the two anonymous referees, John Kultgen and Paul Thompson, for their helpful criticisms and suggestions.

[1] T. S. Eliot, *Four Quartets* (San Diego: Harcourt Brace Jovanovich, 1971), p. 39.
[2] Milton C. Nahm, *Selections from Early Greek Philosophy* (Englewood Cliffs: Prentice-Hall, 1964), p. 70 (fragment 42).

thereby be *proved* to be morally considerable—to have moral standing and to mandate practical respect.[3] The presupposition that some such criterion, or algorithm, for considerability *should* be and *could* be established has shaped a huge amount of the work that has been done in environmental ethics. In this essay, I challenge this presupposition, argue that it is a mistake, and offer a remedy.[4] My remedy is a principle of universal consideration that requires giving moral consideration to anything and everything.

Kenneth Goodpaster introduced the expression "moral considerability" into recent ethics. Here is his general formulation of the moral considerability question:

> However the question gets formulated, the thrust is in the direction of necessary and sufficient conditions on X in
>
> (1) For all A, X deserves moral consideration from A.
>
> where A ranges over rational moral agents and moral 'consideration' is construed broadly to include the most basic forms of practical respect (and so is not restricted to "possession of rights" by X).[5]

As this passage reveals, the question of considerability has been cast, and is still widely understood, in terms of a need for necessary and sufficient conditions which mandate practical respect for whomever or whatever fulfills them. The discovery and formulation of the right conditions are supposed to give us *the* rule (*the* algorithm) for consideration by showing us how X in Goodpaster's formulation is to be instantiated.

---

[3] Tom Regan, "The Nature and Possibility of an Environmental Ethic," in *All That Dwell Therein* (Berkeley: University of California Press, 1982), p. 188. Holmes Rolston, III distinguishes between an ecological ethic and an ethic that is simply *about* the environment in "Is There an Ecological Ethic?" in *Philosophy Gone Wild* (Buffalo: Prometheus Books, 1986), p. 12.

[4] Eric Katz points out, correctly, I think, in his "Searching for Intrinsic Value: Pragmatism and Despair in Environmental Ethics," *Environmental Ethics* 9 (1987): 231-41, that holists in environmental ethics, such as Holmes Rolston, III and J. Baird Callicott, do not make the mistake I am concerned with here because "the idea of intrinsic value loses its sense in a holistic system. An emphasis on intrinsic value, indeed, would preclude the development of a holistic environmental ethic." Thus, Katz suggests that "the creation of a nonindividualistic and nonanthropocentric environmental ethic cannot be based on the search for an intrinsic value that will serve as the ground of moral obligation" (p. 236). Nevertheless, now the usual way of thinking will simply bring up the question of moral considerability again, this time at the level of the whole. Why should the whole planetary ecosystem (or whatever) be given special, moral, consideration? Does it have some special intrinsic value that grounds consideration for it? Although interesting answers have sometimes been attempted at this level, they are not successful in meeting the demand that stimulated them. Among other things, this essay is an attempt to guard holism against the resurgence of the moral considerability question at the level of the whole—by deconstructing the question entirely.

[5] Kenneth E. Goodpaster, "On Being Morally Considerable," *Journal of Philosophy* 75 (1978): 308-25. In this article Goodpaster offers what he calls a "life principle" as the answer to the question of moral considerability. Since his article appeared, there have been many other important treatments of the question, some of which will be mentioned and cited here.

What I take to be the root sense of "giving moral consideration," and the sense in which I mean to use the expression throughout this essay, is as follows:

> To give moral consideration to X is to consider X (to attend to, to look at, to think about, where appropriate to sympathize or empathize with X, etc.) with the goal of discovering what, if any, direct ethical obligations one has to X.[6]

When we take the question of moral considerability *all the way down* to the origins of ethical experience and reflection, to what I call "deontic experience," together with reflection on it, or to the deepest experiential ingredient of what Bernard Williams calls "practical necessity," we will find that the most abiding presuppositions of the question, as it has usually been understood, must be abandoned.[7] These presuppositions are that we can and ought to find, formulate, establish, and institute in our practices, a criterion for (a proof schema of) membership in the class of beings that are moral *consideranda*. These are the beings that have "moral standing," the beings that must be given moral consideration and must not be simply used, or used up, or somehow treated as we might simply please, or treated in a manner uninformed by respect, or treated in a manner that is in no way cognizant of the possibility that we might have direct obligations to them. In short, it has been presupposed (1) that when it comes to moral considerability, there *are*, and *ought* to be, insiders and outsiders, citizens and non-citizens (for example, slaves, barbarians, and women), "members of the club" of *consideranda* versus the rest; (2) that we *can* and *ought* to identify the mark, or marks, of membership; (3) that we *can* identify them in a rational and non-arbitrary fashion; and (4) that we *ought* to institute practices that enforce the marks of membership and the integrity of the club, as well, of course, as maximizing the good of its members.[8]

The problem of finding a criterion of moral considerability is rather clearly a function of imperial power mongering. That is, the question assumes that the

---

[6] A *direct* obligation or duty is an obligation *to* something, in contrast to an *indirect* obligation involving or *regarding* something because of a direct obligation to something else. A standard example is that of endangered species of wild animals. Do we have a direct obligation to them to preserve them (to species and/or individuals)? Or do we have an indirect obligation to preserve them because, say, we have a direct obligation to humans not to deprive humans of the ethical insight they can gain from various relationships with the endangered wild animals? For an account of the difference between direct and indirect duties, see Tom Regan, *The Case For Animal Rights* (Berkeley: University of California Press, 1983), p. 150. The giving of moral consideration centrally involves the attempt to ascertain whether one has direct obligations to the item considered, and what they may be. When the question is one of indirect obligations, then the focus of moral consideration is on that to whom or to which the direct obligation is owed, and from which the indirect obligation derives.

[7] Bernard Williams, *Ethics and the Limits of Philosophy* (Cambridge: Harvard University Press, 1985). There will be further discussion of the notion of practical necessity below.

[8] The pervasive and profound grip of these presuppositions is paradigmatically exhibited by the way the question of moral considerability is put in the introduction to a collection of papers given

moral legitimacy of the Western project of domination and control—and the practical existence of a criterion of moral considerability that legitimates this project. This assumption is an unfortunate, though almost always unintended, subtext of all the texts that address the problem, no matter how well intended they may be. Moral considerability is one of the credentials for membership in the elite club of those humans and nonhumans, like national parks, who are to benefit from the ultimately violent suppression and exploitation of the rest, of the Others (i.e., the objects that are taken as fit for domination and control). Without the assumption that this domination is legitimate, the general question of who and what are to benefit from human practices and institutions, and who and what are to make possible such benefit by being exploited, simply would not arise.

The generally unquestioned presuppositions of the considerability question should thus be understood as part of our imperialist legacy, from the Middle East and Athens to Europe and the Americas. Already at work in 3500 B.C., in *The Epic of Gilgamesh*,[9] these presuppositions eventually became central legitimating threads of the conceptual fabric of our modernist (Cartesian) Western tradition of imperial domination and oppression of humans and nature. That is, the imperial program requires some morally *in*considerable Others (or at least some hierarchy of more and less considerable Others) over which legitimately to exercise power and control, and a criterion of considerability handily creates such Others.

---

at a conference on the topic: "What, if anything, makes something valuable in itself? What loci of value command moral respect? Of what significance within the theory of value are such characteristics as being alive, sentience, having interests, autonomy, the capacity to participate in a moral community, organic or systemic integrity, and flourishing? Answers to these questions have implications for determining the moral standing of embryos and fetuses; endangered species; wilderness areas; ecosystems, etc." L. W. Sumner, Donald Callen, and Thomas Attig, eds., *Values and Moral Standing* (Bowling Green: Bowling Green Studies in Applied Philosophy, 1986), p. v. This is how the question of moral considerability is usually, almost, as it were, "naturally," understood in the terms of our mainstream tradition. For a sound introductory account of the problem of moral considerability as usually approached, see Donald VanDeVeer and Christine Pierce, eds., *People, Penguins, and Plastic Trees* (Belmont: Wadsworth, 1986), especially the "General Introduction," pp. 1-23. For an economical historical account of the evolution of the concept of moral considerability in Western culture, see Roderick Nash, "Do Rocks Have Rights?" *The Center Magazine* 6 (November/December 1977), p. 2. Another important text that reveals the strong grip of the usual way of thinking is Roderick Nash, *The Rights of Nature* (Madison: University of Wisconsin Press, 1989), which argues that even holistically inclined radical environmentalists are closet American-style liberals trying to expand the membership of the club of the morally enfranchised. VanDeVeer and Pierce list and briefly examine the criteria that have been proposed for moral considerability: personhood, potential personhood, rationality, linguistic capacity, sentience, being alive, and being an integral part of an ecosystem. Other criteria that have been proposed, some of which overlap or restate those just listed, include moral agency, possession of a *telos*, autonomy (moral or preference autonomy), having interests, being valuable in itself, having intrinsic value, having inherent value, having inherent worth, and "having an intelligible character."

[9] See *The Epic of Gilgamesh* (London: Penguin Books, 1960). Note also, of course, that the imperialistic assumptions questioned in this paper have not been confined to Western culture.

Without such a criterion, *non-consideranda* (Others) would not be possible. Thus, the institution of *any* practice of *any* criterion of moral considerability is an act of power over, and ultimately an act of violence toward, those others who turn out to fail the test of the criterion and are therefore not permitted to enjoy the membership benefits of the club of *consideranda*. They become "fit objects" of exploitation, oppression, enslavement, and finally extermination. As a result, the very question of moral considerability is ethically problematic itself, because it lends support to the monstrous Western project of planetary domination.

Contrary to our tradition on the question, the truth of the matter seems close to being, in Jim Cheney's words, that "moral regard is appropriate wherever we are *able* to manage it—in light of our sensibilities, knowledge, and cultural/personal histories. . . . [T]he limits of moral regard are set only by the limitations of one's own (or one's species or one's community's) ability to respond in a caring manner. . . ."[10] Certainly in many cultures moral considerability has been afforded to nonhuman beings of various sorts, and even in our own culture there are many people who do give consideration to nonhumans, such as wild animals, trees, mountains, wilderness, and farmland. Of course, their voices are generally marginalized. This essay may be viewed as an attempt to give voice to this marginalized sensibility in a way that mandates its being heard.

Are such cultures and persons confused, irrational, crazy, merely emotional, seriously mistaken, etc.? Consider Andrew Brennan's answer to this question:

> It might be objected that genuine concern for the continued existence, or maintenance, of an item is no evidence that such an item is morally considerable, or a possessor of any kind of moral value.
>
> But suppose my concern shows all the seriousness that is sometimes taken as the hallmark of the moral: my deliberations and actions are constrained in an especially compelling way by concern that the item in question be preserved, be left undisturbed or be maintained in a certain condition. Argument is now needed to show that my concern here is not moral, that the item is not receiving moral consideration from me and that the item does not therefore possess moral value.[11]

Brennan is right, I think, that the burden of proof is on whoever objects to the practice or instance at issue to show that it is somehow inappropriate, mistaken,

---

[10] Jim Cheney, in correspondence and in his "Eco-Feminism and Deep Ecology," *Environmental Ethics* 9 (1987): 144.

[11] Andrew Brennan, *Thinking About Nature* (Athens: University of Georgia Press, 1988), p. 140. Also see Andrew Brennan, "The Moral Standing of Natural Objects," *Environmental Ethics* 6 (1984): 35-56. The point is that if it walks like a duck, quacks like a duck, etc., then it's a duck unless the skeptic can do *his or her* job of showing that it's not really a duck. The burden of proof is on the skeptic. As will become clear below, I part with Brennan's apparent assumption in this passage that a positive valuing of an item is crucially involved in the giving of moral regard.

or not really a case of giving moral consideration. I want to argue that this burden of proof cannot be met without (erroneously) assuming the ethical legitimacy of the imperial modernist project of domination (and the anthropocentrism that goes with it). In sum, I am claiming that it is ethically wrong to suppose that we need and ought to establish a criterion of moral considerability, and that just supposing that we *can* do it means primarily that we intend to persevere in our unethical Western imperial venture. That is, putting any criterion into practice is an act of domination, an arbitrary act of power and violence to the beings that are thereby rendered Other (i.e., constructed as objects for domination and control).

The assumption that we can and ought to establish a criterion of considerability should therefore be abandoned. Once it is, however, we come to the perhaps startling (and to those who are captivated by this assumption, seemingly bizarre) realization that everything must be given moral consideration.[12]

RECONSTRUCTING THE WORK THAT HAS BEEN DONE

In recent years much high quality work in search of a criterion for moral considerability has been done, for example, by Tom Regan, John Rodman, and Paul Taylor.[13] But no one has established *the* criterion of considerability. Because

---

[12] Here is some further evidence of the pervasive grip of the standard moral considerability presuppositions: (1) W. Murray Hunt, in his "Are *Mere Things* Morally Considerable?" *Environmental Ethics* 2 (1980): 59-65, argues against Goodpaster's proposal that "being alive" is the criterion, that "being in existence" is just as plausible on the same arguments that Goodpaster gives for "being alive." He offers this conclusion as a *reductio* of Goodpaster's proposal because it makes "mere things" morally considerable. Nevertheless, without Hunt's unquestioned assumptions that there has to be a criterion and that therefore something, obviously "mere things," couldn't possibly meet it, there is no *reductio*. (2) In "A Refutation of Environmental Ethics," *Environmental Ethics* 12 (1990): 149-60, Janna Thompson argues that an ethics of the environment is virtually impossible because of the virtual impossibility of finding a criterion for intrinsic value in nature (to serve as the needed ground for moral considerability) that is consistent, non-vacuous, and usefully applicable in deciding whether something is intrinsically valuable or not. Her supposition is clearly that there is a need for a criterion if any ethics of the environment is to be developed plausibly. Her argument—that an ethics of the environment cannot be had because this need cannot be met—collapses as soon as we see that there is no need. Also, if we insist, as Thompson insists, that a criterion of moral considerability is necessary for morality to be morality, it follows that morality is morally wrong. To dispel this seeming paradox, we have to conclude that the prevailing instituted "morality system," in Bernard Williams' terms, is ethically flawed. The difficulty we have in justifying moral consideration for nonhumans, in terms acceptable to the prevailing morality system, should be taken as evidence of this flawedness. Note, however, that Thompson's article does a marvelous job of refuting all attempts to establish a criterion of moral considerability.

[13] See Tom Regan, *The Case for Animal Rights* Berkeley: University of California Press, 1983); John Rodman, "Four Forms of Ecological Consciousness Reconsidered," in Scherer and Attig, eds., *Ethics and the Environment* (Englewood Cliffs: Prentice-Hall, 1983); and Paul W. Taylor, *Respect for Nature* (Princeton: Princeton University Press, 1986).

it has already been demonstrated by others, I do not attempt to make this point here.[14] What I want to do instead is show how this work should be understood, what it amounts to, and how it should be reconceived and reconstructed. (This reconstruction itself provides argumentative support for my contention that the usual presuppositions of the moral considerability question are mistaken and for my thesis that universal consideration should be adopted.)

In *The Case for Animal Rights*, Tom Regan says that his criterion for "inherent value" for animals (being the "subject-of-a-life") is sufficient, but not necessary, for inherent value and thus for moral considerability.[15] In an earlier article, Regan held that inherent value is an "objective property" of the things that have it—a "fact about" such things—and that "the inherent value of a natural object is such that toward it the fitting attitude is one of admiring respect."[16] For Regan, moral considerability hinges on, requires, and is a function of inherent value, which is the sort of value that mandates moral consideration and an attitude of admiring respect. After making the case for the moral considerability of animals in terms of his "subject-of-a-life" criterion, he writes that

> ... the very possibility of developing a genuine ethic *of* the environment, as distinct from an ethic *for its use*, turns on the possibility of making the case that natural objects, though they do not meet the subject-of-a-life criterion, can nonetheless have inherent value.[17]

Regan intentionally leaves open the possibility that someone might find some additional justification for inherent value in the further realms of nature, beyond the realm of mammals for which he has established it. The implication, or at least his hope, is that there are other sufficient criteria for inherent value.

I suggest that Occam's razor could well be used to trim out the notion of inherent value from our theorizing about the fundamentals of consideration, and to trim away Paul Taylor's notion of "inherent worth" and other similar notions as well. In Regan's theorizing, the concept of inherent value does no real work because it has no further content beyond what Regan has given it with his "subject-of-a-life" criterion, and perhaps with his notion of admiring respect. In fact, in general, the concept of inherent value and its cousins have no other content than that which comes from the reasons that comprise a successful establishment of a form of moral consideration for whatever can be successfully established as requiring a kind of consideration. (In itself, the notion of inherent value seems to be an empty

---

[14] For example, by Thompson, "A Refutation of Environmental Ethics."

[15] Regan's formulation of the "subject-of-a-life" criterion can be found on p. 243 of his *The Case for Animal Rights*.

[16] Tom Regan, "The Nature and Possibility of an Environmental Ethic," in *All That Dwell Therein*, pp. 199-200.

[17] Tom Regan, *The Case for Animal Rights*, p. 245.

place marker demanded by the logic of our traditional kind of theorizing about moral considerability: the demand for a criterion.) What Regan has actually given us with his "subject-of-a-life" criterion is a justification for the ethical judgment that we ought to give animals moral consideration in a certain way. In a word, he has given us a rule for morally attending to animals.[18]

The upshot is that what Regan and others *have* shown us is that what *can* be found and established is not a criterion, or criteria, for moral considerability, but simply various reasons for practicing moral consideration in various ways in ethically responsive and responsible relationships with various sorts of other entities. That is, Regan seems to be right in holding that one way to practice our moral consideration for certain sorts of animals is to institute a recognition that they have certain sorts of rights. Rodman has shown us that, at least insofar as we can see that some entity has a *telos*, we must show our consideration by worrying about our possible interference with that entity's realization of its *telos*. Taylor has done much the same in the case of entities that can be said to have "a good of their own." Thus, instead of one criterion of moral considerability, there turn out to be various rules for the practicing of various sorts of consideration in relation to various kinds of things. Establishing such various sorts of consideration is what all the sound work on the moral considerability question has accomplished, and how it should now be reconstrued and understood.

METHODS WE MUST AND ALREADY DO USE

In dealing with the moral considerability question we should pay some attention to the philosophical method we use, which seems to be the only method available.

---

[18] Regan's further contention that the way we should show our consideration is to institute a practice of recognizing animal rights is more controversial, though he makes a strong case indeed. Certainly the recognition of rights for animals does not exhaust completely the shape of the ethics of our relating with animals. However that matter turns out, it will not affect the case Regan *has* made in justification of moral considerability for animals (or, more precisely, for mammals, as Callicott has pointed out). It is worth noting that "having rights" is not a necessary condition for moral considerability, and Regan has not made the mistake of saying it is. For him, correctly, the establishment of moral considerability comes first, and the recognition of animal rights is then meant to be the ethically proper, the responsible, response—part, at least, of the way in which this sort of consideration is to be put into practice. In contrast with Regan and Taylor, Rodman, in concluding that "all items having a *telos* are entitled to respectful treatment," does not seem to be offering some general criterion of "inherent value" or "inherent worth," or, in his terminology, of "intrinsic value." Rather he appears to be suggesting that what it *means*, at least in part, to have the kind of "intrinsic value" that is a sufficient warrant for moral considerability is to have a *telos*. His theorizing is thus not cluttered by an empty notion of inherent or intrinsic value. He strikes directly for a general and sufficient criterion for moral considerability, and he argues that having a *telos* is the criterion, at least in the sense of being sufficient. Perhaps the main problem with this suggestion is that there do seem to be items that mandate consideration precisely *because* they have *no* scrutable *telos*, notably items that partake of the sublime. (The same sort of point holds for Taylor's "good of its own.") I am responding here to Rodman's "Four Forms of Ecological Consciousness."

and which is obvious enough to be often overlooked. The central elements in the logic of our method are reasons and arguments to support consideration for certain kinds of things—reasons and arguments that establish policies and rules for considering and attending to others of various kinds in various ways. To understand our method, we must ask what our reasons and arguments are based on, what their grounds are. The grounds are the characteristics, the interests, the *tele*, etc. of the various things we are considering. The method that we cannot help but use is to give them consideration in what I called the root sense above: to attend closely to them with the goal of ascertaining what direct obligations, if any, we may have in relating to them responsibly, what behavior is appropriate in relation to them. Notice, for example, just how closely, carefully, and thoroughly Regan attends to and considers animals. I see no a priori reason why similar close attention should not be given to trees and rocks, and everything else, if our philosophical concern is to see whether or not they should be given practical respect. In fact, it is mandatory. We must do it as responsible philosophers. I suggest that the logic of our philosophical method and practice is, perhaps surprisingly, the logic of consideration in real practical life, or at least something of what that logic should be. What philosophers have to do is what we should recommend to all: universal consideration.[19]

Historically, environmental consciousness and concern have thrown the moral considerability question wide open. From the historical perspective, we see that whenever we have closed off the question with the institution of some practical criterion, we have later found ourselves in error, and have had to open the question up again to reform our practices in a further attempt to make them ethical. The lesson of history is that we must open up the question of moral considerability and *keep it open, not* close it off again by instituting practices based on the latest, and no doubt mistaken, "final" criterion. Of course, keeping the question open *does* mean abandoning the usual presupposition that the question can and should be closed, and resisting the customary impetus toward closure. Keeping the question open requires taking the question *all the way down* and reconstructing our practices of consideration accordingly. When we do so, we will find a new/old logic for a deeper practice of consideration that is open in and by principle and that in part has the shape of what we are right now doing philosophically as we work through recently opened territory. It is basically a matter of owning up explicitly to what we are already doing as ethically concerned environmental philosophers and thoughtful persons.

Raising the question of moral considerability requires us to reconsider our practices in both their general form and in their particular applications. It thus also requires us to *re*-consider the items that are and that are not the objects of

---

[19] Jim Cheney has given us illuminating suggestions about how our method should be more finely tuned, particularly in respect to place, in his "Postmodern Environmental Ethics," *Environmental Ethics* 11 (1989): 117-34.

consideration. Nevertheless, to *reconsider* them is also to *consider* them, to give them moral consideration and the most basic sort of practical respect. Things must be looked at afresh, to see what practices we should institute in relationship to them. This is what it means to give something moral consideration in the root sense, defined above. Our history lesson, thus, is that we should institute practices of universal and continuous consideration, and reconsideration, at least in our present moment in history—and, because we are historically located beings, as far into the future as we can see.[20]

Our historicity is one further reason to hold that the question of moral considerability is perennial and specific to the items with which we are in relationship. Moral practices and institutions (as well as legal institutions) are ethically justifiable only insofar as they facilitate the actual fulfillment of our ethical obligations. Their ethical acceptability is contingent on whether they do so. Obviously, however, our ethical obligations cannot be conscientiously pursued or fulfilled until they are first discovered. Thus, the most fundamental job of the entire business of ethical research is the discovery of our obligations. Nevertheless, it is not possible to discover our obligations to others, of whatever sort, unless and until we give them moral consideration. Only then can we discover what, if anything, is required of us by the relationships in which we live, or ought to live, with them. One of our most fundamental obligations (a *sine qua non* of ethicality) is discovering our obligations. Giving moral consideration to others is necessary to fulfill this most fundamental obligation. If giving universal consideration is necessary to fulfill this most fundamental obligation, then giving universal consideration is itself likewise a most fundamental obligation. (Also, it is a "practical necessity," in Bernard Williams' sense, as I explain below.)

DEONTIC EXPERIENCE AND ITS REQUIREMENTS

By far the most important reasons for reconstructing our practices of moral consideration stem from the shape and place in the logic of practical reason of what I call "deontic experience." Deontic experience is the experience, in response to something or someone, that one *must* do something, that one is called upon to do something. It is the point of origination in our experience of our ethical obligations, of the deontic urge; it is the experiential source and the inspiration of our obligations. Because our values are the ends sought in the doing of what we *must* do, deontic experience is also the original source of our values.[21] Deontic experience is an indispensable part of the testing and proving ground of obliga-

---

[20] Universal consideration, therefore, turns out to be, in the words of Sabina Lovibond, "transcendentally parochial." See her *Realism and Imagination in Ethics* (Minneapolis: University of Minnesota Press, 1983).

[21] My position on the relationship of values to obligations and practical necessity, although perhaps within our deontological tradition in ethics, and present in the thought of Hume, is unorthodox from the viewpoint of our mainstream and teleological tradition, which takes values as

tions, as well as our rules, practices, and institutions. If our deontic experience fails to confirm, or more commonly falsifies, what we have taken to be an obligation, it is time to question and reconsider that now putative obligation. When, for example, our experience does not conform to what our culture has taught us, we properly question these teachings.

The term *deontic* comes from a Greek word meaning "it is binding," or "it behooves." The Greek word has sometimes mistakenly been taken to mean "it is one's duty." As Williams points out, in a remark about *deontological*, the term "is sometimes said to come from the ancient Greek word for duty. There is no ancient Greek word for duty: it comes from the Greek for what one *must* do."[22]

A deontic experience, even of the most powerful sort, is never sufficient to determine an obligation (or a practical necessity), although at some point it is necessary, in the experience of someone. In great measure, the ethical collapse of contemporary Western culture may be attributed to the general deprivation of deontic experience, in favor of canned, preprogrammed experience, and to the neglect and demeaning of it, as well as to the general lack of institutions of any sort, social or personal, for handling it (in contrast, for example, to the social institutions of Native American cultures for dealing with visions and vision quests). A deontic experience is not sufficient to generate an ethical obligation because we may feel strongly called upon to do any number of unethical things. What we take to be an angel calling us to act may turn out, upon reflection, to be Mephistopheles in disguise. What we are strongly called upon to do is not self-justified by the deontic fact of being called upon to do it. It is, therefore, mandatory to assess, to consider, what we feel we must do in terms of our funds of practical ethical knowledge and wisdom, both conceptual and experiential, our knowledge of principle and rules, together with our accumulated practical wisdom. Only then may we discover an ethical obligation, or some other sort of practical necessity. Deontic experience can be generated out of a relationship with any kind of entity: persons, things, systems, ecosystems, other sorts of abstractions, even numbers.

Deontic experiences are things we confess, things about which we testify, things to which we bear witness in our ethical practice, and things about which we tell the stories that constitute our personal and cultural mythologies or worlds. We turn to our deontic experiences and speak of them when we are pressed to explain and justify, and prove to others, our ethical judgments and practices. We turn to them in the course of our own deliberations, to test and prove practical ethical hypotheses.

Although deontic experiences may sometimes be epiphanic, it is doubtful that they are generally so or very often immediately so. That is, the meaning of what

---

the ultimate primitives in ethics. Although I do not explicitly argue for my non-teleological position here, I do want to note that the overall argument on the topic of moral considerability *is* indirect argumentation for it.

[22] Williams, *Ethics and the Limits of Philosophy*, pp. 187-88.

one experiences is not generally so suddenly grasped or instantly clear as in epiphany. More often than not, it takes time, often many years, for such experience to become clarified enough to know its meaning or even to know that we are having it at all.[23] Until we discover that meaning, we often cannot know that we have had the experience. Still, when deontic experience does occur we often know at some level that something has happened that will have to be reckoned with sooner or later. Honing our capacity, our virtue, to recognize such experience consciously and in all the infinite richness of its particularities is a part of what is required to cultivate a deep practice of ethical consideration. Whetted to the point of the ideal, which we might call the point of perfect virtue—a flawless spontaneity—the practice of giving consideration would be the continuous realization of the epiphany of every moment.[24] Even moments of conscious considering, of looking, watching, attending to others of all sorts, to see what is called for in relation to them can become epiphanic moments. When it happens, one realizes part of what it is to be a human being.

Some of my own experience with rocks may clarify these observations. Because I was a child of my culture, climbing and caving were primarily a matter of feats of derring-do, pioneering, going to places where no "man" had gone before, and conquering them. Although at the time I must have known that it meant something, though not what it meant, it has been chiefly in reflection that I have come to identify one of the more important experiences that changed my own mode of consideration of rocks. On one such occasion, I was climbing a route on the First Flatiron overlooking Boulder, Colorado. The route included what we then called an "exciting" hundred-foot friction pitch. A friction pitch has no handholds or footholds, though there are usually slight undulations, slight variations in the steepness of the rock. To climb a friction pitch of this sort, one must first study the entire pitch, planning each placement for hand and foot, and then fully commit oneself to the pitch. There can be no stopping; you cannot miss a placement. If you stop, the momentum that provides the friction that is necessary to stay on the rock

---

[23] In some part, this can be explained as a result of our lack of practice of consideration, particularly in Western culture, which suffers from the modernist and anthropocentric constructions of moral considerability, and the notion that ideal ethicality is algorithmic, the view that I am trying to get beneath in this essay. These notions tend to dull our vision, to blunt the keenness of our attention, and thus impoverish our capacity to recognize and reckon with deontic experience when it occurs.

[24] In this connection see Stanley Cavell's discussion of the central image of chanticleer in Thoreau's *Walden*, in his *The Senses of Walden* (San Francisco: North Point Press, 1981). In his "Walking," Thoreau called it walking, or sauntering. In the Zen tradition it is often called mindfulness. See, for example, Shunryu Suzuki, *Zen Mind, Beginner's Mind* (New York: Weatherhill, 1970), p. 115: "If we are prepared for thinking, there is no need to make an effort to think. This is called mindfulness. Mindfulness is, at the same time, wisdom. By wisdom we do not mean some particular faculty or philosophy. It is the readiness of the mind that is wisdom. Wisdom is not something to learn. Wisdom is something which will come out of your mindfulness. So the point is to be ready for observing things, and to be ready for thinking. This is called emptiness of your mind."

is lost, and you slide off the rock. I do want to emphasize that the pitch I am describing is not one that experienced climbers today would consider very difficult. But for me, at that time, there was no "conquering" of this pitch. It required complete attention to the rock and the manner of participating with the rock in the climbing—complete immersion in, absorption into the relationship with the rock, an absorption of the sort that required complete attention to the rock itself.

Although I didn't consciously realize it at the time, in retrospect, I think that I was stunned in such a way that my normal manner of considering the rock was stopped, and having been cleared and cleansed of it, and, now empty, I was opened to seeing the rock as a being in its own right—and because I had been compelled to extend such complete attention to it, *had* to reconsider it, or *had* to consider it ethically for the first time. I can recall vividly, over thirty years later, many details of slope—the smell, color, temperature, and texture of that rock—as I encountered it at that moment. This experience, together with similar experiences in relation with other rocks, led over the years to a new practice of consideration of rocks generally, a practice that involves attentive appreciation of their particular characteristics, an intentional looking to see what those characteristics are and what the appropriate response to them might be—a practice, moreover, that requires minimizing my impact on rocks, and trying to use them well, when I do use them, with appreciation, respect, and care, hopefully in the manner in which an artisan or artist uses his or her materials.[25]

Most significantly for present purposes, my specific and cumulative deontic experience with rock, in forcing a reconsideration of rock and my practices regarding rock, mandated a reconsideration of moral considerability itself, which, in part, produced this essay. If our concept of moral considerability could block us off from such a thing as proper rock practice (because of the presupposition that there is a general criterion of moral considerability), then it would be sure to taint our practices in relation to plants, trees, forests, animals, and probably with friends, lovers, colleagues and associates, and even with scientific, literary, and philosophical texts and ideas. The entire business of consideration would be called into radical question.

The moral of this story is that when we do take the question of moral considerability "all the way down" to deontic experience, which *is* at the bottom of it all, we discover the need (the practical necessity) to reconstruct our understanding and our basic practices of moral consideration. We need to make

---

[25] Among climbers, deontic experience of the sort I have pointed up in my own case appears to be common. It has led to new climbing practices, such as the use of chocks and nuts rather than pitons in order to lessen impact on the rock. Of course, as we would expect, these new practices are often rationalized away in terms of protection and preservation of the route for other climbers. Nonetheless, I submit that such reductions do not adequately express the ethical realization of many, at least, of those who make them, and that they could be brought to agree that a deeper consideration of the rock is what is centrally at stake.

these practices consistent with the exigencies of our deontic experience. Western culture has misconstrued and largely missed the ethical significance of deontic experience. A close look at our deontic experience shows that an approach to experience with an a priori criterion of moral considerability blinds us to the meaning of our experience. That is, such an approach cuts us off from seeing and reckoning with the experience that is the source and test of all our hypotheses about obligation and value. For this reason, it must therefore be abandoned.

The character of deontic experience, what we see when we do *look* at it, gives us good reasons for developing and instituting a practice, as our most fundamental and continuing practice, of universal consideration. Moreover, the study of deontic experience can show us some of the constituents of the basic reconstructed practice of universal consideration that deontic experience calls for us to cultivate.

RECONSTRUCTING CONSIDERATION

A reconstructed practice of consideration requires, and is required by, *a staying in the place of the origination of practical necessities*.[26] I state this point this way, in terms of practical necessities, in order to accommodate Bernard Williams' illuminating proposition that what lies at the bottom of morality and ethics (which he distinguishes as the broader notion) is "practical necessity":

> When a deliberative conclusion embodies a consideration that has the highest deliberative priority and is also of the greatest importance (at least to the agent), it may take a special form and become the conclusion not merely that one should do a certain thing, but that one *must*, and that one cannot do anything else. We may call this a conclusion of practical necessity.... This is a "must" that is unconditional and *goes all the way down*.[27]

Deontic experience is the source and test of what is of the greatest importance to us. To discover practical necessities, we have to filter what we are called upon to do in deontic experience through our fund of practical wisdom. For instance, Meursault's murder of the man on the beach, in Camus' *The Stranger*, would not be an example of an action that is practically necessary because the deontic urge has not been deliberated upon or run through the filter of practical wisdom. What Williams calls our "peculiar" present institution of morality (which includes our prevailing anthropocentric rules of moral considerability) should be seen as our culture's attempt to structure the practical wisdom that is needed to deliberate on deontic experience in order to ascertain practical necessities. Environmental

---

[26] Thoreau called staying in this place of origination "walking." See also many Native American songs and chants about walking in beauty, in the beauty of the paradise given to us. For just one example, see N. Scott Momaday, *House Made of Dawn* (New York: Harper & Row, 1977), pp. 134-35.

[27] Williams, *Ethics and the Limits of Philosophy*, pp. 186-88.

consciousness, which includes deontic response to such things as the extermination of other species as well as dangers to our own survival, is one of the factors in our postmodern era that is now forcing us to reexamine, reconceive, and reconstruct practical wisdom. Universal consideration, together with the abolition of the old moral considerability question, will be one of the indispensable ingredients, one of the constitutive principles, of a reconstructed understanding of practical wisdom.[28] In other words, whether the others involved have been given adequate consideration will always be a relevant factor in the justification of any practical ethical judgment.

On the basis of entries in the *Oxford English Dictionary*, to consider something is (1) "to view or contemplate it attentively" and/or (2) "to show regard, respect for, to think highly of it, or to give it positive value." Because of our culture's neglect of deontic experience and our imperialist presuppositions about criteria, we have mistakenly taken the second definition (thinking highly of something, valuing something) to be separable from the first and to be the primary meaning. The result has been an ethical practice that requires attentive contemplation only for those entities that are already favorably valued. The rest are then neglected, disregarded, not attended to, without taking the time to determine whether we might have some direct obligation to them. I recommend a recovery of the older priority, according to which the primary meaning of giving consideration is thoughtful, reflective, meditative attentiveness.[29]

Because giving this sort of consideration, or attentiveness, to something does not settle the question of the appropriate obligation or practical necessity, if any, that may arise in relation to what is so considered, it does not produce the paralysis for action that *would* generally follow from having to give everything high regard or positive value (the second definition). The obligation, or practical necessity, may turn out to be to destroy the item considered, or to despise it, or to give it negative value. The need to eradicate AIDS, or perhaps a killer grizzly bear, does not, however, preclude, and actually requires, respecting them. Any competent warrior knows that it is a practical necessity to respect an enemy.[30]

There is a further way in which the first definition, which I am suggesting should be taken as the primary sense of *consider*, involves the second, giving high regard

---

[28] The term *principle* is used very loosely in ethics. Although space does not permit me to develop and explain how and why here, I mean something quite specific by the term, and I contrast ethical principles with rules. For a start on how these terms should be understood see Marcus G. Singer, *Generalization in Ethics* (New York: Alfred A. Knopf, 1961), p. 103. Universal consideration is, I am suggesting, one of a number of ethical principles, which constitute the indispensable guidelines of practical reasonableness, but which will not serve as algorithms for impeccability in judgment. Nevertheless, failing to follow them, failing to give adequate consideration to others, is a partial failing in practical reasonableness.

[29] According to the *Oxford English Dictionary*, this sense of *consider* is somewhat archaic. To the extent that this is (sadly) so, we need then literally to *retrieve* or *recover* something that has nearly been lost.

[30] Such respect is more than simply close attentiveness. It seems also to involve valuing the enemy

or positive value, as in the quote from Brennan above. Acceptance of this connection should help dissolve any remaining puzzles about how something can first be given ethical consideration and then be valued negatively in normal practice (as we attempt to exterminate AIDS), but still be given constant deep consideration. The point is that giving consideration in the first sense of giving attentive contemplation *should* involve an initial generosity of spirit toward all others in terms of their value potential, the possibility that they might have positive value of the usual sort.[31] This attitude can be summed up as an approach to others such that for any X, X is *prima facie* valuable (has positive value). It is a matter of giving others of all sorts a chance to reveal their value, and of giving ourselves a chance to see it, rather than approaching them in hostility as if they have nothing but negative value until they have proved otherwise.

In imperial culture, dedicated to domination, this negative approach to others puts the burden of proof of their positive value on them, the others. Their *prima facie* value has to be negative, and *they* must prove themselves valuable by showing themselves amenable to, or useful for, etc., the project of domination. That is, everything is taken as a *prima facie* threat to the project, which indeed everything does seem to be, at least prior to co-optation, brainwashing, or some other form of sterilization. In this way, the *prima facie* supposition and allegation of negative value is "rational," given the character of the imperial approach. (Even Penelope is to be viewed with suspicion, as potentially a Klytemnestra, until, with twenty years of flawless "faithfulness," she proves herself worthy of Odysseus.)

In contrast, universal consideration requires a reversal of the customary Western placing of the burden of proof. Others are now taken as valuable, even though we may not yet know how or why, until they are proved otherwise. And even after such a proof, the possibility of finding positive value, after reconsideration, is left open, insofar as this is practically feasible. We have recently, for example, reconsidered wildfire, and can now see its positive value even though it may still be practically necessary to act in a way that shuts down similar possibilities, as in the case of extinguishing AIDS, etc. Within the human milieu, because they have cultivated a generosity of spirit, wise and enlightened people already treat other

---

other, though this is perhaps somewhat strained and abstract, for making practical necessities possible for us, for giving us something important and worthwhile to do, for being a part of the whole that makes all this possible. Of course, none of this justifies creating enemies.

[31] The reason why universal consideration should require, and not merely allow, the generosity of spirit that gives *prima facie* positive value, and potential high regard, to others of all sorts is that it seems practically wise. Approaching others with sensitivity to their possible positive value tends to bring that value potential into actuality in practical life. It encourages a climate of trust, of community, of beauty, of love, an atmosphere that fosters the general realization of positive value and of everyone's practical necessities. In contrast, the meanness of the imperial approach generates distrust, hatred, disvalue, and a general social, personal, and ecological disintegration that stifles even the possibility of doing what we must. Universal consideration permits, and therefore requires, us to take up practical residence in the story that we should be telling and trying to inhabit, and to disinhabit the prevailing story or stories of Western culture.

people as *human beings* until it is proved otherwise—there is no a priori requirement for another person to prove his or her worth. Universal consideration requires the extension of the same attitude toward the nonhuman world. Of course, such an extension does not rule out a proper caution and guardedness, or even the sometimes instant recognition of phoniness, deceitfulness, dangerousness, and monstrosity—a monster *is* a *monster!*

In sum, giving all others practical respect or consideration in the primary sense does not involve valuing them positively for purposes of ordinary practice; nor does it make them members of some "moral community" of esteemed *consideranda*. Whether something is to be esteemed or despised in normal practice can only be discovered by giving it consideration in the primary sense. Our reconstruction of consideration must, therefore, recover and put into practice this primary sort of consideration.

With this reconstructive goal in mind, it is helpful to distinguish between (a) *daily*, or normal, practices of consideration and (2) the *deep* practice of consideration. (In Western culture we have confused the priorities of these two in a manner that recapitulates our confusion about the meanings of *considering*, as we have mistakenly absorbed deep consideration into the daily, normal variety.) Daily practices of consideration are the various ways of showing consideration for the various sorts of items that we have found it practically necessary to follow in order to meet the practical necessities that arise out of our relations with these items. Daily practices can be formulated as rules for giving consideration. For example, if Tom Regan is right about animals, then we should institute a daily practice of considering them in terms of their rights. Such a practice would partially characterize the proper, normal way to think about, to attend to, animals and the questions about how we should relate with them. In just this manner, there is widespread agreement that, when it comes to the ethics of relating with other human beings, a daily practice of respecting their rights is a practically necessary institution, because it helps us fulfill at least some of the most important sorts of ethical obligations that seem perennially to arise.

Yet, important though various daily practices may be, and even though it may be practically necessary to have them forever (given our finitude, our fallibility, and the complexity of the world), they are not what is primary. Like other moral rules, they are defeasible, not always relevant, and contingent on historical and cultural location. Also, it is theoretically possible to dispense with them altogether, given a sufficiently honed practice of attentiveness to others, given the perfect virtue mentioned above, given the perhaps infallible spontaneity of enlightenment. They were unnecessary before the Fall, and they would once again become unnecessary after the coming of the Kingdom. (Of course, it would be the usual mistake of fanatics for us usual mortals to believe that we had somehow achieved an enlightenment that permits abrogating, without rational justification, the usual rules of tradition.) What is most important about the rules of daily practices is that they originate in and are reconsidered at a deeper level of consideration. The rules for daily practices are notoriously insufficient and fallible as guides for discovering our obligations. That is, daily practices involve considering other beings in terms of specific categories that may not be sufficiently sensitive to the case at hand. As a result, we must acknowledge the need for, and then institute and cultivate, a deep underlying practice of consideration that requires a complete openness of thought and vision regarding everything and anything. Such openness will be principled, and we will generally, or at least often, be able to tell when we have failed to attend closely enough to others. To be sure, there can be no rules in the sense of effective procedures or algorithms for successful deep consideration. Our overall practices of consideration, and especially our daily practices, must come to be consciously grounded in and constantly informed by deep consideration.

The recurring mistake in Western ethics has been the abandonment of the source of insights about moral considerability in a haste to institute some daily practice of consideration as final and complete and effective as a proof of obligations—to make a rule that is appropriate in limited ways into an overall algorithm. There is no way, or finite set of ways, to consider anything, much less everything. At bottom, there is just the ethically right way, or ways, to respond to each particular entity, being, or person, and at particular times and in specific places. The right way often, of course, may be to do nothing at all. In any case, the right way cannot be found unless each person, each being, each entity is always, insofar as it is humanly possible, considered afresh, and in the context of a fresh look at the whole. Any practice of consideration that is not thoroughly informed with a fundamental openness and attentiveness to others of all sorts is itself a virtual algorithm for eventual ethical failure. This point is nothing new. We have known it in Western culture at least since Socrates, Sophocles, and others called it to our attention. Yet, we have not put this knowledge into practice. It is time we did. It is a practical necessity.[32]

SOME CONSEQUENCES FOR ENVIRONMENTAL ETHICS

There are four consequences that establishing universal consideration has for environmental ethics and for ethics in general.[33] First, it helps make some sense

---

[32] Because universal consideration requires constant, thoughtful attentiveness to others of all sorts, we might well call it the principle of attentiveness. Attentiveness is a prerequisite for making ourselves available to deontic experience. Attentiveness is necessary for the proper exercise of daily practices of consideration, to maintain and further their sensitivity and accuracy of application. Most important, following the principle of attentiveness is practically necessary in order to test our presently instituted and proposed daily practices of consideration, and to determine their ranges of appropriate application.

[33] The fact that universal consideration has these consequences is a further and important argument for it, because illumination and coherence is brought into areas that otherwise stay foggy and confused.

of "biocentric egalitarianism," though the egalitarianism extends farther than just the sphere of life. All things are "equal" in that they must all (including the whole biosphere) be given ethical consideration, both as individuals and as constituents of various wholes.

Second, universal consideration gives us a principled way to begin to practice the ecological feminist concern to make "a central place for values of care, love, friendship, trust, and appropriate reciprocity,"[34] which require approaching others with Marilyn Frye's "loving perception," and with constant and keen attention to what they may themselves be and to what they may require of us, so that we can see what acts are appropriate in given situations. These things cannot be done without giving ethical consideration to all of the beings, entities, etc. involved.

The third is a matter I do not have space to defend here, but which needs to be pointed out anyway, if only in a preliminary manner. Universal consideration requires, and is required by, the preservation of the whole and thus of most of what goes to compose the whole. It is not because nonhuman others of all sorts are to be highly valued or because they have intrinsic or inherent value or worth (most do, but some do not), since the need to prove the intrinsic value of the nonhuman world is no longer central to our theorizing, and we can now delete these vestiges of modernism and foundationalism from it. It is because fully practicing a generosity of spirit toward all others does seem to require a commitment to preserving them. Preservation is required because the nonhuman natural world is necessary for the possibility of giving consideration, of attending, itself necessary for staying in the place of origination of practical necessities—a staying that is a practical necessity. There has to be something to attend to (there has to be a lot to attend to) if our practices of considering are to remain meaningful. These lose their meaning, they become only simulacra, and the real practices are lost if they degenerate into the rote mechanical identification of triggers for rules. With such degeneration, the whole of deep consideration would be lost. The nonhuman, as well as the human, world is *valued* and is preserved, in part, because it does make deep consideration, mindfulness, and attentiveness possible and meaningful. On the assumption that human beings do, or could, have some legitimate place and function on this planet, if something as vast as the radical impoverishment of the planetary ecosystem threatens the continuance of the meaningful attending to others, then it must be given negative value and resisted. One of the greatest practical necessities facing us postmoderns is the need to keep alive the whole business of attending to others, which requires keeping the others in existence, in part, because the whole business is what keeps us alive—is life itself. The alternative seems to be a collapse into a totally introverted, artifactual world of

---

[34] Karen Warren, "The Power and the Promise of Ecological Feminism," *Environmental Ethics* 12 (1990): 143.

simulacra ("copies without originals"),[35] and a resignation of our species from the processes of origination or creation, a flight into the ultimate subjectivity of sheer and groundless imagination.[36]

Fourth, and last, we can now see more clearly what is required to establish what Tom Regan calls an ethics *of* the environment. What we must show is that ethics itself is a practical necessity regarding the environment as well as regarding humans. I hope that this essay's central proposal of universal consideration contributes something toward this task. In my optimistic moments, I think that establishing this practical necessity is the work that all, or most, varieties of environmentally concerned persons are, in fact, in various ways, trying to do.

---

[35] This is Jean Baudrillard's notion of simulacra in Donna Haraway, "A Manifesto for Cyborgs," in Linda J. Nicholson, ed., *Feminism/Postmodernism* (New York: Routlege, 1990), p. 207.

[36] This is by no means the only or the strongest sort of argument for a preservation principle, but the better arguments wait for another essay.

[14]

Radical American Environmentalism and Wilderness Preservation: A Third World Critique

Ramachandra Guha*

I present a Third World critique of the trend in American environmentalism known as deep ecology, analyzing each of deep ecology's central tenets: the distinction between anthropocentrism and biocentrism, the focus on wilderness preservation, the invocation of Eastern traditions, and the belief that it represents the most radical trend within environmentalism. I argue that the anthropocentrism/biocentrism distinction is of little use in understanding the dynamics of environmental degradation, that the implementation of the wilderness agenda is causing serious deprivation in the Third World, that the deep ecologist's interpretation of Eastern traditions is highly selective, and that in other cultural contexts (e.g., West Germany and India) radical environmentalism manifests itself quite differently, with a far greater emphasis on equity and the integration of ecological concerns with livelihood and work. I conclude that despite its claims to universality, deep ecology is firmly rooted in American environmental and cultural history and is inappropriate when applied to the Third World.

Even God dare not appear to the poor man except in the form of bread.
—Mahatma Gandhi

I. INTRODUCTION

The respected radical journalist Kirkpatrick Sale recently celebrated "the passion of a new and growing movement that has become disenchanted with the environmental establishment and has in recent years mounted a serious and sweeping attack on it—style, substance, systems, sensibilities and all."[1] The vision of those whom Sale calls the "New Ecologists"—and what I refer to in this article as deep ecology—is a compelling one. Decrying the narrowly economic goals of mainstream environmentalism, this new movement aims at nothing less

---

*Centre for Ecological Sciences, Indian Institute of Science, Bangalore 560 012, India. This essay was written while the author was a visiting lecturer at the Yale School of Forestry and Environmental Studies. He is grateful to Mike Bell, Tom Birch, Bill Burch, Bill Cronon, Diane Mayerfeld, David Rothenberg, Kirkpatrick Sale, Joel Seton, Tim Weiskel, and Don Worster for helpful comments.
[1] Kirkpatrick Sale, "The Forest for the Trees: Can Today's Environmentalists Tell the Difference," *Mother Jones* 11, no. 8 (November 1986): 26.

than a philosophical and cultural revolution in human attitudes toward nature. In contrast to the conventional lobbying efforts of environmental professionals based in Washington, it proposes a militant defence of "Mother Earth," an unflinching opposition to human attacks on undisturbed wilderness. With their goals ranging from the spiritual to the political, the adherents of deep ecology span a wide spectrum of the American environmental movement. As Sale correctly notes, this emerging strand has in a matter of a few years made its presence felt in a number of fields: from academic philosophy (as in the journal *Environmental Ethics*) to popular environmentalism (for example, the group Earth First!).

In this article I develop a critique of deep ecology from the perspective of a sympathetic outsider. I critique deep ecology not as a general (or even a foot soldier) in the continuing struggle between the ghosts of Gifford Pinchot and John Muir over control of the U.S. environmental movement, but as an outsider to these battles. I speak admittedly as a partisan, but of the environmental movement in India, a country with an ecological diversity comparable to the U.S., but with a radically dissimilar cultural and social history.

My treatment of deep ecology is primarily historical and sociological, rather than philosophical, in nature. Specifically, I examine the cultural rootedness of a philosophy that likes to present itself in universalistic terms. I make two main arguments: first, that deep ecology is uniquely American, and despite superficial similarities in rhetorical style, the social and political goals of radical environmentalism in other cultural contexts (e.g., West Germany and India) are quite different; second, that the social consequences of putting deep ecology into practice on a worldwide basis (what its practitioners are aiming for) are very grave indeed.

II. THE TENETS OF DEEP ECOLOGY

While I am aware that the term *deep ecology* was coined by the Norwegian philosopher Arne Naess, this article refers specifically to the American variant.[2] Adherents of the deep ecological perspective in this country, while arguing intensely among themselves over its political and philosophical implications, share some fundamental premises about human-nature interactions. As I see it, the defining characteristics of deep ecology are fourfold:

---

[2] One of the major criticisms I make in this essay concerns deep ecology's lack of concern with inequalities *within* human society. In the article in which he coined the term *deep ecology*, Naess himself expresses concerns about inequalities between and within nations. However, his concern with social cleavages and their impact on resource utilization patterns and ecological destruction is not very visible in the later writings of deep ecologists. See Arne Naess, "The Shallow and the Deep, Long-Range Ecology Movement: A Summary," *Inquiry* 16 (1973): 96 (I am grateful to Tom Birch for this reference).

First, deep ecology argues, that the environmental movement must shift from an "anthropocentric" to a "biocentric" perspective. In many respects, an acceptance of the primacy of this distinction constitutes the litmus test of deep ecology. A considerable effort is expended by deep ecologists in showing that the dominant motif in Western philosophy has been anthropocentric—i.e., the belief that man and his works are the center of the universe—and conversely, in identifying those lonely thinkers (Leopold, Thoreau, Muir, Aldous Huxley, Santayana, etc.) who, in assigning man a more humble place in the natural order, anticipated deep ecological thinking. In the political realm, meanwhile, establishment environmentalism (shallow ecology) is chided for casting its arguments in human-centered terms. Preserving nature, the deep ecologists say, has an intrinsic worth quite apart from any benefits preservation may convey to future human generations. The anthropocentric-biocentric distinction is accepted as axiomatic by deep ecologists, it structures their discourse, and much of the present discussion remains mired within it.

The second characteristic of deep ecology is its focus on the preservation of unspoilt wilderness—and the restoration of degraded areas to a more pristine condition—to the relative (and sometimes absolute) neglect of other issues on the environmental agenda. I later identify the cultural roots and portentous consequences of this obsession with wilderness. For the moment, let me indicate three distinct sources from which it springs. Historically, it represents a playing out of the preservationist (read *radical*) and utilitarian (read *reformist*) dichotomy that has plagued American environmentalism since the turn of the century. Morally, it is an imperative that follows from the biocentric perspective; other species of plants and animals, and nature itself, have an intrinsic right to exist. And finally, the preservation of wilderness also turns on a scientific argument—viz., the value of biological diversity in stabilizing ecological regimes and in retaining a gene pool for future generations. Truly radical policy proposals have been put forward by deep ecologists on the basis of these arguments. The influential poet Gary Snyder, for example, would like to see a 90 percent reduction in human populations to allow a restoration of pristine environments, while others have argued forcefully that a large portion of the globe must be immediately cordoned off from human beings.[3]

Third, there is a widespread invocation of Eastern spiritual traditions as forerunners of deep ecology. Deep ecology, it is suggested, was practiced both by major religious traditions and at a more popular level by "primal" peoples in non-Western settings. This complements the search for an authentic lineage in Western thought. At one level, the task is to recover those dissenting voices within the Judeo-Christian tradition; at another, to suggest that religious tradi-

---

[3] Gary Snyder, quoted in Sale, "The Forest for the Trees," p. 32. See also Dave Foreman, "A Modest Proposal for a Wilderness System," *Whole Earth Review*, no. 53 (Winter 1986–87): 42–45.

tions in other cultures are, in contrast, dominantly if not exclusively "biocentric" in their orientation. This coupling of (ancient) Eastern and (modern) ecological wisdom seemingly helps consolidate the claim that deep ecology is a philosophy of universal significance.

Fourth, deep ecologists, whatever their internal differences, share the belief that they are the "leading edge" of the environmental movement. As the polarity of the shallow/deep and anthropocentric/biocentric distinctions makes clear, they see themselves as the spiritual, philosophical, and political vanguard of American and world environmentalism.

III. TOWARD A CRITIQUE

Although I analyze each of these tenets independently, it is important to recognize, as deep ecologists are fond of remarking in reference to nature, the interconnectedness and unity of these individual themes.

(1) Insofar as it has begun to act as a check on man's arrogance and ecological hubris, the transition from an anthropocentric (human-centered) to a biocentric (humans as only one element in the ecosystem) view in both religious and scientific traditions is only to be welcomed.[4] What is unacceptable are the radical conclusions drawn by deep ecology, in particular, that intervention in nature should be guided primarily by the need to preserve biotic integrity rather than by the needs of humans. The latter for deep ecologists is anthropocentric, the former biocentric. This dichotomy is, however, of very little use in understanding the dynamics of environmental degradation. The two fundamental ecological problems facing the globe are (i) overconsumption by the industrialized world and by urban elites in the Third World and (ii) growing militarization, both in a short-term sense (i.e., ongoing regional wars) and in a long-term sense (i.e., the arms race and the prospect of nuclear annihilation). Neither of these problems has any tangible connection to the anthropocentric-biocentric distinction. Indeed, the agents of these processes would barely comprehend this philosophical dichotomy. The proximate causes of the ecologically wasteful characteristics of industrial society and of militarization are far more mundane: at an aggregate level, the dialectic of economic and political structures, and at a micro-level, the life style choices of individuals. These causes cannot be reduced, whatever the level of analysis, to a deeper anthropocentric attitude toward nature; on the contrary, by constituting a grave threat to human survival, the ecological degradation they cause does not even serve the best interests of human beings! If my identification of the major dangers to the integrity of the natural world is correct, invoking the bogy of anthropocentricism is at best irrelevant and at worst a dangerous obfuscation.

---

[4] See, for example, Donald Worster, *Nature's Economy: The Roots of Ecology* (San Francisco, Sierra Club Books, 1977).

(2) If the above dichotomy is irrelevant, the emphasis on wilderness is positively harmful when applied to the Third World. If in the U.S. the preservationist/utilitarian division is seen as mirroring the conflict between "people" and "interests," in countries such as India the situation is very nearly the reverse. Because India is a long settled and densely populated country in which agrarian populations have a finely balanced relationship with nature, the setting aside of wilderness areas has resulted in a direct transfer of resources from the poor to the rich. Thus, Project Tiger, a network of parks hailed by the international conservation community as an outstanding success, sharply posits the interests of the tiger against those of poor peasants living in and around the reserve. The designation of tiger reserves was made possible only by the physical displacement of existing villages and their inhabitants; their management requires the continuing exclusion of peasants and livestock. The initial impetus for setting up parks for the tiger and other large mammals such as the rhinoceros and elephant came from two social groups, first, a class of ex-hunters turned conservationists belonging mostly to the declining Indian feudal elite and second, representatives of international agencies, such as the World Wildlife Fund (WWF) and the International Union for the Conservation of Nature and Natural Resources (IUCN), seeking to transplant the American system of national parks onto Indian soil. In no case have the needs of the local population been taken into account, and as in many parts of Africa, the designated wildlands are managed primarily for the benefit of rich tourists. Until very recently, wildlands preservation has been identified with environmentalism by the state and the conservation elite; in consequence, environmental problems that impinge far more directly on the lives of the poor—e.g., fuel, fodder, water shortages, soil erosion, and air and water pollution—have not been adequately addressed.[5]

Deep ecology provides, perhaps unwittingly, a justification for the continuation of such narrow and inequitable conservation practices under a newly acquired radical guise. Increasingly, the international conservation elite is using the philosophical, moral, and scientific arguments used by deep ecologists in advancing their wilderness crusade. A striking but by no means atypical example is the recent plea by a prominent American biologist for the takeover of large portions of the globe by the author and his scientific colleagues. Writing in a prestigous scientific forum, the *Annual Review of Ecology and Systematics*, Daniel Janzen argues that only biologists have the competence to decide how the tropical landscape should be used. As "the representatives of the natural world," biologists are "in charge of the future of tropical ecology," and only they have

---

[5] See Centre for Science and Environment, *India: The State of the Environment 1982: A Citizens Report* (New Delhi: Centre for Science and Environment, 1982); R. Sukumar, "Elephant-Man Conflict in Karnataka," in Cecil Saldanha, ed., *The State of Karnataka's Environment* (Bangalore: Centre for Taxonomic Studies, 1985). For Africa, see the brilliant analysis by Helge Kjekshus, *Ecology Control and Economic Development in East African History* (Berkeley: University of California Press, 1977).

the expertise and mandate to "determine whether the tropical agroscape is to be populated only by humans, their mutualists, commensals, and parasites, or whether it will also contain some islands of the greater nature—the nature that spawned humans, yet has been vanquished by them." Janzen exhorts his colleagues to advance their territorial claims on the tropical world more forcefully, warning that the very existence of these areas is at stake: "if biologists want a tropics in which to biologize, they are going to have to buy it with care, energy, effort, strategy, tactics, time, and cash."[6]

This frankly imperialist manifesto highlights the multiple dangers of the preoccupation with wilderness preservation that is characteristic of deep ecology. As I have suggested, it seriously compounds the neglect by the American movement of far more pressing environmental problems within the Third World. But perhaps more importantly, and in a more insidious fashion, it also provides an impetus to the imperialist yearning of Western biologists and their financial sponsors, organizations such as the WWF and IUCN. The wholesale transfer of a movement culturally rooted in American conservation history can only result in the social uprooting of human populations in other parts of the globe.

(3) I come now to the persistent invocation of Eastern philosophies as antecedent in point of time but convergent in their structure with deep ecology. Complex and internally differentiated religious traditions—Hinduism, Buddhism, and Taoism—are lumped together as holding a view of nature believed to be quintessentially biocentric. Individual philosophers such as the Taoist Lao Tzu are identified as being forerunners of deep ecology. Even an intensely political, pragmatic, and Christian influenced thinker such as Gandhi has been accorded a wholly undeserved place in the deep ecological pantheon. Thus the Zen teacher Robert Aitken Roshi makes the strange claim that Gandhi's thought was not human-centered and that he practiced an embryonic form of deep ecology which is "traditionally Eastern and is found with differing emphasis in Hinduism, Taoism and in Theravada and Mahayana Buddhism."[7] Moving away from the realm of high philosophy and scriptural religion, deep ecologists make the further claim that at the level of material and spiritual practice "primal" peoples subordinated themselves to the integrity of the biotic universe they inhabited.

I have indicated that this appropriation of Eastern traditions is in part dictated by the need to construct an authentic lineage and in part a desire to present deep ecology as a universalistic philosophy. Indeed, in his substantial and quixotic

---

[6] Daniel Janzen, "The Future of Tropical Ecology," *Annual Review of Ecology and Systematics* 17 (1986): 305–06; emphasis added.

[7] Robert Aitken Roshi, "Gandhi, Dogen, and Deep Ecology," reprinted as appendix C in Bill Devall and George Sessions, *Deep Ecology: Living as if Nature Mattered* (Salt Lake City: Peregrine Smith Books, 1985). For Gandhi's own views on social reconstruction, see the excellent three volume collection edited by Raghavan Iyer, *The Moral and Political Writings of Mahatma Gandhi* (Oxford: Clarendon Press, 1986–87).

biography of John Muir, Michael Cohen goes so far as to suggest that Muir was the "Taoist of the [American] West."[8] This reading of Eastern traditions is selective and does not bother to differentiate between alternate (and changing) religious and cultural traditions; as it stands, it does considerable violence to the historical record. Throughout most recorded history the characteristic form of human activity in the "East" has been a finely tuned but nonetheless conscious and dynamic manipulation of nature. Although mystics such as Lao Tzu did reflect on the spiritual essence of human relations with nature, it must be recognized that such ascetics and their reflections were supported by a society of cultivators whose relationship with nature was a far more *active* one. Many agricultural communities do have a sophisticated knowledge of the natural environment that may equal (and sometimes surpass) codified "scientific" knowledge; yet, the elaboration of such traditional ecological knowledge (in both material and spiritual contexts) can hardly be said to rest on a mystical affinity with nature of a deep ecological kind. Nor is such knowledge infallible; as the archaeological record powerfully suggests, modern Western man has no monopoly on ecological disasters.

In a brilliant article, the Chicago historian Ronald Inden points out that this romantic and essentially positive view of the East is a mirror image of the scientific and essentially pejorative view normally upheld by Western scholars of the Orient. In either case, the East constitutes the Other, a body wholly separate and alien from the West; it is defined by a uniquely spiritual and nonrational "essence," even if this essence is valorized quite differently by the two schools. Eastern man exhibits a spiritual dependence with respect to nature—on the one hand, this is symptomatic of his prescientific and backward self, on the other, of his ecological wisdom and deep ecological consciousness. Both views are monolithic, simplistic, and have the characteristic effect—intended in one case, perhaps unintended in the other—of denying agency and reason to the East and making it the privileged orbit of Western thinkers.

The two apparently opposed perspectives have then a common underlying structure of discourse in which the East merely serves as a vehicle for Western projections. Varying images of the East are raw material for political and cultural battles being played out in the West; they tell us far more about the Western commentator and his desires than about the "East." Inden's remarks apply not merely to Western scholarship on India, but to Orientalist constructions of China and Japan as well:

> Although these two views appear to be strongly opposed, they often combine together. Both have a similar interest in sustaining the Otherness of India. The holders of the dominant view, best exemplified in the past in imperial administra-

---

[8] Michael Cohen, *The Pathless Way* (Madison: University of Wisconsin Press, 1984), p. 120.

tive discourse (and today probably by that of 'development economics'), would place a traditional, superstition-ridden India in a position of perpetual tutelage to a modern, rational West. The adherents of the romantic view, best exemplified academically in the discourses of Christian liberalism and analytic psychology, concede the realm of the public and impersonal to the positivist. Taking their succour not from governments and big business, but from a plethora of religious foundations and self-help institutes, and from allies in the 'consciousness industry,' not to mention the important industry of tourism, the romantics insist that India embodies a private realm of the imagination and the religious which modern, western man lacks but needs. They, therefore, like the positivists, but for just the opposite reason, have a vested interest in seeing that the Orientalist view of India as 'spiritual,' 'mysterious,' and 'exotic' is perpetuated.[9]

(4) How radical, finally, are the deep ecologists? Notwithstanding their self-image and strident rhetoric (in which the label "shallow ecology" has an opprobrium similar to that reserved for "social democratic" by Marxist-Leninists), even within the American context their radicalism is limited and it manifests itself quite differently elsewhere.

To my mind, deep ecology is best viewed as a radical trend within the wilderness preservation movement. Although advancing philosophical rather than aesthetic arguments and encouraging political militancy rather than negotiation, its practical emphasis—viz., preservation of unspoilt nature—is virtually identical. For the mainstream movement, the function of wilderness is to provide a temporary antidote to modern civilization. As a special institution within an industrialized society, the national park "provides an opportunity for respite, contrast, contemplation, and affirmation of values for those who live most of their lives in the workaday world."[10] Indeed, the rapid increase in visitations to the national parks in postwar America is a direct consequence of economic expansion. The emergence of a popular interest in wilderness sites, the historian Samuel Hays points out, was "not a throwback to the primitive, but an integral part of the modern standard of living as people sought to add new 'amenity' and

---

[9] Ronald Inden, "Orientalist Constructions of India," *Modern Asian Studies* 20 (1986): 442. Inden draws inspiration from Edward Said's forceful polemic, *Orientalism* (New York: Basic Books, 1980). It must be noted, however, that there is a salient difference between Western perceptions of Middle Eastern and Far Eastern cultures respectively. Due perhaps to the long history of Christian conflict with Islam, Middle Eastern cultures (as Said documents) are consistently presented in pejorative terms. The juxtaposition of hostile and worshiping attitudes that Inden talks of applies only to Western attitudes toward Buddhist and Hindu societies.

[10] Joseph Sax, *Mountains Without Handrails: Reflections on the National Parks* (Ann Arbor: University of Michigan Press, 1980), p. 42. Cf. also Peter Schmitt, *Back to Nature: The Arcadian Myth in Urban America* (New York: Oxford University Press, 1969), and Alfred Runte, *National Parks: The American Experience* (Lincoln: University of Nebraska Press, 1979).

'aesthetic' goals and desires to their earlier preoccupation with necessities and conveniences."[11]

Here, the enjoyment of nature is an integral part of the consumer society. The private automobile (and the life style it has spawned) is in many respects the ultimate ecological villain, and an untouched wilderness the prototype of ecological harmony; yet, for most Americans it is perfectly consistent to drive a thousand miles to spend a holiday in a national park. They possess a vast, beautiful, and sparsely populated continent and are also able to draw upon the natural resources of large portions of the globe by virtue of their economic and political dominance. In consequence, America can simultaneously enjoy the material benefits of an expanding economy and the aesthetic benefits of unspoilt nature. The two poles of "wilderness" and "civilization" mutually coexist in an internally coherent whole, and philosophers of both poles are assigned a prominent place in this culture. Paradoxically as it may seem, it is no accident that Star Wars technology and deep ecology both find their fullest expression in that leading sector of Western civilization, California.

Deep ecology runs parallel to the consumer society without seriously questioning its ecological and socio-political basis. In its celebration of American wilderness, it also displays an uncomfortable convergence with the prevailing climate of nationalism in the American wilderness movement. For spokesmen such as the historian Roderick Nash, the national park system is America's distinctive cultural contribution to the world, reflective not merely of its economic but of its philosophical and ecological maturity as well. In what Walter Lippman called the American century, the "American invention of national parks" must be exported worldwide. Betraying an economic determinism that would make even a Marxist shudder, Nash believes that environmental preservation is a "full stomach" phenomenon that is confined to the rich, urban, and sophisticated. Nonetheless, he hopes that "the less developed nations may eventually evolve economically and intellectually to the point where nature preservation is more than a business."[12]

The error which Nash makes (and which deep ecology in some respects encourages) is to equate environmental protection with the protection of wilderness. This is a distinctively American notion, borne out of a unique social and environmental history. The archetypal concerns of radical environmentalists in

---

[11] Samuel Hays, "From Conservation to Environment: Environmental Politics in the United States since World War Two," *Environmental Review* 6 (1982): 21. See also the same author's book entitled *Beauty, Health and Permanence: Environmental Politics in the United States, 1955–85* (New York: Cambridge University Press, 1987).

[12] Roderick Nash, *Wilderness and the American Mind*, 3rd ed. (New Haven: Yale University Press, 1982).

other cultural contexts are in fact quite different. The German Greens, for example, have elaborated a devastating critique of industrial society which turns on the acceptance of environmental limits to growth. Pointing to the intimate links between industrialization, militarization, and conquest, the Greens argue that economic growth in the West has historically rested on the economic and ecological exploitation of the Third World. Rudolf Bahro is characteristically blunt:

> The working class here [in the West] is the richest lower class in the world. And if I look at the problem from the point of view of the whole of humanity, not just from that of Europe, then I must say that the metropolitan working class is the worst exploiting class in history. . . . What made poverty bearable in eighteenth or nineteenth-century Europe was the prospect of escaping it through exploitation of the periphery. But this is no longer a possibility, and continued industrialism in the Third World will mean poverty for whole generations and hunger for millions.[13]

Here the roots of global ecological problems lie in the disproportionate share of resources consumed by the industrialized countries as a whole *and* the urban elite within the Third World. Since it is impossible to reproduce an industrial monoculture worldwide, the ecological movement in the West must begin by cleaning up its own act. The Greens advocate the creation of a "no growth" economy, to be achieved by scaling down current (and clearly unsustainable) consumption levels.[14] This radical shift in consumption and production patterns requires the creation of alternate economic and political structures—smaller in scale and more amenable to social participation—but it rests equally on a shift in cultural values. The expansionist character of modern Western man will have to give way to an ethic of renunciation and self-limitation, in which spiritual and communal values play an increasing role in sustaining social life. This revolution in cultural values, however, has as its point of departure an understanding of environmental processes quite different from deep ecology.

Many elements of the Green program find a strong resonance in countries such as India, where a history of Western colonialism and industrial development has benefited only a tiny elite while exacting tremendous social and environmental costs. The ecological battles presently being fought in India have as their

---

[13] Rudolf Bahro, *From Red to Green* (London: Verso Books, 1984).

[14] From time to time, American scholars have themselves criticized these imbalances in consumption patterns. In the 1950s, William Vogt made the charge that the United States, with one-sixteenth of the world's population, was utilizing one-third of the globe's resources. (Vogt, cited in E. F. Murphy, *Nature, Bureaucracy and the Rule of Property* [Amsterdam: North Holland, 1977, p. 29]). More recently, Zero Population Growth has estimated that each American consumes thirty-nine times as many resources as an Indian. See *Christian Science Monitor*, 2 March 1987.

epicenter the conflict over nature between the subsistence and largely rural sector and the vastly more powerful commercial-industrial sector. Perhaps the most celebrated of these battles concerns the Chipko (Hug the Tree) movement, a peasant movement against deforestation in the Himalayan foothills. Chipko is only one of several movements that have sharply questioned the nonsustainable demand being placed on the land and vegetative base by urban centers and industry. These include opposition to large dams by displaced peasants, the conflict between small artisan fishing and large-scale trawler fishing for export, the countrywide movements against commercial forest operations, and opposition to industrial pollution among downstream agricultural and fishing communities.[15]

Two features distinguish these environmental movements from their Western counterparts. First, for the sections of society most critically affected by environmental degradation—poor and landless peasants, women, and tribals—it is a question of sheer survival, not of enhancing the quality of life. Second, and as a consequence, the environmental solutions they articulate deeply involve questions of equity as well as economic and political redistribution. Highlighting these differences, a leading Indian environmentalist stresses that "environmental protection per se is of least concern to most of these groups. Their main concern is about the use of the environment and who should benefit from it."[16] They seek to wrest control of nature away from the state and the industrial sector and place it in the hands of rural communities who live within that environment but are increasingly denied access to it. These communities have far more basic needs, their demands on the environment are far less intense, and they can draw upon a reservoir of cooperative social institutions and local ecological knowledge in managing the "commons"—forests, grasslands, and the waters—on a sustainable basis. If colonial and capitalist expansion has both accentuated social inequalities and signaled a precipitous fall in ecological wisdom, an alternate ecology must rest on an alternate society and polity as well.

This brief overview of German and Indian environmentalism has some major implications for deep ecology. Both German and Indian environmental traditions allow for a greater integration of ecological concerns with livelihood and work. They also place a greater emphasis on equity and social justice (both within individual countries and on a global scale) on the grounds that in the absence of social regeneration environmental regeneration has very little chance of succeed-

---

[15] For an excellent review, see Anil Agarwal and Sunita Narain, eds., *India: The State of the Environment 1984–85: A Citizens Report* (New Delhi: Centre for Science and Environment, 1985). Cf. also Ramachandra Guha, *The Unquiet Woods: Ecological Change and Peasant Resistance in the Indian Himalaya* (Berkeley: University of California Press, forthcoming).

[16] Anil Agarwal, "Human-Nature Interactions in a Third World Country," *The Environmentalist* 6, no. 3 (1986): 167.

ing. Finally, and perhaps most significantly, they have escaped the preoccupation with wilderness perservation so characteristic of American cultural and environmental history.[17]

IV. A HOMILY

In 1958, the economist J. K. Galbraith referred to overconsumption as the unasked question of the American conservation movement. There is a marked selectivity, he wrote, "in the conservationist's approach to materials consumption. If we are concerned about our great appetite for materials, it is plausible to seek to increase the supply, to decrease waste, to make better use of the stocks available, and to develop substitutes. But what of the appetite itself? Surely this is the ultimate source of the problem. If it continues its geometric course, will it not one day have to be restrained? Yet in the literature of the resource problem this is the forbidden question. Over it hangs a nearly total silence."[18]

The consumer economy and society have expanded tremendously in the three decades since Galbraith penned these words; yet his criticisms are nearly as valid today. I have said "nearly," for there are some hopeful signs. Within the environmental movement several dispersed groups are working to develop ecologically benign technologies and to encourage less wasteful life styles. Moreover, outside the self-defined boundaries of American environmentalism, opposition to the permanent war economy is being carried on by a peace movement that has a distinguished history and impeccable moral and political credentials.

It is precisely these (to my mind, most hopeful) components of the American social scene that are missing from deep ecology. In their widely noticed book, Bill Devall and George Sessions make no mention of militarization or the movements for peace, while activists whose practical focus is on developing ecologically responsible life styles (e.g., Wendell Berry) are derided as "falling short of deep ecological awareness."[19] A truly radical ecology in the American context ought to work toward a synthesis of the appropriate technology, alternate

---

[17] One strand in radical American environmentalism, the bioregional movement, by emphasizing a greater involvement with the bioregion people inhabit, does indirectly challenge consumerism. However, as yet bioregionalism has hardly raised the questions of equity and social justice (international, intranational, and intergenerational) which I argue must be a central plank of radical environmentalism. Moreover, its stress on (individual) *experience* as the key to involvement with nature is also somewhat at odds with the integration of nature with livelihood and work that I talk of in this paper. Cf. Kirkpatrick Sale, *Dwellers in the Land: The Bioregional Vision* (San Francisco: Sierra Club Books, 1985).

[18] John Kenneth Galbraith, "How Much Should a Country Consume?" in Henry Jarrett, ed., *Perspectives on Conservation* (Baltimore: Johns Hopkins Press, 1958), pp. 91–92.

[19] Devall and Sessions, *Deep Ecology*, p. 122. For Wendell Berry's own assessment of deep ecology, see his "Amplications: Preserving Wildness," *Wilderness* 50 (Spring 1987): 39–40, 50–54.

life style, and peace movements.[20] By making the (largely spurious) anthropocentric-biocentric distinction cental to the debate, deep ecologists may have appropriated the moral high ground, but they are at the same time doing a serious disservice to American and global environmentalism.[21]

---

[20] See the interesting recent contribution by one of the most influential spokesmen of appropriate technology—Barry Commoner, "A Reporter at Large: The Environment," *New Yorker*, 15 June 1987. While Commoner makes a forceful plea for the convergence of the environmental movement (viewed by him primarily as the opposition to air and water pollution and to the institutions that generate such pollution) and the peace movement, he significantly does not mention consumption patterns, implying that "limits to growth" do not exist.

[21] In this sense, my critique of deep ecology, although that of an outsider, may facilitate the reassertion of those elements in the American environmental tradition for which there is a profound sympathy in other parts of the globe. A global perspective may also lead to a critical reassessment of figures such as Aldo Leopold and John Muir, the two patron saints of deep ecology. As Donald Worster has pointed out, the message of Muir (and, I would argue, of Leopold as well) makes sense only in an American context; he has very little to say to other cultures. See Worster's review of Stephen Fox's *John Muir and His Legacy*, in *Environmental Ethics* 5 (1983): 277–81.

[15]

The Relevance of Deep Ecology to the Third World: Some Preliminary Comments

David M. Johns*

> Although Ramachandra Guha has demonstrated the importance of cross-cultural dialogue on environmental issues and has much to tell us about the problems of wilderness preservation in the Third World, I argue that Guha is partly wrong in claiming that deep ecology equates environmental protection with wilderness protection and simply wrong in calling wilderness protection untenable or incorrect as a global strategy for environmental protection. Moreover, I argue that the deep ecology distinction between anthropocentrism and biocentrism is useful in dealing with the two major problems which Guha identifies as undermining the health of the planet—overconsumption and militarism. Although it is true that preservation of wilderness will not be successful unless human social dynamics are taken into consideration, nevertheless, a biocentrism which integrates critical social theory can provide the basis for an ethic that undercuts the environmental degradation from overconsumption and militarism more effectively than a human-centered system.

INTRODUCTION

The appearance in *Environmental Ethics* of work by Third World environmentalists is to be welcomed, and should be greatly encouraged. If the movements for environmental protection anywhere in the world are to be relevent, they must address issues within the global framework. This can only be done in conjunction with and by engaging other movements around the globe. Only through the genuine amalgamation of the various and specific historical experiences can we move toward a new direction(s) for human society. Ramachandra Guha's "sympathetic critique" of deep ecology is an important step and a good example of the necessity of such exchanges, for he raises several issues concerning the tenets of

---

* Department of Political Science, Portland State University, P.O. Box 751, Portland, OR 97207. Johns teaches courses on U.S. and comparative politics. His special interests include the study of relationships among various forms of domination and politico-cultural evolution. He has planted trees in both Oregon and Nicaragua. The author is indebted to the referees who reviewed this paper and made many thoughtful and insightful comments.

deep ecology that are most easily visible from outside the Western industrial world.[1]

In this paper, I comment on two of these issues: wilderness preservation and the usefulness of the anthropocentric/biocentric distinction.

WILDERNESS: ORIGINS AND VALUES

Guha criticizes deep ecology for equating environmental protection with wilderness preservation. This flaw, he argues, is due to deep ecology's lack of awareness of its historical roots and limitations. Adherence to this position, he goes on to argue, actually obscures the real sources of environmental degradation and thus helps to perpetuate the existing order. Moreover, deep ecology fails to recognize the impact of its commitment to wilderness in the Third World. I discuss each of these in turn.

Deep ecology is obviously rooted in the culture of those who espouse it; this is the case of every movement. The very process of transcendence or dialectical working out, moreover, assumes a history; nevertheless, simply pointing out the origins of a particular historical experience does not invalidate it. There is no question that the circumstances of development in the United States—including the pattern of settlement over the huge geographical area available—have helped to shape the response to environmental degradation. e.g., an emphasis on wilderness preservation. There is also little question that in many respects the existence of wilderness may "fit in" with the cultural categories of a consumer society, as a retreat from the insanity of (sub)urban-industrial life—an alternative that only a country like the U.S. can afford, inheriting as it did a virtually unexploited continent still underpopulated compared to the rest of the world, and living off wealth extracted throughout the world.

To the contrary, however, while there may be some cultural fit between wilderness and the existing order that results from the particular experience of material development in the U.S., in most respects it does not "fit." From the very beginning and increasingly so the wilderness system, wildlife refuges, and old-growth forests are under fierce attacks by those who say we cannot afford them because they undermine the viability of an economy based on endless growth.

The real issues are whether Guha's claim that environmental protection means protection of wilderness is an accurate description of deep ecology and whether such a position is wrongheaded in substance. Related to these issues is a larger question: how should humans interact with the rest of the biosphere, and must wilderness preservation stand in opposition to an approach that integrates human livelihood and environmental protection?

---

[1] Ramachandra Guha, "Radical American Environmentalism and Wilderness Preservation: A Third World Perspective," *Environmental Ethics* 11 (1989): 71–83.

I believe that Guha is partly wrong in stating that deep ecology equates environmental protection with wilderness protection and simply wrong in calling wilderness protection untenable or incorrect as a global strategy for environmental protection. The deep ecological support for wilderness is predicated upon two important fact/values: (1) that the Earth can support a limited amount of biomass and the more of it that is composed of humans or turned to human use, the less is available for other life, and (2) that humans do not have the right to so alter the composition of the biomass that there is a resulting destruction, in Leopold's words, of "the integrity, stability and beauty" of the ecosystem. The basis for these values may lie in the experience of Self-realization or through identification with nature as the real community of which one is a part. Whether it is called a transcendence of alienation in its various forms or the healing of a crippled heart, it is supposed to support the claim that human life is no more valuable than any other form of life, life being broadly construed to include plants, animals, ecosystems, rivers, mountains, and the Earth.

Associated with this understanding of the human/Earth relationship is a recognition that in much of the world almost any human impact is destructive of the biosphere. In many ecosystems human livelihood—beyond very minimal numbers and very limited technology—is simply not compatible with maintaining the integrity of the biosphere. Such situations are most obvious when one looks at the fate of other large mammals. Ecosystems must normally be healthy to support them. Their disappearence is a good indication of degradation. Grizzly bears, orangutans, tigers, elephants, and many other species cannot easily coexist with humans in any numbers or with very exploitative technologies. Many ecosystems, moreover, cannot easily accommodate significant human presence without serious deterioration in diversity and balance. Recognition of other species, of ecosystems, and the Earth as valuable in and of themselves, individually or collectively, apart from their usefulness to humans, means that in practice much of the Earth cannot be used for permanent human settlement. Existing devastation, the ever increasing spread of humans into new areas, and the nature of those human incursions, makes the task of protection of areas still in their natural state ever more urgent. Returning large areas to wilderness is only slightly less urgent.

LIVELIHOOD AND WILDERNESS

While preservation of wilderness may seem to be the overriding focus of deep ecology given the ever accelerating destruction of species, ecosystems, and possibly the planet itself, there is a profound recognition that humans have their place *in* nature as well. With regard to places where it is appropriate for humans to settle, how to combine livelihood with environmental integrity is a major emphasis and how to move toward the reestablishment of a real community, embedded in the local ecosystem is a priority of the deep ecology movement.

While it may be a valid criticism that much of the thinking in this area is fuzzy, naive, or falls victim to mystification, it is not true that wilderness is the single goal of deep ecologists. Given the human-nature relationship that deep ecology espouses—that to be effective in allowing nature to heal itself, one must also heal one's own self and community—it seems odd to suggest that deep ecology is unconcerned with human communities and their place in nature.[2]

SOURCES OF ENVIRONMENTAL DEGRADATION

Another criticism that Guha makes of deep ecology is that the focus on humans in general as the problem obscures the real causes of environmental degradation, namely overconsumption and militarization. Although his criticism has much merit, I believe he overstates the case. There can be no doubt that in explaining the particular developments that have resulted in so much destruction over the past two to three hundred years, industrialization, imperialism, overconsumption in the developed world, and the huge commitment of resources to armaments, are paramount. Guha is correct when he says that many in the movement see the problem as simply too many people behaving stupidly, without any regard for the nature of the system in which they live, its dynamics, and the fact that it victimizes most people as well as nature. Because it is probably true that most people who are victimizing nature are themselves victims of the social order, he is right in suggesting that the obstacles to significant changes in the relationship to nature are structural, not simply a matter of altering one's world view.

Yet, for every bit of evidence that this criticism is valid, there is evidence that it is only partially valid. Deep ecology and the German Greens do not see things as differently as Guha suggests. Indeed, I believe the Green movement in Germany and in particular Bahro have informed the deep ecology movement in the U.S.[3] There is a widespread recognition within deep ecology of the great inequality that exists in the world with regard to consumption, great differences

---

[2] Some critics, though not Guha, have accused deep ecology of being fundamentally misanthropic. No doubt there are genuine misanthropes about, but in my reading of the deep ecological literature, both scholarly and popular, I find criticism aimed at human behavior resulting from alienation and disease, not the species per se. Even the angriest statements of those struggling on the front lines against biocide can best be understood in this context. Humans have a place *in* nature; it is when they try to separate themselves that their behavior becomes destructive.

[3] Bahro, one of the leading German Green theorists, comes from a Marxist background, as do many other German Greens. *The Alternative* (London: New Left, 1978), his Marxist critique of "actually existing socialism," is a major contribution to understanding human society. It earned him a lengthy jail sentence in the GDR before he was allowed to emigrate. His later, Green writings are more widely read in the U.S. than his first book, but his historical perspective and concern with human society runs through all of his work, which is read by environmentalists. It is unfortunate, however, that U.S. Marxists are more familiar with his later work than deep ecologists are with his earlier work.

in the existing power of various groups to shape a society's relationship with nature, and a recognition that the solution to ecological problems must address the issues of class, gender, and ethnicity. In addition, there is a recognition that all forms of domination are linked, as evidenced by the ongoing debates between deep ecology and social ecology, between deep ecology and ecofeminism, and between deep ecology and Marxism and other socialisms.

The nature of these linkages is certainly not settled; nonetheless, deep ecology may be distinctive in believing that the resolution of equity issues among humans will not automatically result in an end to the human destruction of the biosphere. One can envision, depending upon the theoretical version chosen, a society without class distinctions, without patriarchy, and with cultural autonomy that still attempts to control or manage the rest of nature in a utilitarian fashion that results in the deterioration of the biosphere. Such social changes would probably lessen the destructiveness of the relationship for two reasons: much of the technology of the last three hundred years is incompatible with a truly egalitarian society and much of the alienation that distorts the expression of human energy into schemes of control would not exist. Thus, it can be argued that although a significant change in the way humans relate to each other is a necessary component in a changed relationship to nature, it is not sufficient to bring about, by itself, the recognition and inclusion of nature as part of the moral community.

Deep ecologists point out, correctly I believe, that in terms of the integrity of an ecosystem, it makes little difference if an old-growth forest and its inhabitants are destroyed to build one house for a North American or fifty simple structures in the Third World. From a strictly human standpoint the latter is much more justifiable than the former. Nevertheless, even if North Americans were to sharply restrict their consumption, the fact remains that it is human numbers as well as levels of consumption that count.[4] There is, I believe, widespread agreement among Greens and deep ecologists that fewer humans (and especially less extensive occupation of the globe) as well as equitable and drastically curtailed consumption are essential to restoring the balance of the planet.

Guha does have much to tell us about the situation in the Third World and the problems of wilderness preservation there. While those of us engaged in political activity in North America are used to confronting the issue of jobs versus environment, it is important to understand that in the Third World "jobs" often equates with actual survival. While sparing old growth in the U.S. within the existing economic structure may cause hardship, sparing tropical forests within

---

[4] Human population pressure combined with human capacity for culture are probably the two most important factors in explaining the dynamic of social evolution and the subsequent alteration of the biosphere. This is discussed more fully below in the section "Values and Culture." In conjunction with this thesis, see note 17 and Mark N. Cohen, *Food Crisis in Prehistory* (New Haven: Yale University Press, 1977), and Marshall Sahlins, *Stoneage Economics* (Chicago: Aldine, 1972).

the existing economic structure may mean immediate hunger. (However, clearing tropical forests may mean eventual hunger as well, depending on the quality of the land cleared.) What Guha is telling us is that efforts to protect the environment by establishing wilderness areas in the Third World hurt the poorest of the poor—they are just more examples of imperialism, the same imperialism that pushes the poor and others into the wilderness in the first place.

The alternative, Guha suggests, is to recognize that wilderness is not appropriate; instead, one must integrate livelihood with environmental protection. Certainly this is the preferred path, when one is discussing how humans should interact with the rest of nature in areas appropriate for human settlement, but it does not address the needs of other species (such as elephants and tigers)—those that cannot coexist in the same area with all but the fewest humans living very simply—or the fact that the integrity of many ecosystems is negatively impacted by the settlement of any humans, even those living at subsistence levels. It is the sheer extensiveness of human settlement in much of Asia (and Europe, and parts of Africa and North America) that is a problem. Humans compete for habitat with other species, threaten their destruction, and otherwise degrade the environment, even diminishing its human carrying capacity.

Wilderness is needed in the Third World as much as it is in Europe and other long settled parts of the globe; nevertheless, it is also important to realize that the structure of imperialism makes the manner in which wilderness is created/protected in the Third World often unjust from a human standpoint. This fact desperately needs to be taken into account by environmentalists. How? First, by understanding how imperialism has created and continues to feed much of the dynamic that threatens ecosystems and species in the Third World from the Amazon to Malaysia; by understanding how countries that have broken or are attempting to break with their historical place in the existing structure in an effort to survive find themselves adopting economic strategies that are environmentally destructive; and by understanding how the wealth extracted from the Third World makes possible the culture of consumption in the First World.[5]

---

[5] We live in a world shaped by the European expansion. Most of Africa, Asia, and Latin America are still bound tightly into an international political economy and state system which keeps them subservient. The wealth continues to flow north. Most of the environmental degradation in the Third World can only be understood in this context. This is also the case with regard to many efforts at conservation, which are often at the expense of the poor. Attempts to break with this international system have sometimes been successful, as with the Russian and later revolutions; nevertheless, the results of such revolutions for the environment have not been impressive. Much more than Marx's nineteenth-century notion of progress (and in direct contradiction with many of his revolutionary goals), it has been the international political environment of hostility that has greeted these revolutionaries, combined, of course, with the inertia of hierarchy inherent in civilized societies, that has resulted in their systems following a very similar road of superexploitation of the environment. Both capitalist and statist societies eat and despoil nature, notwithstanding different internal dynamics. There are similarities as well: increasing competition in weapons and higher levels of consumption.

Second, based upon the understanding just set out, we must come to terms with the severe limits of what can be achieved to protect the environment within the framework of a system based on endless material growth and extreme socioeconomic inequality. We need to grasp the necessity of moving beyond the choices offered by the powers that be; only by pushing beyond the limits of what is acceptable to the existing political-economic order can constraints on ecological-political choices be transcended.

Finally, we must recognize that we cannot alter the existing biocidal order without broad-based support. Only if our thinking and feeling is informed by an understanding of human social relations can we develop successful strategies for protecting the Earth and its diversity. If we are to move beyond the existing order we need to understand who our potential allies are, as well as what the obstacles are. If we treat the poor—who go to the rain forest to farm because they have been driven off the land they formerly cultivated by the wealthy who can make higher profits producing cash crops for the international market—as the problem, rather than the system which constrains their choices, we will fail. If we do not forge alliances with those who oppose the existing order—albeit on the basis of its injury to the poor, to women, to oppressed ethnic groups—we will also fail. The work of EPOCA in Nicaraguan reforestation efforts and in Central America generally the Greenpeace campaign directed at the IMF and the World Bank are both examples of environmental action that reflect at least some elements of what is necessary. Making common cause where possible in pushing for reform and ultimately transformation is essential.

In the short term—given the continued existence of an international political economic system committed to growth and great inequality, and given an international state system in which those who would resist such domination adapt to it to survive—how do we resolve conflicts between particular groups of humans, often the most oppressed, and other species? Even if deep ecologists and other advocates of wilderness creation and protection do attempt to ensure that such activities are not taken at the expense of the oppressed, they will not always be able to achieve both ends: protecting the environment and the poor. By what method do we choose what is to be done? There is no getting around these uncomfortable questions, and previous attempts to ad-

---

More recently there has been a recognition among revolutionaries and even some non-revolutionary governments that protection of the environment is an important value in and of itself and that industrialization and its products are not desirable goals. However, the costs of defending themselves from a hostile world (and the costs of giving in) lead to environmental degradation. Clearly solutions must be global and systemic.

dress them, including Naess's notion of "near and vital," are not adequately developed.[6]

ANTHROPOCENTRIC/BIOCENTRIC DISTINCTION

In attempting to sort out the questions raised in the previous section the biocentric/anthropocentric distinction is critical. If nature and nonhuman life are held by humans to be as valuable as human life, different answers and courses of action certainly follow than if one only regards human life and needs as valuable. Nonetheless, Guha argues that this distinction has little meaning when it comes to addressing two of the major problems undermining the health of the planet: overconsumption and militarization. I argue to the contrary that significantly different practical consequences follow from adherence to a biocentric rather than an anthropocentric system of values: a biocentric world view constrains human activity much more significantly and distinctly than even the most self-enlightened anthropocentric view. Moreover, a biocentric critique of overconsumption and militarization fully takes account of the underlying dynamic which produces these problems. Initially, a few words need to be said concerning the meaning of anthropocentrism.

Guha suggests that the existing social order in the West is not truly anthropocentric in the sense that it does not value and benefit humans per se, but only some humans, e.g., political and economic elites. Humans who do not fit into these categories are often treated much the way other species are—they are valuable only insofar as they are useful to those at the top of the social hierarchy. Such an analysis is certainly accurate; nevertheless, in contrast to the various forms of domination that are salient features of society, the stated values of the elites are in content anthropocentric: all humans have equal dignity and value.

---

[6] Arne Naess has suggested in "Identification as a Source of Deep Ecological Attitudes" (Michael Tobias, ed., *Deep Ecology*, 2d ed. [San Marcos, Calif.: Avant Books, 1980], p. 270) that conflicts between humans and other species can be resolved by balancing the competing interests based upon how "near and vital" the interests are to the species involved. Given the large numbers of Homo sapiens and their extensive settlement, it is difficult to see how this approach could lead to a redress of the current imbalance unless one takes a global perspective. There can be little question, for example, that humans need to give way to tigers, chimps, elephants, grizzlies, and other species. With five billion people and only a few thousand members of certain other species, restoring a balance can only mean a movement in one direction: more room for other species. Of course, the impact on humans of making room for other creatures will not affect all humans equally. Specific humans will have to make way. How are the costs to be spread? If one takes a strictly localized perspective, trying to balance the interests of a local human population only with the interests of a local nonhuman population, an assessment of competing interests gives a result less favorable to nonhuman life. Once one takes the extensive human presence as a given, human interests in their existing livelihood must be weighed without taking into account significant human numbers elsewhere. In this way, the pressure on already diminished populations of other species would continue.

Thus, a social system which benefits the few is justified by arguing that it benefits most or all. Certainly no other justification would do in this age. The contradiction between myth and behavior, however, is most apparent when we look at those who have become superfluous as either producers or consumers: they are "other," objects to be managed by welfare or simply repressed. Widespread toleration of this gap between values and behavior exists for a variety of reasons that we cannot go into here.

Are such systems *anthropocentric*, as deep ecologists argue, or are they better characterized by another term such as *patriarchy* or *class society?* Calling such societies *anthropocentric* seems to miss the fact that only some humans have value (in practice, dominant myths notwithstanding), that control over the human relationship with nature is not shared equally by all, and that the fruits of the exploitation of nature are not shared equally. Yet terms such as *patriarchy* or *class society* seem to ignore the degradation of nonhuman life and the Earth that is so fundamental in almost all existing human societies—and the way in which most humans, regardless of social position, participate in that degradation.[7] Even where critics of the dominant social order recognize the fundamental importance of the human relationship with nature, they continue to share many of the assumptions of the order they criticize: for example, faith in human reason and its ability to solve all problems and the centrality of humans in the universe. Even the phrase "relationship with nature" itself suggests something very different than "place within nature."

My purpose here is not to resolve the issue, but to point out the need for clarification when using these terms as well as the importance of the distinction between values as a world view and as actual behavior. In the discussion that follows anthropocentrism is used to denote any ideology or system of values which is human centered.[8] Such a definition for the purposes of this paper is not meant to slight the reality of the social systems in which these values are embedded—systems which in practice have yet to come anywhere close to living up to their avowed norms. Nor does such a definition imply that the very real and significant differences among human-centered systems of values are unimportant.

---

[7] Under what circumstances and in what situations humans can be treated as a species rather than as classes, genders, and ethnic groups or as individuals is a contentious matter among theoriticians and activists concerned with the environment and human justice. It is an important issue, and is best resolved in regard to specific questions. For some questions a species approach is appropriate; for others another level of aggregation is required. See note 21 below for a related discussion.

[8] After submission of this manuscript Carolyn Marchant published a discussion paper, "Environmental Ethics and Political Conflict: A View from California," *Environmental Ethics* 12 (1990): 45–68, in which she distinguishes egocentric values from anthropocentric values. The dominant values in the U.S. and most other developed societies are most properly termed egocentric rather anthropocentric.

OVERCONSUMPTION

In what ways then is a biocentric system of values meaningful in dealing with overconsumption and militarization? Let's consider overconsumption first. The very meaning of overconsumption differs depending upon whether one takes a biocentric or anthropocentric view. A biocentric view, by giving moral considerability to other species and ecosystems, much more sharply limits human consumption—not only as individuals or groups, but as a species, i.e., it implies a limit on human numbers—than an anthropocentric view which sees value in nature only insofar as it is useful to humans.

If nonhuman nature is valued for itself, if the integrity of the biosphere as a community is valued for itself, then human consumption which disrupts it is wrong: it would constitute overconsumption. Most modern forms of agriculture, forestry, mining, energy extraction and use, housing, transportation, and the like are part of a system of biocidal carnage, and therefore can clearly be called overconsumption.

In a human-centered system of values overconsumption is primarily seen as a social relationship, a problem of distribution between wealthy and poor, a problem of economic ownership.[9] Overconsumption occurs when some consume more than they need at the expense of those who do not have what they need. Generally speaking, material growth and rising levels of consumption are equated with quality of life; the poor can become better off through economic growth and/or more eqalitarian distribution. To this end technology and social organization need to be applied. Such a view does not admit to any finite limit on consumption; nor does it recognize injury to the biosphere as a factor except insofar as it may affect the continued use of the biosphere for human benefit.

Even with most forms of weak anthropocentrism—a view that is sensitive to long-range sustainability—we are still left with a system of values which can and does justify monoculture, high use of energy, massive reclamation projects, conversion of self-regulating ecosystems into cities, and suburbs and agricultural land managed for human use. Such a system continues to view nature as primarily a resource (or a nuisance) and only places limits on consumption which do not affect the sustainability of exploitation. The conversion for human benefit of vast portions of the biosphere is not viewed as wrong even though countless other species are reduced to minimal numbers or to extinction, and ecosystems impoverished or destroyed. Moreover, faith in the centrality of human abilities, particularly as expressed through technology, makes the constraints imposed by

---

[9] Critical theory distinguishes between natural poverty, which is due to a lack of development of productive forces, and social poverty, which is the result of exploitation and inequality. Both are regarded as oppressive. It is fair to say that Marx saw the human/nature schism and subsequent struggle through eyes sensitized to the class struggle. Thus, there can be no essential harmony until the struggle is over, i.e., until the human species brings the rest of nature under its control.

concerns for sustainability so vague as to be ineffective as limits on consumption. In contrast, constraints imposed by regarding the ecosystem and other species as valuable in and of themselves narrow the range of appropriate human behavior very sharply: if it injures the biosphere, don't do it.

The distinction between the two views goes much deeper when we examine the roots and social function of high levels of consumption. On a psychological level much consumption is a result of alienation, not just from nature, but also from self (nature within). Endless accumulation and the distractions it offers are essential features of developed societies and of the elites and middle classes elsewhere in the world. Such pathetic attempts to substitute possession of things in lieu of empowerment, sense of place, and authentic relationships is never satisfactory. A hunger for more always remains.[10] On a social level consumption is used by elites to manage large segments of the population. Give people enough stuff and they will forget their pain and powerlessness. The poor make do with the promise of some adequate level of consumption in the distant future and in the meantime they turn to other forms of distraction, often drugs.

Dominant Western or liberal capitalist views tend to deny that there is such a thing as overconsumption. To liberalism, high levels of consumption are viewed as a true measure of the success of our civilization and the individuals within it. It represents the triumph of control and technique, of humans over nature, the fullest flowering of our human faculties. It embraces dualism, hierarchy, atomism, all the machinery of control: nature is fodder, the "other," something to be mastered and managed. Man (intentionally masculine) is the centerpiece of the universe.

There are many human-centered theories which do recognize the pathological roots and role high levels of consumption play in many societies. The Marxisms of Reich, Marcuse, Gorz, and others are concerned with how factors such as consumption are both the result of and further feed alienation. Nevertheless, most Marxism remains wedded to some kind of control over nature, and thus

---

[10] In *Ecology as Politics* (Boston: South End, 1980), pp. 28–42, Andre Gorz has a very useful discussion on the social definition of poverty and its role in social management. On a psychological level, the existence of narcissistic tendencies (if not clinical narcissism) is important in explaining how individuals become susceptible to the social dynamic of consumption. When people are not aware of their real needs, they often seek socially defined and approved substitutes which necessarily cannot satiate them. Toys, drugs, and even television cannot effectively fill the emptiness left by the inability to experience intimacy, for example. The lack of a developed self leaves individuals particularly vulnerable to manipulation. Christopher Lasch examines this phenomenon in *The Culture of Narcissism* (New York: Norton, 1978). On the psychological aspects of the genesis of narcissism, see the various works of Alice Miller, James F. Masterson, Heinz Kohut, and Otto Kernberg. Alexander Lowen has provided a more popular treatment. The function of the family generally in shaping the malleable young for the roles society requires has long been the subject of some of best psychological writing, including that of Wilhelm Reich, Erik Erikson, Dorothy Dinnerstein, Nancy Chodorow, and others. The phenomenon is not new. Ruskin noted in the last century that the two objects of civilized life are: "Whatever we have—to get more; and wherever we are—to go somewhere else." The only thing we might add today is: to get there faster.

embraces dualism as well as open-ended material growth through progress in technology and social organization.[11] It espouses an unlimited faith in human intelligence and rationality, the ability to understand, control, shape, and improve upon not just human social organization, but all of nature. It is assumed that the evolution of human consciousness will keep pace with any problems, and that we will learn to guide cultural evolution. On the plus side, however, Marxism does reject the view of the world as essentially atomized. As Ollman has so ably demonstrated, Marx saw things as constituted by their relationships and the field of relationships.[12] One can neither change nature without changing oneself nor change an element in a system without changing the system. Despite the limitations of Marxism, a profoundly ecological sort of truth is recognized in such a perspective.

Much radical feminist theory rejects all institutionalized hierarchy and control as being inimical to authenticity.[13] The social problem is not so much who has power, but the power or domination itself. Relationships and community are essential values in this understanding. Both feminists and those concerned with domination based on ethnic differences have been central in pointing out how the category of "the other" runs throughout civilization, justifying oppression, exploitation, and cruelty toward anything that falls within it.[14]

---

[11] Bertell Ollman, *Alienation*, 2d ed. (Cambridge: Cambridge University Press, 1976), pp. 14–32.

[12] For Marx the labor process, by which humans transform nature to make their living, defines the human relationship to nature and is central to human biological and sociocultural evolution. This is linked not only to the conquest or appropriation of nature as central to our "species-being," but to an open-ended notion of human *material* needs. It is in this area that more modern anthropological critiques offer a deeper understanding of the circumstances of human evolution than nineteenth-century theorists had available. One notable point of difference involves the notion of communal ownership which Marx and Engels, following Morgan, viewed as a central feature of egalitarian society. This idea was important because it informed their vision of a future egalitarian society: communism. They believed that the failure of primitive egalitarian society was based upon scarcity; communism would be wealthy, taking advantage of previously developed forces of production. Modern anthropology suggests, nevertheless, that while primitive societies were egalitarian they did not have a sense of ownership, either private or communal. Animals, plants, and nature were part of a community. Much anthropological research also calls into question the notion that there is a deep-seated need in humans to fundamentally transform the environment.

[13] Feminists have offered some of the most cogent criticism of power and hierarchy as such. The work of Susan Griffin, Dorothy Dinnerstein, Kathy E. Ferguson, Mary Daly, and others offers enormously valuable insights into the difference between power (over the other) and empowerment, the nonalienating experience of being a full member in a real community. If one extends the notion of community to include other species and the biosphere, one has, I believe, a version of biocentrism. See Judith Plant, ed., *Healing the Wounds* (Philadelphia: New Society, 1989), for example.

[14] Marjorie Spiegel, in *The Dreaded Comparison* (Santa Cruz: New Society Publishers, 1988), cogently compares the similarities in the arguments that attempt to justify exploitation and domination of and cruelty to humans, animals, and nature. The manner in which some humans separate themselves from other people and from the natural community invariably involves a process in which differences (both real and imagined) are translated into value distinctions simply on the basis of difference. Thus, because Africans or women or wolves are different from Europeans, men, or humans, they are less valuable or of no value.

Although there are several anthropocentric world views which object to Cartesian dualism, liberal atomism, and so on, nature and other species remain excluded from the community either explicitly or by silence. One is left with the gulf between humanity and nature, spirit and body, and with an ungrounded faith in the human mission to make over the planet in its own image.

Some anarchist, Marxist, and feminist theory does suggest that part of realizing one's fullest humanity, i.e., part of the process of transcending alienation, involves embracing one's place in nature. In other words, non-alienated being requires a biocentric view of some sort. In such cases being a "citizen rather than conqueror" is biocentric if the natural as well as the human community is recognized as valuable; however, if one simply values the human interest in non-alienation, then the dualism—and the anthropocentrism—remains and serves as a theoretical foundation for views which rationalize structures of control.

A biocentric view rejects both the enlightenment values which justify the social order that thrives on overconsumption as well as the various forms of faith and dualism which underlie world views critical of much of the existing order. This is not to say that much or even most of feminist, anarchist, or other critical social theory is fundamentally incompatable with biocentrism; nevertheless, where such theory continues to accept the human species as the centerpiece of evolution, with the rest of nature existing solely for humanity's use, it fails to address a central form of domination. As such, even under such critical value systems the biosphere is open to suffer the consequences of the arrogance implied in human attempts to manage nature. If species hierarchy is justified, then hierarchy is justified. Much of what such critiques abhor follows from a human-centered view.

Biocentrism draws a clear line. To reject the human/nature dualism is to reject the "triumph" of the enlightenment attempt to control nature. It is to reject the triumph of knowledge and technique and analysis over Earth wisdom, understanding, and connectedness. It is to reject the focus on things rather than relationships. By rejecting these and valuing nature in and of itself, a biocentric view limits human consumption more fundamentally than any anthropocentric view can; it does so by thoroughly rejecting the roots of such consumption. In its place biocentrism values the web of life, as well as its parts, of which we are one.

MILITARIZATION

As with overconsumption we should ask which system of values will constrain militarism more: the human- or the biosphere-centered? By recognizing the valuableness of nature and other species apart from their usefulness to humans, a significant constraint is imposed on human activity with regard to both the conduct of war and more importantly the economic activity that is essential to preparation for war. Indeed, more than war itself, it is the consumption of "resources" to create and maintain the industrial capacity geared to arms production—for whatever purpose—that is so destructive of the biosphere. All human centered value systems necessarily fall prey to the easy rationalization of militarism.

If one is concerned only with humans, with the perpetuation and protection of particular social systems against internal or external threats, the constraints placed upon the consumption of nature are weak indeed. Even when limits on resources may temper overconsumption generally, there is a real tendency in this sphere of "national security" to literally let the future take care of itself and commit all to the current struggle. Certainly aesthetic regard for nature falls by the wayside. If the machine needs oil, then drill. The Soviet Union, as an example, has some of the strictest environmental legislation in the world. These laws also provide a giant loophole for any endeavor related to the security of the state, virtually negating restrictions.[15] Most countries start with weaker laws to begin with before embracing the exceptions.

There are many human-centered value systems, religious and secular, critical of militarization—and all are largely ineffective. The failure comes in part from the wedding of values to structures of power—be they church or state—that depend upon force for their survival. Insofar as these pacifistic values are taken up by those "outside" these structures they provide some check. But because they are human-centered—the point of opposing militarization is to end human waste and suffering—it is easy to neutralize them by appeal to other human values and to other forms of suffering even worse than war or the costs of deterrence. The other great weakness is that much pacifistic thinking does not address adequately the roots of militarism, something I attempt to do below.

If one values nature in and for itself, then human goals and needs are placed within the context of a larger community. The value placed on the integrity of that community militate heavily against any human-centered rationalization for exploitation. A biocentric view quite simply limits the conversion of ecosystems and biomass to human use to any extensive degree. Although such a view may seem utopian, because it poses a threat to the survival of particular social systems or the system of historical social systems, it does not pose a threat to the survival of the species as some would argue. Quite the opposite, the threat to both us and the planet comes from this system of systems. It is here that biocentrism provides understanding which human-centered approaches cannot, for the latter accept fundamental values which justify the very structures that give rise to the outcomes they criticize.

Consider the roots of militarism. Because modern militarism is particularly

---

[15] See, for example, Boris Kamarov, *Destruction of Nature in the Soviet Union* (White Plains: M. E. Sharpe, 1980).

virulent, attempts to understand and criticize this blight are often limited to the modern period. Certainly the combination of enlightenment arrogance, science, and technology, embedded in the international political economy resulting from the European expansion, has produced a very dangerous world.[16] It is, however, necessary to look more deeply into human history to grasp the underlying dynamic of militarism. While it may have reached new proportions, it is not new, but rather an essential feature of something very old: civilization.[17] It is inseparable from social systems based upon hierarchy (class, gender, and ethnic), control of nature, the denial of self, and the emotions and bonds which constitute the self. It is an essential feature of those societies in which the state exists, the process by which the state attempts to substitute itself for authentic human community is well underway, and conflict between communities has been replaced by the institutionalized conflict of center and periphery and between competing centers.[18] Civilization, and the process of its formation and emergence in the neolithic, is the story of the human attempt to adapt through various strategies of control—control of nature and of people through technology and social organization. It is this attempt to control nature that separates us from it, that constitutes the core of our alienation from life, and that becomes the

---

[16] The literature on imperialism and the world that it has produced is enormous. A good general introduction is Charles K. Wilber, ed., *The Political Economy of Development and Underdevelopment*, 4th ed. (New York: Random House, 1988) or L. S. Stavrianos, *Global Rift* (New York: Morrow, 1981). From there one may pursue more specialized works in economics, politics, conflict, and so on. For those concerned with the environment explicitly, the work of Alfred Crosby stands out; both *The Columbian Exchange* (New York: Greenwood Press, 1973) and *Ecological Imperialism* (Cambridge: Cambridge University Press, 1986) are truly major contributions. Carolyn Merchant's excellent *The Death of Nature* (San Francisco: Harper & Row, 1980) describes the emergence of the scientific world view that formed an essential part of the cultural framework justifying the domination other peoples, nature, and women.

[17] The institutions and processes that constitute civilization are not matters for serious debate. The state, urbanization, extensive division of labor, class structure, patriarchy, militarism, and monumental architecture are among those elements identified by Elman Service, Robert M. Adams. K. C. Chang, Morton Fried, William Sanders, Barbara Price, and others. On the origins and evolution of the problems of agriculture see Wes Jackson, *Altars of Unhewn Stone* (San Francisco: North Point, 1987).

[18] The earliest human communities include nature, but do distinguish the "other," namely, humans belonging to other societies or communities, especially those not involved in some reciprocal relationship. Conflict between communities certainly predates civilization and the neolithic period. Much of this conflict, however, was symbolic and sharply limited in its destructiveness to both persons and the environment. As some communities followed a path toward hierarchical social organization, the ability of the more hierarchical groups to displace, conquer, or otherwise make nonhierarchical groups dependent on them developed. Institutionalized relationships of domination became the order of the day. To escape domination other groups were forced to move, submit, or resist by adopting similar social strategies. The origins of center and periphery and the attendant exploitation, brutalization and conflict between center and periphery and competing centers are thus very old indeed. How egalitarian communities might relate noncompetitively will be a critical issue for humans and others on this planet. Certainly understanding both the psychological dynamic that generates the notion of the "other" and the social/ecological dynamic that creates pressure toward conflict and hierarchical solutions goes a long way toward solving the problem, although it is not sufficient in itself.

foundation for social development that includes patriarchy, class domination, statism, and militarism.

While most, but by no means all human-centered *value* systems eschew militarism, civilization is held as a crowning achievement. Some value systems praise the military spirit, while the majority that condemn it usually do so as a necessary evil, i.e., they simultaneously justify it to one degree or another. The point to be made here is that civilization is based upon and is constituted by relationships of domination that invariably and necessarily produce the conflict and inequality which make militarism inevitable. Certainly some human-centered theory recognizes aspects of the roots of militarism, and it recognizes the terrible price humans have paid, even if ignoring the price nature has paid. Nevertheless, critics maintain a fervent faith in the human mission to manage, in the human ability to disentangle what is inextricably linked. They speak from within the perspective of civilization and cannot see that they must transcend the precarious ground on which they (we) teeter.[19]

Critical theory shares much in common with liberal theory in this area. Some Marxist analysis of the genesis of modern militarism is sound. The notion that many human ills would be solved with the end of class society is also appealing. But the end of class is not the end of the state or of domination, and hence not the end of social systems which produce militarism. (Nor is the end of capitalism the end of class.) The control of nature and the human control of social and cultural evolution are values deeply embedded in most Marxism. Although it has developed useful models for understanding social transformation, the assumptions, perspective, and the content of the transformative vision are very much within the human-centered tradition that is part of the problem.[20]

---

[19] To say that civilization must somehow be fundamentally transcended is to say that the dynamic founded on and constituted by various relationships of domination must be overcome. It may represent a kind of return to the past, but in the service of the future. For the last six thousand years our species has behaved much like one might expect adolescents from a severely dysfunctional family to act. We must go back to where things went wrong—to the origins of our estrangement—and pick up from there. In doing so we make use of all that has occurred in the interim. We have already paid dearly for the lessons.

[20] There are attempts within the Marxist tradition to break out of this approach, but they are limited in degree. For instance, Gorz has argued that socialist industrialization is as bad as capitalist industrialization if it results in the same kind of environmental damage. Nevertheless, in his attempt to integrate environmental constraints into a Marxist framework, he remains concerned with the fate of Homo sapiens, not with the biosphere as such. In the West it is painfully obvious that the Marxist concern with the environment is largely a response to the environmental movement, rather than something generated internally. It remains to be seen how both dialogue and praxis will develop. Recently a journal was founded in this area, *Capitalism, Nature, Socialism*. The analysis of environmental policy in statist or state socialist countries demonstrates that their path reflects a dynamic based on maintaining a bureaucratic oligarchy at home and defending themselves from a hostile world capitalist order. This dynamic, while differing from what moves capitalist societies, offers little in the way of hope for a significant alternative relationship with the environment. Genuinely socialist or communitarian experiments in the Third and First Worlds that reject massive industrialization are another matter.

Some feminism gets much closer to the source of the problem in its critique of hierarchy generally and in particular in its understanding of the central role of patriarchy to militarism and to producing humans amenable to domination. At times, however, feminist theory falls into a kind of intraspecific dualism, i.e., human males are the problem (while at the same time claiming credit for the fact that females created agriculture, which became the economic foundation for the emergence of hierarchy), ignoring that systems adapt to and alter the environment, and individuals adapt to (even while they resist) the roles created by the system's division of labor.[21] Even where this dualism is not at issue, most feminism, like Marxism, remains human-centered. Values such as community, spontaneity, and integration of emotion and intellect militate against the worst features of mainstream human-centered values, but still fail to take account of the relationship with nature as fundamental to all hierarchical systems. Or they remain anthropocentric and fail to address the separation from nature which not only makes possible the superexploitation of the biosphere for the maintenance of the military apparatus, but also underlies the social structures which produce militarism.

While Marxism, feminism, and other critical social theory have contributed

---

[21] The degree of choice that individual humans and human collectives exercise is a matter of serious debate. Factors such as consciousness, social position of particular individuals and groups, and the limits of historical possibility certainly all play a role. In the course of social struggle people necessarily make certain assumptions: that it is possible to realize their goals, that they understand the operation of the social (or natural) world enough to bring about the desired results, and that those opposed to them must be held accountable for their resistance as if they understood what they were doing and recognized the choices available to them. The rather consistent failure in the social realm to realize expectations calls these assumptions into question to varying degrees. Some things, nevertheless, do seem clear: when people make decisions, it is the unintended consequences that tend to outweigh in importance the intended ones—a point which is often never realized. No one set out to invent the state, but a series of decisions over time had that result. There have been "moments" in human history (and prehistory) when real choices about basic human social structure and interaction with nature were possible. At such moments it was possible to switch a train from one track to another. When these moments occur in the future, there must be an awareness that the tracks are available and that awareness must be shared by enough people to make a difference. Most of the time choices are more constrained, with humans squabbling over who gets what seat on the train, rather than where its going. This is not to say, of course, that who sits where is not important to both humans and the rest of nature, for who sits where has much to do with the choices eventually taken when switching tracks is possible. On an individual level, both the powerful and the weak are socialized to roles that neither created, and they are limited in their options by structure and consciousness. Certainly those who benefit more seek to perpetuate and strengthen their position and this contributes to the perpetuation of the system. Even though those who benefit less are more inclined to resist their roles and seek change, they also cooperate in their subjugation because they have been socialized to do so and because they fear repression. The degree to which socialization limits possibilities varies with time and the conjunction of a number of factors. On the other hand, it is important to keep in mind the often overlooked distinction between explaining behavior and excusing it. The former by no means implies the latter. Although people may not choose freely, their actions have consequences and accountability is fundamental to constraining both individuals and collectives. Without it there is no learning.

much to understanding the dynamic of our civilization, they tend to miss the point that if nonhuman life is not valued for itself, then life is not valued for itself. Any system of values that does not transcend nature-as-other cannot limit destruction of the biosphere as effectively as one that embraces nonhuman life as intrinsically valuable. Nor can such a value system help to heal the fundamental split in the human psyche which makes possible civilization and militarism.

Biocentrism is not alone in grasping that the dynamic of human evolution over the last six or seven thousand years may be at a dead end. Certainly the huge growth in human numbers, the displacement of "simpler" societies by more "complex" ones, ones with greater capacity to exploit nature, capture and use energy, and so on suggests that the underlying dynamic is highly adaptive, at least at first glance. What is increasingly clear, however, is that if this dynamic continues we stand a very good chance of killing ourselves along with a good portion of the rest of the planet. The latter is well under way—it's business as usual.

Biocentrism offers a direction for human society based upon a thoroughly fundamental transformation which stresses the centrality of finding our place *in* nature. Such a transformation is as fundamental as the neolithic or industrial revolutions.

A life-centered or planet-centered value system requires that we move toward transcending the split with nature both within our own psyches and in our material relationships: how we consume and alter the biosphere. Far fewer humans, far lower levels of consumption for many, much improved levels for others, the recreation of authentic communities that reintegrate the human into the natural, and the abandonment of the instrumentalities of control—these are a few of the implications of such an ethic.

In contrast, a human-centered approach focuses on wiser if not greater human control. In its more progressive forms we hear words like stewardship rather than ownership; nevertheless, underlying both is the notion that we can replace nature with our intellect, that we can manage our way out of any problems, that we as a species are not only unique (as every species and ecosystem is), but that our uniqueness means we are godlike, better than the others. In short, it is the same arrogance, the same split that has brought us to the current crisis.

VALUES AND CULTURE

All systems of values are part of a broader cultural framework that mediates human behavior by shaping personality and thought. Culture organizes human experience and gives it meaning. Biocentric values are no exception—they are part of a larger cultural framework, albeit an emergent one. Part of that culture includes an understanding of the role of culture generally as well as the critique of particular cultures.

A biocentric approach presents human-centered cultures as both rooted in and perpetuating a split between our species and the rest of nature. This split, which is manifest as both a chronic and debilitating inner tension and as a stressful warlike antagonism toward the environment, is the source of our experience of estrangement. Disembodied, our cortex is a shadow of life, ever busy trying to rationalize the irrational while telling us it is all. Biocentrism offers us back our body by recognizing that the Earth is our real community—that by healing our split from it, by healing the split between cortex and heart, and by healing nature within, we can begin to heal all of nature. It also helps us to understand how that split is possible.

To point to the neolithic as the origin of the culture of control is not enough. A biocentric view places these events in context. It helps us understand how the capacity for culture and the resulting plasticity in human behavior, thought and emotion, and our ability to learn and pass on learning (attitudes and world views as well as technical or social information) enables us to divide ourselves. This capacity for culture allowed humans threatened with localized overpopulation in neolithic times to increase the human carrying capacity by altering both their behavior and the environment in substantial ways. With the new dynamic set in motion the fundamental injury to other needs, the split itself, was probably not very obvious. First, it was cumulative over a long time. Second, the very capacity for culture allowed them (and us) to deny the estrangement, even required such denial for both psychological and social reasons. Third, the emerging social dynamic of hierarchy distributed the costs and benefits of the new adaptive strategies unequally, favoring the decision makers and shapers of a society's values.

Culture has allowed us to trade our place in nature for larger human numbers spread over the entire planet, converting large amounts of the biosphere to our purposes so long as we are willing to pay the price of the various forms of domination and the accompanying anguish with which we are so familiar—and so long as we are willing to deny the value of other life and allow nature to pay the price. The plasticity with which evolution has endowed us allows us to create alienating and biocidal sociocultural systems, but it does not require it; it is not natural in the sense of being necessary or in the sense of being in tune with our deepest nature. (We should not forget that while cancer is a part of nature, it kills its host.) There are other cultural possibilities, including biocentric ones. Indeed, for most of the time humans have been around we have lived in communities which included the rest of nature. We can do so again, this time with full knowledge of what the alternatives are and their price. To limit our biocidal possibilities is not unnatural, as Callicott quite rightly argues, because cultural systems always limit behavior.[22] Culture is always prescriptive.

---

[22] See J. Baird Callicott's discussion on the biosocial role of culture in *In Defense of the Land Ethic* (Albany: SUNY Press, 1989), pp. 63–73. See also Dorothy Lee's *Freedom and Culture* (Englewood Cliffs: Prentice Hall, 1959).

The roots of biocentrism are deep and its emergence in modern form is a result of both the resilience of Earth wisdom and the current crisis—just as surely as human-centered values and cultural systems are a result of the crisis of the neolithic. Both Marxist and feminist anthropology have traced the roots of class and patriarchal domination and have contributed much to understanding the dynamic which emerged. Marx and some Marxists have rightly regarded the split with nature as a decisive milestone in human cultural evolution, albeit a positive one. Some feminist writing has addressed itself to the split with nature, both in the modern epoch and in the distant past, but usually as something ancillary to the development of human forms of domination. A biocentric understanding suggests that the culture of control over nature is part of an adaptation to scarcity and lays the groundwork for other forms of domination.[23]

By accepting biocentric limits upon our behavior we directly undermine the split from nature and the resulting culture of domination which arises from it. In doing so we accept constraints on overconsumption and militarism that no human-centered system of values could impose. Domination and hierarchy, the attempt to control that gives rise to high levels of consumption and militarism, will be unshakable problems until we recognize we cannot substitute our intellect for nature. The degree to which we attempt to do so is a function of our own estrangement.

ALLIANCES

Guha gives us much to think about and we ignore his voice at our peril. Although the Earth and other species need wilderness, we will lose the battle for the planet if we do not realize, as Guha suggests, that imperialism and militarism are our enemies as well as anthropocentrism. We cannot dismiss the struggles over human social structure and realize a deep ecological vision. The land ethic is not compatible with most of the existing order of things. In struggling to alter that order it is necessary to understand how it works, for if we do not, the vision in the hearts of a few will not be enough.

---

[23] See, for example, Peggy Reeves Sanday, *Female Power & Male Dominance* (Cambridge: Cambridge University Press, 1981).

Part IV
Ecology and Feminism

10

WOMEN, HUMANITY AND NATURE

Val Plumwood

There is now a growing awareness that the Western philosophical tradition which has identified, on the one hand, maleness with the sphere of rationality, and on the other hand, femaleness with the sphere of nature, has provided one of the main intellectual bases for the domination of women in Western culture.

There are plenty of good reasons for feminists to distrust both the concept of rationality and the notion of links with nature and the concept of nature. Both of these concepts and their contrasts have been major tools used to inferiorise and exclude women (as well as other groups). The main function of the concept of rationality, which has a confusing array of senses in which it is often hard to discern any precise content, seems to be a self-congratulatory one for the group thought to possess the prized quality and the exclusion and denigration of the contrasting group which does not. Thus the sphere of rationality variously contrasts with and excludes the sphere of the emotions, the body, the passions, nature, the non-human world, faith, matter and physicality, experience and madness. The masculine rational sphere of public life, production, social and cultural life and rational justice is contrasted with the feminine sphere of the private, domestic and reproductive life, the latter representing the natural and individual as against the social and cultural.[1] Again, the rational masculine sphere is a sphere where human freedom and control are exercised over affairs and over nature, especially via science and in active struggle against nature and over circumstances. In contrast, the feminine natural and domestic sphere represents the area of immersion in life, the natural part of a human being, the sphere of passivity, acceptance of unchangeable human nature and natural necessity, of reproduction and necessary and unfree labour.

In these cases there is not merely a contrast but an unfavourable one; the sphere associated with femininity and nature is accorded lower value than that associated with masculinity and freedom. In all the senses of rationality, the 'rational' side of the contrasts is more highly regarded and is part of the ideal human character, so that women, to the extent that they are faithful to the divergent ideals of womanhood, emerge as inferior, impoverished or imperfect human beings, lacking or possessing in a reduced form the admired characteristics of courage, control, rationality and freedom which make humans what they are, and which, according to this view, distinctively mark them off from nature and the animal. Feminine 'closeness to nature' in this sense is hardly a compliment. The ideals of the masculine sphere and those of humanity are identical or are convergent. Those of femininity and humanity are divergent. To put the point another way, the ideals of the rational sphere give us a character model of the human which is masculine.

The concept of nature too has been and remains a major tool in the armoury of conservatives intent on keeping women in their place and supporting a rigid division of sexual spheres, or worse. It is allegedly nature, not contingent and changeable social arrangements, which determines that the lot of women will be that of reproduction and domestic arrangements and which justifies inequality. Women have been seen as connected with nature in both its two major contrasted senses, that of nature in contrast to culture or society, the realm of necessity in contrast to that of freedom, of controllable human cultural and social arrangements, and that of nature in contrast to the *human* world, or what is distinctively human in the world. The first sense, in which what is natural is what is not open to explanation or change, inspires the following conservative comment:

> Nature isn't fair, and never will be – it is not concerned with justice. Nature has made Man with more Assertion, so that he will not willingly let Woman take first place. If she tries to he will always feel his manhood affronted, and he will not like her so much. It isn't fair, but it is a fact.

> Without women men will always fight and drink and live like crows – they are really little savages. It's women who are the homemakers, the civilisers, the gentle, the beautiful

WOMEN, HUMANITY AND NATURE

ones – and all they require of men is Security and Love. But they get more enjoyment out of the Arts, more fun out of being creative, more love out of little children, more depth out of life. To ask to be equal as well – is it really fair?[2]

(No, it's not a contemporary of Rousseau's. That appeared in a book published in 1985.)

As Genevieve Lloyd has noted in her book *The Man of Reason*, however, the attitude to both women and nature resulting from the identification has not always been a simple one, and as Carolyn Merchant notes, it has not always been purely negative.[3] The connection has sometimes been used to provide a *limited* affirmation of both women and nature, for example, in the romantic tradition. But the dominant tradition has been one in which the connection with nature accords women a lower status (even if one that is sometimes accorded some virtue as a 'complement'), and has been used to confine them to limited and impoverished lives.

Given this background, it is not surprising that many feminists regard with some suspicion a recent view, expressed by a growing number of writers in the ecofeminist camp, that there may be something to be said *in favour* of feminine connectedness with nature, and that there are important connections between the oppression of women and the domination and destruction of the natural world which feminism cannot afford to ignore. The very idea of feminine connection with nature seems to many to be regressive and insulting, summoning up images of women as passive, reproductive animals, contented cows immersed in the body and in unreflecting experiencing of life.

It is both tempting and common therefore for feminists to view the traditional connection between women and nature as no more than in instrument of oppression, a relic of the bad old days which should simply wither away once its roots in an oppressive tradition are exposed. After all, this is 1990. It seems obvious enough that women must now claim full and equal participation in the sphere of humanity and rationality from which they have been excluded, and to which their traditional sphere of nature has been opposed. Freed of traditional prejudice and of the traditionally enforced tie to the natural, women can at last take their place simply as equal human beings. The connection with nature is best forgotten. Women

VAL PLUMWOOD

(especially modern women) have no more real connection with nature than men.

What I want to argue in this paper is that there are several reasons why this widespread, 'common-sense' approach to the issue is unsatisfactory. There are several reasons why the question of a woman/nature connection can't just be set aside, why the question should be examined carefully by feminists. The first of these, which is developed in the first part of the paper, is that it is essential to give critical examination to the issue both because of its repercussions for the model of humanity and for the treatment of nature.

The second reason, which is developed in the later part of the paper, is that it is essential for feminism to address the issue because the ecofeminist argument reveals an important ambiguity in feminist theory itself. Examination of the ecofeminist argument can throw valuable light on questions at the heart of feminism itself, and has significant implications for distinguishing different strains of feminism and different associated strategies.

The common-sense approach might better be called the 'naive' approach on analogy with naive realism in epistemology, since like naive realism it takes to be unproblematic what is not unproblematic. According to the naive view, the connection of women with nature should simply be set aside as a relic of the past, the problem for both women and men being that of becoming simply unproblematically and fully *human*. But the question of what is human is itself now highly problematic, and one of the areas in which it is most problematic is in the relation of humans to nature, to the non-human world.

Another problem is that what is in question is not just a model of *feminine* connectedness with and passivity towards nature, but also a contrasting and complementary one of masculine *disconnectedness* from and domination of nature. But the assumptions in the masculine model are not seen as such because the masculine model is taken for granted as simply a *human* model and the feminine as a deviation from that. Hence to simply repudiate the old tradition of feminine connection with nature and to put nothing in its place, usually amounts to implicitly endorsing an alternative *masculine* model of the human and of human relations to nature and to implicitly endorsing also female absorption into this model. It is not, as it might at first appear, a neutral position, because unless the question of relation to

214

WOMEN, HUMANITY AND NATURE

nature is explicitly put up for consideration and renegotiation, it is already settled – and settled in an unsatisfactory way – by the dominant Western model of humanity into which women will be fitted. This is a model of domination and transcendence of nature, in which freedom and virtue are construed in terms of control over, and distance from, the natural sphere. The critique of the domination of nature developed by environmental philosophers in the last ten years has shown I think that there are excellent reasons to be critical of this model.[4] Unless there is some critical re-evaluation of this masculine model in the area of relations to nature, the old female/nature connection will be replaced by a dominant model of distance from, transcendence and control of nature which is masculine. Some critical examination of the question then has to have a place, and an important one, on the feminist agenda if a masculine model of the human and of human relations to nature is not to triumph by default.

There is another reason then why the issue cannot be set aside in the way the naive view assumes. As a number of ecofeminists have observed, feminism needs to put its *own* house in order on this issue. If women do not have to fight the battles of other groups in a display of traditional altruism and self-abnegation, to carry the world's ills in recognition of motherly duty, as some arguments from peace and environmental activists suggest, it is also true that they can't base their own freedom on endorsing the continued lowly status of the sphere from which they have lately risen. Moves upwards in human groups are often accompanied by the vociferous insistence that those new recruits to the privileged class are utterly disassociated from the despised group from which they have emerged – hence the phenomenon of lower middle-class respectability and the officer risen from the ranks. Arguments for women cannot convincingly be based on a similar put-down of the non-human world.

But much of the traditional argument has been so based. For Mary Wollstonecraft, for example, what is valuable in the human character ideal to which women must aspire is defined in contrast to the inferior sphere of brute creation. Thus she begins her *Vindication* by asking: 'In what does man's pre-eminence over the brute creation, consist? The answer is as clear as that a half is less than a whole, in Reason.' And she goes on:

VAL PLUMWOOD

> For what purpose were the passions implanted? That man by struggling with them might attain a degree of knowledge denied to the brutes.
>
> Consequently the perfection of our nature and capability of happiness must be estimated by the degree of reason, virtue and humanity that distinguish the individual and that from the exercise of reason, knowledge and virtue naturally flows.[5]

In her argument that women do have the capacity to join men in 'superiority to the brute creation', the inferiority of the natural order is simply taken for granted. It is certainly no longer acceptable for feminists to argue for equality in this way.

* * *

Several critiques converge to necessitate reconsideration of the model of feminine connectedness with nature and masculine distance from and domination of it and to problematise the concept of the human. They are:

a) the critique of masculinity and the valuing of traits associated with it traditionally;

b) the critique of rationality; relevant here is not only the masculine and instrumental character of rationality, but also its over-valuation and use as a tool for the exclusion and oppression of the contrasting classes of the non-human (since rationality is often taken as the distinguishing mark of the human) and of women (because of its association with maleness). The over-valuation of rationality is deeply entrenched in Western culture and intellectual traditions, not always taking the extreme form of some of the classical philosophers (for example the Platonic view that the unexamined life was worthless, or the Augustinian one that rationality was the ultimate value to which all others were instrumental)[6] but appearing in many more subtle modern forms, e.g, the limitation of consideration to rational moral agents.

c) the critique of the human domination of nature, human chauvinism, speciesism, of the treatment of nature in purely instrumental terms and the low valuation placed on it in relation to the human and cultural spheres. Included in this is a critique

WOMEN, HUMANITY AND NATURE

of the model of the ideal human character and of human virtue, which points out that the Western human ideal is one which maximises difference and distance from the animal and the natural; the traits thought distinctively human, and valued as a result are not only those associated with masculinity but those unshared with animals.[7] Usually these are taken to be mental characteristics. An associated move is the identification of the human with the higher, mental capabilities and of the animal or natural with lower bodily ones, and the identification of the authentic human individual with the mental sphere.[8]

The critiques converge for several reasons. A major one is that the characteristics traditionally associated with masculinity are also those used to define what is distinctively human, e.g, rationality (and selected mental characteristics and skills), transcendence and activity, i.e. domination and control of nature as opposed to passive immersion in it (consider the characterisation of 'savages' as lower orders of humanity on this account), productive labour, sociability and culture. These last characteristics are assumed to be confined to humans but also associated with the masculine sphere of public life as opposed to the private, domestic, and reproductive sphere assigned to women. *Masculine* virtues are also taken to be *human* virtues, what distinguishes humans from the sphere of nature, especially the qualities of rationality, transcendence and freedom. Some traditional feminist arguments also provide striking examples of this identification of the human and the masculine. Thus Mary Wollstonecraft in the *Vindication* appeals strongly to the notion of an ungendered human character as an ideal for both sexes ('the first object of laudable ambition is to obtain a character as a human being'),[9] but this human character is implicitly masculine. The human character ideal she espouses diverges sharply from the feminine character ideal, which she rejects, 'despising that weak elegancy of mind, exquisite sensibility, and sweet docility of manners'. Instead she urges that women become 'more masculine and respectable'. The complementary feminine character ideal is rejected – both sexes should participate in a common human character ideal (p. 23) which despite some minor modifications (men are to become more modest and chaste and in that respect to take on feminine characteristics) coincides in its specifications with the masculine character. A single 'unsexed' character ideal is substituted for the old two-sexed one, where the old feminine ideal was perceived as subsidiary and sexed.

The key concepts of rationality (or mentality) and nature then form a crucial link between the human and the masculine, so that to problematise masculinity and rationality is at the same time to problematise the human, and with it, the relation of the human to the contrasted non-human sphere. The naive approach mistakenly takes the concept of the human to be unproblematic and fails to observe its masculine bias. This dual connection is then another reason why the issue of the traditional connection of women and nature can't simply be ignored, why the problems raised must be considered.

The concept of the human is itself very heavily normative. The notion of being fully or properly human carries enormous positive weight, and usually with little examination of the assumptions behind this, or the inferiorisation of the class of non-humans this involves. Things are deplored or praised in terms of conformity to a concept of 'full humanity'. But the dignity of humanity, like that of masculinity, is maintained by contrast with an excluded inferior class.[10]

The concept of the human plays an important but somewhat shadowy role in the problem, and assumptions about the ideal nature of the human often stand silently in the background in discussion on masculinity and femininity, as well as in other areas. Thus for example behind the view that there is something insulting or degrading about linking women and nature stands an unstated set of assumptions about the inferior status of the non-human world. Behind the view that the traditional connection between women and nature can be forgotten stands the assumption that women can now be fitted unproblematically into the current concept of the human, and, again, that this concept itself is unproblematic.

Once these assumptions are made explicit, the connection between the stance adopted on the issue of the woman/nature connection and the different possibilities for feminism becomes clearer. In terms of this framework the main traditional position – the point of departure for feminism – can be seen as one in which the ideal of human character is not, as it often pretends to be, gender-neutral, but instead coincides or converges with that of masculine character, while the ideals of womanhood

WOMEN, HUMANITY AND NATURE

diverge. Included, and indeed having pride of place in this character ideal are the ideals of rationality, self-expression, freedom and control via transformation and domination of the natural. Womanly character ideals of emotionality, passivity acceptance and nurturance stand in contrast. Thus, as Simone de Beauvoir has so powerfully stated, the tragedy of being a woman consisted in not only having one's life and choices impoverished and limited, but also in the fact that to be a good woman was to be a second-rate human being.[11] So that to the extent that these 'neutral' human character ideals were subscribed to and absorbed and the traditional feminine role also accepted, women must forever be forced to see themselves as inferiors and to be so seen. Because women were excluded from the activities and characteristics which were highly valorised and seen as *distinctively human*, they were forced to be satisfied with being mere spectators of what the *distinctively human* business of life was all about, the real business of the struggle with nature.

De Beauvoir's solution to this tragic dilemma was stated with great force and clarity – change was to come about by women fitting themselves and being *allowed* to fit themselves into the dominant model of the human, and women were thus to become *fully human*. The model itself, and the model of freedom via the domination of nature it is especially based on, are never themselves brought into question, and indeed women's eagerness to participate in it confirms and supports the superiority of the model. Similarly for others, e.g. Harriet Taylor and Mary Wollstonecraft. As this earlier feminism saw it, the tragedy of women was that they were treated as less than fully human, or that, prevented from becoming fully human, they were kept at the level of the brutes.

This has been called the first, masculinising, wave of feminism.[12] The problem for women was to claim full humanity, i.e. to conform to the main human character ideal, defined by traits characteristic also of the masculine, and to fit into, adapt themselves to, the corresponding social institutions of the public sphere. These might require some minor modification but basically it was women who were to change and adapt (sometimes with help), and women (or what society had made of them) who were the problem. The position can be summed up as that of demanding participation by women in a masculine concept or ideal of humanity, and the associated activist strategy as that of demanding equal admittance for women to a masculine-defined sphere and masculine institutions.

Central to these was the domination of nature. Women, in this strategy, are to join men in participation in areas which especially exhibit human freedom, such as science and technology, from which they have been especially strongly excluded. These areas are especially strongly masculine not only because their style strongly involves the highly valorised masculine traits of objectivity, abstractness, rationality and suppression of emotionality, but also because of their function which exhibits most strongly the masculine virtues of transcendence of, control of and struggle with nature. In the equal admittance strategy, women enter science, but science itself and its orientation to the domination of nature remain unchanged.

This masculinising strategy is the one which is being implicitly adopted when the problem of the woman/nature connection is simply side-stepped or set aside. It is assumed that the solution is for women to fit into a masculine model of human relations to nature which does not require change or challenge.

In the last decade this first, masculinising strategy of feminism has come under strong criticism from several feminist quarters and a number of its problems identified. One problem is that the masculine model of the human and corresponding social institutions has been arrived at precisely by exclusion and devaluation of women and feminine characteristics. Because it has been defined by *exclusion*, it is loaded against women in a variety of subtle and less subtle ways and women will not benefit from admittance to it as much as they think. As Genevieve Lloyd notes, 'Women cannot easily be accommodated into a cultural ideal that has defined itself in opposition to the feminine.'[13] Absorption into the masculine model is not likely to be successful.

Other major criticisms come from those who see the need to reject or modify the masculine character ideal as well as (or in some cases instead of) the feminine character ideal rejected or modified in the masculinising strategy. There are several different angles from which this criticism is directed. One is from difference theorists, who reject the masculine character ideal as a model, at least for women and in some cases for both men and women. Another is from ecofeminists, who reject the masculine model especially in the area of human relations to nature, and argue more directly that this masculinising strategy amounts to having

WOMEN, HUMANITY AND NATURE

women join men in belonging to a privileged class in turn defined by excluding the inferior class of the non-human; that is, it is a strategy of having women equally admitted to a now wider dominating class, without questioning the structure of, or the necessity for, domination. The conceptual apparatus relating superior to inferior orders remains intact and unquestioned.[14] What is achieved is a broadening of the dominating class, without changing or challenging the basis of domination itself. And the attempt to simply enlarge the privileged class by extending to and including women not only ignores a crucial moral dimension of the problem, it ignores the way in which different kinds of domination act as models for and as support and reinforcement for one another, and the way in which the same conceptual structure of domination reappears in very different inferiorised groups, e.g. women, inferior humans, slaves, manual labourers, 'savages', people of colour – all 'closer to the animals'.[15]

What seems to be involved here is often not so much an affirmation of feminine connectedness with and closeness to nature[16] as distrust and rejection of the masculine character model of disconnectedness from and domination of the natural order. The masculine character ideal is similarly rejected by the broader ecofeminists and by some theorists of non-violence, who link the masculine character ideal (and in some cases biological maleness) to aggression against fellow humans, especially women, as well as against nature. They reject the absorption of women into this mould, which is perceived as yielding a culture not of life but of death.[17]

* * *

One thing that has emerged from the disscussion so far is that a critical and thoroughgoing contemporary feminism is and must be engaged in a lot more than merely challenging and revising ideals of feminine character, that it is and must be engaged in revising and challenging as well the ideals both of *masculine* and of *human* character. The masculinising strategy is unsatisfactory and superficial precisely because it does not do this. In the light of this understanding it seems worthwhile to try to compare and evaluate some alternative strategies for revising the human character ideal and to try to spell out more clearly what alternative model the ecofeminist argument is really appealing to, and especially how it differs from conservative positions it is often confused with. It seems clear that the basic common ground of the ecofeminist and non-violence argument is rejecting the masculine model of the human as a character ideal, at least for women, but beyond that there is confusion, ambiguity and indeterminacy, and a number of different alternatives are possible.

Perhaps the most obvious way to interpret the ecofeminist argument is as one which replaces the masculine model of the human character by a new feminine model. That is, if the masculinising strategy rejected the feminine character ideal and affirmed a masculine one for both sexes, this feminising strategy rejects the masculine character ideal and affirms a feminine one for both sexes. The masculinising wave of feminism is succeeded by a new feminising wave. Several slogans sum up this feminising strategy, e.g. 'the future is female', 'Adam was a rough draft, Eve is a fair copy' (courtesy Macquarie University toilet door). There are several different forms the assertion of a feminine character ideal can take, and it is important to be clear about the difference.

First, a feminine character ideal can be affirmed *not* as a rival to the masculine character ideal but as a *complement*. The masculine model is not really challenged at all in this strategy and may in fact be affirmed and supported, although there may be some degree of upward revaluation of the relative worth of feminine traits. For example, the romantic tradition often does this, affirming the value of the feminine but in a way that does not really challenge the masculine ideal, but rather complements it or adds a separate feminine model.

An associated strategy is that of affirming a traditional model of feminine character obtained by reversing the values, so that traits previously regarded as lowly and despised become instead virtues and are given a high value: e.g. closeness to nature, previously used to put women down, is recast as a virtue.

There is a fairly strong tendency for a position which thus simply reverses the value of traditional feminine traits to collapse into a complementary position, and conversely for a complementary affirmation of feminine character to affirm traditional traits. One reason for this is that really traditional feminine traits include appropriate attitudes of subservience or self-abnegation which require a masculine complement. Thus where feminine virtues

WOMEN, HUMANITY AND NATURE

are developed in a situation of exclusion and complementation there is a problem about how they can stand on their own.[18] The associated social change strategy is that of separate spheres – recognising and revalorising traditional femininity as a complement to masculinity. This is a conservative pre-feminist or anti-feminist strategy, and is included here for completeness of alternatives, and so that it can be seen in relation to other positions.

A different strategy is that of affirming a feminine character ideal as a *rival* ideal, attempting to replace the masculine ideal, not merely to complement it. To be a genuine rival, it has to be affirmed as a rival model of the *human*, displacing or competing with the masculine model of the human. The human ideal then becomes a feminine rather than, as traditionally, a masculine one, and human virtues are now feminine virtues and character traits rather than masculine ones. Thus a feminine ideal is seen as desirable for both sexes, although there may be doubts as to how far biological males can ever approximate to it. Thus, according to Sally Miller Gearhart,

> it is time to dare to admit that some of the sex-role mythology is in fact true and to insist that the qualities attributed to women (specifically empathy, nurturance and co-operativeness) *be affirmed as human qualities* capable of cultivation by men even if denied them by nature.[19]

The 'primacy of the female' (i.e. of feminine character traits, not necessarily biological femaleness) would be acknowledged 'as primary, the source of all life'.[20]

What has come to be called 'difference theory' can involve the celebration and articulation of woman's difference from the ideal and actuality of masculine character, and, in some forms, can represent another strand of this feminising strategy. In contrast to the sort of position discussed above, which assumes that the identification of feminine traits is clear and that they can be known to include such traditional traits as nurturance and empathy, this alternative strand takes the form of the celebration of what is proclaimed as the genuinely feminine, which may be 'a feminine principle not to be defined'.[21] The project of the discovery and emergence of the genuinely feminine, is conceptualised not as something whose character has been formed by exclusion from the masculine sphere, but as an independent force, silenced and

223

unable to reach expression under patriarchy, but ready and able to emerge once the barriers of phallocentric society to its expression are removed. Women's bodily experience is often taken as the starting point in the attempt to give expression to the silenced and unknown feminine.[22]

If the strategy associated with the first, masculinising model is that of equality (in masculine institutions), and the strategy of the second, complementary feminine form is that of separate (but equally valued) spheres, the strategy of the third, feminising form is that of separatism, in which feminine virtues can be developed and come to dominate, or the unknown and yet to be discovered true feminine can emerge.

There seem to be numerous difficulties in both strands of the position. This paper is mainly concerned with understanding the motivation for and structure of the ecofeminist argument, the range of options available and which account of the human character ideal it appeals to, rather than with the detailed elaboration and critique of these positions which is undoubtedly required in feminist theory. Nevertheless, some critical comment on the third 'feminising' approach seems to be in order, if only to motivate the examination of alternatives.

Much of the problem turns on the question of what the characteristics of the alternative feminine ideal are, and of how the desired traits can be identified as feminine. If the position of the first strand is adopted, some virtues (e.g. nurturance, empathy in Gearhart) are identified as feminine or feminine-associated and put forward as the new ideal for the human. But how is this identification of these traits as feminine arrived at? Are the traits in question taken to be characteristic of all women in all circumstances (which is not very convincing), or only under traditional and complementary circumstances, in which case how can we know that they will survive translation to a different non-traditional and non-complementary context? Or is there some other alternative? Are they really traits of all *actual* women (or only some?) arrived at by examining what actual women are like, or are they traits simply traditionally attributed to women? So if traditional traits are affirmed, there is the problem Lloyd points to as to how traits developed in a complementary context (e.g. nurturance) can stand alone as a human ideal.

Gearhart skirts the problem by referring to the relevant traits as 'feminine-associated', an expression which is neatly four-way

WOMEN, HUMANITY AND NATURE

ambiguous between 'attributed to women', 'attributed traditionally to women', 'occurring with women', and 'occurring with women in the traditional context'. The ambiguity enables her to assume that those traits attributed to women in fact occur with them unproblematically in a non-traditional context.

Gearhart also conveniently overlooks numerous negative traits associated with women under patriarchy and in the traditional feminine, such as subservience, and does not explain what ensures that we will get the desirable characteristics but not the undesirable ones. Are the undesirable ones assumed to be produced by a patriarchal context, and the desirable ones somehow not? There are a host of problems.

If we examine the second difference theory strand we encounter a different set of equally serious problems, now turning on specifying what the characteristics of the alternative feminine ideal are. Independent criteria for selection and identification of feminine traits are lacking.

Since these are not *traditional* virtues or character traits associated with the feminine, what are they? There has to be some way of determining which are to be affirmed in opposition to masculine traits. Usually they are not identified or taken as identifiable (e.g. because of silencing), or are treated as to be discovered.

The genuinely feminine is either unknowable or as yet unknown, to be brought into existence. In this case there seems no way of showing whether the desired characteristics, e.g. alternatives to domination of nature, will or will not be present among the group of traits. Arguments from psychoanalysis may suggest that they will be but are hardly conclusive as they stand; and as Claire Duchen suggests, relying solely on them appears to involve denying the importance of other non-individual and social influences and bases of character.[23]

The problem, then, is how to say what this concept of 'the feminine' is, and what the ideal human character being affirmed is like. Obviously its character cannot be determined by examining the sorts of characteristics actual women *now* display, since these have been determined by exclusion under patriarchy. Thus for example it is hardly convincing to suggest that passivity, insecurity, and the poorly developed sense of self and of independence many women are obliged to develop under patriarchal conditions are genuine but unrecognised human virtues. Again it seems impossible not to recognise that the oppression of women has produced undesirable as well as desirable character traits.

So, since it cannot be *actual* existing women whose character forms the basis for the ideal, this position sets off a search for some sort of feminine essence which eludes expression in present societies, but appears as an unrealised potential, so much unrealised that it is, in some versions, almost essentially inexpressible. Since it seems that this character can never be instantiated by actual women in existing oppressive societies, the position has difficulty in explaining exactly how the ideal character appealed to 'belongs to' women, and which women it belongs to, i.e. what makes it feminine. And it seems inevitably either nebulous or circular, since we are asked to undertake a remaking of the human in the mould of a set of 'feminine' characteristics which cannot be specified unless and until that remaking is achieved, and whose relation to actually existing women is, at best, unclear. And the suggestion that we should thus blindly swear allegiance to the nation of the female body, and to whatever characteristics it may develop or display, seems a mere piece of nationalism.

The body is sometimes thus introduced in an attempt to solve the problem of identifying the feminine, in what appears to be a form of reverse dualism. The position apparently accepts the mind/body division and its correspondence to masculinity and femininity, but replaces the masculine notion of identity as based in the mind or in consciousness with the supposedly feminine one of identity as based in – and apparently reducible to – the sexed body. To the extent that bodily difference is taken as determining of the feminine, that the feminine is endorsed as the ideal of human character, and that what is involved is the assertion of a rival human ideal which men will *necessarily* never be able to participate in, the position seems to have built into it another hierarchy, another exclusion. There may be difference here, but too much remains the same.

In brief the position, whether interpreted according to strand 1 or strand 2, faces a dilemma as a base for the ecofeminist argument. If it follows strand 1 and specifies the traits, selecting only desirable ones such as nurturance, it faces the problem of explaining how these relate to existing women and how they are feminine. If it fails to do so, specifying them only in their relation to female bodies or to the emergence of an unspecified 'genuine

WOMEN, HUMANITY AND NATURE

femininity', it needs to provide a basis for believing, what is needed for the ecofeminist argument, that the desirable traits are included or will emerge. In neither case, it seems, can the ecofeminist argument be adequately based on position 3. Is the argument therefore to be abandoned?

I want to argue that it doesn't have to be, although this particular form of it needs to be. Initially it seems obvious that the ecofeminist and peace argument is grounded on accepting a special feminine connectedness with the natural or with peaceful characteristics, and then affirming this as a rival ideal of the human (or as part of such an ideal). But on closer examination this is not so clear.

The argument doesn't have to take this form. The ecofeminist argument basically involves the rejection of the masculine model of the human and of the aggression towards and domination of nature seen as part of that model. But to reject the masculine model of the human is not necessarily to affirm a rival feminine ideal, nor to accept any other special connection between nature and the feminine. To free the concept of the human from the connection to the masculine which has lain behind its guise of neutrality doesn't mean that it has to be replaced by a rival feminine ideal specified in reaction to the masculine ideal.

The choice between the masculine model of the human and its feminine rival is, fortunately then, a false choice. This can be seen clearly if we examine the logical options for the human ideal and its relation to a masculine or feminine ideal. They can be set out as follows (using the symbol 'R' to mean 'reject' and the symbol 'A' to mean 'accept'):

1) A masculine model, R feminine model
2) A masculine model, A feminine model
3) A feminine model, R masculine model
4) R masculine model, R feminine model

This set of alternatives is exclusive and exhaustive of the possibilities for an ideal if the categories are treated as wholes, but of course a further set of options can be generated if they are not or if the necessity for a human character ideal itself is questioned, viz. *no* character ideal at all.[24] It is apparent from this set of alternatives that the assumption that an alternative to 1) (the traditional model) or 2) (the romantic complementary or separate spheres model) must be 3) (the feminine model) is wrong.

Thus it is open to an ecofeminist to agree in part with the common-sense view and assert that women are in fact no more significantly or essentially connected to nature than men (except in so far as an alleged connection has been used to inferiorise both and has involved exclusion of women from technology and culture) but that what is needed is an account of the human ideal *for both sexes*, which accepts the undesirability of the domination of nature associated with masculinity. This would be a strategy which rejected the masculine concept of the human, but because it denied any special significant connection between nature and the feminine, was not committed to a rival feminine ideal. The fact that the concept of the human is up for remaking doesn't mean that it *has* to be remade in the mould of either the masculine or the feminine.

Not only *can* an ecofeminist argument appeal more satisfactorily to the fourth model than the third, that is clearly what it often does. For example, Rosemary Ruether, one of the pioneers of the position, is clearly appealing to model 4, not model 3, when she writes

> Both men and women must be resocialised from their traditional distorted cultures of masculinity and femininity in order to find that humanized culture that is both self-affirming and other-affirming. It is precisely in this creation of a humanity that is truly affirming of all life, both one's own and that of others, that the writers seek to find the deepest connections between feminism and non-violence.[25]

In some writers the adherence to model 4 over 3 is even more explicit, e.g.

> If the masculine character ideal supports militarism, what can support peace? Femininity? No, for that character ideal also has been shaped by patriarchy and includes along with virtues such as gentleness and nurturance a kind of dependency which breeds the passive-aggressive syndrome of curdled violence.[26]

The rejection of the masculine character ideal does not imply acceptance of corresponding feminine traits, and a critique of both masculinity and femininity and their complementary characters

WOMEN, HUMANITY AND NATURE

may be involved. Further, the rejection of both the masculine and feminine character ideals is linked with the rejection of the traditionally associated dualisms of mind/body, rationality/emotionality, public/private, and so on, which are also rejected as false choices, so that the *transcendence* of the traditional gendered characters becomes part of, is linked with the systematic transcendence of this wider set of dualisms.[27] These dualisms are subject to independent criticism in the ecofeminist literature. From this perspective, the model which simply replaces the masculine by the feminine is a reactive model which fails to take adequate account of the way in which gender is structured as a dualism. It shares the general inadequacy of the 'reverse-value' strategy for dealing with dualisms which simply affirms the underside – that to do so is implicitly to accept and to preserve, rather than to challenge, the dualistic structure.

The fourth model for demasculinising the human character may be developed in various different ways. One of the most obvious and popular ways to develop it is in terms of androgyny. Thus Kokopeli and Lakey continue:

> We are encouraged by the vision of androgyny, which acknowledges that the best characteristics now allocated to the two genders indeed belong to both; gentleness, intelligence, nurturance, courage, awareness of feelings, co-operativeness . . . . Many of these characteristics are now allocated to the feminine role which has led some men to conclude that the essential liberating task is to become effeminate. We don't agree, since some desirable characteristics are now allocated to the masculine role (for example initiative, intelligence).[28]

But androgyny is not the only construction to place on the fourth model, and it in turn has its problems. The concept of androgynous human character suggests a recipe analogy, in which the new human ideal is put together from existing ingredients: take good points of each gender and place in bowl, mix gently, throw bad points into dustbin.

But such a model is far too simple and shallow, ignoring relations of exclusion, complementation and so on between traits and suggesting that their allocation to their respective sex is arbitrary. It treats the problem as if it could be solved by an amalgam of certain *existing* characteristics thrown together, just as the androgynous human is pictured as a physical composite of male and female organs. Similarly the androgynous terminology suggests that no significant character differences should remain between masculine and feminine characteristics, that there will be a single model for both sexes composed of the same set of character traits.

These assumptions are both unsatisfactory and unnecessary, and are not an inevitable part of the fourth model. The androgynous way of developing the fourth model should be distinguished from other ways, e.g. where what is involved is not an amalgam of genders leading to identical gender roles, but a transcendence of the dualistic gender characteristics to produce a third set of characteristics that will often be different from either. The androgynous model overlooks the fact that the gender contrasts of existing character traits are often false contrasts. In fact the gender categories and associated institutions can be seen as a systematic and related network of false choices. A good example is provided by the egoism/altruism contrast, associated with masculinity and femininity respectively. If egoism is taken as consisting in pursuit of a person's own selfish interest, and altruism to consist in a person's denying or setting aside their own interest in favour of that of others, the false contrast standardly presented between the two overlooks the alternative of interdependence of interest, the situation where interests are not discrete and disconnected but where a person's interest essentially involves the interests of others.[29] Similar points can be made for most of the other gender-related dualisms.

There are several further ways of developing the fourth model. One of them is *degendered*, in that selection of characteristics to be affirmed is not based on association with one sex or the other or exclusively on traditional gender characteristics, but aims to transcend them. Another one might be thought of as *regendered*[30] in that it does not aim to eliminate gender and gender difference as such, but rather to reconstruct it so as to free it from dualistic construction and in particular to dissolve particular dualistically-paired traits such as the pairs dominant/submissive and over-emphasised/under-emphasised ego boundaries. The regendering alternative would not need to deny difference or assume the neutrality of the body, to deny that differently sexed bodies might give rise to different experiences and different orientations to the world, although it would resist

WOMEN, HUMANITY AND NATURE

the attempt to treat such difference as lacking a social context and as giving rise to fixed essences not open to change.[31] Nor does it have to try to create a unique human character ideal, as opposed to a multiplicity of such ideals.

Such a position can also allow room for certain ways of affirming or valuing the feminine, in at least some senses of this highly ambiguous phrase, that is for affirming as humanly valuable certain traits previously treated as of little consequence, confined to women and excluded from the masculine character or from more prestigious areas of human life. Such upward revaluing of traits such as nurturance is an important part of the ecofeminist position. But it is important to note the ambiguities and the difference between, first, valuing all and only feminine traits and valuing them *because* or on the grounds that they are feminine, regardless and in ignorance of what they will actually turn out to be (the position corresponding to the feminising strategy discussed earlier), and, second, revaluing, on a selective basis, certain important traits which have been devalued because of their association with the feminine and with nature.

It is important to note the difference too between affirming feminine traits as part of a dualised and patriarchal structure, and affirming them in a way which challenges such a structure. For example affirming feminine traits as *confined* to women and as exclusively possessed by them leads to an essentially conservative position, and it is important for a radical ecofeminism to show how its position differs from this conservative affirmation of feminine traits and roles, which may involve showing how in affirming them it gives them a different significance from the one they had in a dualised, patriarchal context. Nurturance, for example, a trait often affirmed in ecofeminism, has been devalued because of its links to both femininity and animality, and in a patriarchal context where it has been confined to women and the private sphere, is such that women's exclusive nurturance confirms and supports male control of the world. But a critical and non-conservative ecofeminism must affirm it in ways which do not do this, which remove it from the dualised context, making it a virtue for men as well as women, and giving it the significance of nurturing the natural world and others not confined to the nuclear family. By such transformations of context and significance, feminine traits arising from a patriarchal context can develop into real and radical strengths in a non-patriarchal context.[32] But although such traits can *develop* from traditional feminine traits given an appropriate transformation of social context, they are not *identical* with them, and representing them as the same traits (as 'affirming the feminine' suggests) can be misleading about the sense in and extent to which they are feminine. For they are neither the same as those in the traditional context, and hence 'feminine' in that sense of being part of the traditional characteristics and virtues of womankind, nor feminine in the sense that it is not possible for men to share them or aspire to them.

Where does all this leave the ecofeminist argument and the notion that women are 'closer to nature'? Women have been treated historically as aligned with nature, and this has shaped the characteristics of feminine identity. But there are problems in embracing such a feminine alternative, to the extent that it has involved the confinement of women to activities such as reproduction and denial to them of capacities for reason, intelligence and control of life conditions, that is, their exclusion from the valued features of human life and culture. So a *different* concept of closeness to nature from the traditional one has to be invoked.

But on the other side of the dualism, women's alignment with nature has been matched by the development of a masculine identity centring around distance from nature and such 'natural' areas in human life as reproduction, and around control, domination and inferiorisation of the natural sphere. Such distance is obtained by the location of value in the area of human character and culture that has been taken as both masculine and distinguishing of humans from the non-human world. This is the model of human life ecofeminists reject, as a model both for men and for women liberated from the constraints of the traditional position.

Neither model is acceptable, and an acceptable 'androgynous' solution cannot be obtained by somehow combining the two unsatisfactory ones. What is needed is a regendered model, which realigns the gender power structure, reconstructs the gender identities and challenges the dualisms on which they have been based. Thus a sophisticated ecofeminism based on the feminist critique of gender and of dualism challenges nature/culture dualism, and the dominant masculine model of human culture and the human self as separated and maximally distanced from the natural world, and from features of the human self shared

WOMEN, HUMANITY AND NATURE

with the natural world – 'nature within'. On such an alternative model of the human we would not *overemphasise* or *overvalue* the characteristics that set humans apart from the natural world nor attempt obsessively to maximise the differences as the main source of virtue. We would be able to see value not only in the natural world but in the characteristics which we as humans share with it, which have been allocated to the feminine and treated as culturally problematic.

NOTES

This chapter, now revised, was first published in *Radical Philosophy* 48 (Spring 1988)

1. The sex/gender distinction is important in stating ecofeminist positions but has been under attack recently. It remains, I believe, both defensible and useful, although often loaded with *additional* less defensible assumptions. As used here it carries no dualistic implications, e.g. that the character traits involved are purely mental (physical and integrated characteristics can and normally will be included), or that biological sex is a brute fact involving no element of social or cultural determination. For a defence of the distinction and elaboration of some of the issues surrounding it see my 'Do We Need A Sex/Gender Distinction?', *Radical Philosophy* 51, spring 1989.
2. p. 297 Dr C. McT. Hopkins, *As You Take It*, Neptune Press, Geelong, 1985.
3. Genevieve Lloyd, *The Man of Reason*, Methuen, 1984, and Carolyn Merchant, *The Death of Nature*, Wildwood House, 1980.
4. See e.g. R. and V. Routley, 'Against the Inevitability of Human Chauvinism', in K.E. Goodpaster and K.M. Sayre, eds, *Ethics and Problems of the 21st Century*, University of Notre Dame Press, 1979.
5. Mary Wollstonecraft, *A Vindication of the Rights of Woman*, Dent, London, 1982, p. 15.
6. For some account of this see Susan Moller Okin, *Women in Western Political Thought*, Princeton, New Jersey, 1979.
7. See, for example, John Rodman, 'Paradigm Change in Political Science', *American Behavioural Scientist*, Vol. 24, No. 1 (1980). Also Mary Midgley, *Beast & Man*, p. 40, Methuen, 1979, Ch. 11.
8. These points are developed in more detail in V. Plumwood, 'Ecofeminism: an Overview and Discussion of Positions and Arguments', in Janna L. Thompson, ed., *Women and Philosophy*, *Australasian Journal of Philosophy*, Supplement to Vol. 64, June 1986, pp. 120–38.
9. Wollstonecraft, op. cit., p. 5.
10. For examples see Keith Thomas, *Man and the Natural World*, Penguin, 1983, p. 41ff.

11 Simone de Beauvoir, *The Second Sex*, Foursquare Books, 1965.
12 Ariel Salleh, 'Contribution to the Critique of Political Epistemology', *Thesis Eleven*, 1984, No. 8.
13 Lloyd, op. cit., p. 104.
14 For example Elizabeth Dodson Gray, *Why the Green Nigger: Remything Genesis*, Roundtable Press, Wellesley, Mass. 1979.
15 See Rosemary Radford Ruether, *New Woman New Earth*, Seabury Press, New York, 1975, and Susan Griffin, *Women and Nature*, Harper & Row, New York, 1978.
16 Although this may be involved in some cases, e.g. Elizabeth Dodson Gray, op. cit.
17 See e.g. Leonie Caldecott and Stephanie Leland, eds, *Reclaim the Earth: Women speak out for Life on Earth*, The Women's Press, London, 1983, and Pam McCallister, ed., *Reweaving the Web of Life: Feminism and Non-violence*, New Society Publishers, Philadelphia, 1982.
18 On complementation see Genevieve Lloyd, op. cit., Ch. 7.
19 Sally Miller Gearhart, 'The Future – If There is One – Is Female', in McCallister, op. cit., p. 271.
20 Ibid., p. 272.
21 Christiane Makward, 'To Be or Not to Be . . . A Feminist Speaker', in Hester Eisenstein and Alice Jardine, eds, *The Future of Difference*, Rutgers University Press, New Brunswick, 1985, p. 96.
22 Chantal Chawaf, cited in Makward, op. cit., p. 96. in E. Marks and I. de Courtivron, eds, *New French Feminisms*, Harvester, Brighton, 1980, p. 103.
23 Claire Duchen, *Feminism in France*, Routledge 1986, Ch. 5.
24 This seems to be what is being suggested in Moira Gatens, 'Feminism, Philosophy and Riddles Without Answers', in C. Pateman and E. Gross, eds, *Feminist Challenges*, pp. 28–9.
25 Quoted in P. McCallister, op. cit.
26 Bruce Kokopeli and George Lakey, 'More Power Than We Want: Masculine Sexuality and Violence', in P. McCallister, op. cit., p. 239.
27 For an account of these linked dualisms and of how they are treated in the ecofeminist argument see V. Plumwood, 'Ecofeminism: an Overview and Discussion of Positions and Arguments', op. cit.
28 Kokopeli and Lakey, op. cit., p. 239.
29 For more details see R. and V. Routley, 'Against the Inevitability of Human Chauvinism' in K. Goodpaster and K. Sayre eds, *Ethics and Problems of the 21st Century*, Notre Dame University Press, Notre Dame, 1979.
30 The difference between regendering and degendering is explained in more detail in my article 'Do We Need a Sex/Gender Distinction?', op. cit. For telling criticisms of androgyny see Hester Eisentein, *Contemporary Feminist Thought*, Unwin Paperbacks, London, 1984, Ch. 6.
31 See Plumwood, op. cit.
32 As explained in Jean Baker Miller, *Toward a New Psychology of Women*, Penguin Books, London, 1986.

[17]

Ecological Feminism and Ecosystem Ecology[1]

KAREN J. WARREN and JIM CHENEY

Ecological feminism is a feminism which attempts to unite the demands of the women's movement with those of the ecological movement. Ecofeminists often appeal to "ecology" in support of their claims, particularly claims about the importance of feminism to environmentalism. What is missing from the literature is any sustained attempt to show respects in which ecological feminism and the science of ecology are engaged in complementary, mutually supportive projects. In this paper we attempt to do that by showing ten important similarities which establish the need for and benefits of on-going dialogue between ecofeminists and ecosystem ecologists.

Ecological feminism is a feminism which attempts to unite the demands of the women's movement with those of the ecological movement in order to bring about a world and worldview that are not based on socioeconomic and conceptual structures of domination. Many ecological feminists have claimed that what is needed is a feminism that is ecological and an ecology that is feminist (see King 1983, 1989). They have shown ways in which ecology, understood in its broadest sense as environmentalism, is a feminist issue.[2] What has yet to be shown is that ecology, understood in its narrower sense as "the science of ecology" (or, scientific ecology) also is or might be a feminist issue. Establishing *that* claim involves showing that ecological feminism makes good scientific ecological sense.[3]

In this paper we discuss ten noteworthy similarities between themes in ecological feminism and ecosystem ecology—similarities that show the two are engaged in complementary, mutually supportive projects. Our goal is modest and suggestive. We are *not* arguing for the stronger claims that ecosystem (or, more generally, scientific) ecology must be feminist, that feminists must be ecologists, or that these similarities establish that ecosystem ecology is feminist. To establish these claims, much more would be needed than is provided in this paper.[4] Rather, we are identifying theoretical points of intersection between ecofeminism and ecosystem ecology in the interest of furthering discussion on the nature and direction of future bridge-building between the two.[5]

ECOLOGICAL FEMINISM AND ECOFEMINIST ETHICS

We take ecological feminism to refer "to a sensibility, an intimation, that feminist concerns run parallel to, are bound up with, or, perhaps, are one with concern for a natural world which has been subjected to much the same abuse and ambivalent behavior as have women" (Cheney 1987, 115). Although there are a variety of ecofeminist positions (Warren 1987), the common thread that runs through ecofeminist scholarship is that the domination of women and the domination of nature are "intimately connected and mutually reinforcing" (King 1989, 18). All ecofeminists endorse the view that an adequate understanding of the nature of the connections between the twin dominations of women and nature requires a feminist theory and practice informed by an ecological perspective and an environmentalism informed by a feminist perspective (Warren 1987, 4-5).

Much of ecofeminist scholarship concerns the ethical nature of human relationships to the nonhuman natural world. Like feminist ethics generally, "ecofeminist ethics" includes a variety of positions. What makes ecofeminist ethics feminist is a twofold commitment to critique male bias in ethics and to develop analyses which are not male-biased (see Jaggar 1990, 23). However, ecofeminist ethics extends feminist ethical critiques of sexism and other social "isms of domination" to include critiques of "naturism," i.e., the unjustified domination of nonhuman animals and nature by humans. As such, ecofeminist ethics critiques not only androcentric but also anthrocentric and naturist bias in ethics. Ecofeminist ethics is grounded in the assumption that the dominations of women and of nature are morally wrong and ought to be eliminated. Like feminist ethics (see Jaggar 1990, 24-5), the practical import of ecofeminist ethics is as a guide to action on issues in the pre-feminist, patriarchal present. This guidance is aimed at assisting persons in resisting sexist, naturist, and interconnected racist, classist, heterosexist practices, and in envisioning and creating morally desirable alternatives. The women-initiated non-violent Chipko movement begun in 1974 in Reni, India is one such alternative action (see Shiva 1988 and Warren 1988).

One way to image ecofeminist ethics is as a quilt-in-the-making (see Warren 1988, 1990). Like the AIDS Names Project Quilt, ecofeminist ethics is a quilt-in-process, constructed from "patches" contributed by persons located in different socioeconomic, cultural, historical circumstances. Since these patches will reflect the histories of the various quilters, no two patches will be just the same. Nonetheless, the quilts-in-process will each have borders that not only delimit the spatiotemporal dimensions of the quilt, but also put some necessary conditions, "boundary conditions," on what can become part of the quilt. What these boundary conditions do *not* do is delimit the interior of the quilt, what the design or actual pattern of the quilt will be. That design will emerge out of the life experiences, ethical concerns, and specific socio-economic historical contexts of the quilters (see Warren 1990).

What are some of the boundary conditions of ecofeminist ethics? Just what does, and what does not, belong on the quilt? Since ecofeminism is a critique of interrelated social systems of domination, no "isms of domination" (for example, sexism, racism, classism, heterosexism, naturism) belong on the quilt (Warren 1990). This means that any conceptual framework (or, set of basic beliefs, values, attitudes, and assumptions which grow out of and reflect one's view of oneself and one's world) which sanctions, justifies, or perpetuates these "isms of domination"— oppressive and patriarchal conceptual frameworks— does not belong on the quilt. What *does* belong on the quilt are those descriptions and prescriptions of social reality that do not maintain, perpetuate, or attempt to justify social "isms of domination" and the power-over relationships used to keep them intact. These will include patches that make visible and challenge local and global forms of environmental abuse, the disproportional effects of environmental pollution on women, children, the poor, dislocated indigenous persons, and peoples in so-called less developed countries; patches that provide present-day alternatives to environmental exploitation; patches that document and celebrate the morally respectful dimensions of women's experiences with the nonhuman world; and patches that include the experiences of indigenous people, when those experiences are neither sexist nor naturist. Taken together, the patches on the quilt provide the ethical theorist with concrete, pictoral ways of understanding the nature of a morality which treats both women's moral experiences and human interactions with the nonhuman natural world respectfully.

Ecosystem Ecology

Many controversies in modern ecosystem ecology about the nature of ecosystems can be understood as arguments between two approaches to the study of ecosystems: the "population-community" approach and the "process-functional" approach.[6] The population-community approach focuses on the growth of populations, the structure and composition of communities of organisms, and the interactions among individual organisms. It is grounded in Darwinian theory of natural selection. It "tends to view ecosystems as networks of interacting populations whereby the biota *are* the ecosystem and abiotic components such as soil or sediments are external influences" (O'Neill et al. 1986, 8). The population-community approach typically is identified with the work of such ecologists as Clemens, Lotka, Gauss, and Whittaker.

In contrast, the process-functional approach is based on a quantitative, mathematical, thermodynamic, biophysical model which emphasizes energy flows and nutrient cycling. It assumes that the fundamental units of ecosystems include both organisms and physical components, biotic and abiotic components. The process-functional approach was developed during this century by such ecologists as Tansley, Lindeman, and Odum.

Although discussions of ecosystems ecology often present "the ecological perspective" as if there were only one perspective, debates arising from differences between the population-community and process-functional approaches to ecosystems ecology reveal that there currently is *no single model* of ecosystems.[7] Furthermore, there is a third alternative way to conceive ecosystems. That alternative is "hierarchy theory" or what, for important feminist reasons, we prefer to refer to as "observation set theory."[8] We understand hierarchy theory to be the most viable attempt to date by scientific ecologists to provide an inclusive theoretical framework for the variety of ecosystem analyses. Ecologists such as O'Neill, DeAngelis, Waide and Allen are among its main advocates (O'Neill et al. 1986).

Central to hierarchy (observation set) theory is the notion of an *observation set*. O'Neill et al. describe an observation set as "a particular way of viewing the natural world. It includes the phenomena of interest, the specific measurements taken, and the techniques used to analyze the data" (1986, 7). Although specific problems always call for particular observation sets, theory making calls for consideration of multiple observation sets:

> Each of these points of view emphasizes different phenomena and quite different measurements. But since neither encompasses all possible observations, neither can be considered to be more fundamental. When studying a specific problem, the scientists must always focus on a single observation set. However, when developing theory, many observation sets must be considered (O'Neill et al. 1986, 7).

According to hierarchy theory, both an adequate conception of the complexity of ecosystems and meaningful ecosystem comparisons require that one consider multiple observation sets.

Spatiotemporal scale is an important characteristic of an observation set both because it changes as the ecological problem changes and because "ecological principles often do not translate well across these scales" (O'Neill et al. 1986, 20). The meanings of such basic ecological concepts as "stability," "equilibrium," "temporary," "enduring," "local," and "global," are relative to some particular scale. Depending on the spatiotemporal scale used in any given observation set, "ecosystems have been seen as static or dynamic, as steady-state or as fluctuating, as integrated systems or as collections of individuals" (O'Neill et al. 1986, 20). For example, a forest stand can be looked at from an organismic standpoint (e.g., as enduring, stable individual trees or populations of trees) *or* from an energy flow and nutrient cycling standpoint (e.g., as fluxes and flows of carbon and oxygen recycled through photosynthesis). Because the forest stand may accurately be viewed in either way, it is incorrect, in fact impossible, "to designate *the* components of *the* ecosystem"—the designation depends on the spatiotemporal scale and changes as that scale changes (O'-Neill et al. 1986, 83).

The basic contribution of hierarchy (observation set) theory is to call attention to the importance of observation sets and spatiotemporal scales to ecosystem ecology. The complexity of natural systems is overlooked or discounted when one focuses on a single observation set. An exclusivist "either-or" approach to describing or studying ecosystems (e.g., an exclusivist population-community or functional process approach) is thereby viewed as based on a false dichotomy which results in an inadequate, because incomplete, *theory* of ecosystems (O'Neill et al. 209).[9]

Similarites Between Ecofeminism and Ecosystems Ecology

We are now in a position to show some of the similiarities between ecofeminism (particularly ecofeminist ethics) and ecosystem ecology seen through the lens of hierarchy (observation set) theory. These similarities suggest various ways in which ecofeminism and ecosystem ecology inform and support one another.[10]

First, central to hierarchy theory is the view that space-time dependent observation sets provide different vantage points or frameworks from which one makes ecological observations and engages in ecological theory building. It is through the notion of multiple observation sets that the idea of one single model of ecosystems is rejected. In this respect, hierarchy theory rules out any notion of an observation set free or *decontextualized* science: how one views ecosystems will depend on the observation sets one employs.

One is immediately struck by the similarity between the hierarchy theorist's emphasis on observation sets, "windows through which one views the world," and the ecofeminist's emphasis on "ways of thinking," "world-views," and "conceptual frameworks," especially oppressive and patriarchal ones (see Warren 1987, 1990). The notion of a patriarchal and oppressive conceptual framework is as central to ecofeminism and ecofeminist ethics as the notion of an observation set is to hierarchy theory in ecosystem ecology: one could not generate the observations and conclusions of each without them. An attention to observation sets is also an acknowledgment of the importance of the contexts in and through which one observes, measures, and theorizes. One's observation set, like one's conceptual framework, will quite literally shape and affect what one sees; both provide a *context* for theorizing.

There are at least three interrelated reasons why attention to context is of importance to ecofeminist ethics. First, *what* a thing (person, community, population, species, animal, river) is, is in part a function of *where* it is, a function of the relationships in which it stands to other things and to its history, including (where applicable) its evolutionary history. It is this attention to *place* that fuels bioregionalist ecofeminism (see Plant, 1990) and the importance many ecofeminists give to narratives, myth, and ritual in the construction of ecofeminist ethics (see Cheney 1987, 1989a; Diamond and Orenstein 1990; Warren 1990). Second, an understanding of context is important in assessing the putatively universal claims of reason and ethical deliberation. Feminist worries about ahistorical and allegedly gender-neutral conceptions of reason and rationality in the Western philosophical tradition provide one way of understanding the importance of context—historical location and gender identity in theory building in the pre-feminist present.[11] Ecofeminist theory building seeks to rid prevailing conceptions of reason, rationality, and morality of whatever male and naturist bias they have.

More than this, however, and this is our third point, an attention to context permits one to stress the idiographic dimension of our ethical journeys through this world and of ethics itself. Holmes Rolston has been a strong advocate for recognizing this aspect of ethical thought in environmental ethics, and this advocacy derives from his understanding that a thing is what it is in part because of where it is. As Rolston puts it:

> An ethics should be rational, but rationality inhabits a historical system. The place that is to be counted morally has a history; the ethics that befits such a place will take on historical form; the ethics will itself have a history. The place to be mapped ... will have twin foci. One focus will be nomothetic, recurrent; the other will be idiographic, uniquely particular. ...
>
> The rationality of the ethic, as well as the area to be mapped, will be historical. That is, logic will be mixed with story. The move from *is* to *ought* ... is transformed into movement along a story line (Rolston 1988, 341-42).

An attention to context does not split off the idiographic as what ethics permits, provided that the universal demands of morality are met. Instead, the ethically idiographic is the very center of each individual's ethical life; it is the place from which we not only test the claims of the "universal" and the "rational," but from which we construct the "universal" claims of "rationality." In this way, the "universal" and the "rational" are always in some manner or other inflected with historicity. The "universal" and the "rational" are themselves moments in a story, reflecting some observation set.

The ecological dimension of ethical reflection stems in large part from the fact that ecology is context (or observation set) dependent. We agree with Brennan that:

what ecology shows is not simply that the context makes a difference to the kind of action we engage in. It shows, rather, that what kinds of things we are, what sort of thing an individual person is, and what sort of options for fulfillment and self-realisation are open, are themselves very much context-dependent (Brennan 1988, 162).

One way ecofeminist ethics centralizes this context-dependent feature of ethical discourse is by conceiving of ethics as growing out of what Cheney (1987, 144) calls defining relationships, that is, relationships understood as in some sense defining who one is. These relationships include those of moral agents with the nonhuman natural world, including animals.

Second, hierarchy theory provides a methodological means of investigating ecological problems. According to hierarchy theory, the "ontology" that emerges from any particular investigation is relative to the observation set that produces it. This does not make that ontology "subjective" in any pernicious sense; but it does mean that to accept a solution to a particular problem is not thereby to make any ontological commitments in any absolute (i.e., non-observation set dependent) sense. Thus, the methodology of hierarchy theory makes it imperative that the epistemological requirements of particular problems, given in terms of observer-affected observation sets, dictate to ontology (rather than the converse); ecology does not determine that an ecological problem must be pressed into the shape of a preferred ontology. According to hierarchy theory, it would be quixotic to think in terms of striving for an articulation of *the* structure (even *the* hierarchical structure) of an ecosystem.

As a methodological stance, hierarchy theory rejects the view that there is only one way to describe ecological phenomena. Which description is appropriate will depend upon the observation set and on what it is one is attempting to describe, explain, or predict. In this respect, hierarchy theory privileges methodological and epistemological considerations over ontology, the attempt to specify what is "really" in the world. The ontology embedded in both explanation and phenomena being explained is always a function of the appropriate observation set. Any grand attempt to provide one metaphysics of morals seems doomed because misguided: it puts the metaphysical/ontological cart before the epistemological/ methodological horse.

Like hierarchy theory, ecofeminism makes no attempt to provide *the* point of view, one single model, an "objective" (i.e., value-neutral, unbiased) point of view—none, that is, beyond the very "boundary conditions" of ecofeminism itself. Ecofeminists criticize up-down, value-hierarchical, value dualistic thinking which they say characterizes Western philosophical thinking about women and nature as being both patriarchal and insular—as if what is observed, prescribed, and theorized are independent of *any* conceptual framework (Gray 1981; Griffin 1978; King 1983b, 1989; Ruether 1975; Warren 1987, 1988,

1990). Ecofeminists acknowledge up front their basic feminist value commitments: the twin dominations of women and nature exist, are wrong, and ought to be eliminated. Ecofeminists see these twin dominations as social problems rooted in very concrete, historical, socioeconomic conditions, as well as in oppressive, patriarchal conceptual frameworks that maintain and sanction these conditions.

As a methodological and epistemological stance, all ecofeminists centralize, in one way or another, the "voices" and experiences of women (and others) with regard to an understanding of the nonhuman natural world. Like hierarchy theory, this is not to say that an ecofeminist "ontology" does not include material objects—real trees, rivers, and animals. It does! But it acknowledges that these objects are in important senses both materially given and socially constructed: what counts as a tree, river, or animal, how natural "objects" are conceived, described, and treated, must be understood in the context of broader social and institutional practices. Centralizing women's voices is important methodologically and epistemologically to the overall critique and revisioning of the concept of nature and the moral dimensions of human-nature relationships.

Third, hierarchy theory is antireductionist. Population-community based observation sets cannot be reduced to process-functional based observation sets (or vice versa). Consequently, a functional-process understanding of organisms does not render an "object ontology" of discrete organisms (trees, rivers, animals) obsolete, or render organisms mere conduits or configurations of energy, as environmental ethicist J. Baird Callicott has claimed (1986). There is no ontologically prior or privileged or fundamental description of nature. Hierarchy theory rules out a view of individual entities (e.g., animals) as ontologically parasitic on something more fundamental (e.g., energy flows or nutrient cycles), a point we return to shortly. If hierarchy theory is correct, then in contemporary scientific ecology, there is no place for a notion of degrees of reality. *Both* individuals and energy flows are real.

Because it is antireductionist, hierarchy theory centralizes diversity; it takes difference or diversity to be a fundamental feature of phenomena, not reducible to talk of the "sameness" of organisms or the "oneness" of energy flows. That would be the case only if one approach had epistemological, metaphysical, or ontological priority over the other. In fact, one of the most interesting features of hierarchy theory is that it privileges the notion of diversity or difference when studying interactions between different subsystems ("holons") of ecosystems, *and* the notion of commonalities among members of the same subsystem. Hierarchy theory is therefore a framework which provides for both an ecology of differences and an ecology of commonalities, depending on the context and observation set.[12]

Ecofeminist ethics is also antireductionist. It is a structurally pluralistic framework that centralizes both diversity or difference (e.g., among women, among people of color, between humans and nonhumans) *and* commonalities (e.g., among women, among people of color, between humans and nonhumans). A nonreductionist ecofeminist stance acknowledges differences between humans and members or elements of nonhuman nature, while nonetheless affirming that humans are animals and members of an ecological community. An ecology of differences and commonalities fits well with an ecofeminist politics and ethics of differences and commonalities.

Fourth, hierarchy theory is an inclusivist theory that offers a framework for mediating between historically opposed approaches to ecosystem ecology, making a central place for the insights of each without inheriting the defects of either when viewed exclusively as the right or correct way to study ecosystems. Hierarchy theory suggests that the future of at least ecosystem ecoogy may well lie in successfully integrating these two approaches into a model that centralizes the importance of observation sets and locates any particular ecosystem analyis in or relative to a particular observation set.

Similarly, ecofeminst ethics is an inclusivist ethic (see Warren 1990) that offers a framework for mediating between two historically opposed approaches in environmental ethics: deontological rights-, virtues-, or holistic-based ethics and consequentialist-based ethics. Warren has argued that ecofeminism "involves a shift *from* a conception of ethics as primarily a matter of rights, rules, or principles [whether deontological or consequentialist] determined and applied in specific cases to entities viewed as competitors in the contest of moral standing," *to* one which "makes a central places for values . . . that presuppose that our relationships to others are central to our understanding of who we are" (Warren 1990, 143). An ecofeminist ethic may involve a commitment to rights in certain contexts and for certain purposes (for example, in the protection of individual animals against unnecessary pain or suffering); it may use consequentialist considerations in other contexts and for other purposes (for example, when considering behavior toward ecosystems). Like hierarchy theory, ecofeminist ethics is one possible framework for developing such an inclusivist alternative.[13]

As a fifth and related point, hierarchy theory provides a framework for viewing historically opposed approaches as complementary. Dualisms fade into the complexity of multiple vantage points and find complementarity where once there was only oppositionality (e.g., stability or instability, diversity or sameness, energy flow or discrete organism). This rejection of oppositional polarities is accomplished *not* by reducing population-community to process-functional accounts, or vice versa, or by reducing both to a still more basic or primitive ontological framework; it is accomplished by providing a unifying framework for studying and relating to one another various analyses, each with their own epistemology and context-dependent ontology. As a "unified theory," it is a unity which does not erase difference.

The earliest ecofeminist literature was grounded in a rejection of oppositional value dualisms (see Gray 1981; Griffin 1978). Ecofeminist ethics needs to follow suit[14] by emphasizing difference in a way that does not reduce difference to the terms of some (reductionist) privileged discourse.

Sixth, because it centralizes diversity, hierarchy theory complexifies rather than simplifies the variety of ways natural phenomena can be described. It does this by emphasizing the sorts of interrelationships that exist among organisms and the relevance of scalar and other dimensions to the observations made. It rejects exclusivist models of ecosystems (i.e., population-community or process-functional models) that simplify rather than complexify the nature of ecosystems, typically by an imposed naive reductionism that focuses on sameness, similarity, or shared traits. Interrelationships among biotic and abiotic nature that are based on a single, unitary model of ecosystems are viewed as misrepresentations of the variety of relationships in nature.

Similarly, as a context-dependent, inclusivist framework that centralizes difference, ecofeminism complexifies the variety of ways in which ethics is conceived and practiced, in which humans may be in relationship with others (including the nonhuman natural environment), and in which human-nature, women-nature connections may be described. As we have argued elsewhere (Warren 1988, 1990; Cheney 1987, 1989a, 1990), ecofeminist ethics complexifies the moral arena by making a central place for values often lost or overlooked in mainstream ethics (e.g., values of care, love, friendship, diversity, appropriate reciprocity) in the context of human-nonhuman relationships. This includes taking seriously the sort of "indigenous technical knowledge" that women and others who work closely with the land have (see Warren 1988).

Seventh, and perhaps most importantly for ethics, hierarchy theory permits meaningful ecological talk of "individual" and "other" without the caveat that these are nonprimitive notions, ultimately reducible to notions of energy flow and pattern. At the same time, it also permits meaningful talk of "whole-system" behavior in both population-community and process-functional terms, neither of which is reducible to the other. Hierarchy theory thus permits meaningful discussion of discrete (and, in varying degrees and modes, autonomous) individual objects as well as of whole systems. Hierarchy theory shows that "object theory" is not obsolete; it is an acceptable and alternative way to describe organisms—appropriate for some observation sets and not others.

This alternative way of describing ecosystems is accomplished in hierarchy theory in part by an eighth characteristic, one shared by ecofeminism and ecofeminist ethics: it encourages a network or relational view of organisms, whether conceived as "knots in a biospherical web of relationships" or as separate (although not isolated or solitary), discrete individuals, members of species, populations, or communities. In both cases, ecosystems are networks, either networks of interacting individuals, populations, and communities or of interacting energy and nutrient flows and cycles.

This dual acknowledgment of the autonomous existence of individuals (characteristic seven) and the relational existence of individuals in webs of relationships (characteristic eight) fits nicely with those feminist ethics which insist that it is of primary importance to acknowledge and foster individual autonomy (after all, oppressed persons are still trying to have their autonomy recognized) *and* to recognize that people exist in webs of relationships that are to some extent constitutive of who they are. Much work in feminist ethics (often strongly influenced by the work of Gilligan 1982) has emphasized the centrality of relationships in women's ethical thinking. Others (e.g., Friedman 1989 and Young 1986) have critiqued communitarian ideals and stressed the importance for women of autonomy and a politics of difference in a world in which the penchant for defining oneself relationally can easily be turned into sacrifice of the self. Many feminists have been concerned to develop conceptions of self and society that avoid the problems of what Alison Jaggar calls abstract individualism, that is, the position that it is possible to identify a human essence or human nature that exists independently of any particular historical context (Jaggar 1980, 29).

This concern carries over into ecofeminist ethical reflection on nature. An ecofeminist ethic that emphasizes the nature of individuals or "others" as beings-in-relationships permits meaningful ecological discussion of *both* "self" and "other," of "individuals" (populations, communities) and "webs of relationships." For ecofeminists the contexts and relationships that help construct "the self" include ecological contexts and relationships with nonhuman nature. For an ecofeminist one cannot give an adequate account of what it is to be human in terms that do not acknowledge humans as members of ecological communities.

That hierarchy theory provides for meaningful discussion of "self" and "other" suggests one reason ecofeminists are and ought to be suspicious of some of the claims about scientific ecology made by other, allegedly "minority position" environmental ethics. For example, in "The Metaphysical Implications of Ecology," Callicott argues that scientific ecology "undermines the concept of a separable ego and thus renders obsolete any ethics which involves the concepts of 'self' and 'other' as primitive terms" (1986, 301). Callicott's overarching conclusion is that scientific ecology ontologically subordinates matter and living natural objects (e.g., humans, deer, trees) to energy flows, making an "object ontology" inappropriate as an ecological description of the natural environment.

Views such as Callicott's are not borne out by state-of-the-art hierarchy theory in ecosystem ecology. Hierarchy theory shows that even if at some level of inquiry it is plausible to hold that the universe and everything in it are constituted of energy, that everything is a perturbation in an all-encompassing energy field, this does *not* imply that entities revealed through other observation sets (e.g., as individual organisms, populations, or communities) are not "primitive," that they are reducible to the ontology of some other observation set. Hierarchy theory not only permits but demands meaningful ecological discussion of "self" and "other" on the one hand *and* of "whole-system behavior" on the other. Certain ecological observation sets relevant to ethics yield an ontology of autonomous individual organisms interacting with one another. Other observation sets paint a holistic picture of ecosystem function. But there is no a priori or ecological reason (other than a misguided reductionism) to give (ethical or metaphysical) pride of place to the latter.[15]

What *is* crucial is our particular mode of access to the objects of our moral concern. We need to formulate our "ethical ontology" and ethical theory in light of an understanding of our epistemological relationship to the objects of moral concern. In terms of actual practice, we certainly can say things, significant and important things, about individuals without drawing in the rest of the universe. We can gain at least certain kinds of knowledge of individuals without an analysis of the relations that constitute or produce the individual as the individual it is; that is, we can come to know the individual without knowing anything much about the shaping factors.

Ninth, hierarchy theory makes a place for whatever "hard" scientific data scientific ecology produces regarding the natural environment, although it always contextualizes that data relative to a given observation set with specific scalar dimension. It is *always* scientifically relevant to ask about particular observation sets within which and from which the "hard" data are gathered. According to hierarchy theory, all scientific data and questions of ecology come with and have a context; proper scientific theorizing involves making visible the observation sets (contexts) within which one conducts the observations and analyses. Hierarchy theory thereby leaves open the door for saying that whatever ecologists learn about organisms or ecosystems from computer modeling techniques, mathematical or statistical models, or data projections conducted within the closed system of a laboratory may not tell us all there is to know, or even the most relevant information and material we need to know, about terrestrial organisms and ecosystems—i.e., nature outside the laboratory. But we may need to know it, nonetheless, to solve pressing environmental problems.[16]

Ecofeminism welcomes appropriate ecological science and technology. Environmental problems demand scientific and technological responses as part of the solution. These "data" represent a piece of the ecological pie. What ecofeminists insist on is that the perspectives of women and indigenous peoples with regard to the natural environment also be recognized as relevant "data." As a *feminism*, ecofeminism insists that relevant "data" about the historical and interconnected twin exploitations of women (and other oppressed peoples) and nature be included in solutions to environmental problems; as an *ecological* feminism, ecofeminism insists upon the inclusion of appropriate insights and "data" of scientific ecology. What ecological feminism opposes is the practice of one without the other.

Lastly, hierarchy theory invites a reconceiving of ecosystems research and methodololoy, objectivity, and knowledge. In its rejection of the view that there is one ahistorical, context-free, neutral observation stance, in its incorporation of multiple observation sets and its refusal to privilege the ontology of one over the ontology of any other, in its acceptance of multiple understandings of ecosystems and the complexity of the relationships that exist within them, hierarchy theory exemplifies, to some extent, what Donna Haraway (1988) has called embodied objectivity. What is obviously absent in hierarchy theory is an ethical and political dimension, however, which is present in Haraway's notion.

Objectivity, as Haraway puts it, is "about particular and specific embodiment and definitely not about the false vision promising transcendence of all limits and responsibility" (Haraway 1988, 582-83). Because all knowledge is "situated knowledge" (Haraway 1988, 581), no knowledge is innocent; all knowledge involves risks and implies responsibility. As Haraway argues:

> admitted or not, politics and ethics ground struggles over knowledge projects in the exact, natural, social, and human sciences. Otherwise, rationality is simply impossible, an optical illusion projected from nowhere comprehensively (1988, 587).

The ethical and political dimensions of knowledge and objectivity suggest an important contribution that ecofeminism can offer hierarchy theory. The "partial knowledges" that emerge from various observation sets do not constitute an innocent plurality of bodies of knowledge. Both the positions taken (with their resultant situated knowledges) and the connections made are "power-sensitive" (Haraway 1988, 589). Situated knowledges are partial knowledges,

> not partiality for its own sake but, rather, for the sake of the connections and unexpected openings situated knowledges make possible. Situated knowledges are about communities, not about isolated individuals. The only way to find a larger vision is to be somewhere in particular. (Haraway 1988, 590)

Since ecofeminism sees theory building, objectivity, and knowledge as historically situated, illuminated, and created, theory is not something static—it is both "situated" (in Haraway's sense) and "in process," emerging from people's different experiences and observations and changing over time. It *is* like quilting.

Are there, then, any ethical implications of ecosystem ecology? It depends. The ethical implications of ecosystem ecology, like the hierarchy theory that might be used to support them, only have axiological status within and from the vantage points of certain observation sets. As ecologist Mark Davis claims of any ecological model, "any set of ethical implications derived or inspired from the model must always be regarded as only one of many possible such sets" (Davis 1988, 4).

The contextualist conception of objectivity at work in hierarchy theory is consistent with the notion of objectivity being developed in some feminist postmodernist theorizing. The problem faced by postmodern science, as Haraway puts it, is "how to have *simultaneously* an account of radical historical contingency for all knowledge claims and knowing subjects . . . *and* a no-nonsense commitment to faithful accounts of a 'real' world" (Haraway 1988, 579). But just as Haraway would insist upon an ethical and political basis for objectivity in the sciences, so she would add the idea of the "object" of knowledge as an active agent in the construction of knowledge. She rightly points out that feminists have been suspicious of scientific accounts of objectivity that portray the "object" of knowledge as passive and inert. Haraway's view in response to this passive understanding of the object of scientific inquiry is as follows:

> Situated knowledges require that the object of knowledge be pictured as an actor and agent, not as a screen or a ground or a resource, never finally as slave to the master that closes off the dialectic in his unique agency and his authorship of "objective" knowledge. The point is paradigmatically clear in critical approaches to the social and human sciences. . . . But the same point must apply to the other knowledge projects called sciences. (Haraway 1988, 592-93).

If we understand the objects of scientific inquiry as actors and agents *and* insist upon an ethical and political basis for objectivity, accounts of the world based "on a logic of 'discovery' " give way to "a power-charged social relation of 'conversation.' The world neither speaks itself nor disappears in favor of a master decoder" (Haraway 1988, 593). In this regard, Haraway herself calls attention to the promise of ecofeminism:

> Ecofeminists have perhaps been most insistent on some version of the world as active subject. . . . Acknowledging the agency of the world in knowledge makes room for some unsettling possibilities, including a sense of the world's independent sense of humor. . . . There are . . . richly evocative figures to promote feminist visualizations of the world as witty agent. We need not lapse into appeals to a primal mother resisting her translation into resource. The Coyote or Trickster . . . suggests the situation we are in when we give up mastery but keep searching for

fidelity, knowing all the while that we will be hoodwinked....
We are not in charge of the world. We just live here and try to
strike up noninnocent conversations (Haraway 1988, 593-94).

We agree with Haraway's concluding words: "Perhaps our hopes for accountability, for politics, for ecofeminism, turn on revisioning the world as coding trickster with whom we must learn to converse" (Haraway 1988, 596). The significance of the finding that ecofeminism and ecosystem ecology are involved in complementary, mutually reinforcing projects would then lie in what they can contribute together to our conversation with the world as "coding trickster."

NOTES

1. We gratefully acknowledge the helpful comments received on an earlier draft of this paper from Roxanne Gudeman, Donna Haraway, Sandra Harding, Alison M. Jaggar, Ruthanne Kurth-Schai, Toby McAdams, Michal McCall, Lindsay Powers, Truman Schwartz, Geoff Sutton, Nancy Tuana, Leslie Vaughn, Anthony Weston, and Cathy Zabinski.

2. See ecofeminist critiques of environmental practices cross-culturally in Caldecott and Leland (1983), Diamond and Orenstein (1990), Merchant (1980), Peterson and Merchant (1986), Plant (1989), Shiva (1988), and Warren (1988).

3. Showing that scientific ecology is a feminist issue is not as easy as one might expect. As scientific ecologists are quick to point out, there is a difference between the ecology movement (or, popular environmentalism) and the science of ecology. Even if the women's movement and the ecology movement are inextricably connected, and even if understanding the connections between the domination of women and the domination of nature is crucial to an adequate feminism, environmentalism, or ethic, still, none of this shows any respects in which the *science* of ecology must be feminist. In this paper, we attempt to put into place some considerations which bear on *that* issue.

4. In helpful comments on an earlier draft of this paper, Sandra Harding pointed out that even if there are striking similarities between ecological feminism and ecosystem ecology, there might be very good reasons for feminists to reject some claims of ecosystem ecology, and vice versa. One such reason would be the inattention to issues of power in ecosystem theory construction and practice. Since an analysis of power is central to feminist critiques of socially constructed "isms of domination," one would need very good reasons for accepting *as feminist* any theory or practice in scientific ecology which did not include an analysis of power and power-over relationships.

5. Our discussion of ecological feminism is limited to emerging themes in ecofeminism and ecofeminist ethics which are not tied to any one feminism. This is because there is not *one* ecological feminism anymore than there is *one* feminism; the varieties of ecofeminisms will reflect differing feminist commitments of liberal, marxist, radical, socialist feminisms as well as feminisms of women of color (nationally and internationally). Similarly, our discussion of scientific ecology is limited to ecosystem ecology, since it is ecosystem analysis that is the focus of much of the current literature

in environmental ethics on the ethical or metaphysical implications, if any, of ecology (see Brennan 1988; Callicott 1986; Cheney 1991b; Golley 1987; Rolston 1988, 1989).

6. We express our gratitude to Mark Davis, Department of Biology and Director of Environmental Studies at Macalester College, for the information he provided about the population-community and functional-process approaches to ecosystem ecology and hierarchy theory. Much of that information is presented in his unpublished article "Should Moral Philosophers Be Listening to Ecologists?" (1988).

7. There are also feminist reasons to worry about construing these two approaches as the only approaches to studying ecosystems, reasons having to do both with a general concern about theoretical descriptions of material reality in terms of mutually exclusive polarities. (See, e.g., Gray, 1981.)

8. Insofar as so much feminist, including ecofeminist, theory has focused on a critique of value hierarchical thinking and its function in creating, maintaining, and perpetuating social systems of domination, the name "hierarchy theory" is most unfortunate from a feminist point of view. In her comments on an earlier draft of this paper, Alison Jaggar suggested that the name is "toxic" and could well predispose feminists to be antagonistic towards hierarchy theory from the outset. Since, as will be shown, the notion of an "observation set" is central to hierarchy theory and yet does not connote problematic value hierarchies, we have chosen to refer to hierarchy theory frequently throughout this paper as "hierarchy (observation set) theory." (We do not discuss here that aspect of hierarchy theory which gives it its name, though we do in our forthcoming book *Ecological Feminism*, Westview Press.) If it were not for the established usage of the expression "hierarchy theory" within the scientific ecological community, we would refer to the theory simply as "observation set theory."

9. O'Neill et al. stress that they have exaggerated the differences between the population-community and process-functional approaches and that "few ecologists would hold to either extreme of the spectrum" (1986, 10). The distinction between the two approaches is better viewed on a continuum, with the population-community and process-functional approaches at each end and ecologists "drawn in one direction or the other by the specific problems that interest them" (1986, 10).

10. That the discussion format moves from hierarchy (observation set) theory to ecofeminism is not intended to privilege either perspective. Furthermore, more space is provided below to ecosystem ecology when discussing similarities than to ecofeminism for two main reasons: first, there is a virtual absence in the literature of ecofeminism of any attempt to spell out the details of just how ecological feminism might draw support for its position from, or impart its own insights to, ecological science. To begin to remedy this omission, we deliberately have chosen to focus on ecosystem ecology (rather than on ecological feminism) and *then* show important similarities between the two— similarities that are more detailed and specific about "ecology" than are general appeals to the importance of ecosystems, interconnectedness among life forms, or ecological well-being to the survival of the planet. Second, we have presented elsewhere our views on ecological feminism and ecofeminist ethics (Cheney 1987, 1989a, 1991a; Warren 1987, 1988, 1990) and did not want to duplicate those efforts here.

11. For essays and a literature overview on this issue, see the American Philosophical Association *Newsletter on Feminism and Philosophy* Special Issue on "Reason, Rationality and Gender," edited by Nancy Tuana and Karen J. Warren, vol 88, no. 2 (March 1989).

12. We develop this argument in more detail in our forthcoming book *Ecological Feminism*.

13. Warren has argued that ecofeminist ethics needs to evaluate ethical claims partly in terms of a condition of inclusiveness: *Those claims are morally and epistemologically favored (preferred, better, less partial, less biased) which are more inclusive of the perspectives and felt, lived experiences of the most amount of people, particularly including the perspectives and experiences of oppressed persons* (Warren 1988, 1990).

14. We say "needs to" because some ecofeminists have been criticized for substituting a value-hierarchical "women *are* closer to nature than men" ontology and ethics for an unacceptable patriarchal value-hierarchical schema which puts nature and what is female gender-identified together as inferior and opposed to that which is male gender-identified. The criticism is that the very oppositional dualism which prompts the question "Are women closer to nature than men?" is itself the problem. Switching the answer by elevating women and nature (in opposition to men) only perpetuates the problem. (See Griscom, 1981; King, 1981; Ortner, 1974; Warren, 1987.)

15. The implication is clear: just as "it is quite feasible and even reasonable to maintain an individualistic (i.e., Gleasonian) concept of the community and a holistic concept of ecosystem function" (O'Neill et al. 1986, 189) so too it is quite feasible and reasonable to understand the moral community as consisting, in part, of autonomous agents with properties in their own right while at the same time treating that community as in some respects holistic.

16. A popular environmentalist slogan, sometimes endorsed by ecofeminists, is that everything is connected: a tug on any part of the system has an effect on every other part of the system. This image of ecosystems is one that O'Neill et al. take great pains to dispel (86). Critical to the stability of an ecosystem is the relative insulation or *disconnection* of sub-systems ("holons") from one another (with strong interaction within holons and weak interaction between holons). Overconnectedness in a system, where tugs on any part of the system produce effects on all parts of the system, are *unstable* (94). This perspective renders problematic the oft-repeated remark that ecology demonstrates that everything is connected with everything else— the interconnection is only within holons, not between holons.

An adequate ecofeminist ecology, then, must acknowledge that the world, so to speak, "strives" to organize itself into discrete and relatively autonomous holons as a condition of its own stability. This is at least as important a feature of our world as is its connectedness. And, indeed, individuals still come into their own with the same sterling ontological credentials as the energy flow patterns which emerge from process-functional analyses. Everything may be tied to everything else in *some* sense, but hierarchy (observation set) theory suggests that it is not in any metaphysically reductionist, holistic sense.

REFERENCES

Brennan, Andrew. 1988. *Thinking about nature: An investigation of nature, value and ecology*. Athens, GA: University of Georgia Press.

Caldecott, Léonie and Stephanie Leland, eds. 1983. *Reclaim the earth: Women speak out for life on earth*. London: The Women's Press.

Callicott, J. Baird. 1986. The metaphysical implications of ecology. *Environmental Ethics* 8: 301-16.

Cheney, Jim. 1987. Eco-feminism and deep ecology. *Environmental Ethics* 9: 115-45.

———. 1989a. Postmodern environmental ethics: Ethics as bioregional narrative. *Environmental Ethics* 11: 117-34.

———. 1989b. The Neo-stoicism of radical environmentalism. *Environmental Ethics* 11: 293-325.

———. 1991a. Review of Arne Naess, Ecology, community and life-style. *Environmental Ethics.* Forthcoming.

———. 1991b. Callicott's "Metaphysics of morals." *Environmental Ethics.* Forthcoming.

Davis, Mark. 1988. Should moral philosophers be listening to ecologists? Unpublished manuscript.

Diamond, Irene and Gloria Femen Orenstein, eds. 1990. *Reweaving the world: The emergence of ecofeminism.* San Fransisco: Sierra Club Books.

Friedman, Marilyn. 1989. Feminism and modern friendship: Dislocating the community. *Ethics* 99: 275-90.

Gilligan, Carol. 1982. *In a different voice: Psychological theory and women's development.* Cambridge: Harvard University Press.

Golley, Frank B. 1987. Deep ecology from the perspective of environmental science. *Environmental Ethics* 9(1): 45-55.

Gray, Elizabth Dodson. 1981. *Green paradise lost.* Wellesley, MA: Roundtable Press.

Griffin, Susan. 1978. *Women and nature: The roaring inside her.* San Francisco: Harper and Row.

Griscom, Joan L. 1981. On healing the nature/history split in feminist thought. In *Heresies #13: Feminism and Ecology* 4 (1): 4-9.

Haraway, Donna. 1988. Situated knowledges: The science question in feminism and the privilege of partial perspective. *Feminist Studies* 14: 575-99.

Jaggar, Alison M. 1990. Feminist ethics: Problems, projects, problems. *Feminist Ethics.* Claudia Card, ed. Forthcoming.

King, Ynestra. 1981. Feminism and the revolt of nature. *Heresies #13: Feminism and Ecology* 4(1): 12-16.

———. 1983. Toward an ecological feminism and a feminist ecology. In *Machina ex dea: Feminist perspectives on technology.* Joan Rothschild, ed. New York: Pergamon Press.

———. 1989. The ecology of feminism and the feminism of ecology. In *Healing the wounds: The power of ecological feminism.* Judith Plant, ed. Philadelphia and Santa Cruz: New Society Publishers.

Leopold, Aldo. 1970. *A sand county almanac.* New York: Ballantine Books.

Merchant, Carolyn. 1980. *The death of nature: Women, ecology, and the scientific revolution.* San Francisco: Harper and Row.

Murphy, Patrick, ed. 1988. Feminism, ecology and the future of the humanities. Special issue of *Studies in the Humanities.* 15(2).

O'Neill, R. V., D. L. DeAngelis, J. B. Waide, and T. F. H. Allen. 1986. *A hierarchical concept of ecosystems.* Princeton: Princeton University Press.

Ortner, Sherry B. 1974. Is female to male as nature is to culture? In *Women, culture, and society.* Michelle Rosaldo and Lousie Lamphere, eds. Stanford, CA: Stanford University Press.

Peterson, Abby and Carolyn Merchant. 1986. Peace with the earth: Women and the environmental movement in Sweden. *Women's Studies International Forum.* 9(5-6): 465-79.

Plant, Judith. 1990. Searching for common ground: Ecofeminism and bioregionalism. In *Reweaving the world: The emergence of ecofeminism.* Irene Diamond and Gloria Feman Orenstein, eds. San Francisco: Sierra Club Books.

Plant, Judith, ed. 1989. *Healing the wounds: The power of ecological feminism.* Philadelphia and Santa Cruz: New Society Publishers.

Rolston, Holmes, III. 1988. *Environmental ethics: Duties to and values in the natural world.* Philadelphia: Temple University Press.

———. 1989. Review of Andrew Brennan, "Thinking about nature: An investigation of nature, value and ecology." *Environmental Ethics* 11: 259-67.

Ruether, Rosemary Radford. 1975. *New woman/new earth: Sexist ideologies and human liberation.* New York: Seabury Press.

Salleh, Ariel Kay. 1984. Deeper than deep ecology: The eco-feminist connection. *Environmental Ethics* 6: 339-45.

Shiva, Vandana. 1988. *Staying alive: Women, ecology, and development.* London: Zed Books.

Warren, Karen J. 1987. Feminism and ecology: Making connections. *Environmental Ethics* 9: 3-21.

———. 1988. Toward an ecofeminist ethic. *Studies in the Humanities* 15 (2): 140-156.

———. 1990. The power and promise of ecological feminism. *Environmental Ethics* 12: 125-146.

Young, Iris. 1986. The ideal of community and the politics of difference. *Social Theory and Practice* 12: 1-26.

Zimmerman, Michael. 1987. Feminism, deep ecology, and environmental ethics. *Environmental Ethics* 9: 21-44.

[18]

The Power and the Promise of Ecological Feminism

Karen J. Warren*

Ecological feminism is the position that there are important connections—historical, symbolic, theoretical—between the domination of women and the domination of nonhuman nature. I argue that because the conceptual connections between the dual dominations of women and nature are located in an oppressive patriarchal conceptual framework characterized by a logic of domination, (1) the logic of traditional feminism requires the expansion of feminism to include ecological feminism and (2) ecological feminism provides a framework for developing a distinctively feminist environmental ethic. I conclude that any feminist theory and any environmental ethic which fails to take seriously the interconnected dominations of women and nature is simply inadequate.

INTRODUCTION

Ecological feminism (ecofeminism) has begun to receive a fair amount of attention lately as an alternative feminism and environmental ethic.[1] Since Francoise d'Eaubonne introduced the term *ecofeminisme* in 1974 to bring atten-

---

*Philosophy Department. Macalaster College. 1600 Grand Avenue, St. Paul, MN 55105. Warren's main research and teaching interests are in feminism, environmental ethics, and philosophical psychology. She also teaches philosophy and critical thinking to teachers and students, grades K to 12. Warren is currently writing a book with Jim Cheney, *Ecological Feminism: What It Is and Why It Matters*, to be published by Westview Press. Earlier versions of this paper were presented at the American Philosophical Association Meeting in New York City, December 1987, and at the University of Massachusetts, April 1988. The author wishes to thank the following people for their helpful comments and support: Bob Ackerman, Kim Brown, Jim Cheney, Mahmoud El-Kati, Eric Katz, Michael Keenan, Ruthanne Kurth-Schai, Greta Gaard, Roxanne Gudeman, Alison Jaggar, H. Warren Jones, Gareth Matthews, Michael McCall, Patrick Murphy, Bruce Nordstrom, Nancy Shea, Nancy Tuana, Bob Weinstock-Collins, Henry West, and the anonymous referees of *Environmental Ethics*.

[1] Explicit ecological feminist literature includes works from a variety of scholarly perspectives and sources. Some of these works are Leonie Caldecott and Stephanie Leland, eds., *Reclaim the Earth: Women Speak Out for Life on Earth (London: The* Women's Press, 1983); Jim Cheney, "Eco-Feminism and Deep Ecology," *Environmental Ethics* 9 (1987): 115–45; Andrée Collard with Joyce Contrucci, *Rape of the Wild: Man's Violence against Animals and the Earth* (Bloomington: Indiana University Press, 1988); Katherine Davies, "Historical Associations: Women and the Natural World," *Women & Environments* 9, no. 2 (Spring 1987): 4–6; Sharon Doubiago, "Deeper than Deep Ecology: Men Must Become Feminists," in *The New Catalyst Quarterly*, no. 10 (Winter 1987/88): 10–11; Brian Easlea, *Science and Sexual Oppression: Patriarchy's Confrontation with Women and Nature* (London: Weidenfeld & Nicholson, 1981); Elizabeth Dodson Gray, *Green Paradise Lost* (Wellesley, Mass.: Roundtable Press, 1979); Susan Griffin, *Women and Nature: The Roaring Inside Her* (San Francisco: Harper and Row, 1978); Joan L. Griscom, "On Healing the Nature/History Split in Feminist Thought," in *Heresies #13: Feminism and Ecology* 4, no. 1 (1981): 4–9; Ynestra King,

tion to women's potential for bringing about an ecological revolution,[2] the term has been used in a variety of ways. As I use the term in this paper, ecological feminism is the position that there are important connections—historical, experiential, symbolic, theoretical—between the domination of women and the domination of nature, an understanding of which is crucial to both feminism and environmental ethics. I argue that the promise and power of ecological feminism is that *it provides a distinctive framework both for reconceiving feminism and for developing an environmental ethic which takes seriously connections between the domination of women and the domination of nature.* I do so by discussing the nature of a feminist ethic and the ways in which ecofeminism provides a feminist and environmental ethic. I conclude that any feminist theory *and* any environmental ethic which fails to take seriously the twin and interconnected dominations of women and nature is at best incomplete and at worst simply inadequate.

FEMINISM, ECOLOGICAL FEMINISM, AND CONCEPTUAL FRAMEWORKS

Whatever else it is, feminism is at least the movement to end sexist oppression. It involves the elimination of any and all factors that contribute to the

---

"The Ecology of Feminism and the Feminism of Ecology," in *Healing Our Wounds: The Power of Ecological Feminism,* ed. Judith Plant (Boston: New Society Publishers, 1989), pp. 18–28; "The Eco-feminist Imperative," in *Reclaim the Earth,* ed. Caldecott and Leland (London: The Women's Press, 1983), pp. 12–16, "Feminism and the Revolt of Nature," in *Heresies #13: Feminism and Ecology* 4, no. 1 (1981): 12–16, and "What is Ecofeminism?" *The Nation,* 12 December 1987; Marti Kheel, "Animal Liberation Is A Feminist Issue," *The New Catalyst Quarterly,* no. 10 (Winter 1987–88): 8–9; Carolyn Merchant, *The Death of Nature: Women, Ecology and the Scientific Revolution* (San Francisco, Harper and Row, 1980); Patrick Murphy, ed., "Feminism, Ecology, and the Future of the Humanities," special issue of *Studies in the Humanities* 15, no. 2 (December 1988); Abby Peterson and Carolyn Merchant, "Peace with the Earth': Women and the Environmental Movement in Sweden," *Women's Studies International Forum* 9, no. 5–6. (1986): 465–79; Judith Plant, "Searching for Common Ground: Ecofeminism and Bioregionalism," in *The New Catalyst Quarterly,* no. 10 (Winter 1987/88): 6–7; Judith Plant, ed., *Healing Our Wounds: The Power of Ecological Feminism* (Boston: New Society Publishers, 1989); Val Plumwood, "Ecofeminism: An Overview and Discussion of Positions and Arguments," *Australasian Journal of Philosophy,* Supplement to vol. 64 (June 1986): 120–37; Rosemary Radford Ruether, *New Woman/New Earth: Sexist Ideologies & Human Liberation* (New York: Seabury Press, 1975); Kirkpatrick Sale, "Ecofeminism—A New Perspective," *The Nation,* 26 September 1987): 302–05; Ariel Kay Salleh, "Deeper than Deep Ecology: The Eco-Feminist Connection," *Environmental Ethics* 6 (1984): 339–45, and "Epistemology and the Metaphors of Production: An Eco-Feminist Reading of Critical Theory," in *Studies in the Humanities* 15 (1988): 130–39; Vandana Shiva, *Staying Alive: Women, Ecology and Development* (London: Zed Books, 1988); Charlene Spretnak. "Ecofeminism: Our Roots and Flowering," *The Elmswood Newsletter,* Winter Solstice 1988; Karen J. Warren, "Feminism and Ecology: Making Connections," *Environmental Ethics* 9 (1987): 3–21; "Toward an Ecofeminist Ethic," *Studies in the Humanities* 15 (1988): 140–156; Miriam Wyman, "Explorations of Ecofeminism," *Women & Environments* (Spring 1987): 6–7; Iris Young, " 'Feminism and Ecology' and 'Women and Life on Earth: Eco-Feminism in the 80's'," *Environmental Ethics* 5 (1983): 173–80; Michael Zimmerman, "Feminism, Deep Ecology, and Environmental Ethics," *Environmental Ethics* 9 (1987): 21–44.

[2] Francoise d'Eaubonne, *Le Feminisme ou la Mort* (Paris: Pierre Horay, 1974), pp. 213–52.

continued and systematic domination or subordination of women. While feminists disagree about the nature of and solutions to the subordination of women, all feminists agree that sexist oppression exists, is wrong, and must be abolished.

A "feminist issue" is any issue that contributes in some way to understanding the oppression of women. Equal rights, comparable pay for comparable work, and food production are feminist issues wherever and whenever an understanding of them contributes to an understanding of the continued exploitation or subjugation of women. Carrying water and searching for firewood are feminist issues wherever and whenever women's primary responsibility for these tasks contributes to their lack of full participation in decision making, income producing, or high status positions engaged in by men. What counts as a feminist issue, then, depends largely on context, particularly the historical and material conditions of women's lives.

Environmental degradation and exploitation are feminist issues because an understanding of them contributes to an understanding of the oppression of women. In India, for example, both deforestation and reforestation through the introduction of a monoculture species tree (e.g., eucalyptus) intended for commercial production are feminist issues because the loss of indigenous forests and multiple species of trees has drastically affected rural Indian women's ability to maintain a subsistence household. Indigenous forests provide a variety of trees for food, fuel, fodder, household utensils, dyes, medicines, and income-generating uses, while monoculture-species forests do not.[3] Although I do not argue for this claim here, a look at the global impact of environmental degradation on women's lives suggests important respects in which environmental degradation is a feminist issue.

Feminist philosophers claim that some of the most important feminist issues are *conceptual* ones: these issues concern how one conceptualizes such mainstay philosophical notions as reason and rationality, ethics, and what it is to be human. Ecofeminists extend this feminist philosophical concern to nature. They argue that, ultimately, some of the most important connections between the domination of women and the domination of nature are conceptual. To see this, consider the nature of conceptual frameworks.

A *conceptual framework* is a set of *basic* beliefs, values, attitudes, and assumptions which shape and reflect how one views oneself and one's world. It is a socially constructed lens through which we perceive ourselves and others. It is affected by such factors as gender, race, class, age, affectional orientation, nationality, and religious background.

Some conceptual frameworks are oppressive. An *oppressive conceptual framework* is one that explains, justifies, and maintains relationships of domination and subordination. When an oppressive conceptual framework is *pa-*

---

[3] I discuss this in my paper, "Toward An Ecofeminist Ethic."

triarchal, it explains, justifies, and maintains the subordination of women by men.

I have argued elsewhere that there are three significant features of oppressive conceptual frameworks: (1) value-hierarchical thinking, i.e., "up-down" thinking which places higher value, status, or prestige on what is "up" rather than on what is "down"; (2) value dualisms, i.e., disjunctive pairs in which the disjuncts are seen as oppositional (rather than as complementary) and exclusive (rather than as inclusive), and which place higher value (status, prestige) on one disjunct rather than the other (e.g., dualisms which give higher value or status to that which has historically been identified as "mind," "reason," and "male" than to that which has historically been identified as "body," "emotion," and "female"); and (3) logic of domination, i.e., a structure of argumentation which leads to a justification of subordination.[4]

The third feature of oppressive conceptual frameworks is the most significant. A logic of domination is not *just* a logical structure. It also involves a substantive value system, since an ethical premise is needed to permit or sanction the "just" subordination of that which is subordinate. This justification typically is given on grounds of some alleged characteristic (e.g., rationality) which the dominant (e.g., men) have and the subordinate (e.g., women) lack.

Contrary to what many feminists and ecofeminists have said or suggested, there may be nothing *inherently* problematic about "hierarchical thinking" or even "value-hierarchical thinking" in contexts other than contexts of oppression. Hierarchical thinking is important in daily living for classifying data, comparing information, and organizing material. Taxonomies (e.g., plant taxonomies) and biological nomenclature seem to require *some* form of "hierarchical thinking." Even "value-hierarchical thinking" may be quite acceptable in certain contexts. (The same may be said of "value dualisms" in non-oppressive contexts.) For example, suppose it is true that what is unique about humans is our conscious capacity to radically reshape our social environments (or "societies"), as Murray Bookchin suggests.[5] Then one could truthfully say that humans are better equipped to radically reshape their environments than are rocks or plants—a "value-hierarchical" way of speaking.

The problem is not simply *that* value-hierarchical thinking and value dualisms are used, but *the way* in which each has been used *in oppressive conceptual*

---

[4] The account offered here is a revision of the account given earlier in my paper "Feminism and Ecology: Making Connections." I have changed the account to be about "oppressive" rather than strictly "patriarchal" conceptual frameworks in order to leave open the possibility that there may be some patriarchal conceptual frameworks (e.g., in non-Western cultures) which are *not* properly characterized as based on value dualisms.

[5] Murray Bookshin, "Social Ecology versus 'Deep Ecology'," in *Green Perspectives: Newsletter of the Green Program Project*, no. 4–5 (Summer 1987): 9.

frameworks to establish inferiority and to justify subordination.[6] It is the logic of domination, *coupled with* value-hierarchical thinking and value dualisms, which "justifies" subordination. What is explanatorily basic, then, about the nature of oppressive conceptual frameworks is the logic of domination.

For ecofeminism, that a logic of domination is explanatorily basic is important for at least three reasons. First, without a logic of domination, a description of similarities and differences would be just that—a description of similarities and differences. Consider the claim, "Humans are different from plants and rocks in that humans can (and plants and rocks cannot) consciously and radically reshape the communities in which they live; humans are similar to plants and rocks in that they are both members of an ecological community." Even if humans are "better" than plants and rocks with respect to the conscious ability of humans to radically transform communities, one does not *thereby* get any *morally* relevant distinction between humans and nonhumans, or an argument for the domination of plants and rocks by humans. To get *those* conclusions one needs to add at least two powerful assumptions, viz., (A2) and (A4) in argument A below:

(A1) Humans do, and plants and rocks do not, have the capacity to consciously and radically change the community in which they live.
(A2) Whatever has the capacity to consciously and radically change the community in which it lives is morally superior to whatever lacks this capacity.
(A3) Thus, humans are morally superior to plants and rocks.
(A4) For any X and Y, if X is morally superior to Y, then X is morally justified in subordinating Y.
(A5) Thus, humans are morally justified in subordinating plants and rocks.

Without the two assumptions that *humans are morally superior* to (at least some) nonhumans, (A2), and that *superiority justifies subordination*, (A4), all one has is some difference between humans and some nonhumans. This is true *even if* that difference is given in terms of superiority. Thus, it is the logic of domination, (A4), which is the bottom line in ecofeminist discussions of oppression.

Second, ecofeminists argue that, at least in Western societies, the oppressive conceptual framework which sanctions the twin dominations of women and nature is a patriarchal one characterized by all three features of an oppressive conceptual framework. Many ecofeminists claim that, historically, within at least the dominant Western culture, a patriarchal conceptual framework has sanctioned the following argument B:

---

[6] It may be that in contemporary Western society, which is so thoroughly structured by categories of gender, race, class, age, and affectional orientation, that there simply is no meaningful notion of "value-hierarchical thinking" which does not function in an oppressive context. For purposes of this paper, I leave that question open.

(B1) Women are identified with nature and the realm of the physical; men are identified with the "human" and the realm of the mental.
(B2) Whatever is identified with nature and the realm of the physical is inferior to ("below") whatever is identified with the "human" and the realm of the mental; or, conversely, the latter is superior to ("above") the former.
(B3) Thus, women are inferior to ("below") men; or, conversely, men are superior to ("above") women.
(B4) For any X and Y, if X is superior to Y, then X is justified in subordinating Y.
(B5) Thus, men are justified in subordinating women.

If sound, argument B establishes *patriarchy*, i.e., the conclusion given at (B5) that the systematic domination of women by men is justified. But according to ecofeminists, (B5) is justified by just those three features of an oppressive conceptual framework identified earlier: value-hierarchical thinking, the assumption at (B2); value dualisms, the assumed dualism of the mental and the physical at (B1) and the assumed inferiority of the physical vis-à-vis the mental at (B2); and a logic of domination, the assumption at (B4), the same as the previous premise (A4). Hence, according to ecofeminists, insofar as an oppressive patriarchal conceptual framework has functioned historically (within at least dominant Western culture) to sanction the twin dominations of women and nature (argument B), both argument B and the patriarchal conceptual framework, from whence it comes, ought to be rejected.

Of course, the preceeding does not identify which premises of B are false. What is the status of premises (B1) and (B2)? Most, if not all, feminists claim that (B1), and many ecofeminists claim that (B2), have been assumed or asserted within the dominant Western philosophical and intellectual tradition.[7] As such, these feminists assert, as a matter of historical fact, that the dominant Western philosophical tradition has assumed the truth of (B1) and (B2). Ecofeminists, however, either deny (B2) or do not affirm (B2). Furthermore, because some ecofeminists are anxious to deny any ahistorical identification of women with nature, some ecofeminists deny (B1) when (B1) is used to support anything other than a strictly historical claim about what has been asserted or assumed to be true

---

[7] Many feminists who argue for the historical point that claims (B1) and (B2) have been asserted or assumed to be true within the dominant Western philosophical tradition do so by discussion of that tradition's conceptions of reason, rationality, and science. For a sampling of the sorts of claims made within that context, see "Reason, Rationality, and Gender," ed. Nancy Tuana and Karen J. Warren, a special issue of the American Philosophical Association's *Newsletter on Feminism and Philosophy* 88, no. 2 (March 1989): 17–71. Ecofeminists who claim that (B2) has been assumed to be true within the dominant Western philosophical tradition include: Gray, *Green Paradise Lost;* Griffin, *Woman and Nature: The Roaring Inside Her;* Merchant, *The Death of Nature;* Ruether, *New Woman/New Earth.* For a discussion of some of these ecofeminist historical accounts, see Plumwood, "Ecofeminism." While I agree that the historical connections between the domination of women and the domination of nature is a crucial one, I do not argue for that claim here.

within patriarchal culture—e.g., when (B1) is used to assert that women properly are identified with the realm of nature and the physical.[8] Thus, from an ecofeminist perspective, (B1) and (B2) are properly viewed as problematic though historically sanctioned claims: they are problematic precisely because of the way they have functioned historically in a patriarchal conceptual framework and culture to sanction the dominations of women and nature.

What *all* ecofeminists agree about, then, is the way in which *the logic of domination* has functioned historically within patriarchy to sustain and justify the twin dominations of women and nature.[9] Since *all* feminists (and not just ecofeminists) oppose patriarchy, the conclusion given at (B5), all feminists (including ecofeminists) must oppose at least the logic of domination, premise (B4), on which argument B rests—whatever the truth-value status of (B1) and (B2) *outside of* a patriarchal context.

That *all* feminists must oppose the logic of domination shows the breadth and depth of the ecofeminist critique of B: it is a critique not only of the three assumptions on which this argument for the domination of women and nature rests, viz., the assumptions at (B1), (B2), and (B4); it is also a critique of patriarchal conceptual frameworks generally, i.e., of those oppressive conceptual frameworks which put men "up" and women "down," allege some way in which women are morally inferior to men, and use that alleged difference to justify the subordination of women by men. Therefore, ecofeminism is necessary to *any* feminist critique of patriarchy, and, hence, necessary to feminism (a point I discuss again later).

Third, ecofeminism clarifies why the logic of domination, and any conceptual framework which gives rise to it, must be abolished in order both to make possible a meaningful notion of difference which does not breed domination and to prevent feminism from becoming a "support" movement based primarily on shared experiences. In contemporary society, there is no one "woman's voice," no *woman* (or *human*) *simpliciter*: every woman (or human) is a woman (or human) of some race, class, age, affectional orientation, marital status, regional or national background, and so forth. Because there are no "monolithic experiences" that all women share, feminism must be a "solidarity movement" based on shared beliefs and interests rather than a "unity in sameness" movement based on shared experiences and shared victimization.[10] In the words of Maria

---

[8] Ecofeminists who deny (B1) when (B1) is offered as anything other than a true, descriptive, historical claim about patriarchal culture often do so on grounds that an objectionable sort of biological determinism, or at least harmful female sex-gender stereotypes, underlie (B1). For a discussion of this "split" among those ecofeminists ("nature feminists") who assert and those ecofeminists ("social feminists") who deny (B1) as anything other than a true historical claim about how women are described in patriarchal culture, see Griscom, "On Healing the Nature/History Split."

[9] I make no attempt here to defend the historically sanctioned truth of these premises.

[10] See, e.g., Bell Hooks, *Feminist Theory: From Margin to Center* (Boston: South End Press, 1984), pp. 51–52.

Lugones, "Unity—not to be confused with solidarity—is understood as conceptually tied to domination."[11]

Ecofeminists insist that the sort of logic of domination used to justify the domination of humans by gender, racial or ethnic, or class status is also used to justify the domination of nature. Because eliminating a logic of domination is part of a feminist critique—whether a critique of patriarchy, white supremacist culture, or imperialism—ecofeminists insist that *naturism* is properly viewed as an integral part of any feminist solidarity movement to end sexist oppression and the logic of domination which conceptually grounds it.

ECOFEMINISM RECONCEIVES FEMINISM

The discussion so far has focused on some of the oppressive conceptual features of patriarchy. As I use the phrase, the "logic of traditional feminism" refers to the location of the conceptual roots of sexist oppression, at least in Western societies, in an oppressive patriarchal conceptual framework characterized by a logic of domination. Insofar as other systems of oppression (e.g., racism, classism, ageism, heterosexism) are also conceptually maintained by a logic of domination, appeal to the logic of traditional feminism ultimately locates the basic conceptual interconnections among *all* systems of oppression in the logic of domination. It thereby explains at a *conceptual* level why the eradication of sexist oppression requires the eradication of the other forms of oppression.[12] It is by clarifying this conceptual connection between systems of oppression that a movement to end sexist oppression—traditionally the special turf of feminist theory and practice—leads to a reconceiving of feminism as *a movement to end all forms of oppression*.

Suppose one agrees that the logic of traditional feminism requires the expansion of feminism to include other social systems of domination (e.g., racism and classism). What warrants the inclusion of nature in these "social systems of domination"? Why must the logic of traditional feminism include the abolition of "naturism" (i.e., the domination or oppression of nonhuman nature) among the "isms" feminism must confront? The conceptual justification for expanding feminism to include ecofeminism is twofold. One basis has already been suggested: by showing that the conceptual connections between the dual dominations of women and nature are located in an oppressive and, at least in Western

---

[11] Maria Lugones, "Playfulness, 'World-Travelling,' and Loving Perception," *Hypatia* 2, no. 2 (Summer 1987): 3.

[12] At an *experiential* level, some women are "women of color," poor, old, lesbian, Jewish, and physically challenged. Thus, if feminism is going to liberate these women, it also needs to end the racism, classism, heterosexism, anti-Semitism, and discrimination against the handicapped that is constitutive of their oppression as black, or Latina, or poor, or older, or lesbian, or Jewish, or physically challenged women.

societies, patriarchal conceptual framework characterized by a logic of domination, ecofeminism explains how and why feminism, conceived as a movement to end sexist oppression, must be expanded and reconceived as also a movement to end naturism." This is made explicit by the following argument C:

(C1) Feminism is a movement to end sexism.
(C2) But Sexism is conceptually linked with naturism (through an oppressive conceptual framework characterized by a logic of domination).
(C3) Thus, Feminism is (also) a movement to end naturism.

Because, ultimately, these connections between sexism and naturism are conceptual—embedded in an oppressive conceptual framework—the logic of traditional feminism leads to the embracement of ecological feminism.[13]

The other justification for reconceiving feminism to include ecofeminism has to do with the concepts of gender and nature. Just as conceptions of gender are socially constructed, so are conceptions of nature. Of course, the claim that women and nature are social constructions does not require anyone to deny that there are actual humans and actual trees, rivers, and plants. It simply implies that *how* women and nature are conceived is a matter of historical and social reality. These conceptions vary cross-culturally and by historical time period. As a result, any discussion of the "oppression or domination of nature" involves reference to historically specific forms of social domination of nonhuman nature by humans, just as discussion of the "domination of women" refers to historically specific forms of social domination of women by men. Although I do not argue for it here, an ecofeminist defense of the historical connections between the dominations of women and of nature, claims (B1) and (B2) in argument B, involves showing that within patriarchy the feminization of nature and the naturalization of women have been crucial to the historically successful subordinations of both.[14]

If ecofeminism promises to reconceive traditional feminism in ways which include naturism as a legitimate feminist issue, does ecofeminism also promise to reconceive environmental ethics in ways which are feminist? I think so. This is the subject of the remainder of the paper.

---

[13] This same sort of reasoning shows that feminism is also a movement to end racism, classism, age-ism, heterosexism and other "isms," which are based in oppressive conceptual frameworks characterized by a logic of domination. However, there is an important caveat: ecofeminism is *not* compatible with all feminisms and all environmentalisms. For a discussion of this point, see my article, "Feminism and Ecology: Making Connections. What it *is* compatible with is the minimal condition characterization of feminism as a movement to end sexism that is accepted by all contemporary feminisms (liberal, traditional Marxist, radical, socialist, Blacks and non-Western).

[14] See, e.g., Gray, *Green Paradise Lost;* Griffin, *Women and Nature;* Merchant, *The Death of Nature;* and Ruether, *New Woman/New Earth.*

CLIMBING FROM ECOFEMINISM TO ENVIRONMENTAL ETHICS

Many feminists and some environmental ethicists have begun to explore the use of first-person narrative as a way of raising philosophically germane issues in ethics often lost or underplayed in mainstream philosophical ethics. Why is this so? What is it about narrative which makes it a significant resource for theory and practice in feminism and environmental ethics? Even if appeal to first-person narrative is a helpful literary device for describing ineffable experience or a legitimate social science methodology for documenting personal and social history, how is first-person narrative a valuable vehicle of argumentation for ethical decision making and theory building? One fruitful way to begin answering these questions is to ask them of a particular first-person narrative.

Consider the following first-person narrative about rock climbing:

> For my very first rock climbing experience, I chose a somewhat private spot, away from other climbers and on-lookers. After studying "the chimney," I focused all my energy on making it to the top. I climbed with intense determination, using whatever strength and skills I had to accomplish this challenging feat. By midway I was exhausted and anxious. I couldn't see what to do next—where to put my hands or feet. Growing increasingly more weary as I clung somewhat desperately to the rock, I made a move. It didn't work. I fell. There I was, dangling midair above the rocky ground below, frightened but terribly relieved that the belay rope had held me. I knew I was safe. I took a look up at the climb that remained. I was determined to make it to the top. With renewed confidence and concentration, I finished the climb to the top.
>
> On my second day of climbing, I rappelled down about 200 feet from the top of the Palisades at Lake Superior to just a few feet above the water level. I could see no one—not my belayer, not the other climbers, no one. I unhooked slowly from the rappel rope and took a deep cleansing breath. I looked all around me—really looked—and listened. I heard a cacophony of voices—birds, trickles of water on the rock before me, waves lapping against the rocks below. I closed my eyes and began to feel the rock with my hands—the cracks and crannies, the raised lichen and mosses, the almost imperceptible nubs that might provide a resting place for my fingers and toes when I began to climb. At that moment I was bathed in serenity. I began to talk to the rock in an almost inaudible, child-like way, as if the rock were my friend. I felt an overwhelming sense of gratitude for what it offered me—a chance to know myself and the rock differently, to appreciate unforeseen miracles like the tiny flowers growing in the even tinier cracks in the rock's surface, and to come to know a sense of *being in relationship* with the natural environment. It felt as if the rock and I were silent conversational partners in a longstanding friendship. I realized then

that I had come to care about this cliff which was so different from me, so unmovable and invincible, independent and seemingly indifferent to my presence. I wanted to be with the rock as I climbed. Gone was the determination to conquer the rock, to forcefully impose my will on it; I wanted simply to work respectfully with the rock as I climbed. And as I climbed, that is what I felt. I felt myself *caring* for this rock and feeling thankful that climbing provided the opportunity for me to know it and myself in this new way.

There are at least four reasons why use of such a first-person narrative is important to feminism and environmental ethics. First, such a narrative gives voice to a felt sensitivity often lacking in traditional analytical ethical discourse, viz., a sensitivity to conceiving of oneself as fundamentally "in relationship with" others, including the nonhuman environment. It is a modality which *takes relationships themselves seriously*. It thereby stands in contrast to a strictly reductionist modality that takes relationships seriously only or primarily because of the nature of the *relators* or parties to those relationships (e.g., relators conceived as moral agents, right holders, interest carriers, or sentient beings). In the rock-climbing narrative above, it is the climber's relationship with the rock she climbs which takes on special significance—which is itself a locus of value—in addition to whatever moral status or moral considerability she or the rock or any other parties to the relationship may also have.[15]

Second, such a first-person narrative gives expression to a variety of ethical attitudes and behaviors often overlooked or underplayed in mainstream Western ethics, e.g., the difference in attitudes and behaviors toward a rock when one is "making it to the top" and when one thinks of oneself as "friends with" or "caring about" the rock one climbs.[16] These different attitudes and behaviors suggest an ethically germane contrast between two different types of relationship humans or climbers may have toward a rock: an imposed conqueror-type relationship, and

---

[15] Suppose, as I think is the case, that a necessary condition for the existence of a moral relationship is that at least one party to the relationship is a moral being (leaving open for our purposes what counts as a "moral being"). If this is so, then the Mona Lisa cannot properly be said to have or stand in a moral relationship with the wall on which she hangs, and a wolf cannot have or properly be said to have or stand in a moral relationship with a moose. Such a necessary-condition account leaves open the question whether *both* parties to the relationship must be moral beings. My point here is simply that however one resolves *that* question, recognition of the relationships themselves as a locus of value is a recognition of a source of value that is different from and not reducible to the values of the "moral beings" in those relationships.

[16] It is interesting to note that the image of being friends with the Earth is one which cytogeneticist Barbara McClintock uses when she describes the importance of having "a feeling for the organism," "listening to the material [in this case the corn plant]," in one's work as a scientist. See Evelyn Fox Keller, "Women, Science, and Popular Mythology," in *Machina Ex Dea: Feminist Perspectives on Technology*, ed. Joan Rothschild (New York: Pergamon Press, 1983), and Evelyn Fox Keller, *A Feeling For the Organism: The Life and Work of Barbara McClintock* (San Francisco: W. H. Freeman, 1983).

an emergent caring-type relationship. This contrast grows out of, and is faithful to, felt, lived experience.

The difference between conquering and caring attitudes and behaviors in relation to the natural environment provides a third reason why the use of first-person narrative is important to feminism and environmental ethics: it provides a way of conceiving of ethics and ethical meaning as *emerging out of* particular situations moral agents find themselves in, rather than as being *imposed on* those situations (e.g., as a derivation or instantiation of some predetermined abstract principle or rule). This emergent feature of narrative centralizes the importance of *voice*. When a multiplicity of cross-cultural *voices* are centralized, narrative is able to give expression to a range of attitudes, values, beliefs, and behaviors which may be overlooked or silenced by imposed ethical meaning and theory. As a reflection of and on felt, lived experiences, the use of narrative in ethics provides a stance from which ethical discourse can be held accountable to the historical, material, and social realities in which moral subjects find themselves.

Lastly, and for our purposes perhaps most importantly, the use of narrative has argumentative significance. Jim Cheney calls attention to this feature of narrative when he claims, "To contextualize ethical deliberation is, in some sense, to provide a narrative or story, from which the solution to the ethical dilemma emerges as the fitting conclusion."[17] Narrative has argumentative force by suggesting *what counts* as an appropriate conclusion to an ethical situation. One ethical conclusion suggested by the climbing narrative is that what counts as a proper ethical attitude toward mountains and rocks is an attitude of respect and care (whatever that turns out to be or involve), not one of domination and conquest.

In an essay entitled "In and Out of Harm's Way: Arrogance and Love," feminist philosopher Marilyn Frye distinguishes between "arrogant" and "loving" perception as one way of getting at this difference in the ethical attitudes of care and conquest.[18] Frye writes:

> The loving eye is a contrary of the arrogant eye.
>
> The loving eye knows the independence of the other. It is the eye of a seer who knows that nature is indifferent. It is the eye of one who knows that to know the seen, one must consult something other than one's own will and interests and fears and imagination. One must look at the thing. One must look and listen and check and question.
>
> The loving eye is one that pays a certain sort of attention. This attention can require a discipline but *not* a self-denial. The discipline is one of self-knowledge, knowledge of the scope and boundary of the self. . . . In particular, it is a matter of

---

[17] Cheney, "Eco-Feminism and Deep Ecology," 144.
[18] Marilyn Frye, "In and Out of Harm's Way: Arrogance and Love," *The Politics of Reality* (Trumansburg, New York: The Crossing Press, 1983), pp. 66–72.

being able to tell one's own interests from those of others and of knowing where one's self leaves off and another begins. . . .

The loving eye does not make the object of perception into something edible. does not try to assimilate it, does not reduce it to the size of the seer's desire, fear and imagination, and hence does not have to simplify. It knows the complexity of the other as something which will forever present new things to be known. The science of the loving eye would favor The Complexity Theory of Truth [in contrast to The Simplicity Theory of Truth] and presuppose The Endless Interestingness of the Universe.[19]

According to Frye, the loving eye is not an invasive, coercive eye which annexes others to itself, but one which "knows the complexity of the other as something which will forever present new things to be known."

When one climbs a rock as a conqueror, one climbs with an arrogant eye. When one climbs with a loving eye, one constantly "must look and listen and check and question." One recognizes the rock as something very different, something perhaps totally indifferent to one's own presence, and finds in that difference joyous occasion for celebration. One knows "the boundary of the self," where the self—the "I," the climber—leaves off and the rock begins. There is no fusion of two into one, but a complement of two entities *acknowledged* as separate, different, independent, yet *in relationship;* they are in relationship *if only* because the loving eye is perceiving it, responding to it, noticing it, attending to it.

An ecofeminist perspective about both women and nature involves this shift in attitude from "arrogant perception" to "loving perception" of the nonhuman world. Arrogant perception of nonhumans by humans presupposes and maintains *sameness* in such a way that it expands the moral community to those beings who are thought to resemble (be like, similar to, or the same as) humans in some morally significant way. Any environmental movement or ethic based on arrogant perception builds a moral hierarchy of beings and assumes some common denominator of moral considerability in virtue of which like beings deserve similar treatment or moral consideration and unlike beings do not. Such environmental ethics are or generate a "unity in sameness." In contrast, "loving perception" presupposes and maintains *difference*—a distinction between the self and other, between human and at least some nonhumans—in such a way that perception of the other as other *is* an expression of love for one who/which is recognized at the outset as independent, dissimilar, different. As Maria Lugones says, in loving perception, "Love is seen not as fusion and erasure of difference but as incompatible with them."[20] "Unity in sameness" alone is an *erasure of difference*.

---

[19] Ibid., pp. 75–76.
[20] Maria Lugones, "Playfulness," p. 3.

"Loving perception" of the nonhuman natural world is an attempt to understand what it means *for humans* to care about the nonhuman world, a world *acknowledged* as being independent, different, perhaps even indifferent to humans. Humans *are* different from rocks in important ways, even if they are also both members of some ecological community. A moral community based on loving perception of oneself *in relationship with* a rock, or with the natural environment as a whole, is one which acknowledges and respects difference, whatever "sameness" also exists.[21] The limits of loving perception are determined only by the limits of one's (e.g., a person's, a community's) ability to respond lovingly (or with appropriate care, trust, or friendship)—whether it is to other humans or to the nonhuman world and elements of it.[22]

If what I have said so far is correct, then there are very different ways to climb a mountain and *how* one climbs it and *how* one narrates the experience of climbing it matter ethically. If one climbs with "arrogant perception," with an attitude of "conquer and control," one keeps intact the very sorts of thinking that characterize a logic of domination and an oppressive conceptual framework. Since the oppressive conceptual framework which sanctions the domination of nature is a patriarchal one, one also thereby keeps intact, even if unwittingly, a patriarchal conceptual framework. Because the dismantling of patriarchal conceptual frameworks is a feminist issue, *how* one climbs a mountain and *how* one narrates—or tells the story—about the experience of climbing also are *feminist issues*. In this way, ecofeminism makes visible why, at a conceptual level, environmental ethics is a feminist issue. I turn now to a consideration of ecofeminism as a distinctively feminist and environmental ethic.

ECOFEMINISM AS A FEMINIST AND ENVIRONMENTAL ETHIC

A feminist ethic involves a twofold commitment to critique male bias in ethics wherever it occurs, and to develop ethics which are not male-biased. Sometimes this involves articulation of values (e.g., values of care, appropriate trust, kinship, friendship) often lost or underplayed in mainstream ethics.[23] Sometimes it involves engaging in theory building by pioneering in new directions or by revamping old theories in gender sensitive ways. What makes the critiques of old theories or conceptualizations of new ones "feminist" is that they emerge out of sex-gender analyses and reflect whatever those analyses reveal about gendered experience and gendered social reality.

As I conceive feminist ethics in the pre-feminist present, it rejects attempts to conceive of ethical theory in terms of necessary and sufficient conditions, because it assumes that there is no essence (in the sense of some transhistorical,

---

[21] Cheney makes a similar point in "Eco-Feminism and Deep Ecology," p. 140.
[22] Ibid., p. 138.
[23] This account of a feminist ethic draws on my paper "Toward an Ecofeminist Ethic."

universal, absolute abstraction) of feminist ethics. While attempts to formulate joint necessary and sufficient conditions of a feminist ethic are unfruitful, nonetheless, there are some necessary conditions, what I prefer to call "boundary conditions," of a feminist ethic. These boundary conditions clarify some of the minimal conditions of a feminist ethic without suggesting that feminist ethics has some ahistorical essence. They are like the boundaries of a quilt or collage. They delimit the territory of the piece without dictating what the interior, the design, the actual pattern of the piece looks like. Because the actual design of the quilt emerges from the multiplicity of voices of women in a cross-cultural context, the design will change over time. It is not something static.

What are some of the boundary conditions of a feminist ethic? First, nothing can become part of a feminist ethic—can be part of the quilt—that promotes sexism, racism, classism, or any other "isms" of social domination. Of course, people may disagree about what counts as a sexist act, racist attitude, classist behavior. What counts as sexism, racism, or classism may vary cross-culturally. Still, because a feminist ethic aims at eliminating sexism and sexist bias, and (as I have already shown) sexism is intimately connected in conceptualization and in practice to racism, classism, and naturism, a feminist ethic must be anti-sexist, anti-racist, anti-classist, anti-naturist and opposed to any "ism" which presupposes or advances a logic of domination.

Second, a feminist ethic is a *contextualist* ethic. A contextualist ethic is one which sees ethical discourse and practice as emerging from the voices of people located in different historical circumstances. A contextualist ethic is properly viewed as a *collage* or *mosaic*, a *tapestry* of voices that emerges out of felt experiences. Like any collage or mosaic, the point is not to have *one picture* based on a unity of voices, but a *pattern* which emerges out of the very different voices of people located in different circumstances. When a contextualist ethic is *feminist*, it gives central place to the voices of women.

Third, since a feminist ethic gives central significance to the diversity of women's voices, a feminist ethic must be structurally pluralistic rather than unitary or reductionistic. It rejects the assumption that there is "one voice" in terms of which ethical values, beliefs, attitudes, and conduct can be assessed.

Fourth, a feminist ethic reconceives ethical theory as theory in process which will change over time. Like all theory, a feminist ethic is based on some generalizations.[24] Nevertheless, the generalizations associated with it are themselves a pattern of voices within which the different voices emerging out of concrete and alternative descriptions of ethical situations have meaning. The

---

[24] Marilyn Frye makes this point in her illuminating paper, "The Possibility of Feminist Theory," read at the American Philosophical Association Central Division Meetings in Chicago, 29 April–1 May 1986. My discussion of feminist theory is inspired largely by that paper and by Kathryn Addelson's paper "Moral Revolution," in *Women and Values: Reading in Recent Feminist Philosophy*, ed. Marilyn Pearsall (Belmont, Calif.: Wadsworth Publishing Co., 1986) pp. 291–309.

coherence of a feminist theory so conceived is given within a historical and conceptual context, i.e., within a set of historical, socioeconomic circumstances (including circumstances of race, class, age, and affectional orientation) and within a set of basic beliefs, values, attitudes, and assumptions about the world.

Fifth, because a feminist ethic is contextualist, structurally pluralistic, and "in-process," one way to evaluate the claims of a feminist ethic is in terms of their *inclusiveness:* those claims (voices, patterns of voices) are morally and epistemologically favored (preferred, better, less partial, less biased) which are more inclusive of the felt experiences and perspectives of oppressed persons. The condition of inclusiveness requires and ensures that the diverse voices of women (as oppressed persons) will be given legitimacy in ethical theory building. It thereby helps to minimize empirical bias, e.g., bias rising from faulty or false generalizations based on stereotyping, too small a sample size, or a skewed sample. It does so by ensuring that any generalizations which are made about ethics and ethical decision making include—indeed cohere with—the patterned voices of women.[25]

Sixth, a feminist ethic makes no attempt to provide an "objective" point of view, since it assumes that in contemporary culture there really is no such point of view. As such, it does not claim to be "unbiased" in the sense of "value-neutral" or "objective." However, it does assume that whatever bias it has as an ethic centralizing the voices of oppressed persons is a *better bias*—"better" because it is more inclusive and therefore less partial—than those which exclude those voices.[26]

Seventh, a feminist ethic provides a central place for values typically unnoticed, underplayed, or misrepresented in traditional ethics, e.g., values of care, love, friendship, and appropriate trust.[27] Again, it need not do this at the exclusion of considerations of rights, rules, or utility. There may be many contexts in which talk of rights or of utility is useful or appropriate. For instance, in contracts or property relationships, talk of rights may be useful and appropriate. In deciding what is cost-effective or advantageous to the most people, talk of

---

[25] Notice that the standard of inclusiveness does not exclude the voices of men. It is just that those voices must cohere with the voices of women.

[26] For a more in-depth discussion of the notions of impartiality and bias, see my paper, "Critical Thinking and Feminism," *Informal Logic* 10, no. 1 (Winter 1988): 31–44.

[27] The burgeoning literature on these values is noteworthy. See, e.g., Carol Gilligan, *In a Different Voice: Psychological Theories and Women's Development* (Cambridge: Harvard University Press, 1982); *Mapping the Moral Domain: A Contribution of Women's Thinking to Psychological Theory and Education*, ed. Carol Gilligan, Janie Victoria Ward, and Jill McLean Taylor, with Betty Bardige (Cambridge: Harvard University Press, 1988); Nel Noddings, *Caring: A Feminine Approach to Ethics and Moral Education* (Berkeley: University of California Press, 1984); Maria Lugones and Elizabeth V. Spelman, "Have We Got a Theory for You! Feminist Theory, Cultural Imperialism, and the Women's Voice," *Women's Studies International Forum* 6 (1983): 573–81; Maria Lugones, "Playfulness"; Annette C. Baier, "What Do Women Want In A Moral Theory?" *Nous* 19 (1985): 53–63.

utility may be useful and appropriate. In a feminist *qua* contextualist ethic, whether or not such talk is useful or appropriate depends on the context; *other values* (e.g., values of care, trust, friendship) are *not* viewed as reducible to or captured solely in terms of such talk.[28]

Eighth, a feminist ethic also involves a reconception of what it is to be human and what it is for humans to engage in ethical decision making, since it rejects as either meaningless or currently untenable any gender-free or gender-neutral description of humans, ethics, and ethical decision making. It thereby rejects what Alison Jaggar calls "abstract individualism," i.e., the position that it is possible to identify a human essence or human nature that exists independently of any particular historical context.[29] Humans and human moral conduct are properly understood essentially (and not merely accidentally) in terms of networks or webs of historical and concrete relationships.

All the props are now in place for seeing how ecofeminism provides the framework for a distinctively feminist and environmental ethic. It is a feminism that critiques male bias wherever it occurs in ethics (including environmental ethics) and aims at providing an ethic (including an environmental ethic) which is not male biased—and it does so in a way that satisfies the preliminary boundary conditions of a feminist ethic.

First, ecofeminism is quintessentially anti-naturist. Its anti-naturism consists in the rejection of any way of thinking about or acting toward nonhuman nature that reflects a logic, values, or attitude of domination. Its anti-naturist, anti-sexist, anti-racist, anti-classist (and so forth, for all other "isms" of social domination) stance forms the outer boundary of the quilt: nothing gets on the quilt which is naturist, sexist, racist, classist, and so forth.

Second, ecofeminism is a contextualist ethic. It involves a shift *from* a conception of ethics as primarily a matter of rights, rules, or principles predetermined and applied in specific cases to entities viewed as competitors in the contest of moral standing, *to* a conception of ethics as growing out of what Jim Cheney calls "defining relationships," i.e., relationships conceived in some sense as defining who one is.[30] As a contextualist ethic, it is not that rights, or rules, or principles are *not* relevant or important. Clearly they are in certain

---

[28] Jim Cheney would claim that our fundamental relationships to one another as moral agents are not as moral agents to rights holders, and that whatever rights a person properly may be said to have are relationally defined rights, not rights possessed by atomistic individuals conceived as Robinson Crusoes who do not exist essentially in relation to others. On this view, even rights talk itself is properly conceived as growing out of a relational ethic, not vice versa.

[29] Alison Jaggar, *Feminist Politics and Human Nature* (Totowa, N.J.: Rowman and Allanheld, 1980), pp. 42–44.

[30] Henry West has pointed out that the expression "defining relations" is ambiguous. According to West, "the 'defining' as Cheney uses it is an adjective, not a principle—it is not that ethics defines relationships; it is that ethics grows out of conceiving of the relationships that one is in as defining what the individual is."

contexts and for certain purposes.[31] It is just that what *makes* them relevant or important is that those to whom they apply are entities *in relationship with others*.

Ecofeminism also involves an ethical shift *from* granting moral consideration to nonhumans *exclusively* on the grounds of some similarity they share with humans (e.g., rationality, interests, moral agency, sentiency, right-holder status) *to* "a highly contextual account to see clearly what a human being is and what the nonhuman world might be, morally speaking, *for* human beings."[32] For an ecofeminist, *how* a moral agent is in relationship to another becomes of central significance, not simply *that* a moral agent is a moral agent or is bound by rights, duties, virtue, or utility to act in a certain way.

Third, ecofeminism is structurally pluralistic in that it presupposes and maintains difference—difference among humans as well as between humans and at least some elements of nonhuman nature. Thus, while ecofeminism denies the "nature/culture" split, it affirms that humans are both members of an ecological community (in some respects) and different from it (in other respects). Ecofeminism's attention to relationships and community is not, therefore, an erasure of difference but a respectful acknowledgement of it.

Fourth, ecofeminism reconceives theory as theory in process. It focuses on patterns of meaning which emerge, for instance, from the storytelling and first-person narratives of women (and others) who deplore the twin dominations of women and nature. The use of narrative is one way to ensure that the content of the ethic—the pattern of the quilt—may/will change over time, as the historical and material realities of women's lives change and as more is learned about women-nature connections and the destruction of the nonhuman world.[33]

Fifth, ecofeminism is inclusivist. It emerges from the voices of women who experience the harmful domination of nature and the way that domination is tied to their domination as women. It emerges from listening to the voices of indigenous peoples such as Native Americans who have been dislocated from

---

[31] For example, in relationships involving contracts or promises, those relationships might be correctly described as that of moral agent to rights holders. In relationships involving mere property, those relationships might be correctly described as that of moral agent to objects having only instrumental value, "relationships of instrumentality." In comments on an earlier draft of this paper, West suggested that possessive individualism, for instance, might be recast in such a way that an individual is defined by his or her property relationships.

[32] Cheney, "Eco-Feminism and Deep Ecology," p. 144.

[33] One might object that such permission for change opens the door for environmental exploitation. This is not the case. An ecofeminist ethic is anti-naturist. Hence, the unjust domination and exploitation of nature is a "boundary condition" of the ethic; no such actions are sanctioned or justified on ecofeminist grounds. What it *does* leave open is some leeway about what counts as domination and exploitation. This, I think, is a strength of the ethic, not a weakness, since it acknowledges that *that* issue cannot be resolved in any practical way in the abstract, independent of a historical and social context.

their land and have witnessed the attendant undermining of such values as appropriate reciprocity, sharing, and kinship that characterize traditional Indian culture. It emerges from listening to voices of those who, like Nathan Hare, critique traditional approaches to environmental ethics as white and bourgeois, and as failing to address issues of "black ecology" and the "ecology" of the inner city and urban spaces.[34] It also emerges out of the voices of Chipko women who see the destruction of "earth, soil, and water" as intimately connected with their own inability to survive economically.[35] With its emphasis on inclusivity and difference, ecofeminism provides a framework for recognizing that what counts as ecology and what counts as appropriate conduct toward both human and nonhuman environments is largely a matter of context.

Sixth, as a feminism, ecofeminism makes no attempt to provide an "objective" point of view. It is a social ecology. It recognizes the twin dominations of women and nature as social problems rooted both in very concrete, historical, socioeconomic circumstances and in oppressive patriarchal conceptual frameworks which maintain and sanction these circumstances.

Seventh, ecofeminism makes a central place for values of care, love, friendship, trust, and appropriate reciprocity—values that presuppose that our relationships to others are central to our understanding of who we are.[36] It thereby gives voice to the sensitivity that in climbing a mountain, one is doing something in relationship with an "other," an "other" whom one can come to care about and treat respectfully.

Lastly, an ecofeminist ethic involves a reconception of what it means to be human, and in what human ethical behavior consists. Ecofeminism denies abstract individualism. Humans are who we are in large part by virtue of the historical and social contexts and the relationships we are in, including our relationships with nonhuman nature. Relationships are not something extrinsic to who we are, not an "add on" feature of human nature; they play an essential role in shaping what it is to be human. Relationships of humans to the nonhuman environment are, in part, constitutive of what it is to be a human.

By making visible the interconnections among the dominations of women and nature, ecofeminism shows that both are feminist issues and that explicit acknowledgement of both is vital to any responsible environmental ethic. Feminism *must* embrace ecological feminism if it is to end the domination of women because the domination of women is tied conceptually and historically to the domination of nature.

---

[34] Nathan Hare, "Black Ecology," in *Environmental Ethics*, ed. K. S. Shrader-Frechette (Pacific Grove, Calif.: Boxwood Press, 1981), pp. 229–36.

[35] For an ecofeminist discussion of the Chipko movement, see my "Toward an Ecofeminist Ethic," and Shiva's *Staying Alive*.

[36] See Cheney, "Eco-Feminism and Deep Ecology," p. 122.

A responsible environmental ethic also *must* embrace feminism. Otherwise, even the seemingly most revolutionary, liberational, and holistic ecological ethic will fail to take seriously the interconnected dominations of nature and women that are so much a part of the historical legacy and conceptual framework that sanctions the exploitation of nonhuman nature. Failure to make visible these interconnected, twin dominations results in an inaccurate account of how it is that nature has been and continues to be dominated and exploited and produces an environmental ethic that lacks the depth necessary to be truly *inclusive* of the realities of persons who at least in dominant Western culture have been intimately tied with that exploitation, viz., women. Whatever else can be said in favor of such holistic ethics, a failure to make visible ecofeminist insights into the common denominators of the twin oppressions of women and nature is to perpetuate, rather than overcome, the source of that oppression.

This last point deserves further attention. It may be objected that as long as the end result is "the same"—the development of an environmental ethic which does not emerge out of or reinforce an oppressive conceptual framework—it does not matter whether that ethic (or the ethic endorsed in getting there) is feminist or not. Hence, it simply is *not* the case that any adequate environmental ethic must be feminist. My argument, in contrast, has been that it *does* matter, and for three important reasons. First, there is the scholarly issue of accurately representing historical reality, and that, ecofeminists claim, requires acknowledging the historical feminization of nature and naturalization of women as part of the exploitation of nature. Second, I have shown that the conceptual connections between the domination of women and the domination of nature are located in an oppressive and, at least in Western societies, patriarchal conceptual framework characterized by a logic of domination. Thus, I have shown that failure to notice the nature of this connection leaves at best an incomplete, inaccurate, and partial account of what is required of a conceptually adequate environmental ethic. An ethic which *does not* acknowledge this is simply *not* the same as one that does, whatever else the similarities between them. Third, the claim that, in contemporary culture, one can have an adequate environmental ethic which is *not* feminist assumes that, in contemporary culture, the label *feminist* does not add anything crucial to the nature or description of environmental ethics. I have shown that at least in contemporary culture this is false, for the word *feminist* currently helps to clarify just *how* the domination of nature is conceptually linked to patriarchy and, hence, how the liberation of nature, is conceptually linked to the termination of patriarchy. Thus, because it has critical bite in contemporary culture, it serves as an important reminder that in contemporary sex-gendered, raced, classed, and naturist culture, an unlabeled position functions as a privileged and "unmarked" position. That is, without the addition of the word *feminist*, one presents environmental ethics as if it has no bias, including male-gender bias, which is just what ecofeminists deny: failure to notice the connections between the twin oppressions of women and nature *is* male-gender bias.

One of the goals of feminism is the eradication of all oppressive sex-gender (and related race, class, age, affectional preference) categories and the creation of a world in which *difference does not breed domination*—say, the world of 4001. If in 4001 an "adequate environmental ethic" is a "feminist environmental ethic," the word *feminist* may then be redundant and unnecessary. However, this is *not* 4001, and in terms of the current historical and conceptual reality the dominations of nature and of women are intimately connected. Failure to notice or make visible that connection in 1990 perpetuates the mistaken (and privileged) view that "environmental ethics" is *not* a feminist issue, and that *feminist* adds nothing to environmental ethics.[37]

CONCLUSION

I have argued in this paper that ecofeminism provides a framework for a distinctively feminist and environmental ethic. Ecofeminism grows out of the felt and theorized about connections between the domination of women and the domination of nature. As a contextualist ethic, ecofeminism refocuses environmental ethics on what nature might mean, morally speaking, *for* humans, and on how the relational attitudes of humans to others—humans as well as nonhumans—sculpt both what it is to be human and the nature and ground of human responsibilities to the nonhuman environment. Part of what this refocusing does is to take seriously the voices of women and other oppressed persons in the construction of that ethic.

A Sioux elder once told me a story about his son. He sent his seven-year-old son to live with the child's grandparents on a Sioux reservation so that he could "learn the Indian ways." Part of what the grandparents taught the son was how to hunt the four leggeds of the forest. As I heard the story, the boy was taught, "to shoot your four-legged brother in his hind area, slowing it down but not killing it. Then, take the four legged's head in your hands, and look into his eyes. The eyes are where all the suffering is. Look into your brother's eyes and feel his pain. Then, take your knife and cut the four-legged under his chin, here, on his neck, so that he dies quickly. And as you do, ask your brother, the four-legged, for

---

[37] I offer the same sort of reply to critics of ecofeminism such as Warwick Fox who suggest that for the sort of ecofeminism I defend, the word *feminist* does not add anything significant to environmental ethics and, consequently, that an ecofeminist like myself might as well call herself a deep ecologist. He asks: "Why doesn't she just call it [i.e., Warren's vision of a transformative feminism] deep ecology? Why specifically attach the label *feminist* to it . . .?" (Warwick Fox, "The Deep Ecology-Ecofeminism Debate and Its Parallels," *Environmental Ethics* 11, no. 1 [1989]: 14, n. 22). Whatever the important similarities between deep ecology and ecofeminism (or, specifically, my version of ecofeminism)—and, indeed, there are many—it is precisely my point here that the word *feminist* does add something significant to the conception of environmental ethics, and that any environmental ethic (including deep ecology) that fails to make explicit the different kinds of interconnections among the domination of nature and the domination of women will be, from a feminist (and ecofeminist) perspective such as mine, inadequate.

forgiveness for what you do. Offer also a prayer of thanks to your four-legged kin for offering his body to you just now, when you need food to eat and clothing to wear. And promise the four-legged that you will put yourself back into the earth when you die, to become nourishment for the earth, and for the sister flowers, and for the brother deer. It is appropriate that you should offer this blessing for the four-legged and, in due time, reciprocate in turn with your body in this way, as the four-legged gives life to you for your survival." As I reflect upon that story, I am struck by the power of the environmental ethic that grows out of and takes seriously narrative, context, and such values and relational attitudes as care, loving perception, and appropriate reciprocity, and doing what is appropriate in a given situation—however that notion of appropriateness eventually gets filled out. I am also struck by what one is able to see, once one begins to explore some of the historical and conceptual connections between the dominations of women and of nature. A *re-conceiving* and *re-visioning* of both feminism and environmental ethics, is, I think, the power and promise of ecofeminism.

[19]

The Ecofeminism/Deep Ecology Debate: A Reply to Patriarchal Reason

Ariel Salleh *

> I discuss conceptual confusions shared by deep ecologists over such questions as gender, essentialism, normative dualism, and eco-centrism. I conclude that deep ecologists have failed to grasp both the epistemological challenge offered by ecofeminism and the practical labor involved in bringing about social change. While convergencies between deep ecology and ecofeminism promise to be fruitful, these are celebrated in false consciousness, unless remedial work is done.

I. ECOFEMINISM AND ENVIRONMENTAL ETHICS

THE AUTONOMY OF ECOFEMINISM

Environmental ethics is an emerging field in philosophy that examines "man's relation to nature," as distinct from socialism, which treats "man's relation to man," and feminism, which takes up the question of "man's relation to woman." Ecofeminism, like Green politics—at least in its ideal form, expresses a synthesis of all three concerns. Within the field of environmental ethics itself, several alternative paradigms coexist: the mystical wilderness tradition of Thoreau; the ecological stance of Naess and others with its increasingly psychological emphasis; an aesthetic, playful, humanist tendency manifest in Goodman; the agrarian approach of Mollison; Singer's legalistic ethical extensionism; and Ehrlich's managerial focus on resource conservation. These various formulations range roughly across a continuum of human attitudes toward the natural world from "soft" to "hard," from "let it be" to "let's get the most out of it." Accordingly, a number of them have been crossed in debate: such deep ecologists as Devall have criticized resource conservationists for the shallowness of their environmentalism and such ethical extensionists as Regan have accused ecological holists, for

---

* P.O. Box 133, Glebe, New South Wales 2037, Australia. Salleh is an ecofeminist activist and theorist. She writes on convergencies and contradictions between socialism, feminism, and ecology and is currently completing a book about ecofeminism and Green politics. She is a convener of the Women's Environmental Education Centre in Sydney and is an occasional visiting scholar in the Environmental Conservation Education Program at New York University.

example, Callicott, of ecocentric fascism! On the other hand, there is not always a clear distinction between one ethical position and another. The agrarian approach as articulated by Berry has mystical elements, while some rights theorists, for example, Stone, are just as committed to seeing a paradigm shift away from anthropocentrism as their deep ecology critics are.[1]

Contrary to the assumption of many environmental ethicists, ecofeminism did not evolve out of these philosophical inquiries, which for the most part reflect the endeavors of academic men and middle-class nature lovers. Even so, although ecofeminism cannot be subsumed by environmental ethics, tendencies within the feminist tradition can be found that parallel the various eco-ethical standpoints. Those who take pleasure in mystical identification with a re-enchanted nature, for example, wilderness environmentalists and deep ecologists, are matched by such ecofeminists from the U.S. West Coast as Starhawk or Deena Metzger, who celebrate the cycles of nature and the communion of women's bodies with these changes, and weave powerful spiritually affirmative feminist rituals around this link to the source of our being. In England and West Germany, through the work of Monica Sjoo, pagan nature worship and the lost wisdom of "witches" are being rediscovered. A parallel to the artistic and agrarian ethic appears among Canadian feminists, associated with the journal *Women and Environments*, who design domestic and city spaces for mothers with small children, the disabled, and the aged. Practitioners of organic farming include Lea Harrison in Australia and Margrit Kennedy in Germany. Women bioregionalists and communitarian socialists also converge on an agrarian ethic. The ethical extensionists find their equivalent among feminist animal liberation and wilderness preservation activists. Nevertheless, most ecofeminists have moved away from arguments about rights toward a radical-feminist-inspired depth analysis of interspecies exploitation. Andree Collard and Connie Salamone in the U.S. are notable instances. Meanwhile, resource conservationists can be found within feminism as well. In Kenya, Wangari Maathai and women of the Greenbelt Movement recently received an Alternative Nobel Prize for their work in reforesting the marginal areas of their farm lands. In the industrial nations, women interested in resource conservation per se tend to be liberal feminists, who recognize no particular connection between this and women's immediate political concerns. Like urban based Marxist feminists, they tend to envisage environmental questions as management problems with technological solutions.

The major difference between ecofeminism and the field of environmental ethics is that none of the latter's paradigms succeed in integrating a social analysis. For this reason, while many Greens and eco-political activists are now interested in the highly popularized environmental ethics position called deep ecology, at the

---

[1] Roderick Nash's *The Rights of Nature* (Madison: University of Wisconsin Press, 1989) is a good introduction to the field.

same time, they are encountering a number of trenchant critiques of the stand. As far as I am aware, the debate between ecofeminism and deep ecology began with a paper I presented at a conference in Australia in 1983, "Deeper than Deep Ecology: The Eco-Feminist Connection," which was published in *Environmental Ethics* a year later.[2] Nevertheless, feminist and ecofeminist difficulties with deep ecology continue to be uncovered as essays by Janet Biehl, Ynestra King, and Sharon Doubiago demonstrate. In addition to the reservations expressed by women, left-wing radicals as theoretically diverse as Tim Luke, Murray Bookchin, and George Bradford also regularly contest deep ecology's philosophical coherence and political adequacy.[3] One thing is sure: the passion that deep ecology arouses testifies to the fact that it has hit a significant nerve in modern Western societies—and, of course, the same can be said for ecofeminism.

A DIALECTICAL PROCESS

Ecofeminism has a specific history of its own, shaped by the day-to-day efforts of ordinary women to survive with their families. In highlighting the ecological dimension and drawing on the grass-roots experiences of women in both developed and so-called developing countries, ecofeminism opens up the feminist movement itself to a new cluster of problems and challenges urban-based theoretical paradigms—liberal, Marxist, radical, post-structuralist—that have dominated feminist politics over the last two decades. By pitting new empirical concerns against established feminist analyses, ecofeminism is encouraging a new synthesis in feminist political thought. In seeking a review of "man's relation to nature," ecofeminism certainly addresses the same project as environmental ethics. As a feminism, however, ecofeminism takes on its project in a compound sense, since it simultaneously calls for a review of "man's relation to woman" as it goes along. Unlike environmental ethics in general, and deep ecology in particular, ecofeminism does not go after its object with a simple linear critique. It is obliged to engage in a zig-zag dialectical course between (a) its feminist task of establishing the right of women to a political voice; (b) its ecofeminist task of undermining the patriarchal basis of that political validation by dismantling the

---

[2] Ariel Salleh, "Deeper that Deep Ecology: The Eco-Feminist Connection," *Environmental Ethics* 6 (1984): 335-41. The critique addresses two texts: Arne Naess, "The Shallow and the Deep. Long Range Ecology Movement," *Inquiry* 16 (1973): 95-100, and Bill Devall, "The Deep Ecology Movement," *Natural Resources Journal* 20 (1980): 299-322.

[3] See Janet Biehl, "It's Deep But is It Broad?" *Kick it Over*, Winter 1987, pp. 2A-4A; Ynestra King, "What is Eco-Feminism?" *The Nation*, 12 December 1987, pp. 702, 730-731; Sharon Doubiago, "Mama Coyote Talks to the Boys," in J. Plant, ed., *Healing the Wounds* (Philadelphia: New Society Publishers, 1989); Murray Bookchin, "Social Ecology versus Deep Ecology," *Socialist Review* 18 (1988); Tim Luke, "The Dreams of Deep Ecology," *Telos* 76 (1988): 65-92; George Bradford, *How Deep is Deep Ecology?* (Ojal, Calif.: Times Change Press, 1989).

patriarchal relation of man to nature; and (c) its ecological task of demonstrating how women have been able to live differently in relation to nature.

The need for a multilevered and reflexive epistemological stance is not often recognized by men, for whom patriarchal social reality is a relatively straightforward affair, in large part because the underbelly of their social life is held together by psychosocial maintenance work that women of most races are socialized to do. Women are thus already sensitized to meta-levels of communication as a result of the "master-slave dialectic" that they find built into patriarchy.[4] Furthermore, as women begin forging new cultural meanings of their own and such shared realities as feminist understanding, many men find themselves left behind—which can be a very disturbing experience. To a large extent, some of the category mistakes and misrepresentation of ecofeminism made by deep ecologists are a result of their attempts to cope with this dialectical process. The deep ecologists and their comrades at arms, Earth First!, were quick to come to the aid of their environmental ethic. Unfortunately, many of their replies have been mere reassertions of the standpoint or attempts to "shoot the (feminist) messenger," rather than efforts to assimilate the criticism and nurture a shared feminist ecological perspective. Hopefully, the following review of deep ecological responses to ecofeminism, based largely on material published in *Environmental Ethics* since 1986, will help locate the obstacles to that common political understanding. Broadly, there are three kinds of difficulties: *philosophical* points at issue: the *sociopolitical* grounding of deep ecology itself; and the *psychosexual dynamic* which runs through the exchange. Although these problems are interlocked and each works to reinforce the others, in the present paper I focus only on the first of these areas.[5]

Talk about examining the "relative merits" of deep ecology and ecofeminism, a competitive either/or approach common to several defenders of deep ecology, misses the intent of the original ecofeminist criticism, which was not to dismiss deep ecology, but to urge it to sharpen up its political awareness. As I affirmed right at the beginning of "Deeper than Deep Ecology":

> ... deep ecology is already an attempt to transcend the shortsighted instrumental pragmatism of the resource-management approach to the environmental crisis. It argues for a new metaphysics and an ethic based on recognition of the intrinsic worth of the nonhuman world. It abandons the hardheaded scientific approach to reality in favor of a more spiritual consciousness. It asks for voluntary simplicity in living and a nonexploitative steady-state economy. *The appropriateness of these attitudes as*

---

[4] My position, temporarily and strategically privileging women's standpoint, is similar to Sandra Harding's formulation in *The Science Question in Feminism* (Ithaca: Cornell University Press, 1986). See also the earlier epistemological essay, "On the Dialectics of Signifying Practice," *Thesis Eleven* 5-6 (1982): 72-84, where I discuss the role of suffering as a phenomenological lever behind political perception.

[5] The other two difficulties will be covered in Ariel Salleh, "Class, Race, and Gender Discourse in the Ecofeminism/Deep Ecology Debate," *Environmental Ethics*, forthcoming.

expressed in Naess and Devall's seminal papers on the deep ecology movement is indisputable.[6]

For some reason, none of the respondents to that ecofeminist position paper has acknowledged this endorsement of the deep ecological project. Each proceeds "as if" I had dispensed with deep ecology itself, by indicating where it falls short from a viewpoint based on women's experiences. Needless to say, this continuing oversight has been destructive of future unity between deep ecology and ecofeminist aims. At least, the need for some kind of synthesis has now been acknowledged by deep ecologists—and that is an achievement. But there is still a lot of flailing around over what can only be called "scholastics."

It is possible that the wry title phrase "Deeper that Deep Ecology" has been felt as a threat by deep ecologists, particularly, among them, men not used to having their ideas tested by women. If so, then it negates the reason for the paper in the first place. If we are to forge a politics based on a radically new appreciation of the potentials of all beings, then men's openness to the views of women is an essential part of the program. Women and men are at a point in history when each is learning to find and use the parts of themselves previously suppressed by patriarchal lore. The same rediscovery is a major facet of deep ecology, and again, one acknowledged by my own statement: "This is the self-estranged male reaching for the original androgynous natural unity within himself." Conversely, women are learning to express themselves in public with a new confidence and assertiveness. As committed radicals who are working on themselves, deep ecological men should ideally be able to accept women speaking out, and relate to them appropriately as equals. Women should not have to continue pandering to men's need for authority, as they have been trained to do under patriarchy, by tiptoeing around and dressing up their objections to what men do in euphemisms that safeguard the masculine ego. The recognition of this point is a crucial part of any political work in the twentieth century. Hence, when a male friend read the newly finished "Deeper than Deep Ecology" manuscript and suggested that I change the title to "Deepening Deep Ecology" as a "strategic move," I thought it over and decided not to. First, it seemed to me that to do so would mean accepting the traditional pattern of protecting men from something they might not want to hear from a woman, a reactionary move. Second, his proposed title implied that ecofeminism was not a discrete politics "in its own right," but rather a contribution to a basically male defined environmental ethic. Third, by undertaking the task of "deepening" deep ecology, ecofeminists would simply have slipped back into the role of doing men's theoretical housework for them, tidying up their concepts— unrecognized, unacknowledged, just as mother's work was. Ironically, the fact that a meta-discursive preamble to this paper is called for suggests that women are

---

[6] Salleh, "Deeper than Deep Ecology," p. 339 (emphasis added).

still required to attend to the fabric of social relations that sustains patriarchal discourse, including the academic sort.

II. UNFAMILIARITY WITH FEMINIST HISTORY AND THEORY

Liberal, Marxist, Radical, Post-Structural Paradigms

Looking at the *philosophical* points of argument pursued by deep ecologists, it is plain that each man comes to accept the reality of patriarchy in his own time. In this respect, the contrast between Alan Wittbecker's reactive outrage toward ecofeminist criticism in "Deep Anthropology, Ecology and Human Order" in 1986 and Michael Zimmerman's more cautious response in "Feminism, Deep Ecology, and Environmental Ethics" in 1987 is instructive. Zimmerman does make an effort to acquaint himself with the feminist literature and has a veritable think tank of women advisers. The bulk of his long essay, in fact, is quite a useful summary of women's writing on ethics and epistemology. Jim Cheney's piece, "Eco-Feminism and Deep Ecology," in the next issue goes even further. Sympathetic to the synthesis of feminism and ecology, Cheney leaves behind the patriarchal mindset altogether and works creatively with feminist theory. Don Davis' "Ecosophy: The Seduction of Sophia" in 1986 is in a similar vein. Both Davis and Cheney agree that deep ecology responds to criticism by simply attaching ecofeminist insights to a basically masculine ethical orientation. However, Warwick Fox's extended tract in 1989, "The Deep Ecology-Ecofeminism Debate and its Parallels," is, like Zimmerman's, well out of its depth concerning the shifts in consciousness that mark paradigms within feminism—liberal, Marxist, radical, and ecofeminist, in this case. Clearly, the finer points of feminist political thought are not something academics or environmentalists necessarily know about. For that reason, though, some intellectual modesty might have protected deep ecologists here.[7]

Feminism's earliest liberal phase has been and still is concerned mainly with equality for women in a system designed by men—access to educational opportunities and jobs, without discrimination and harassment. Part of the drive for equality, as much of contemporary society understands it, is to ensure that women's reproductive activities do not impede their progress alongside male peers. Accordingly, availability of contraception and the right to abortion are companion political issues in this struggle. The objectives of Marxist feminists have been similar to those following the liberal agenda, but rest on such structural changes as the full-scale entry of women into the waged sector and the socializa-

---

[7] Alan Wittbecker, "Deep Anthropology, Ecology, and Human Order," *Environmental Ethics* 8 (1986): 268-70; Donald Davis, "The Seduction of Sophia," *Environmental Ethics* 8 (1986): 151-62; Michael Zimmerman, "Feminism, Deep Ecology, and Environmental Ethics," *Environmental Ethics* 9 (1987): 21-44; Jim Cheney, "Eco-Feminism and Deep Ecology," *Environmental Ethics* 9 (1987): 115-45; Warwick Fox, "The Deep Ecology-Ecofeminism Debate and its Parallels," *Environmental Ethics* 11 (1989): 5-25.

tion of domestic functions. The radical feminist consciousness of women's "difference," and its later convergence with post-structuralism, has broadened the agenda yet again, introducing a substantial challenge to the patriarchal terms of reference on which equality itself has been sought. More recently, an international movement of *ecofeminists* is applying this politics of difference in a full-scale analysis of the global crisis.

Although Zimmerman gives footnote acknowledgment to the fact that the terms *ecofeminism* and *feminism* cannot necessarily be used interchangeably, he forgets this point in constructing his deep ecological reply to ecofeminism. Arguments grounded in liberal or Marxist feminism, for example, cannot be placed against ecofeminism in order to demonstrate the latter's internal inconsistency. Trying to do so is like using arguments from Burke against J. S. Mill in order to prove that Western political thought is contradictory. Ecofeminist theorists in the 1980s worked through the liberal, social, and radical arguments and have digested these concerns in the light of women's lived experiences in a way that is much broader than liberal or Marxist feminism alone. Drawing together the insights of earlier analyses, ecofeminism now moves on to include an ecological sensibility. Despite Karen Warren's helpful exposition along these lines, neither Zimmerman nor Cheney nor Fox appear to be entirely clear about the distinction between liberal feminism and ecofeminism.[8] Broadly, it can be said that the former focuses on the distribution of power and resources in society, while the latter involves both cultural and structural revolution and spiritual search. Paraphrasing the words of Australian "re-sister" Becca Milier, we do not want a piece of the patriarchal pie: we want to bake a new pie altogether. Fox shows uncertainty about Marxist, radical, and ecofeminist paradigms as well, when he tries to chastise ecofeminists by quoting Warren to the effect that the radical feminists who have influenced them have paid "too little attention to the historical and material features of women's oppression." In fact, Fox gets it upside down. Warren's text actually reads: ". . . there are noteworthy worries about radical feminism from an ecofeminist perspective. First, since radical feminism generally pays too little attention to the historical and material features of women's oppression. . . ."[9] What can be said is that ecofeminists in the United States, along with Greens and environmentalists there, have paid relatively little attention to historical and material forces. This lack of attention reflects the general suppression of Marxist scholarship and labor history in that society. The same observation does not apply to ecofeminism internationally, though, and a perusal of the European, Asian, and Australian literature makes this point evident.[10]

---

[8] Karen Warren, "Feminism and Ecology: Making Connections," *Environmental Ethics* 9 (1987): 3-20. Although this article is a very useful, Warren's synchronic philosophical analysis has the effect of setting up static categorical boundaries between paradigms. Actually, a continual process of learning and revisioning has gone on among women within the feminist movement.

[9] Fox, "The Deep Ecology-Ecofeminism Debate," p. 14, and Warren, "Feminism and Ecology," p. 15.

[10] Maria Mies, *Patriarchy and Accumulation* (London: Zed, 1987); Vanadana Shiva, *Staying*

The women of many racial backgrounds who are engaged in ecofeminist activities, may not themselves identify with the tradition of feminist ideas produced by educated middle-class Westerners over the past two decades. Nevertheless, they will have their own sense of the underlying patriarchal power at work behind violence against nature and the degradation of women. The Shibokusa grannies of Japan and the Roman Catholic housewives coping with industrial toxins in Seveso, Italy are cases in point. In fact, peace activism and environmental struggle may be a first step toward developing an interest in feminism as an ideology per se. Similarly, in the so-called advanced nations, where women domestic consumers, according to Zimmerman and Fox, are accomplices in environmental exploitation and reap "the advantages" of it, many women become ecofeminist activists without having been feminists first. Some even move on to an understanding of how they have been manipulated into consumerism by a capitalist patriarchy (hence, the organic, grass-roots nature of ecofeminist politics). At the same time, intellectually inclined women have been using the feminist tradition to weave an interpretative literature around woman-nature links. Consider the more strictly feminist contributions of say, Susan Griffin in the U.S., Hilkka Pietila in Finland, or Giovanna Merola in Venezuela. Some ecofeminism, then, is contiguous with radical or socialist paradigms, but by going back to women's lived experiences in a time of global crisis, it brings fresh understandings to these movement ideologies.

Now the question arises: why use the word *ecofeminism* at all? Fox certainly believes it is redundant, given what he sees as the broader shoulder of deep ecology. Of course, from the perspective of historians of ideas, sociologists of knowledge and social movements, the term usefully situates a very particular direction in feminist politics. Marking the spontaneous appearance of a new consciousness among women, it is remarkable how it cropped up in several places around the globe during the 1970s. The term *ecofeminism* is a logical combination, integrating and transcending both feminist and environmental concerns alike.

DIFFERENCE AND ASYMMETRY

Consistent with the patriarchal subsumption of women's labor and ideas, deep ecologists miss the point when they propose women simply call ecofeminism "deep ecology." There is an urgent feminist political moment embodied in this little word: the need for lessons from a different cultural experience to be aired, listened to, taken seriously, and acted upon. This difference becomes even plainer as the dialogue between deep ecology and ecofeminism goes on. The number of

---

Alive: Women, Ecology and Development in India (London: Zed, 1988); Gail Omvedt et al., *Women and Struggle* (New Delhi: Kali for Women, 1988); Ariel Salleh, "Eco-Socialism/Ecofeminism," *Capitalism, Nature, Socialism* 2 (1990): 129-34; Mary Mellor, "Eco-Feminism and Eco-Socialism: Dilemmas of Essentialism and Materialism," *Capitalism, Nature, Socialism* 3 (1992): 43-62.

deep ecologists who have offered to join women in advancing an ecofeminist sensibility are few. My experience of men in workshops with the U.S. Green movement is more promising. Many see the logic of our analysis, appreciate how it fits with their own emancipation as bearers of patriarchy, and look forward to working with women to restore the balance. Such Green political figures as Per Gharton, leader of the Grona in the Swedish parliament, have even publicly endorsed a period of matriarchy to redress the horrors of the patriarchal millennium. Nevertheless, although he means well, Gharton's interpretation of ecofeminism is itself patriarchal. Feminism has never been about gaining power over the rest of society, while ecofeminism, specifically, is about a transvaluation of values, such that the repressed feminine, nurturant side of our culture can be woven into all social institutions and practices. Zimmerman observes that the ecofeminist position and deep ecology are at least superficially in agreement over their opposition to a rights-based reform environmentalism. However, again his text slips into using "feminists" here, which nullifies his argument because, although liberal feminists do endorse a rights-based politics, they would be loath to recognize the intrinsic value of nature. Moreover, in the light of the deeper radicalism which an ecofeminist analysis using the theory of difference offers, deep ecology itself looks shallow and reformist.

Zimmerman's "Feminism, Deep Ecology, and Environmental Ethics" starts out with a fair synopsis of my criticism of deep ecology, apart from a couple of points. Keeping in mind that the sentence and quote marks are his, he writes, "Male deep ecologists should consult women who are more in tune with natural world than men."[11] This paraphrase disregards the historical process at work here, namely, that it is patriarchal domination that puts women close to nature, while men are seen to be active in the sphere of culture. This process causes women's experiences and identities to be linguistically mediated by reference to nature. Not only is the feminine psyche constructed differently by this means, but the work roles that women are assigned also revolve around nature, "putting the dirt back where it should be." These roles, in turn, reinforce women's hands-on knowledge of natural processes. Zimmerman's "critical observations" expose a very limited understanding of these ideological dynamics, although this failing is not surprising, given the liberal feminist analysis that he so often relies on. Hence, he writes, "After having gone through the phase of seeking to dissolve differences between men and women, many feminists began to affirm those differences—and to conclude that woman is better than man." This cryptic treatment of feminist theory does little to open up understanding of the specificities of women's experience for men of good faith who want to understand them. Even though Zimmerman's text acknowledges the historicity of feminism by noting its "phases," his account fails to amplify the tragic fact that the sameness women initially sought alongside men as liberal feminists, and the radical difference they later asserted, are both

---

[11] Zimmerman, "Feminism, Deep Ecology, and Environmental Ethics," p. 38.

conceptually tied to patriarchal logic. If there is any doubt about the oppressive misogynist web that we are dealing with, remember that the very origin of the word *feminine* itself is *feminus*, a Latin word meaning "without faith."

Having set up a straw ecofeminist argument about "difference" meaning that women are "better than men," Zimmerman pits the feminist argument over essentialism against it.[12] Unfortunately, student of Heidegger though he is, Zimmerman has his own proclivity to essentialism. Mixing up sex and gender categories, he asks, "Would authentic female experience be formed by a feminist culture? And what then would happen to authentic male experience?" Here his thinking gets caught in the old dualist—either/or—grid. Feminism at large does not aim to be a blueprint for some purist matriarchal dictatorship—the mirror image of patriarchy. Feminism is a catalyst in the ongoing development of human self-consciousness. Ecofeminists are now waiting for men to take the corresponding next step in their emancipation from patriarchy so that together we can "negotiate" a fair and human "contract" with "nature," as it were.[13] Zimmerman is mistaken in supposing that patriarchal culture is just an "interpretative framework" like feminism. The problem is that he has not understood patriarchy as a system of power relations. *Patriarchy* does not simply exist as an idea; rather, the term stands for a solid set of oppressive facts. Recall, for instance, the International Labor Organization statistics showing that women are fifty percent of the world's population, do sixty-five percent of the world's work, get less than ten percent of all wages paid, and own less that one percent of all property. Again, the International Women's Tribune Center in New York reports that because Islam forbids the execution of virgins, women activists in Iran are raped first and then executed. These may be disturbing items of information to educated Westerners, but for millions of women, East and West, daily life under patriarchy is hell. Incredulously, Zimmerman can ask: "Does feminism pretend to provide a non-distorted, impartial way of interpreting experience?"

Under patriarchal culture, the program of repression that has treated women and colored peoples as resources, from the beginning of recorded history, has also been the ideology that plunders nature. This association of women and minorities with nature means that if there is to be any chance of political change in attitudes toward the environment, there will have to be a shift in gendered and racial attitudes at the same time. Although it is encouraging to hear men such as the deep

---

[12] Ibid., p. 34. He is guided here by Hester Eisenstein and Alice Jardine, *The Future of Difference* (New Brunswick: Rutgers University Press, 1985). Jardine's more recent attitude toward difference is more sympathetic, in line with new feminist explorations of the deconstructive potential of stances dismissed earlier as essentialist. A recent paper by Zimmerman in Irene Diamond and Gloria Orenstein, eds., *Reweaving the World* (San Francisco: Sierra Club, 1990) repeats these confusions and ambivalences.

[13] I am tempted to use the more spiritual word *covenant* here, but both it and *contract* have patriarchal baggage attached. At least *contract* conveys a sense of *the process* of negotiation between women and men that we look forward to.

ecologists wanting to speak on behalf of nature, the deep ecology movement looks rather like a young man driving a car who shifts impatiently from first to fourth gear. All levels of oppression on the "Great Chain of Being"—speciesism, racism, sexism, classism—are interlinked and must be attended to. At this point in history, women, a global majority, both dominated and empowered, are well equipped to take up the case for "other" beings. Nevertheless, it is not a matter of "speaking for": men have always spoken for women and it has not helped much. Rather, it is a matter of unraveling the conceptual roots of an exploitative white male dominant multinational corporate system that continues to take the integrity of other life forms away.

III. CATEGORY MISTAKES AND CONCEPTUAL CONFUSIONS

SEX VERSUS GENDER, FEMINIST VERSUS MASCULINIST

Cheney, for one, is keenly aware of the pitfalls of confusing sex and gender terms. As he notes, *feminine* and *masculine* are gender categories and are culturally defined in such a way that they cannot be expected to overlap in any thoroughly systematic way with biological sex. Further, as those trained in the social sciences recognize, neither are these categories historically or culturally invariant. Wittbecker, a forest conservationist and computer consultant, is less successful in dealing with these issues. His un-self-conscious text still refers to "Man," even while our government directives, publishers, and professional associations promulgate new guidelines for non-sexist language. His use of the term *androgyny* also shows no familiarity with the controversy over the concept in the last decade of feminist literature. Regarding ecofeminism, Wittbecker has three axes to grind—that deep ecology is said to be "ignorant of feminism," that it is accused of being "non-feminine," and that ecofeminism is a "reductionist femocentrism." While the first point may yet be substantiated, it is doubtful that the other two will be. Wittbecker not only overlooks my acknowledgment in "Deeper than Deep Ecology" that deep ecologists are genuinely reaching for the feminine, but he also fails to note my objection that they reappropriate the feminine in an unexamined and politically naive way. There is no hint in his discussion that the feminine role constellation may already contain a deep ecological sense. The deep ecologists' lack of insight into these matters may be connected to the psychosexual context in which the patriarchal ego forms. It is here that the *feminist*, as opposed to *feminine*, argument enters the picture. Had the original deep ecology formulations been more politically reflexive so as to acknowledge the different experiences of women and men, and the unique environmental potential of women's orientation, they would not have drawn criticism.

Wittbecker's lack of assimilation of ecofeminist theory is confirmed by his views on "reductionist femocentrism." As Elizabeth Dodson-Gray's "Great Chain of Being" shows, a fundamental cleavage in consciousness exists between ruling-class men, God's stewards, and "all others" in descending echelons of the hierarchy—white men over black men, white women over black women, and finally children, animals, plants, and rocks.[14] Significantly, it is easier for white men to acknowledge their exploitation of black men than to acknowledge exploitation of their own women. The psychosexual dynamic of racial "otherness" is not so bound up with the primal structure of masculine identity. Ecofeminism certainly recognizes the rigid patriarchal dualism between male and female as a key political problem, but in going after the pyschosexual drive on which all domination feeds, it also embraces children's and animal liberation, and caring for plants and rocks. Remember that the campaign against child sexual abuse was first put on the agenda by the women's movement. As politics, then, our perspective is not a "reductionist femocentrism," as Wittbecker complains, when he says that "Salleh limits the center to a feminine principle." Rather, the paradigm attempts to remedy the way in which women and others have been historically and discursively marginalized by a patriarchal center. In doing so, it makes a major contribution to the same "decentering" to which deep ecology claims it is committed.

Cheney, by contrast, is open to examining whether or not deep ecology embodies a feminist liberatory potential. He could, of course, have surveyed the political activities of men and women in the movement, including what has been until recently a rather macho and racist Earth First! Instead, however, he undertakes a penetrating analysis of how the deep ecological identification with the ecosphere is constituted. Sagely, Cheney clarifies the parameters of his own understanding of ecofeminism at the outset. It refers to ". . . a sensibility, an intimation, that feminist concerns run parallel to, are bound up with, or perhaps are one with concern for a natural world which has been subjected to much the same abuse and ambivalent behavior as have women."[15] However, although he is aware that there are different paradigmatic phases within feminism, his use of the phrase—"that feminist concerns run parallel to"—glosses over these differences, and could lead to the same misunderstanding that Zimmerman produces, especially when he unknowingly pulls out differences between various feminist paradigms in an attempt to illustrate that ecofeminism itself is internally inconsistent. Given the substantial growth of ecofeminism outside movement ideology, Cheney's definition would also be more appropriate if he had said "women's concerns" rather that "feminist concerns."

One further clarification is also needed. We have to make use of the words we have, while still trying to shift their political sense. Ecofeminists do not want to deny ontological continuity between the so-called "natural" and the "social" or historical spheres, as the prevailing liberal paradigm's nature-nurture cliché does.

---

[14] Elizabeth Dodson-Gray, *Green Paradise Lost* (Wellesley, Mass.: Roundtable, 1981).
[15] Cheney, "Eco-Feminism and Deep Ecology," p. 115.

Nevertheless, any route is vulnerable to the fact that current language is embedded in patriarchal assumptions about the superiority of culture over nature. Deep ecologists should be more attentive to this problem in their writing. In everyday talk, the word *female* is usually used to denote biological functions. If it is used to denote social attributes, the allusion may be a derogatory one. Similarly, use of the word *female* as a noun is invariably a dismissive sense because of the biological connotation. *Male*, as opposed to *masculine*, can also have an undertone of sexual hostility to it. Even so, there are times when it is unavoidably correct to use it. A discerning reader stands advised by the context of a piece of writing as to the intent behind a usage. The inadequacy of the current lexicon crops up again when Cheney writes that "concern for nature in the modern world can be described as a 'feminization' of masculine attitudes toward nature." There is no way of knowing whether he means feminization coming from *feminine* or feminization coming from *feminist*. Cheney's text compounds this ambiguity by using the word *feminist* in parallel to the adjective *masculine*. This semantic asymmetry—*masculine* properly pairs with *feminine* rather than *feminist*—simply reflects modern gender inequalities. *Feminist*, implying a rejection of patriarchal values and a reconstructed sensibility, has no parallel term in the asymmetrically gendered experience of men. Again, the term *masculinist* is incorrectly used by Zimmerman as an equivalent to *feminist;* this designation was devised to express the antagonism of men to the arrival of feminism, and is a renewal of patriarchal attitudes. I wonder how long it will take before some linguistic parity is arrived at?[16]

AUTHENTICITY AND ESSENTIALISM

If men and women alike are deformed by patriarchal social relations, how then, Zimmerman wonders, can ecofeminism claim to represent the "authentic" voice of women? Zimmerman assumes some pure archaic essence of "the feminine" in the ecofeminist stand. In other words, he projects an essentialist position onto the argument in "Deeper than Deep Ecology," ignoring my dialectical arguments on overdetermination and the deconstruction of masculine and feminine categories in "Contribution to the Critique of Political Epistemology" and other cited papers.[17] In fact, the many voices of women that we are asking deep ecologists,

---

[16] An excellent treatment of gender in patriarchal language is Dale Spender, *Man Made Language* (London: Routledge, 1980).

[17] Ariel Salleh, "Contribution to the Critique of Political Epistemology," *Thesis Eleven* 8 (1984): 23-43. For the discussion on essentialism, see pp. 30-33 especially, and for more on overdetermination, see "From Feminism to Ecology," *Social Alternatives* 4 (1984): 8-12. See also "Essentialism and Eco-Feminism," *Arena*, no. 94 (1991): 167-73. A more recent paper by Roger King, "Caring about Nature," *Hypatia* 6 (1991): 75-89, suffers from the same problems as Zimmerman's, as does Catherine Roach's insufficiently researched essay, "Caring for Your Mother," *Hypatia* 6 (1991): 46-59, which uses the label "nature feminism" for the "Deeper than Deep Ecology" position.

socialists, and others to pay attention to are precisely those that love and labor under patriarchy, the real empirical voices of living women now. As I wrote then, "... if women's lived experience were recognized as meaningful and were given legitimation in our culture, it could provide an immediate 'living' social basis for the alternative consciousness which the deep ecologist is trying to formulate and introduce as an abstract ethical construct."[18] Surely this majority is the minority tradition par excellence! The abstract hypothetical authenticity that Zimmerman as a male academic phenomenologist projects here is symptomatic of how thought is shaped in the context of a class, race, and gender stratified division of labor. Under capitalist patriarchy, all people, white or colored, men or women, are proscribed from knowing the full range of their own and each other's capacities. In a related vein, a literal-minded reader might also want to object to my phrase, "we women need to be allowed to love what we are," on the grounds that it naively suggests a "nature feminism" with an unproblematic essentialist notion of authenticity and an inadequate appreciation of gender construction. In my understanding of how ideological forces impact on identity formation, however, it is largely women's historically contrived, or workaday "second Nature" that has political relevance in today's crisis.

Zimmerman also objects to my use of the word *woman* as a shortcut for the litany of traditionally ascribed feminine characteristics. At this stage in history, what other word are we to use? Nevertheless, on this basis, he levels another charge of essentialism at my thesis: a "genetic doctrine" is his phrase. Wittbecker also falls back on this terminology. Again, where I write "women already ... flow with the system of nature," Fox adds the coda, "by their essential nature."[19] First, because the word *essence* is not used in my text, and second, because the debate over essentialism turns out to have been a spurious byproduct of the dualist thought frame of patriarchal liberalism, Fox's remarks are particularly off target. It is nonsense to assume that women are any closer to nature than men. The point is that women's reproductive labor and such patriarchally assigned work roles as cooking and cleaning bridge men and nature in a very obvious way, and one that is denigrated by patriarchal culture. Mining or engineering work similarly is a transaction with nature. The difference is that this work comes to be mediated by a language of domination that ideologically reinforces masculine identity as

---

[18] Salleh, "Deeper than Deep Ecology," p. 340. Compare my findings in "Environment: Consciousness and Action," *Journal of Environmental Education* 20 (1988): 26-31.

[19] See Zimmerman, "Feminism, Deep Ecology, and Environmental Ethics," p. 265; Fox, "The Deep Ecology-Ecofeminism Debate," p. 17. On the question of essentialism, it is interesting to compare the glib assertions of deep ecologist Dolores LaChappelle: "What came natural to me, as a woman, was to see that everyone was functioning at their peak level. ... This larger corpus collosum gives women the advantages of feeling connected to nature. ... Many women are paranoid for two years after the birth of each child. ... Such a person cannot make valid decisions for a group. ..." Dolores LaChappelle, "No, I'm Not an Eco-Feminist: A Few Words in Defence of Man," *Earth First!*, 21 March 1989.

powerful, aggressive, and separate over and above nature. The language that typifies a woman's experience, in contrast, situates her along with nature itself. She is seen, and accordingly sees herself, as somehow part of it. Although men and women both wear historically manufactured identities, in times of ecological devastation, the feminine one is clearly the more wholesome human attitude. As Ynestra King has put it, ecofeminists would like to see men give up their attempts to control women and nature and join women in their identity with nature. The fact that deep ecologists have embarked on this revolutionary process, but are yet still uneasy with it, is evidenced by their profound discomfort at accepting what women have to say.

UNIVERSAL OPPRESSION AND NORMATIVE DUALISM

Even if the categories of masculine and feminine are individually and culturally variable, the *universality* of women's oppression is still up for debate. However, this debate is riddled with methodological difficulties over empirical evidence and over interpretation. Here the social construction of knowledge meets the social construction of gender head on. Socialists such as Engels and ecofeminists such as Adrienne Rich and Charlene Spretnak have postulated a prehistorical matriarchy in which social relations were organized around loving, sensuous, life-affirming activities rather than on competition and power. But verifiability is impossible, and the matriarchal image is usually depicted simply in the way that a negative utopia might evolve in the future. In my view, this debate over the universality of feminine oppression is sheer scholasticism. Looking at the real world, can one name a single modern society not governed by men or by a token woman operating within patriarchal values?

Citing Ellen Messer-Davidow as his source, Zimmerman points out that according to feminist anthropologists—it is not clear whether these are liberal, radical, or Marxist—the universality of the woman-nature, male-culture divide does not hold cross-culturally. He identifies the polarity as a byproduct of Western Enlightenment thought, but provides no references. Rosemary Ruether, on the other hand, has documented the pervasiveness of this alignment of women and nature since early Judaeo-Christian times.[20] Indian and Japanese traditions manifest it as well. Surely, however, the point of the argument is that it is Western patriarchy that is becoming globally dominant through the neo-colonial development process, and that it is this culture that men and women of all races are going to have to contest. Meanwhile, Davidow's own examples of classification by inclusion-exclusion actually support a thesis for the ubiquity of patriarchal logic. The schemas mentioned, "clearing" versus "bush" and degree of language

---

[20] Rosemary Ruether, *New Women, New Earth* (New York: Seabury, 1975); also Mary Daly, *Beyond God the Father* (Boston: Beacon, 1973); Marylin French, *Beyond Power* (New York: Summit, 1985).

competence, closely parallel Levi-Strauss' deep structural categories of "raw" versus "cooked." A related oversight is Davidow's apparently superficial defence of the father of liberalism, John Locke. Here, she is said to claim that since Locke recognized "several categories" of being—children, animals, women, idiots and "other classes"—he cannot be seen as operating within a rigid dichotomizing thought mode. Plainly, the oppressive grid—normative dualism—is still at work here, the self-other, white-black, rational-irrational, valued-nonvalued way of organizing the world.

While I am obviously impatient over our patriarchal division of labor and its psychological consequences for men and women, and clearly will be glad once this way of life is obsolete, Wittbecker, nevertheless, asserts that I divide "the sexes as if they were two species and seem to think women have no masculine aspects."[21] Of course, it is patriarchal ideology that creates this dichotomy. However, a reader will not recognize the origin of this division unless he or she first acknowledges the reality of patriarchy itself. I have been researching this masculine propensity to dualism in logic, mathematics, philosophy, language, and other social institutions for some time—for gender is not the only popular set of two. The fascinating thing is that even empirical findings suggest the "two sexes" are, in fact, a continuum of assorted potentials. As I have argued in "Contribution to the Critique of Political Epistemology," male-sexed bodies may have feminine personality attributes and vice versa. The past fifteen years of feminism have encouraged women to acknowledge and draw on their own so-called "masculine" capacities in their workplace, sex life, and so on. White men, rewarded as they already are by the status quo, may have little incentive to find their missing other half. Some may be too damaged to know where to begin. Others, working-class or minority men, may have little but the option of their masculinity to get by with. True, some in the gay movement have been remaking sex and gender, although even here there are factions who have used the gay experience to shore up the patriarchal stereotype of masculinity. Patrick Murphy helps clarify this admittedly difficult area when he writes:

> Thus the 'other' is always implicated in psychical activities, and indicates that the 'self' itself is not singular, unified, or total, but is multiple, through the non-identity of the conscious and unconscious and self-conceptions and drive. It is precisely this recognition of non-identity and the need for inner-dialogue, specifically between 'masculine' and 'feminine' aspects of the psyche, that Salleh sees missing from the propositions of Deep Ecology and seriously impairs its subversion of patriarchy's hegemony.[22]

---

[21] Wittbecker, "Deep Anthropology, Ecology, and Human Order," p. 265. Compare Salleh, "Contribution to the Critique of Political Epistemology," pp. 30-33.

[22] Patrick Murphy, "Prolegomenon for an Eco-Feminist Dialogics," in D. Bauer and S. McKinstry, eds., *Feminism and the Dialogic* (Albany: SUNY Press, 1990). Murphy's reading of my work here corrects the interpretation in his earlier paper, "Sex-Typing the Planet," *Environmental Ethics* 10 (1988): 155-68. For further work on nonidentity, see arguments in Salleh, "On the Dialectics of

Ignoring the dialectic that has women and men moving in opposite directions as they try to reclaim skills unavailable to them by patriarchal convention. Wittbecker thinks he sees bad faith in my use of the masculine medium of academic argument to get an ecofeminist message across. The situation, however, is even more complex. While ecofeminism works to ecologize the feminist movement, and to feminize the consciousness of the environmental movement, it also challenges the very validity of the patriarchal legitimating structures that are its instruments in this process. Conversely, masculine identity meshes with patriarchal structures in an unproblematic way, and only a rare man is motivated to examine this relationship. Wittbecker suggests that I cite my earlier publications in "Deeper than Deep Ecology" as a form of academic "one-upmanship." While it is certainly true that women operating in a male-dominated ecosystem need all the status validation they can get, my reason for referring to this material was to keep my argument short, while yet amplifying more contentious points and showing how these have been debated in a wider context of socialist and feminist politics. All of the papers cited were interlocking. Considering ecofeminist theory's evolution in dialogue with other movements, Wittbecker's concern over its "narrow filter" is especially inappropriate. He would have served deep ecology better by following the normal practice of examining an author's position fully before taking it on.

Part of the trouble is that even in the twentieth century, women under capitalist patriarchy are "to be seen and not heard." As "others," objects in a system of domination, women are looked at, manipulated, used, and finally abused. With political change we may arrive at a society without the engendered differentiations that support this domination. Yet, some among the respective sexes would still be able to grow new members of the species within their bodies. It is symptomatic of the masculine dilemma over this asymmetry that a veritable taboo has been placed on talk about the reproductive side of human relations—especially the birth act. Symptomatic, too, is the fact that in recent times technology has provided patriarchy with new means of controlling this remnant of women's generative power—through the gynecological profession, government-sponsored population programs, and the harvesting of women through *in vitro* fertilization and surrogate motherhood.[23] But, leaving the deeper psychodynamic aside, even though biology provides us with a range of genetic differences which combine randomly with a range of body types, there is no reason in the world why these variations of capacity should be "valued" differently. Race and sex differences can be acknowledged without ascribing a hierarchy of political rights and social privileges to them. Valuing them differently involves a slide from the ontological to the normative again—the very same category mistake that patriarchal thinking both outlaws and perpetrates.

Signifying Practice," and Salleh, "Contribution to the Critique of Political Epistemology."
[23] Rita Arditti et al., eds., *Test Tube Women* (London: Pandora, 1984).

ECO-EGALITARIANISM VERSUS ANDROCENTRISM

In "The Deep Ecology-Ecofeminism Debate and its Parallels," Fox is concerned that many ecofeminists "do not make their (presumed) commitment to an ecocentric egalitarianism particularly explicit," with the result that they reinforce anthropocentrism rather than overcome it.[24] Ecocentrism should not need to be made explicit; it is fundamental to an ecofeminist deconstruction of the nature-humanity split and our very first premise. Griffin's prologue to *Women and Nature* comes to mind here:

> We are the birds' eggs. Birds' eggs, flowers, butterflies, rabbits, cows, sheep; we are caterpillars, leaves of ivy and sprigs of wallflower. We are women. We rise from the wave. We are gazelle and doe, elephant and whale, lilies and roses and peach, we are air, we are flame, we are oyster and pearl, we are girls. We are women and nature. And he says he cannot hear us speak.[25]

Wittbecker similarly fails to comprehend my comment regarding "women's special potency," which is about revaluing life itself and nurture, not a denial of egalitarianism. Let us look more deeply into this misunderstanding, for it seems to have an existential basis.

The literature on "the woman-nature" link suggests that the relationship of a man to his mother, a dependency which is very exclusive under patriarchal conditions, is deeply problematic for the masculine sense of self. The originative power of his mother appears to leave a residue of psychosexual insecurity and unresolved resentment, one that expresses itself in modern attempts to better Mother Nature through dam construction, intergalactic travel, genetic engineering. One of the political aims of ecofeminists and their allies is to replace this intense female nurturing, destructive to women as much as to their sons, by setting up communal forms of child care. Liberated men are already involving themselves as equals in this nurturant labor. Wittbecker should be reassured by these moves toward egalitarianism. In the meantime, however, the deep ecologists' own approach to mothering, via population "control" as a panacea for environmental preservation, remains patronizing and managerial, sitting badly with their professed egalitarianism. Given that women, no longer chattels, are now supposed to be treated with "intrinsic worth and dignity" and have "the freedom to unfold in their own way," they, not men, must be allowed to make decisions about how they use their own bodies. They must also have the opportunity to make their decisions in an informed way. Historically, the time has come when the issue of fertility should pass back into women's hands. As we build a new society based on a new relation to nature, men of all races are going to have to relinquish some of the

---

[24] Fox, "The Deep Ecology-Ecofeminism Debate," p. 13.
[25] Susan Griffin, *Made from this Planet* (London: Women's Press, 1982), p. 83.

privileges they have held for centuries, including the "appropriation" of life through monogamous marriage and the arbitrary "elimination" of life through war. A first step, however, is for thinking men to recognize what their privileges are and to ask whether they, as individuals, really need these props anymore.

Symptomatically, while he can proclaim that "anthropocentrism is the natural centering of human experience" or that "the foot is a unit of measurement based on the length of the human foot," Wittbecker denies that "woman" is "other," because, like Zimmerman and Fox, he seems unable to recognize the factuality of patriarchal relations. Instead, he reaches for a spiritual or transcendent "other," a projection of our material selves as "ultra human nature." These remarks sound dangerously like the kind of transcendental projection of human essence that Ruether, Marilyn French, and Mary Daly describe in their conjectural accounts of patriarchy's formative stages. Surely, we want reintegration with our natural, material base, not abstract, disembodied, transcendence out of it? Nevertheless, Fox lobbies for "biocentrism" and Wittbecker an "ultra-human nature." Yet, as opposed to Fox, Wittbecker's own evolutionist document is also favorably disposed to "anthropomorphism," "anthropocentrism," and "anthropometrism," while deep ecology itself is characterized as "polycentric." This sesquipedalian logic is best left to the deep ecologists themselves to sort out.

Meanwhile, as Fox vacillates back and forth about whether deep ecology is truly "androcentric itself" or simply "androcentric in focus," he strongly objects to Green political claims that draw on such "androcentric" causes as class and race as "the root problem," treating the nonhuman "natural" world simply as a backdrop. He argues correctly that it is quite possible to conceive of a nonandrocentric and egalitarian society that is nevertheless anthropocentric and exploitative of nature. Ecofeminism, however, attempts a synthesis of all these levels. Although my own commitment to ecocentrism goes back to arguments about instrumental reason as early as 1979 and surfaces in several later papers,[26] Fox's superficial acquaintance with that writing leads him to classify my perspective as "thoroughly interhuman." Further, he does not understand why women should use the epithet *feminist* at all, if they are genuinely committed to establishing a biocentric egalitarianism.[27] Like Wittbecker before him, Fox mistakes the focus of my critique for its philosophical vision. The former, although it is merely remedial and transitional, nonetheless, has a very necessary knot to unravel. What is needed in the deep ecologists' response to ecofeminism is a sense of the

---

[26] Ariel Salleh, "Of Portnoy's Complaint and Feminist Problematics," *Australian and New Zealand Journal of Sociology* 17 (1981): 4-13. Alternatively, see Ariel Salleh, "Epistemology and the Metaphors of Production," *Studies in the Humanities* 15 (1988): 130-39, and Ariel Salleh, "At the Interface," *Australian Society* 7 (1984).

[27] Fox's status quo liberalism is echoed in a recent paper by Robyn Eckersley, who also asks whether ecofeminism really adds anything. See "The Paradox of Eco-Feminism," in Ken Dyer and John Young, eds., *Ecopolitics IV Proceedings*, Adelaide, 1989.

complex interlocking issues, economic and ideological, that have to be dealt with, and a sense of the "labor" involved in bringing about social change. There are many centers to work from—speciesism, sexism, racism, classism. Gender is strategically crucial to dismantling each one of these. Without this political work, the deep ecologist's "biocentrism" may become a premature and reductionist closure, failing to help us out of the global impasse.

CONCLUSION

There is nothing uniquely patriarchal about deep ecology. The exchange between deep ecology and ecofeminism is merely an exemplar for women's critiques of other standpoints within the field of environmental ethics. As far as deep ecology goes, we do not question the ultimate intentions of the project, and especially its aim to break down the ontological dualism of humanity versus nature. Nevertheless, the movement's unconscious androcentrism continues to be a very real obstacle to that "self-realization," as they call it. While feminism has moved on, society in general and many men in particular, have not yet assimilated the message of even its first stage—liberal feminism. As a result, as ecofeminists put it, the very ground that society stands on is in danger of being washed away by the flood tides of reaction. Women's views are still not listened to unconditionally, whether they use patriarchal tools like parliament or academia, befitting the first stage of feminist consciousness, or alternative homespun methods of protest. Women's efforts to share their experience outside the narrow confines of a supportive movement are still ridiculed by those with a vested interest in keeping them quiet. What is missing from the deep ecologists' caricature of ecofeminist politics is a comprehension of this historical context together with the fact that feminism is not a static bloc, but a fluid, subtle weave of intentions and opportunities. The feminist consciousness in politics has learned to be aware not only of where "the master" is at, viz., government preference for a rights-based ethic, but when to use liberal strategies, or socialist analyses, or ecofeminist arguments. Many men fail to see that there is no symmetry between men's and women's use of patriarchal discourse and institutions. Women must handle these things self-consciously and reflexively; for most men, it is simply "the way it's done." Having spent a good decade learning to appear "like men," and later rejecting this tack, the feminist political consciousness is complex in contrast to the relatively unproblematic one-dimensional social reality of patriarchal relations. Accordingly, many of the weaknesses and inconsistencies that deep ecologists have picked out in feminism are simply reflections of deep ecology's inability to grasp feminism's transformative energies dialectically.

In some retrospective comments on "Deep Ecology and Its Critics," Kirkpatrick Sale has noted that

> ... it is probably accurate to say that ecologists think primarily in biotic rather than social terms. They regard the fundamental issue to be the destruction of nature and

the suffering of rapidly dying species and eco-systems as distinct from those who regard the basic issue as the absence of justice and the suffering of human populations.[28]

This is a concise summation of the matter as deep ecologists perceive it: you either struggle "for nature" or "for man." Ecofeminists are believed to belong to the second category of activists, those who give human concerns priority.

However, there are four important issues embedded in this statement. First, the dichotomizing either/or conceptualization of the problem serves to replicate the patriarchal logic at the very root of the environmental crisis. The "rational" severance of humanity-nature, man-woman, remains intact. Accordingly, ecofeminists are judged to be injecting exclusively "women's questions" into Green politics, thereby deflecting energy from the main culprit, anthropocentric attitudes. What is lost by marginalizing the ecofeminist project in this conventional way is its broad epistemological challenge. Ecofeminism confronts not only social institutions and practices, but the language and logics by which Western patriarchy constructs its relation to nature. In doing so, it has already traveled a long way down the very same road that deep ecological opponents of anthropocentrism are looking for.

Instead of perpetuating the polarized mindset of "man" versus "nature," a social versus a biocentric emphasis, ecofeminism demands to know how and why the cultural dichotomy has become established at all. This observation introduces the second point at issue: there is a certain naive realism in the claim that matters are either social or biocentric. Ecofeminists are acutely aware that the discursive and institutional medium through which our political debates are being hammered out is not itself neutral or transparent. Ecofeminism takes on a critical examination and deconstruction of that "reason" by which the anthropocentric man-nature split is always regenerated. Instead of thinking man vis-à-vis woman or man vis-à-vis nature, we invite deep ecologists to reorient their static, dualistic thought patterns around, in, and through a several-dimensioned formula, "woman-nature-man."

Woman, or rather the social fabrication of feminine identity under patriarchal domination, serves as a prism through which radical ecologists can come to see how and why they themselves have been constituted as men against nature. Herein lies the third embedded issue. By working through this broader constellation of relationships, deep ecology can gain an actively historical sense of itself, a deeper realization.

Fourth, as deep ecologists come to appreciate how "man's relation to nature" is constructed by means of his relation to woman, they will help build the movement bridges necessary for the emergence of holistic Green politics. To conclude, the convergencies between ecofeminism and deep ecology promise to

---

[28] Kirkpatrick Sale, "Deep Ecology and its Critics," *The Nation*, 14 May 1989, p. 672.

be fruitful. Yet, if we are to celebrate these without false consciousness, careful remedial work needs to be done. By dealing with ecological resistances to the appearance of ecofeminism, I have here attempted to catalyze that deeper insight and understanding.

Part V
Are Humans Part of Nature or Separate From it?

[20]

Can and Ought We to Follow Nature?

Holmes Rolston, III*

"Nature knows best" is reconsidered from an ecological perspective which suggests that we ought to follow nature. The phrase "follow nature" has many meanings. In an absolute law-of-nature sense, persons invariably and necessarily act in accordance with natural laws, and thus cannot but follow nature. In an artifactual sense, all deliberate human conduct is viewed as unnatural, and thus it is impossible to follow nature. As a result, the answer to the question, whether we can and ought to follow nature, must be sought in a relative sense according to which human conduct is sometimes more and sometimes less natural. Four specific relative senses are examined: a homeostatic sense, an imitative ethical sense, an axiological sense, and a tutorial sense. Nature can be followed in a homeostatic sense in which human conduct utilizes natural laws for our well-being in a stable environment, but this following is nonmoral since the moral elements can be separated from it. Nature cannot be followed in an imitative ethical sense because nature itself is either amoral or, by some accounts, immoral. Guidance for inter-human ethical conduct, therefore, must be sought not in nature, but in human culture. Nevertheless, in an axiological sense, persons can and ought to follow nature by viewing it as an object of orienting interest and value. In this connection, three environments are distinguished for human well-being in which we can and ought to participate—the urban, the rural, and the wild. Finally, in a tutorial sense, persons can and ought to follow nature by letting it teach us something of our human role, our place, and our appropriate character in the natural system as a whole. In this last sense, "following nature" is commended to anyone who seeks in his human conduct to maintain a good fit with the natural environment—a sense of following nature involving both efficiency and wisdom.

INTRODUCTION

"Nature knows best" is the third law of ecology according to Barry Commoner and the gravity of his claim is underlined by its ranking with the first two, that everything is interconnected and that nothing is ever destroyed, only recycled.[1] But this third law is curiously normative, not merely describing what nature does, but evaluating it, and implying that we

---

* Department of Philosophy, Colorado State University, Fort Collins, CO 80523. Rolston teaches philosophy including a course on environmental ethics. His research interests are concentrated on philosophical, religious, and scientific conceptions of nature. He is a backpacker and amateur botanist with an interest in bryophytes.

1. Barry Commoner, *The Closing Circle: Nature, Man & Technology* (New York: Alfred A. Knopf, 1972), p. 41.

ought to follow nature. Such following may ordinarily be more prudential than moral for Commoner, but for others, if not for him too, the deepest commands of nature reach the ethical level. Radcliffe Squires writes of Robinson Jeffers, "To direct man toward a moral self by means of the wise, the solemn lessons of Nature: that has been Jeffers' life work."[2]

But there are dissenting voices. We have for too long thought of "Mother Nature" as "sensitive, efficient, purposeful, and powerful," laments Frederick E. Smith, a Harvard professor of resources and ecology. She does not exist; nature is adrift. "This absence of 'goal' in the world systems is what makes the concept of Mother Nature dangerous. In the final analysis nothing is guiding the ship."[3] This, of course, exempts us from following nature—to the contrary, we must take control of our aimless ecosystem. And, again, if this is for Smith more a matter of prudence than of morality, another earlier Harvard professor noted with intensity the moral indifference of nature. Coining a memorable phrase, William James called us to "the moral equivalent of war" in our human resistance to amoral nature:

Visible nature is all plasticity and indifference,—a moral multiverse... and not a moral universe. To such a harlot we owe no allegiance; with her as a whole we can establish no moral communion; and we are free in our dealing with her several parts to obey or to destroy, and to follow no law but that of prudence in coming to terms with such of her particular features as will help us to our private ends.[4]

Those with a philosophical memory will see that the environmental debate reconnects with a longstanding problem in the ethics of nature, and recognize the two camps into which those before us have so often divided, the one setting human conduct morally and valuationally in essential discontinuity with our environment, the other finding continuity there. John Stuart Mill stands within one paradigm: "Conformity to nature has no connection whatever with right and wrong."[5] Ralph Waldo Emerson represents the other: "Right is conformity to the laws of nature so far as they

---

2. Radcliffe Squires, *The Loyalties of Robinson Jeffers* (Ann Arbor: University of Michigan Press, 1956), p. 134.

3. Frederick E. Smith, "Scientific Problems and Progress in Solving the Environmental Crisis" (Address delivered at conference on "Environment, the Quest for Quality," Washington, D.C., February 19, 1970), pp. 3, 5.

4. William James, "The Moral Equivalent of War," in *Memories and Studies* (New York: Longmans, Green, and Co., 1911), pp. 267-96; "Is Life Worth Living?" in *The Will to Believe* (New York: Longmans, Green, and Co., 1896), pp. 43-44.

5. John Stuart Mill, "Nature," in *Collected Works*, 19 vols. (Toronto: University of Toronto Press, 1963-77), 10:400.

are known to the human mind."6 Sometimes old debates can be thrown into fresh perspective by more recent insights and discoveries. Of late, having become ecologically aware, can we say anything more about the question, "Can and ought we to follow nature?"

Much of the puzzle is in the way we use that grand word *nature* and here an analysis of our language is necessary. Still, it is not a sufficient answer to the question. The issue will finally turn on one's sensitivities to value, and to what degree this can be found in the environment we address. We shall try here to disentangle the phrase "follow nature," reaching in conclusion limited but crucial senses in which we both can and ought to follow nature. *Nature* is an absolutely indispensable English word, but there are few others with such a tapestry of meanings. In this respect it is like other monumental words round which life turns to such a high degree that we often capitalize them—*Freedom, the Good, the Right, Beauty, Truth, God, my Country, Democracy, the Church*—words that demand an ethical response, words that we cannot altogether and at once keep in logical perspective, but can only attack piecemeal, always reasoning out of the personal backing of our responsive perceptual experience. Earlier and in the foreground, we will put "following nature" into logical focus. But, later on and in the background, we can only invite the reader to share our moral intuitions. In ethics, Aristotle remarked, "The decision rests with perception."7

Nature is whatever is, all in sum, and in that universal sense the word is quite unmanageable. Even the sense of the physical universe going back to the Greek *physis* is both too broad and too simple. We reach the meaning we need (which also recalls the sense of *physis*) if we refer to our complex earthen ecosphere—a biosphere resting on physical planetary circulations. Nature is most broadly whatever obeys natural laws, and that also includes astronomical nature. Used in this way the word has a contrast only in the supernatural realm, if such there is. But nevertheless we restrict the word to a global, not a cosmic sense, as our typical use of the word *nature* still retains the notion, coming from the Latin root *natus* and also present in *physis*, of a system giving birth to life. No one urges that we follow physicochemical nature—dead nature. What is invariably meant features that vital evolutionary or ecological movement we often capitalize as *Nature* and sometimes personify as *Mother Nature*.

In the present state of human knowledge we are not in any position to estimate the cosmic rarity or frequency of this motherhood on our planet. Perhaps it has regularly appeared wherever nature has been given proper opportunity to organize itself; if so, that would tell us a great deal about the tendency of nature. But it may be that all this vitality is but an eddy in the all-

---

6. Ralph Waldo Emerson, *Journals* (Cambridge, Mass.: Riverside Press, 1910), 3:208.

7. Aristotle *Nicomachean Ethics* 2. 8. 1109^{b}23.

consuming stream of entropy. Although it seems that the stars serve as the necessary furnaces in which all the chemical elements except the very lightest are forged—elements foundational to any biosystem—we nevertheless know little about the contributions of astronomical nature to our local ecosystem. We draw many conclusions about universal nature based on our knowledge of physics and chemistry, but we are reluctant to do so with biology, for we do not like to project from only one known case. Furthermore, profound and mysterious though it is, astronomical nature is too simple. We know nature in its most sophisticated organization on Earth; so, we speak now only of that face of nature which has yielded our own flourishing organic community—eco-nature.

In what follows we distinguish seven senses in which we may follow nature—first, in general terms, an *absolute* sense, an *artifactual* sense, and a *relative* sense, and then, in more detail, four specific relative senses, a *homeostatic* sense, an *imitative ethical* sense, an *axiological* sense, and finally a *tutorial* sense. We answer our basic question, whether we can and ought to follow nature, in terms of each.

FOLLOWING NATURE IN AN ABSOLUTE SENSE

Everything which conducts itself or is conducted in accordance with the laws of nature "follows nature" in a broad, elemental sense, and here it is sometimes asked whether human conduct does or ought to follow these laws. The human species has come into evolutionary nature lately and yet dramatically and with such upset that we are driven to ask whether persons are some sort of anomaly, literally apart from the laws that have hitherto regulated and otherwise still regulate natural events. No doubt our bodies have very largely the same biochemistries as the higher animals. But in our deliberative and rational powers, in our moral and spiritual sensitivities, we do not seem to run with the same mechanisms with which the coyotes and the chimpanzees so naturally run. These faculties seem to "free" us from natural determinisms; we transcend nature and escape her clutches.

Perhaps it is true that in their cultural life humans are not altogether subject to the laws of evolutionary nature. But we may immediately observe that humans are, in a still more basic sense, subject to the operation of these natural laws which we sometimes seem to exceed. If nature is defined as the aggregate of all physical, chemical, and biological processes, there is no reason why it should not *include* human agency. The human animal, as much as all the others, seems to be subject to all the natural laws that we have so far formulated. Although we live at a higher level of natural organization than any other animal, and even though we act as intelligent agents as perhaps no other animal can, there does not seem to be any law of nature that we violate either in our biochemistry or in our psychology. It is, however, difficult to get clear on the logical connections, to say nothing of the psychosomatic connections, of agency with causation. In any case, insofar as we operate as agents on the world, we certainly do so by using rather than by exempting ourselves from laws of nature. No one has ever broken the laws of gravity, or those of electricity, nutrition, or psychology. All human conduct is natural inasmuch as the laws of nature operate in us and on us willy-nilly. We cannot help but follow nature, and advice to do so in this basic law-of-nature sense is idle and trivial even while some high-level questions about the role of human deliberation in nature remain open.

FOLLOWING NATURE IN AN ARTIFACTUAL SENSE

Still, within this necessary obedience to the laws of nature humans do have options through agentive capacities. Submit we must, but we may nevertheless sometimes choose our route of submission. Something remains "up to us." We alter the course of spontaneous nature. That forces us to a second extreme—asking whether, in what we may call an *artifactual* sense, we can follow nature. The feeling that deliberation exempts us from the way that nature otherwise runs suggests the possibility that all agentive conduct is unnatural. Here nature is defined as the aggregate of all physical, chemical, and biological processes *excluding* those of human agency. What we most commonly mean by a natural course of events lies not so much in a scientific claim about our submission to natural laws as it does in a contrast of the natural with the artificial, the artifactual. Nature runs automatically and, within her more active creatures, instinctively; but persons do things by design, which is different, and we for the most part have no trouble distinguishing the two kinds of events. A cabin which we encounter hiking through the woods is not natural, but the rocks, trees, and the stream that form its setting are. A warbler's nest or a beaver's skull are natural while a sign marking the way to a lake or an abandoned hiking boot are not. These things differ in their architecture. The one kind is merely caused. The other kind is there for reasons.

By this account no human has ever acted deliberately except to interfere in the spontaneous course of nature. All human *actions* are in this sense unnatural because they are artifactual, and the advice to follow nature is impossible. We could not do so if we tried, for in deliberately trying to do so we act unnaturally.[8]

---

8. We take notice here of a common usage of *nature* in order to set it aside. The word is sometimes used in the sense of "not affected, spontaneous" and applied to conduct that is not studied or strained. Such conduct is not deliberated, not a result of intentional effort, and, hence, natural like the spontaneous course of non-deliberative nature. Notice that our senses of "follow" shift, although they all unfold from the basic sense of "going in the track of." The senses of "follow" which mean to replace or to succeed in a chronological or causal sequence are not used here.

Each extreme—the absolute and the artifactual—so strongly appeals to part of our usage of the word *nature* that some inquirers are stalled here and can go no further. Yet even Mill, whose celebrated essay on "Nature" begins with these as the only two options, continues to ask at length about following nature as though it is possible and optional, an inquiry which cannot arise in terms of either of the above senses of the phrase. Are there not some other intermediate and reasonably distinct senses in which we can follow nature?

FOLLOWING NATURE IN A RELATIVE SENSE

There is a relative sense in which we may follow nature. Although always acting deliberately, we may conduct ourselves more or less continuously or receptively with nature as it is proceeding upon our entrance. Man is the animal with options who, when he acts, chooses just how natural or artificial his actions will be. All human agency proceeds in rough analogy with the sailing of a ship, which, if it had no skipper, would be driven with the natural wind. But the skipper may set the sails to move crosswind or even tack against the wind using the natural wind all the while. There are no unnatural energies. Our deliberative agency only manages to shift the direction of these natural forces, and it is that intervention which we call unnatural. But our interventions are variously disruptive, and, having admitted these senses in which they are all both natural and unnatural, we recognize further a range across which some are more and some are less natural.

Any parents who "plan" their children act unnaturally in the artifactual sense. Yet marriage, mating, and the rearing of children proceed with the laws of nature. In between, we debate just how natural or artificial birth control methods really are. Some moralists and some medical persons dislike methods that greatly tamper with natural cycles. In contrast to the natural love of man and woman, homosexual conduct is unnatural, "queer," which is one of the strongest reasons why many condemn it. All childbirth is natural, all medically attended childbirth is unnatural, and in between we speak of natural childbirth as opposed to a more medically manipulative childbirth.

All landscaping is artificial. On the other hand, no landscaping violates the laws of nature. Some landscaping, which blends with natural contours and uses natural flora or introduced plants compatible with it, is considered natural; however, landscaping which involves bulldozing out half a hill and setting a building and artificial shrubbery against a scarred landscape is unnatural. All farming is unnatural, against spontaneous nature, but some farming practices fit in with the character of the soil and climate while others do not. Bluegrass does well in Kentucky and in the Midwest, but the Southern farmer is foolish to plant it; and who would plant cotton in New England? On millions of acres found on every continent our unnatural agricultural practices strain fragile semi-desert ecosystems with the fate of millions of persons at stake. Highly manipulative industrial agriculture seems increasingly unnatural with its hybrid "strains," herbicides and pesticides, monocultures, factory farming of chickens, and hormone lacing of beef cattle on feedlots. Some lakes are natural while others are man-made, but among the latter a pond with a relatively fixed shoreline which permits natural flora to flourish there seems more natural than a drawdown reservoir with barren edges.

All clothing is unnatural; only nudists go *au naturel*. We are usually oblivious to whether style and color have any connection with our environment, but still, when the issue arises, we may prefer "the natural look." The traditional Scots plaids come almost literally from the landscape; "earth tones" are in. The iridescent, gaudy colors of modern chemistry are unnatural. Some prefer furniture with a "natural finish" to having the wooden grain hidden beneath DuPont's latest exotic colors. We hardly object to trails for hikers in our natural areas, but if humans go there with motors and highways the wildness is spoiled. Even along interstate highways we prohibit billboards lest they pollute the countryside.

It is sometimes thought that with increasing amendation and repair of spontaneous nature the degree of unnaturalness is roughly the same as the degree of progress—the successful shift from nature to culture. But our ecological perspective has forced us to wonder whether modern life has become increasingly out of kilter with its environment, lost to natural values that we ought to conserve. Big city life in a high rise apartment—to say nothing of the slum—as well as a day's work in a windowless, air-conditioned factory represent synthetic life filled with plastic everything from teeth to trees. They are foreign to the earthen element from which we were reared. We have lost touch with natural reality; life is, alas, artificial.

This relative sense of following nature has to do with the degree of alteration of our environment, with our appreciative incorporation of this environment into our life styles, and with our nearness to nature. But is it not natural for us to be cultured? Consider our hands, each composed of four dexterous fingers and an opposable thumb. Their natural homologues run back through the primates and even to the birds and reptiles. Consider our brain evolving for speech with the jaw released from prehensile functions, and our eyes moving round to frontal focus on hands that enable us to be agents in the world. What are we to say when we deliberately use this natural equipment? That we act unnaturally? Surely not more so than when we use our eyes and ears. Yet with the brain and hand what are we to do? To follow nature? To build a culture that opposes it? Or is there room for the pursuit of both?

With these questions in mind we now examine four specific relative senses of following nature.

FOLLOWING NATURE IN A HOMEOSTATIC SENSE

The ecological crisis has introduced us to what we may call the homeostatic sense of following nature: "You ought not to upset the stability of the ecosystem." Here human welfare and survival depend upon our following nature, but in a sense so basic and rudimentary that we wonder whether it is moral. Human conduct may run through a spectrum from what is minimally to maximally disruptive of natural cycles. In its primitive state the human race had only local and relatively inconsequential environmental impact, but technological humanity has at its option powers capable of massive environmental alteration. We use these clumsily and wrongly, partly out of ignorance, partly because of the erratic, unplanned growth of society, but significantly too because of our defiant refusal to participate in our environment, to accept it, and to fit into it. Environmental rebels, we seek to exploit nature and become misfits. Our modern conduct is thus unnatural.

Ecology awakens us to these unnatural actions. Natural systems fluctuate dynamically and sometimes dramatically, but there is also a resilience and recuperative capacity built into them. Still, they may be pushed to the point of collapse. Ordinarily, if a species becomes much of a misfit, it perishes while the system continues. But humankind may push the system to collapse, perish taking nearly everything else down with it, and thus wreck all. This danger is especially clear in the case of hundreds of soil/water/air interactions. What will supersonic jets or aerosol cans do to the ozone layer? Where does all the DDT go, or the strontium 90? What becomes of the pollutants from coal-fired generators, or from nuclear plants? Where we use natural chemicals, we sling them around in unnatural volumes allowing lead from gasoline, arsenic from pesticides, mercury from our batteries, and nitrogen from fertilizers to find their way into places where they are more disruptive than most people imagine. Worse, so much of our chemistry is exotic, not biodegradable, unnatural in the sense that nature cannot break it down and recycle it, or does so very slowly. Every rock made underground can be eroded at the surface; every compound organically synthesized has some enzyme that will digest it, and so on. But our artificial products choke up the system. Alas, not only our technology, but our whole profiteering, capitalistic, industrial system may be "unnatural" in that it cheats by incurring an environmental debt which moves us ever onward toward reduced homeostasis.

Should we then behave naturally? Humans are the only animals with deliberate options and these options do enable us to command nature, the more so with the advance of science. This capacity to command nature is indeed a sort of escape from obeying nature, but of the sort that must remain in intimate contact with nature if it is to continue. We can no more escape from nature than we can from human nature, than the mind can from the body, but we can bring all these increasingly under our deliberative control.

Technology does not release us from natural dependencies; it only shifts the location and character of these, releasing us from some dependencies while immediately establishing new ones. A tree escapes above the soil pushing ever higher only by rooting ever more deeply. On the one hand, we are driven back to our original observation that we can never escape the laws of nature, but must obey them willy-nilly. The only sense in which we can ever break natural laws is to neglect to consider their implication for our welfare. We might even say that any creature acts unnaturally whose behavior is such that the laws of nature run to the detriment of that organism, and when that happens such an unnatural creature soon becomes extinct.

But then, on the other hand, we must not forget our second observation, that all our human actions are unnatural. According to this viewpoint, our successful actions relieve us from the need of following nature—in the sense of submitting to narrow natural constraints—by enlarging our sphere of deliberate options. Room for the homeostatic sense of following nature must be found somewhere between these extremes. The key point we need to consider seems to be that among our deliberate options some will help retain stability in the ecosystem and in our relationship to it while others will not. In this sense it seems perfectly straightforward to say that we may or may not follow nature, and that we both can and ought to do so. To follow nature means to choose a route of submission to nature that utilizes natural laws for our well-being.

It may be objected that the advice to *follow* nature has been subtly converted into the injunction to *study* nature—conduct with which no rational person will quarrel. According to this objection, *studying* nature has nothing to do with *following* nature. To the contrary, its purpose is to repair nature, to free us from conforming to its spontaneous course, by examining just how much alteration we can get by with. This objection has force, but its scope is too narrow, for we study nature to manipulate only parts of it, always within the larger picture of discovering our organic, earthen roots, the natural givens to which we have to submit and with which we have to work. We study cancer in order to eradicate it; we study diabetes in order to repair a natural breakdown in insulin production; but we study the laws of health in order to follow them. We study the causes of floods in order to prevent them, but we study the laws of ecosystemic health in order to follow them. Those who study nature find items they may alter, but they also discover that the larger courses of nature are always to be obeyed. This applies not only in the strong sense in which we have no option, but also in the weak optional sense of intelligently fitting ourselves into their pattern of operation; and in that sense we do study nature, in the end, in order to follow nature.

But is any of this moral? There are a great many ways in which morality readily combines with the injunction to find a life style compatible with our planetary ecosystemic health. The jet set who have insisted on flying in SSTs, should these planes prove to deplete the protective ozone in the atmosphere, would be acting immorally against their fellow humans, as would farmers who continue long-term poisoning of the soil with non-biodegradable pesticides in order to achieve short-term gains. But it is relatively easy to isolate out the *moral* ends here—respect for the welfare of others—and to see the natural means—conformity to the limitations of our ecosystem—as *nonmoral*.[9] So, we are forced to conclude that there is nothing moral about following nature in and of itself; our relations with nature are always technical or instrumental; and the moral element emerges only when our traffic with nature turns out to involve our inter-human relations. We establish no moral communion with nature, but only with other persons. It is not moral to repair a ship nor immoral to sink it except if it happens to be one that we and our fellow travelers are sailing in. We have reached then a homeostatic sense in which we both can and ought to follow nature only to find it submoral or premoral because the morality surrounding it can be separated off from it.

FOLLOWING NATURE IN AN IMITATIVE ETHICAL SENSE

It is difficult to propose that we ought to follow nature in an imitative ethical sense because our usual estimate—and here we vacillate—is that nature is either amoral or immoral. We call nature amoral because morality appears in humans alone and is not, and has never been, present on the natural scene. Human conduct may be moral or immoral, but the "conduct" of nature, if indeed it can be called that, is simply amoral. The moral dimension in human nature has no counterpart in mother nature. No being can be moral unless he is free deliberatively; something must be "up to him"; and nothing else in nature has sufficient mental competence to be moral. Mother nature simply unfolds in creatures their genetic programming, like the developing seed, and they respond to their environments driven like the leaf before the wind. Even if there are erratic, indeterminate elements in nature, these provide no moral options; they just happen. Biological and evolutionary processes are no more moral than the laws of gravity or electricity. Whether something does or must happen has nothing to do with whether it ought to happen. Out of this estimate arises the basic cleavage that runs through the middle of the modern mind dividing every study into the realm of the *is* and the realm of the *ought*. No study of nature whether physical, biological, or even social can tell us what ought to happen, and following nature where it is possible and optional is something that is never in itself moral. Nature is blind to this dimension of reality. It is a moral nullity.

9. See Holmes Rolston, III, "Is There an Ecological Ethic?" in *Ethics: An International Journal of Social, Political, and Legal Philosophy* 85 (1975): 93-109.

We immediately grant that there are no other moral agents in nature, whether orangutans, butterflies, wind, or rain; nor is nature as a whole a moral agent even when personified as "Mother Nature." We have no evidence that any natural species or forces do things deliberately, choosing the most moral route from less moral options. If anyone proposes that we "follow nature" in something like the ethical sense in which Christians "follow Jesus," or the Buddhists, Buddha, he has very much gone astray, and the blind does indeed lead the blind. Such a person ignores the emergent sphere of deliberative morality in humans for which there is no precedent in birds or field mice. In this sense, Mill is undoubtedly right when he protests that conformity to nature has no connection with right and wrong. There is no way to derive any of the familiar moral maxims from nature: *"One ought to keep promises." "Tell the truth." "Do to others as you would have them do to you." "Do not cause needless suffering."* There is no natural decalogue to endorse the Ten Comandments; nature tells us nothing about how we should be moral in this way, *even if* it should turn out that this is approximately the morality ingrained by natural selection in human nature.

But this does not end the matter, for there may nevertheless be some good or goods in nature with which we morally ought to conform even if these goods have not been produced by the process of deliberative options necessary to us if we are to be moral. The resolution of this form of our question will prove more difficult. Because nature has no moral agency, and because inter-human relations are clearly moral, it has been easy to suppose that there is nothing moral in our relations with nature. It has also been easy to conclude that morality is not "natural," but rather belongs to our "super-natural" nature. But to grant that morality appears with the emergence of human beings out of nonmoral nature does not settle the question whether we, who are moral, should follow nature.

When the issue of good in nature is raised, we are at once confronted with the counterclaim that the course of nature is bad—one which, if we were to follow it, would be immoral. Nature proceeds with an absolute recklessness that is not only indifferent to life, but results in senseless cruelty which is repugnant to our moral sensibilities. Life is wrested from her creatures by continual struggle, usually soon lost; and those "lucky" few who survive to maturity only face more extended suffering and eventual collapse in disease and death. With what indifference nature casts forth to slaughter ten thousand acorns, a thousand grasshoppers, a hundred minnows, and a dozen rabbits, so that one of each might survive. Things are no sooner sprouted, hatched, or born than they are attacked; life is unrelieved stress, until sooner or later, swiftly or by inches, fickle nature crushes out the life she gave, and the misery is finally over. All we can be sure of from the hands of nature is calamity. We are condemned to live by attacking other life. Nature is a gory blood bath; she permits life only in agony. The world's last word is what the Buddhists call *duhkha*, suffering. Few persons can read Mill's essay on "Nature" without being chastened in their zeal for following nature:

> In sober truth, nearly all the things which men are hanged or imprisoned for doing to one another, are nature's everyday performances.... Nature impales men, breaks them as if on the wheel, casts them to be devoured by wild beasts, burns them to death, crushes them with stones like the first christian martyr, starves them with hunger, freezes them with cold, poisons them by the quick or slow venom of her exhalations, and has hundreds of other hideous deaths in reserve, such as the ingenious cruelty of a Nabis or a Domitian never surpassed.... A single hurricane destroys the hopes of a season; a flight of locusts, or an inundation, desolates a district.... Everything, in short, which the worst men commit either against life or property is perpetrated on a larger scale by natural agents.[10]

The Darwinian paradigm of nature in the nineteenth century strongly reinforced that of Mill. Nature became a kind of hellish jungle where only the fittest survive, and these but barely. The discovery of the genetic basis of Darwin's random mutations only added to the sense of nature's rudderless proceedings, law-like to be sure in the sense that natural selection conserves beneficial mutations, but still aimless since natural selection operates blindly over mutations which are mostly worthless, irrelevant, or detrimental. There seemed a kind of futility to it all, certainly nothing worthy of our moral imitation. This portrait of nature affected several generations of ethicists who frequently concluded that ethics had nothing to do with the laws of nature unless it was to alter and overcome our natural instincts and drives, lest we too behave "like beasts." The *is/ought* cleavage became entrenched in earlier twentieth-century philosophy in large part because of this nineteenth-century portrait of nature. G.L. Dickinson expresses with great force the protest of this period:

> I'm not much impressed by the argument you attribute to Nature, that if we don't agree with her we shall be knocked on the head. I, for instance, happen to object strongly to her whole procedure: I don't much believe in the harmony of the final consummation... and I am sensibly aware of the horrible discomfort of the intermediate stages, the pushing, kicking, trampling of the host, and the wounded and dead left behind on the march. Of all this I venture to disapprove; then comes Nature and says, "but you ought to approve!" I ask why, and she says, "Because the procedure is mine." I still demur, and she comes down on me with a threat—"Very good, approve or no, as you like; but if you don't approve you will be eliminated!" "By all means," I say, and cling to my old opinion with the more affection that I feel myself invested with some-

---

10. Mill, *Collected Works*, 10:385-86.

thing of the glory of a martyr.... In my humble opinion it's Nature, not I, that cuts a poor figure!¹¹

Here we have undoubtedly reached a moral sense of following nature, but one we cannot recommend. Virtually none of us, except perhaps ethical mavericks like Nietzsche, will recommend that this pushing, kicking, and trampling be taken as a moral model for inter-human conduct. So, offered this imitative ethical sense of following nature, we observe that nature is not a moral agent and therefore really cannot be followed, and secondly that there are elements in nature which, if we were to transfer them to inter-human conduct, would be immoral, and therefore ought not to be imitated. But does it follow that nature is therefore bad, a savage realm without natural goods? Is this ferocity and recklessness all that is to be said, or even the principal thing to be said, or can this be set in some different light?

FOLLOWING NATURE IN AN AXIOLOGICAL SENSE

In order to develop an axiological sense in which human conduct may be natural, let us make a fresh start and postpone answering the question we have just posed until we can come at it from another side. Three environments—the urban, the rural, and the wild—provide three human pursuits—culture, agriculture, and nature. All three are vocations which ought to be followed and environments which are needed for our well-being. We are concerned for the moment with human activity collectively and will examine individual responsibility later. When Aristotle observed that "Man is by nature a political animal,"¹² he was speaking in terms of the Greek word *polis*, city-state, of which Athens is such a memorable example. Here *city* refers indiscriminately to village, town, and city. We are social animals and the story of civilization is largely the growth of our capacity for building a cultured state. We are both *Homo sapiens* and *Homo faber*; the brain and the hand combine in wisdom and in craft to construct the enormous world of artifacts which is our urban environment. All these products are unnatural in the sense that they are independent of nature's spontaneous production. It cannot, on the other hand, be unnatural for us to build cities, for, after all, nature has supplied us with the brain and the hand as well as the social propensities for community. Humans are the creatures whom nature did not specialize, but rather equipped with marvelous faculties for culture and craft. We ought to use them, both prudentially and morally, for is not wasted talent a sin? In this sense it is not unnatural for man to be urban even though, as soon as we do anything deliberately, we alter spontaneous nature.

11. Goldworthy Lowes Dickinson, *The Meaning of Good* (New York: McClure, Phillips and Co., 1907), p. 46.

12. Aristotle *Politics* 1. 2. 1253ª2.

We reach the paradox that "Man is the animal for whom it is natural to be artificial."[13]

In culture we allow a discontinuity between human life and nature, but this discontinuity is still an extension out of the ultimate natural environment. Nature releases us to develop our culture; here she offers no model; we are on our own; the mores of the human city are up to us, albeit judged by a culturing of those native endowments we call reason and conscience. The city is in some sense our *niche*; we belong there, and no one can achieve full humanity without it. Cultured human life is not possible in the unaltered wilderness; it is primitive and illiterate if it remains at a merely rural level. The city mentality provides us with literacy and advancement, whether through the market with its trade and industry, or through the library and laboratory, out of which so much of our knowledge of nature has come.

By the term *rural environment* we mean nature as domesticated for the life support of the human population, primarily the cultivated landscape, the field, the woodlot, the pasture, the groved road, the orchard, the ranch. The farm feeds the city, of course, and that may be taken as a metaphor for the whole support of society in soil, water, and air—for the organic circulations of the city in nature. The rural environment is the one in which humans meet nature in productive encounter, where we command nature by obeying her. Here there is a judiciously mixed sense of discontinuity and continuity: by human agency we adapt the natural course—yet we adopt it too; we alter nature—yet accept its climates and capacities. We both get into nature's orbit and bring nature into our orbit. We direct nature round to our goals; yet, if we are intelligent, we use only those disruptions that nature can absorb, those appropriate to the resilience of the ecosystem under cultivation. In the urban environment, no burden of proof rests on a person proposing an alteration whether or not the change is natural (so long as it does not spill over to disrupt rural or wild areas). But in the rural environment, a burden of proof does rest upon the proposer to show that the alteration will not deteriorate the ecosystem. Within our agricultural goals our preference is for those alterations that can be construed as "natural," those most congenial to the natural environment; and we prohibit those that disfigure it.

The rural environment is an end in itself as well as an instrument for the support of the city. It has beauty surpassing its utility. If we ask why there are gardens, we answer "for food," only to recall that there are also flower gardens. The English garden combines both the rose and the berry bush. Both the farm and the park belong in the pastures of the Shenandoah Valley, the blue grass farms of Kentucky, and the cornfields of Iowa, where there is a form of beauty not possible either in the city or in the wilderness. We love the green, green grass of

---

13. Lucius Garvin, *A Modern Introduction to Ethics* (Cambridge, Mass.: Houghton Mifflin, 1953), p. 378.

home, the tree in the meadow, the forested knobs behind the church, and the walk down by the pond. We are deeply satisfied by the rural environment. Although we appreciate our modern freedom from the drudgery of the farm, many still cherish, within limits, experiences that can only be had in the country—sawing down an oak tree, shelling peas, drawing a bucket of water from a well.

The rural environment is, or ought to be, a place of *symbiosis* between humankind and nature, for we may sometimes improve a biosystem. The climax forest of an ecosystemic succession is usually not suited for the maximum number and kinds of fauna and flora, and this succession can be interrupted by agriculture with benefit to those natural species which prefer fields and edging. There are more deer in Virginia now than when the Indians inhabited its virtually unbroken forests, and that is probably true of cottontails, bobwhites, and meadowlarks. Suitable habitat for all but a few of the wildest creatures can be made consistent with the rural use of land. With pleasant results humans have added the elm and the oak to the British landscape, the Russian olive to the high plains, the eucalyptus to California, the floribunda rose to interstate highway roadsides, and the ring-necked pheasant to the prairies. In his idyllic love of nature, Emerson did not write of the wilderness so much as of the domestic New England countryside. When we sing "America the Beautiful," we sing largely of this gardened nature.

We may even speak of a micro-rural environment—an urban garden, a city park, an avenue of trees with squirrels and rabbits, a suburban fence row with cardinals and mockingbirds, a creekside path to a school. Anyone who flies over all but the worst of our Eastern cities will be impressed by how much nature is still there. We love something growing about us if only trees and lawns, and everyone would consider a city improved if it had more green space, more landscape left within it. We prefer our homes, bridges, streets, offices, and factories to be "in a natural setting." We want our cities graced with nature, and that alone suffices to undermine Mill's claim that "All praise of Civilization, or Art, or Contrivance, is so much dispraise of Nature."[14] The wood fire on the stone hearth or the gentle night rain on a tin roof recall for us this natural element; even our plastic trees vicariously return us to nature.

Our requirements for wild nature are more difficult to specify than those for tamed nature, but nonetheless real. The scarcest environment we now have is wilderness, and, when we are threatened with its possible extinction, we are forced to think through our relationships to it. Do we preserve wild nature only as a potential resource for activity that humans may someday wish to undertake in terms of urban or rural nature? Or are there richer reasons, both moral and prudential, why we ought to maintain some of our environment in a primitive state?

14. Mill, *Collected Works*, 10:381.

It is beyond dispute that we enjoy wild places, that they fill a *recreational* need, but that word by which we typically designate this fulfillment seems a poor one until we notice a deeper etymology. Something about a herd of elk grazing beneath the vista of wind and sky, or an eroded sandstone mesa silhouetted against the evening horizon, *re-creates* us. We have loved our national parks almost to death, the more so because they are kept as close to spontaneous nature as is consistent with their being extensively visited. Worried about park overuse, we are now struggling to preserve as much wilderness area as possible, resolving to keep the human presence there in lower profile. We set aside the best first—the Yellowstone, the Grand Canyon, the High Sierra, the Great Smokies, the Everglades—but later found that there was really no kind of landscape for which we did not wish some preservation—the desert, the pine barrens, the grasslands, the wild rivers, the swamps, the oak-hickory forests. We began by preserving the buffaloes and lady-slippers, and soon became concerned for the toads and mosses. But why is it that sometimes we would rather look for a pasqueflower than see the latest Broadway hit?

Wild nature is a place of encounter where we go not to act on it, but to contemplate it, drawing ourselves into its order of being, not drawing it into our order of being. This accounts for our tendency to think of our relationship to wild nature as recreational, and therefore perhaps idle, since we do not do any work while there. We are at leisure there, often, of course, an active leisure, but not one that is economically productive. In this respect our attitude toward wilderness will inevitably be different from that of our grandfathers who for the most part went into it to reduce the wild to the rural and urban. Their success forces us to the question of the worth of the wild. But, when the answer has to be given in non-resource terms, it is not the kind or level of answer to which we are accustomed in questions about nature. For in important senses wild nature is not for us a commodity at all. Even when the answer is given in terms of some higher, noneconomic value, our philosophical apparatus for the analysis and appraisal of wild value is, frankly, very poorly developed, for we have too much fallen into the opinion that the only values that there are, moral or artistic or whatever, are human values, values which we have selected or constructed, over which we have labored. Modern philosophical ethics has left us insensitive to the reception of nonhuman values.

We need wild nature in much the same way that we need the other things in life which we appreciate for their intrinsic rather than their instrumental worth, somewhat like we need music or art, philosophy or religion, literature or drama. But these are human activities, and our encounter with nature has the additional feature of being our sole contact with worth and beauty independent of human activity. We need friends not merely as our instruments, but for what they are in themselves, and, moving one order beyond this, we need wild nature precisely because it is a realm of values which are independent of us Wild nature has a kind of integrity, and we are the poorer if we do not recognize it and enjoy it. That is why seeing an eagle or warbler, a climbing fern or a blue spruce is a stirring experience. The Matterhorn leaves us in awe, but so does the fall foliage on any New England hillside, or the rhododendron on Roan Mountain. Those who linger with nature find this integrity where it is not at first suspected, in the copperhead and the alligator, in the tarantula and the morel, in the wind-stunted banner spruce and the straggly box elder, in the stormy sea and the wintry tundra. Such genuine nature precedes and exceeds us despite all our dominion over it or our uniqueness within it, and its spontaneous value is the reason why contact with nature can be re-creating.

We are so indisposed to admit the possibility of wild value that the cautious naturalist, finding himself undeniably stimulated by his outings, will still be inclined to locate these values within himself—values which he believes he has somehow constructed or unfolded out of the raw materials of natural encounter. These encounters provide him with an account of why only some of nature has value for him. If he has successfully used it, it has value. The rest of nature, left unused, has no value, not yet at least. Wild nature, then, according to this account, serves only as an occasion of value; it triggers dormant human potential. Even such a naturalist, however, needs wild nature for the triggering of these values, and he will have to reckon with why nature has this capacity to occasion value, being necessary if insufficient for it. But what makes this account peculiarly unsatisfying is its persistent anthropocentrism and its artificiality in actual natural encounter. It takes considerable straining, even after studying philosophy, to accept the idea that the beauty of the sunset is only in the eye of the beholder. The sensitive naturalist is again and again surprised by nature, being converted to its values and delighted by it just because he has gone beyond his previous, narrowly-human values. It is the autonomous otherness of the natural expressions of value that we learn to love, and that integrity becomes vain when this value secretly requires our composing.

This value is often artistic or aesthetic, and is invariably so if we examine a natural entity at the proper level of observation or in terms of its ecological setting. An ordinary rock in microsection is an extraordinary crystal mosaic. The humus from a rotting log supports an exquisite hemlock. But this value also has to do with the intelligibility of each of the natural members; and here natural science, especially ecology, has greatly helped us. This intelligibility often leads to a blending of the autonomy of each of the natural kinds creating a harmony in the earthen whole. A world in which there are many kinds of things, the simple related to the complex, is a valuable world, and especially so if all of them are intelligibly related. Everything has its *place*, and that justifies it. Natural value is further resident in the vitality of things, in their struggle and zest, and it is in this sense that we often speak of a reverence for life, lovely or not. Or should we say that we find all life beautiful, even when we sometimes must sacrifice it? We love the natural mixture of consistency and freedom; there is something about the word *wild* that goes well with the word *free*, whether it is the determined freedom of the wild river or the more spontaneous freedom of the hawk in the sky. In this splendor, sublimity, and mystery the very word *wild* is one of our value words. Simply put, we find *meanings* in wild things.

In this context we may offer yet another answer to our question. We may be said to follow that which is the object of our orienting interest, as when we follow sports, medicine, or law, or the latest news developments. Many scientists, perhaps all the "pure" ones, "follow nature" in that they find its study to be of consuming interest—intrinsically worthwhile—and those who are also naturalists go on in varying senses to say that they appreciate nature, find great satisfaction in it, and even love it. We follow what we "participate in," especially goals we take to be of value. This sense of "follow" is less than "ethical imitation," but it is significantly more than the notion that our conduct toward nature is not moral. For we look to nature as a realm of natural value beyond mere natural facts, which, maintained in its integrity, we may and ought to encounter. The notion of "following" nature, in addition, is deeper than following art, music, or sports, in that, when encountering nature, we are led by it through sensitive study to the importation of nonhuman kinds of meaning. When I delight in the wild hawk in the wind-swept sky, that is not a value that I invent, but one that I discover. Nature has an autonomy which art does not have. We must follow nature to gain this meaning—in the sense of leaving it alone, letting it go its way. We take ourselves to it and listen for and to its natural forms of expression, drawn by a range and realm of values which are not of our own construction. We ought not to destroy this integrity, but rather preserve it and contemplate it, and in this sense our relations with nature are moral. Even G.E. Moore, who so much lamented the "naturalistic fallacy," by which we mistakenly move from a natural *is* to an ethical *ought*, still finds that appreciation of the existence of natural beauty is a good.[15] But morality is the science of the good; so, as soon as we move from a natural *is* to a natural *is good*, our relations with that natural good are moral. We follow what we love, and the love of an intrinsic good is always a moral relationship. We thus find it possible to establish that moral communion with nature which James thought impossible. In this axiological sense, we ought to follow nature, to make its value one among our goals; and, in so doing, our conduct is here guided by nature.

How far is this value so distributed that each individual is obligated to moral conduct towards nature? There is no person who ought not to be concerned with the preservation of natural goodness, if only because others undeniably do find values there. Nevertheless, we allow individuals to weight their preferences, and there may be differing vocations, some seeking the social goods more than the natural ones. But a purely urban person is a one-dimensional person; only those who add the rural and the wild are three-dimensional persons. As for myself, I consider life morally atrophied when respect for and appreciation of the naturally wild is absent. No one has learned the full scope of what it means to be moral until he has learned to respect the integrity and worth of those things we call wild.

FOLLOWING NATURE IN A TUTORIAL SENSE

In positing a tutorial sense in which human conduct may follow nature, I admit that I can only give witness and invite the sharing of a gestalt, rather than provide a reasoned conceptual argument. I find I can increasingly "draw a moral" from reflecting over nature—that is, gain a lesson in living. Nature has a "leading capacity"; it prods thoughts that educate us, that lead us out (*educo*) to know who and where we are, and what our vocation is. Take what we call natural symbols—*light and fire, water or rock, morning and evening, life and death, waking and sleeping, the warmth of summer and the cold of winter, the flowers of spring and the fruits of fall, rain and rivers, seeds and growth, earth and sky*. How readily we put these material phenomena to "metaphorical" or "spiritual" use, as when we speak of life's "stormy weather," of strength of character "like a rock," of insecurity "like shifting sand," of the "dark cloud with the silver lining," or of our "roots" in a homeland. Like a river, life flows on with persistence in change. How marvelously Lanier could sing of the watery marshes of Glynn—and the darkey, of Old Man River! How profound are the psychological forces upon us of the grey and misty sky, the balmy spring day, the colors we call bright or somber, the quiet of a snowfall, the honking of a skein of wild geese, or the times of natural passage—birth, puberty, marriage, death! How the height of the mountains "elevates" us, and the depths of the sea stimulates "deep" thoughts within!

Folk wisdom is routinely cast in this natural idiom. The sage in Proverbs admonishes the sluggard to consider the ways of the ant and be wise. The farmer urges, "Work, for the night comes, when man's work is done." "Make hay while the sun shines." The Psalmist notices how much we are like grass which flourishes but is soon gone, and those who understand the "seasonal" character of life are the better able to rejoice in the turning of the seasons and to do everything well in its time. Jesus asks us, in our search for the goods of life, to note the natural beauty of the lilies of the field, which the affected glory of Solomon could not surpass, and he points out birds to us, who, although hardly lazy, are not anxious or worried about tomorrow.

---

15. George Edward Moore, *Principia Ethica* (Cambridge: Cambridge University Press, 1903), pp. 36-58, 188, 193, 195-98, 200, 206.

"What you sow, you reap." "Into each life some rain must fall." "All sunshine makes a desert." "By their fruits shall you know them." "The early bird gets the worm." "Time and tide wait for no man." "The loveliest rose has yet its thorns." "The north wind made the Vikings." "The tree stands that bends with the wind." "White ants pick a carcass cleaner than a lion." "Every mile is two in winter." "If winter comes, can spring be far behind." It is no accident that our major religious seasons are naturally scheduled: Christmas comes at the winter solstice, Easter with the bursting forth of spring, and Thanksgiving with the harvest. Encounter with nature integrates me, protects me from pride, gives a sense of proportion and place, teaches me what to expect, and what to be content with, establishes other value than my own, and releases feelings in my spirit that I cherish and do not find elsewhere.

Living well is the catching of certain natural rhythms. Those so inclined can reduce a great deal of this to prudence, to the natural conditions of value; and we may be particularly prone to do this because nature gives us no ethical guidance in our inter-human affairs. But human conduct must also be an appropriate form of life toward our environment, toward what the world offers us. Some will call this mere efficiency, but for some of us it is a kind of wisdom for which prudence and efficiency are words that are too weak. For we do not merely accept the limits that nature thrusts upon us, but endorse an essential goodness, a sufficiency in the natural fabric of life which encompasses both our natural talents and the constitution of the world in which, with our natural equipment, we must conduct ourselves. What I call a larger moral virtue, excellence of character, comes in large part, although by no means in the whole, from this natural attunement; and here I find a natural ethic in the somewhat old-fashioned sense of a way of life—a life style that should "follow nature," that is, be properly sensitive to its flow through us and its bearing on our habits of life. A very significant portion of the *meaning* of life consists in our finding, expressing, and endorsing its naturalness. Otherwise, life lacks propriety.

We have enormous amounts of nature programmed into us. The protoplasm that flows within us has flowed naturally for over a billion years. Our internal human nature has evolved in response to external nature for a million years. Our genetic programming—which largely determines what we are, making each of us so alike and yet so different—is entirely natural. It is difficult to think that we do not possess a good natural fit in the wellsprings of our behavior. Our cultural and our agentive life must be, and, so far as it is optional, ought to be consistent with that fit—freeing us no doubt for the cities we build, permitting our rural adaptations, and yet in the end further fitting us for life within our overarching natural environment. We are not, in the language of geographers, environmentally determined, for we have exciting options, and these increase with the advance of culture. But we are inescapably environmentally grounded as surely as we are mortal. This *is* the case, and hence our optional conduct *ought* to be commensurately natural; and, if we can transpose that from a grudging prudential *ought* to a glad moral *ought*, we shall be the happier and the wiser for finding our "place under the sun." Life moves, we are saying, not so much against nature as with it, and that remains true even of cultured human life which never really escapes its organic origins and surroundings. Our ethical life *ought* to maintain for us a good natural fit in both an efficient and a moral sense. This is what Emerson means when he commends moral conduct as conformity to the laws of nature. There is in this communion with nature an ethic for life, and that is why exposure to natural wildness is as necessary for a true education as the university.

Someone may complain, and perhaps fiercely, that in this ethic nature only serves as an occasion for the construction of human virtues; that the natural wisdom we have cited shows only the virtues that develop *in us* when we confront nature; and that thus there is no following of nature, but rather a resistance to it, a studied surmounting in which we succeed despite nature. But this anthropocentric account is too one-sided. Evolution and ecology have taught us that every kind of life is what it is not autonomously but because of a natural fit. We are what I call *environmental reciprocals* indebted to our environment for what we have become in ways which are as complementary as they are oppositional. Nature is, I think, not sufficient to produce all these virtues in us, and that allows for our own integrity and creativity—but nature is necessary for them. Admittedly, we must attain these virtues before we find and establish natural symbols for them—we must undergo the natural course in order to understand it—but I do not think that this ethical strength is merely and simply inside us. It is surely relational, at a minimum, arising out of the encounter between humans and nature. At the maximum, we are realizing and expressing in this strong and good life which we live something of the strength and goodness which nature has bequeathed us.

Nature is often enigmatic. Human life is complex. Each contains many times and seasons. The danger here is that any secretly desired conduct can somehow be construed as natural and found virtuous. Nature gives us little help concerning how we are to behave toward one another. In these matters we are free to do as we please, although nature has endowed us with reason and conscience out of which ethics may be constructed. Especially suspicious are arguments which assign human roles to nature, as is sometimes done with women or blacks, for we easily confuse the natural with the culturally conventional.

There may also be cases where we learn what is bad from nature. In rare cases, we may unwisely elect to follow some process in nature which in itself is indefensible—as some say the bloodthirsty conduct of the weasel is. I do not wish to defend the course of nature in every particular, but most of these cases involve learning something bad—an ethic of selfishness, a dog-eat-dog attitude, or a might-makes-right life style—by inappropriately projecting into moral inter-human conduct, and thereby making bad what is quite appropriate at some lower, nonmoral level—for example, the principle of the survival of the fittest or the self-interest programmed into the lower life forms. We cannot assume that the way things work at lower, nonmoral levels is the way that they ought to work at human, moral levels, for the appearance of the capacity for moral deliberation makes a difference. This is what is correct about the *is-ought* distinction. Our moral conduct exceeds nature, and we must deliberate with an ethic based on reason and conscience which supplants instinct. It is our conduct or mores insofar as it fits us to our environment—our ethic of bearing toward the natural world, not toward other persons—that I refer to in the tutorial sense, and which I here defend. Moreover, I call this conduct moral too in the sense that it contributes to our wisdom and our excellence of character.

In catching these natural rhythms, we must judiciously blend what I call *natural resistance* and *natural conductance*. Part of nature opposes life, increases entropy, kills, rots, destroys. Human life, like all other life, must struggle against its environment, and I much admire the human conquest of nature. However, I take this dominion to be something to which we are naturally impelled and for which we are naturally well-equipped. Furthermore, this struggle can be resorbed into a natural conductance, for nature has both generated us and provided us with life support—and she has stimulated us into culture by her resistance. Nature is not all ferocity and indifference. She is also the bosom out of which we have come, and she remains our life partner, a realm of otherness for which we have the deepest need. I resist nature, and readily for my purposes amend and repair it. I fight disease and death, cold and hunger—and yet somehow come to feel that wildness is not only, not finally, the pressing night. Rather, that wildness with me and in me kindles fires against the night.

I am forced, of course, to concede that there are gaps in this account of nature. I do not find nature meaningful everywhere, or beautiful, or valuable, or educational; and I am moved to horror by malaria, intestinal parasites, and genetic deformities. My concept of the good is not coextensive with the natural, but it does greatly overlap it; and I find my estimates steadily enlarging that overlap. I even find myself stimulated positively in wrestling with nature's deceits. They stir me with a creative discontent, and, when I go nature one better, I often look back and reflect that nature wasn't half bad. I notice that my advanced life depends on nature's capacity to kill and to rot, and to make a recycling and pyramidal use of resources. Nature is not first and foremost the bringer of disease and death, but of life, and with that we touch the Latin root, *natus*. When nature slays, she takes only the life she gave as no murderer can; and she gathers even that life back to herself by reproduction and by re-enfolding organic resources and genetic materials, and produces new life out of it.

Environmental life, including human life, is nursed in struggle; and to me it is increasingly inconceivable that it could, or should, be otherwise. If nature is good, it must be both an assisting and a resisting reality. We cannot succeed unless it can defeat us. My reply, then, to G.L. Dickinson's lament over the kicking and pushing in nature is that, although I do not imitate it, certainly not in human ethics, I would not eliminate it if I could, not at least until I have come to see how life could be better stimulated, and nobler human character produced without it. Nature is a vast scene of birth and death, springtime and harvest, permanence and change, of budding, flowering, fruiting, and withering away, of processive unfolding, of pain and pleasure, of success and failure, of ugliness giving way to beauty and beauty to ugliness. From the contemplation of it we get a feeling for life's transient beauty sustained over chaos. There is as it were a music to it all, and not the least when in a minor key. Even the religious urges within us, though they may promise a hereafter, are likely to advise us that we must for now rest content with the world we have been given. Though we are required to spend our life in struggle, yet we are able to cherish the good earth and to accept the kind of universe in which we find ourselves. It is no coincidence that our ecological perspective often approaches a religious dimension in trying to help us see the beauty, integrity, and stability of nature within and behind its seeming indifference, ferocity, and evils.

Dickenson's portrait can give an account of only half of nature, natural resistance, and even that is an enigmatic account of human life set oddly, set for martyrdom, in a hostile world. He can give no account of natural conductance; indeed, he cannot even see it, and thus he has mistakenly taken the half for the whole. But the account which I am seeking contains both elements, and not merely as a nonsensical mixture of goods and evils—each is a surd in relation to the other. A world in which there is an absurd mixture of helps and hurts is little better than a world of steady hostility. Neither could tutor us. What one needs is a nature where the evils are tributary to the goods, or, in my language of philosophical ecology, where natural resistance is embraced within and made intelligible by natural conductance. It is not death, but life, including human life as it fits this planetary environment, which is the principal mystery that has come out of nature. For several billion years, the ongoing development and persistence of that life, culminating in human life, have been the principal features of eco-nature behind which the element of struggle must be contained as a subtheme. Our conduct morally ought to fit this natural conductance. Life follows nature because nature follows life.

I do endorse in principle, though not without reservations, the constitution of the ecosystem. I do not make any long-range claims about the invariable, absolute law of evolution, about who is guiding the ship, or about the overall record of cosmic nature. There is beauty, stability, and integrity in the evolutionary ecosystem that we happen to have. There is a natural, an earthen, trend to life, although we cannot know it as a universal law. We ought to preserve and to value this nature, if only because it is the only nature that we know in any complexity and detail. If and when we find ourselves in some other nature, of a sort in which we earthlings can still maintain our sanity, we can then revise our ethic appropriately. In the meantime, however, we can at least sometimes "seek nature's guidance" in a tutorial sense almost as one might seek guidance from the Bible, or Socrates, or Shakespeare, even though nature, of course, does not "write" or "speak." None of us lives to the fullest who does not study the natural order, and, more than that, none of us is wise who does not ultimately make his peace with it.

When Mill faces the prospect of an unending expansion of the urban and rural environments, his attitude toward nature shifts, and, rather surprisingly, we find him among the defenders of nature. Suppose, God forbid, he writes, that we were brought by our industry to some future "world with nothing left to the spontaneous activity of nature; with every rood of land brought into cultivation, which is capable of growing food for human beings; every flowery waste or natural pasture ploughed up, all quadrapeds or birds which are not domesticated for man's use exterminated as his rivals for food, every hedgerow or superfluous tree rooted out, and scarcely a place left where a wild shrub or flower could grow without being eradicated as a weed in the name of improved agriculture." Such a world without "natural beauty and grandeur," Mill asserts, "is not good for man." Wild nature "is the cradle of thoughts and aspirations which are not only good for the individual, but which society could ill do without." [16] Thus, in the end, we enlist even this celebrated opponent of our morally following nature among those who wish to follow nature in our axiological sense.

For a closing statement on the tutorial sense of following nature, however, we do better to consult a poet rather than an ecologist or an ethicist. "I came from the wilderness," remembers Carl Sandburg as he invites us to reflect on the wilderness—how it tries to hold on to us and how, in our tutorial sense, we ought not to be separated from it:

> There is an eagle in me and a mockingbird... and the eagle flies among the Rocky Mountains of my dreams and fights among the Sierra crags of what I want... and the mockingbird warbles in the early forenoon before the dew is gone, warbles in the underbrush of my Chattanoogas of hope, gushes over the blue Ozark foothills of my wishes—And I got the eagle and the mockingbird from the wilderness.[17]

16. John Stuart Mill, *Principles of Political Economy*, in *Collected Works*, 3:756. Mill also records that reading Wordsworth's poetry reawoke in him a love of nature after his analytic bent of mind had caused a crisis in his mental history. See John Stuart Mill, *Autobiography* (Boston: Houghton Mifflin, 1969), pp. 88-90.

17. Carl Sandburg, "Wilderness," in *Complete Poems* (New York: Harcourt, Brace, Jovanovich, 1970), p. 100. Ellipsis in original.

Letting in the Jungle

MICHAEL F. SMITH

ABSTRACT *The destruction of the environment is a matter for moral concern and cannot be halted in the long term by appeals to human utility. However, the inadequacy and naïvety of humanist styles of ethical argument become apparent when attempts are made to extend them to environmental issues. They usually abstract certain supposed features of natural objects, e.g. sentience, and reify these as essential characteristics which operate to carry or ground ethical values. These arguments necessarily lead to the exclusion of objects which are, in fact, ethically valued or entail an unacceptably expansive egalitarianism. Such egalitarianism is often followed by a return to human-centred prejudices opposed to the originally stated aims of 'biocentric' ethicists like Taylor. Similarly, those physical and ecological holisms which rely not upon shared 'natural' features, but upon sharing in nature itself cannot solve this dilemma as they are incapable of explaining differential ethical values. The attempt to place boundaries on moral considerability should be abandoned in favour of an ethical pluralism which places emphasis on the context of valuations.*

It should not be necessary to begin an essay in environmental ethics with a litany of ecological disasters. I assume that we are all well aware of the earth's current predicament and its human causes. Suffice it to say, that the jungle has nearly all gone.

The argument of this paper is that the survival of the remaining few wilderness areas is a matter for moral concern. Their preservation will not be urged in the usual terms of human utility. For example, rain forests are often referred to as gene banks, potential resources for sustainable development, oxygen factories and so on. Though in some sense they may be all of these things, to justify their preservation by reference to these roles is to accept the language and rationale of their exploitation. These are expressions of human-centred attitudes towards nature and concrete examples of the imposition of managerial and financial constraints upon nature. Just as in our present bureaucratic/consumer society all has to be managerially approved and financially profitable, it is often argued that wilderness too needs to justify its continued existence on the same grounds. Though the defence of wild places by such means may be successful as a short-term expedient, to justify their preservation only in these terms is tacitly to accept the status quo and the ultimate hegemony of human self-interest.

The long-term consequences of such a policy are likely to be disastrous. If the fundamental reasons given for preserving habitats are those of human utility, then whenever and wherever the balance of utility favours habitat destruction this will occur. Once destroyed it can rarely be replaced. Bit by bit the wilderness is eroded until all that remains are a few curios, remnants of what once was, to be stared at and picked over. This is not just idle speculation, but describes what is happening now [1].

If arguments stemming from human utility cannot stop this degradation, and they are after all one of its primary causes, can ethical concerns help to preserve the wilderness? There are after all, ethical traditions of forms of ethics stretching back many hundreds of years which have allowed some place for non-human well-being. (Attfield [2], for example, following Passmore [3], traces these concerns in Western philosophy back to the Judeo-Christian heritage.)

A calculus of moral utilitarianism homologous to the concept of economic utility would fare no better. Indeed, it would be subject to identical drawbacks as far as giving a rationale for preserving wilderness areas is concerned. (This problem would not be resolved by advocating a 'rule' rather than an 'act' utilitarianism so long as the rules formulation took into account only human interests.) This paper argues that the forms of ethics that might help preserve wilderness are not those well-developed traditions, including utilitarianism, which might for want of a better word be labelled *humanist*. 'Humanist' is not used in contrast to 'religious' ethics, but rather to label a particular way of 'doing' ethics which would include Attfield's own style of argument. Just what it means will, I hope, become clear from the examples that follow.

One such humanist conception of morality might be described in these terms: as we look at the history of western society there seems to have been a uni-directional expansion of the bounds of moral considerability, from the immediate social group to ever widening categories of moral objects [4]. Peter Singer has noted that a popular metaphor for describing this broadening of ethical horizons is that of the expanding circle. A typical example he quotes comes from Lecky's *History of European Morals*, first published in 1869.

> ... benevolent affections embrace merely the family, soon the circle expanding includes first a class, then a nation, then a coalition of nations, then all humanity and finally its influence is felt in the dealings of man with the animal world. [5]

Singer's own book *The Expanding Circle* both elucidates and epitomises this humanist approach [6]. His thesis is that ethics originated in forms of biological behaviour such as kin selection and reciprocal altruism, whereby apparently altruistic acts of individuals are explained by their role in increasing the genetic contribution of that individual's genome to the gene pool of the next generation. A mother shares 50% of her genes on average with an offspring. Put in its crudest form, those mothers who die saving more than two offspring will be selected for. Thus, altruism as a feature seems amenable to explanation in terms of so called 'selfish genes'.

The altruistic faculty, according to Singer, comes to take on a new form for humans because of our endowment with language and rationality. Justifications of actions affecting the wider community come to be given in terms of reasons. For example, I may justify my claim to a greater than average share of the food on the basis that I do more work than most. This might be accepted, but then someone points out that Freda does more work that I do and so is entitled, on this basis, to more food still.

Once utilised, this form of rational argument suggests that we "cannot get away with different ethical judgments in apparently identical situations" [7]. Certain rational considerations can call into question previously held prejudices about the limits of moral considerability. For example, if it is right to help person A in a given situation then why not person B?

Altruistic tendencies had, in the first instance, only extended to the immediate family or our own group, but the 'autonomy of reasoning' entails a logic whereby the boundaries of 'our own' expand to the next largest community with which we identify [8]. Perhaps this community is a social class or a race, but once such an extension of moral considerability has been justified then its boundary too is, in turn, open to questioning. Why, for example, should one skin colour be preferable to another? Viewed in this way, the history of morals comes to be seen as an increasingly enlightened view about those we conceive of as having affinities to ourselves. Like the layers of an onion the boundaries of moral considerability come to overlie each other as the rational justification for each is formulated and then challenged.

Eventually, we reach a stage where claims to moral status are justified in terms of features of something called 'human nature'; perhaps the possession of a rational faculty itself. This stage is equivalent to the roughly Kantian position, that if I am morally considerable because of my rationality then all rational beings must be so considered. This being so, individual members of different races, sexes and so forth apparently obtain equal moral status (unless of course we can find reasons for doubting that all sections of humanity are equally rational). In connection with the issue of moral extensionism, it is worth noting the importance that has been attached at different times to phrenology, I.Q. tests and other 'scientific' methods of discrimination [9].

However, why should the policy of extension stop at the level of species? In a famous quotation, Bentham points out both the drawbacks in relying on rationality or language to delimit moral considerability and suggests instead that ability to feel pain or pleasure is the appropriate moral arbiter.

> It may one day be recognised, that the number of legs, the villosity of the skin, or the termination of the os sacrum, are reasons equally insufficient [to skin colour] for abandoning a sensitive being... What else is it that should trace the insuperable line? Is it the faculty of reason, or perhaps, the faculty of discourse? But a full-grown horse or dog is beyond comparison a more rational, as well as a more conversable animal, than is an infant of a day, or a week or even a month old... the question is not, can they reason? nor can they talk? but, can they suffer? [10]

This is, indeed, the position that Singer takes, claiming that this is the outer layer of the onion. The difference between sentience and non-sentience is not, says Singer, a morally arbitrary boundary in the way that species differences are. A line drawn at sentience does, however, exclude most of the animal kingdom, and certainly plants, waterfalls and whole ecosystems from moral considerability.

However, we can fill out this notion of rationally argued affinities in still other ways. Instead of using shared natural characteristics—genes, skin colour, sex, human nature or even life itself to dictate moral boundaries one could refer to shared interests. For example, our affinity with a particular class might be a common interest in overcoming the exploitation we suffer due to our social position. The less alike our social circumstances the less likely we may be to consider someone as part of our moral community.

Now, if interests are of critical importance, the outer layer of moral considerability will be bounded by an ability to possess interests. Singer believes that the capacity to possess interests is co-extensive with sentience, but others have a wider perspective. Why should plants not have interests in obtaining enough water and nutrients? Thus, for philosophers like Robin Attfield, plants too find a place within the expanded circle. For him the interests of non-sentient beings lie in "their flourishing or their capacity for flourishing after the manner of their kind..." [11]

Paul Taylor proposes a 'biocentric' theory of environmental ethics where the outer layer of moral considerability is to be determined by a thing's ability to possess a good-of-its-own. To have a good-of-its-own the object must be capable of being harmed or benefited as a teleological centre of life, having its own species-specific goals. The goals of an organism are realised when it has successfully maintained "the normal biological functions of its species", thus developing to its full potential. A butterfly species, for example, has a life cycle from egg to caterpillar to chrysalis to imago. To stop any individual butterfly from playing each of these roles would constitute a harm to it. Having a good-of-one's own is then a necessary condition of moral considerability (of having inherent worth in Taylor's terminology), but is not sufficient. This distinction between things which have and which lack a good-of-their-own equates, according to Taylor, to that between the living and non-living, and constitutes the justification principle which marks the outer boundary of moral standing [12].

To summarise the argument so far: the humanist presents us with a succession of features or capacities that are supposed to determine the bounds of moral considerability. All previous boundaries as they become superseded are seen to have been mistaken, their core justifying principle being too limited in scope. They were based on the wrong objective essential characteristic; that characteristic which has the role of carrying, or at least grounding, value. However, now the humanist faces a serious dilemma. For, as the boundary principles become less and less specific to take account of the wider categories of ethical objects we wish to countenance, this form of rational argument brings with it a new and more expansive egalitarianism.

If, for example, possession of interests is the criterion used there seems to be no over-arching reason why the interests of one type of organism should have more importance than any other. All things capable of having their interests benefited or harmed are equally considerable whether aphids, dandelions or humans. This extreme position would be held by very few. On the other hand, to relate everything to similarity of interests with humans seems unjustifiably prejudiced. Faced with the possibility of widespread natural egalitarianism most humanists backtrack and busy themselves constructing rational justifications for their prejudice in much the same way as others had previously tried to exclude various sections of humanity from equal consideration [13].

Not all ethicists are equally culpable in these human-centred prejudices. Taylor is specifically concerned to promote this natural egalitarianism as the heart of his 'biocentric' perspective. He states, "All animals however dissimilar to humans they may be are beings that have a good-of-their-own" and "... all plants are likewise beings that have a good-of-their-own" [14]. And again:

> The first thing we do when we accept the biocentric outlook is to take the fact of our being members of a biological species to be a fundamental feature of our existence. We do not deny the differences between ourselves and other species, any more than we deny the differences among other species themselves. Rather, we put aside these differences and focus our attention upon our nature as biological creatures.
>
> ... we keep in the forefront of our consciousness the characteristics we share with all forms of life on Earth. Not only is our common origin in one evolutionary process fully acknowledged, but also the common environmental circumstances that surround us all. We view ourselves as one with them, not as set apart from them. We are then ready to affirm our fellowship with them as equal members of the whole Community of Life on Earth. [15]

Such is his theoretical standpoint, but when it comes to the practical implications of this policy for human interaction with the environment he is less candid. All that this egalitarianism practically requires is that "*certain* habitats used by wild-species populations are not destroyed, and some wildlife is given a chance to survive alongside the works of human culture" [my emphasis]. "Animals" says Taylor, "are not of *greater* worth [than humans] so there is no obligation to further their interests at the cost of basic interests to humans" [original emphasis] [16].

However, surely there are cases where if equal moral status is to count for anything, the basic interests of animals and plants will outweigh those of humans. Indeed, since Taylor's theory gives inherent worth to microscopic individuals, almost every act we perform becomes of immense moral importance, harming and destroying millions of our fellow citizens. In spraying a crop we destroy vast quantities of insects, fungi, etc., all supposedly on an equal footing with ourselves. Taylor chooses to ignore the potentially restrictive nature of his thesis and instead makes some extremely bland generalisations about living in harmony with nature.

The approaches I have called humanist are, I suggest, ethically inappropriate, unworkable and vastly over-simplistic. Similarities of faculty are reified into universal demarcation principles in an attempted emulation of the natural sciences. The only empirical evidence admitted is scientific evidence on the distribution of the chosen demarcating faculty in the natural world. Thus, intelligence testing, biological taxonomy and sociobiology are all admissible as evidence for the possible moral considerability of a class of objects. What is not admissible, though, is evidence about whether people actually do so regard an object. What is positively dismissed is the massive plurality of reasons why people can and do value things morally. The humanist mania for objective theoretical criteria leads to a monolithic reductionism combined with an unwarranted mystification of one particular faculty as somehow bearing moral value [17].

Perhaps the clearest way to see the problems this view creates is by looking at those things that are drawn out of moral bounds, things beyond the periphery of the expanded circle. In discussing his concept of the good of a being Taylor contrasts a child with a pile of sand. The sand, he writes, has "... no good of its own. It is not the sort of thing that can be included in the range of application of the concept entity-that-has-a-good-of-its-own" [18]. This being so it is excluded from moral considerability.

Yet we certainly can extend moral considerability even to piles of sand in certain contexts (the context is all important). The barchans, great crescent-shaped dunes found in the sand deserts of Arabia's 'empty quarter' have inspired the imagination of many travellers and moulded the lives of people like the Bedouin who have lived amongst them. The sandstone of regions like Exmoor or, more impressively, the Pakaraima mountains on the border of Guyana and Venezuela, containing some of the world's highest waterfalls, is directly responsible for their particular ambience. The feelings and forms of life these 'piles of sand' have generated can lead and have led to their being valued for their own sake in ways that can best be described as ethical. To take a different example, when oil spills from tankers onto sandy beaches we think such avoidable occurrences morally reprehensible, not just because they are unaesthetic, but because it makes sense to talk of desecration of the beaches.

Before pursuing these points further I want to look at some closely related ways of expanding ethical consideration. These too could be labelled 'humanist'. The rationale for extending such consideration has so far depended upon the sharing of certain features or capacities judged to be of moral importance. A different method might depend not on sharing anything with nature, but upon sharing in nature itself. One example of this form of argument is provided by J. Baird Callicott. He takes a radical reductionist stance based on his interpretation of quantum physics. According to Callicott, our apparent individuality and isolation from nature is mistaken, for at the level of quanta we are actually continuous with the world: Callicott endorses Alan Watt's sentiment that "the world is your body" [19].

If this is the case then Callicott thinks we can dismiss arguments about the intrinsic value of different attributes, we need only posit that the self is valuable: "nature is intrinsically valuable to the extent that the self is intrinsically valuable". Environmental degradation is thus to be seen as an attack on my extended person: "the injury *to me* of environmental destruction is primarily and directly to my extended self, to the larger body and soul with which I am continuous" [20].

This form of argument can be given short shrift. Humans do not operate ethically at the level of quanta. The fact, if it is a fact, that we are one with nature at this level gives us no ethical guidance at all, for so too are murderers, logging companies and industrialists. This is not to say that the perception that we form a part of a greater whole will always be morally insignificant. Such a view may, for example, lead to the valuing of nature as a whole system. The acknowledgement of holism may be central to particular ethical ideals in other ways, as it was for the stoics [21] and Spinoza. (An ethical approach based upon Spinoza's philosophy by Gilles Deleuze lends itself to environmental applications [22].) However, by itself, Callicott's holism cannot give us any ethical guidance. To live in the world we need to act differentially to parts of it. We have to relate on a human scale with whales, mountains and other humans. Ethics is about the resolution of conflicts between ourselves and about our relationship with our environment. The existence of physical links does not necessitate that conflict will cease.

Furthermore, in terms of quantum physics it is very difficult to talk about environmental destruction at all. The destruction entailed when a beefburger is produced via a circuitous route from forest, to grassland, to cattle, can not be expressed in terms of quanta. It can only be expressed in terms of the forest itself. Fundamental ethical dilemmas are left entirely unaltered by this egocentric holism. Egoism is not itself unproblematic. An egocentric holism like Callicot's could just as easily support a complete disrespect for the surrounding world on the ground that as the world is a part of *my* body, it is mine to do with as I wish.

Callicott's holism is only partly based on quantum physics. It also rests upon what might be termed eco-holism. Here, the science appealed to is ecology rather than physics. Ecology, it is claimed, reveals our place in nature as a locus in an interdependent network of organisms [23]. This interdependency should lead us to re-evaluate the worth of other natural things and see ourselves as just one amongst many. Again, this may be true, and have important metaphorical and practical implications, but it seems far from clear that it has any *necessary* ethical implications. One could clearly grasp an ecological understanding of our place in nature and yet still treat other organisms as mere means to human ends.

The eco-holist view reaches its apex in the Gaia hypothesis of Lovelock [24] which sees the earth as one giant self-regulating organism. Lovelock is concerned to stress the scientific nature of his theory and in his later work at least regards his theory as having no *necessary* ethical implications. He states that "there is no prescription for living with Gaia, only consequences" [25]. As Lovelock remarks, with mild approbation, some of his 'followers' have, however, taken his theory on a different level as a mystical concept entailing specific modes of treating the world. Lovelock's tacit approval for the mystical interpretation of his work seems to stem from a pragmatic approach towards influencing others to care for their environment. For example, Christians might be persuaded to see the Virgin Mary as embodied in Gaia and thus come to change environmentally destructive practices [26]. This mystical interpretation of Lovelock's hypothesis is closely akin to the pantheistic holism of Romantics like William Wordsworth. In an essay entitled "Wordsworth in the tropics" Aldous Huxley makes points relevant to the argument against humanism in this paper. He writes:

> It is only very occasionally that he [Wordsworth] admits the existence in the world around him of those 'unknown modes of being' of which our immediate intuitions of things make us so disquietingly aware. Normally what he does is to pump the dangerous unknown out of nature and refill the emptied forms of hills and woods, flowers and waters with something more reassuringly familiar—with humanity, with Anglicanism. He will not admit that a yellow primrose is simply a yellow primrose—beautiful but essentially strange, having its own alien life apart. He wants it to possess some sort of a soul, to exist humanly, not simply flowerily... But the life of vegetation is radically unlike the life of man.
>
> The jungle is marvellous, fantastic, beautiful, but it is also terrifying, it is also profoundly sinister. There is something in what, for want of a better word, we must call the character of great forests... which is foreign, appalling, fundamentally and utterly inimical to intruding man. [27]

Later, he writes:

> A few months in the jungle would have convinced him [Wordsworth] that the diversity and utter strangeness of nature are at least as real and significant as its *intellectually discovered* unity [my emphasis]. [28]

We can now reconsider the general humanist rationale I have outlined, bearing Huxley's comments in mind. It is only now that I want to refer directly to the title of the paper. It comes, of course, from Kipling's *Second Jungle Book* [29]. Jungle epitomises, or at least epitomised in Kipling's era, wildness, ferocity, the power of nature, the unexplained, the untamable, that part of nature over which humans lacked control. It is a particular historical representation of the 'otherness' of nature, the alien character described by Huxley. In our own case, let it refer to the deserts, the oceans and all the remnants of wilderness left in the world, however small.

I want to suggest that the reasons for valuing the jungle or the primrose, the desert or Antarctica are manifold and often concerned more with our perception of their disparity from humanity than any affinities, whether these are natural or intellectually contrived. It would, however, be wrong to take the present argument as putting forward 'alien otherness' as itself a criterion for moral considerability. The perception of something as 'other' does, in certain contexts lead to its being ethically valued (e.g. Iris Murdoch [30]), but it is by no means necessary that it should do so. As Stephen Clark points out [31], our valuation of the environment as 'other' may be dinted when we realise that it now survives only by human protection. How alien is an area enclosed in barbed wire or patrolled by uniformed rangers? This sort of question raises problems for anyone wishing to treat 'otherness' as a criterion of moral considerability. However, the point of this paper is to subvert the whole enterprise of providing necessary criteria, or grounds for moral considerability. The methods of humanist philosophy which depend upon shared common features (even shared 'otherness') reach the end of the road where the jungle begins. They were never logically compelling in any case. They serve only to impose too rigid a structure upon our moral beliefs and values. If moral consideration is to be extended to non-humans this has to be done not on the contrived and spurious basis of shared properties, but on due recognition of our natural phenomena for their differences as well as their similarities, and the many and varied ways we can relate to them.

The thesis of the expanding circle provides a graphic representation of anthropocentrism. Humanity sits at the centre of a concentrically ordered nature, as the archetype of ethical value, both the measure and the measurer of all things. In its theoretical development the humanist rationale develops an abstract and unworkable egalitarianism. In practice, the greater the difference between 'us' and 'them' the less is the gravitational pull on our moral faculties. In reality, the periphery of moral considerability is determined by whatever arbitrary feature or concept we are happiest with in any given historical and cultural circumstance. The continual discovery of new and 'better' demarcation principles is a fiction, a 'Just So Story' to use another of Kipling's phrases. In its dependence upon an a-contextual rationality it ignores the vital place that context plays in our ascription of values.

This is no way to treat nature with respect. Nor will it provide any better a barrier to habitat destruction than human utility. It is a parody of our ethical valuation of nature on much the same level that Walt Disney's *Jungle Book* is a parody of Kipling's book, a cartoon fulfilling our inclinations to anthropomorphize everything: the pyromaniac apes who want to be 'men' [sic], the bear that wears a grass skirt. These sanitised symbols show nature with a human face, obeying human rationales. The Walt Disney of humanist philosophy allows us both to subjugate the 'otherness' of nature, and to simplify ethical complexities into categories of the morally considerable and those beyond the pale.

A striking feature of Kipling's jungle is its otherness. The wolves are wolves and have their own world-view, the law of the jungle. Perhaps the only long-term chance for the survival of the jungle lies in our coming to see it as being of intrinsic value on its own terms. The jungle offers us a chance to escape a world where all we see reflects 'humanity' back at us. The appropriateness of using ethical language in discussions of environmental concerns lies not in the similarity of the moral objects to ourselves, but in morality's ability to express concerns about a wider community, a community not of equals, but of inter-relationships. What we need to do is to let the jungle into our moral considerations.

If we have a passion for wilderness it will not be stemmed by the humanist who calls us unreasonable. If it is unreasonable to value rivers, if mountains are not morally considerable and deserts not intrinsically valuable to humanists, that is because they have too narrow a vision. Their eyes are closed. They have failed to grasp what the jungle is and what it can represent.

To quote David Hume: "... reason is and only should be the slave of the passions" [32]. To quote Bagheera: "We of the jungle know that man is the wisest of all. If we trusted our ears we should know that of of all things he is most foolish".

Michael F. Smith, Department of Philosophy, University of Stirling, Stirling FK9 4LA, United Kingdom.

Acknowledgements

I would like to thank Andrew Brennan and Stephen Clark for comments on this paper, a shortened version of which was given at the inaugural British meeting of the International Society for Environmental Ethics at the joint session of Mind and the Aristotelian Society, 13-16 July, 1990.

NOTES

[1] See, for example, the Worldwatch Institute Report (1990) *The State of the World 1990* (London, Unwin).
[2] ROBIN ATTFIELD (1983) *The Ethics of Environmental Concern* (Oxford, Blackwell).
[3] JOHN PASSMORE (1974) *Man's Responsibility for Nature* (London, Duckworth).
[4] Papers on moral considerability of particular relevance to environmental ethics are CHRISTOPHER STONE (1974) *Should Trees Have Standing?* (Los Angeles, Kaufman); KENNETH GOODPASTER (1978) On Being Morally Considerable, *Journal of Philosophy*, 75, pp. 308-324; ANDREW BRENNAN (1984) The Moral Standing of Natural Objects, *Environmental Ethics*, 6, pp. 35-36.
[5] The Lecky quotation prefaces PETER SINGER (1981) *The Expanding Circle, Ethics and Sociobiology* (Oxford, Clarendon Press).
[6] Ibid.
[7] Ibid., p. 93.
[8] Ibid., p. 113. Singer is proposing that autonomous rational argument is the primary cause of the historical extension of moral considerability beyond kin and reciprocal altruism to wider society. But, we might note that the development of a language complex enough to produce and express such rational arguments might itself require a fairly stable and complex society, presumably including some moral norms. The historical and causal primacy of rational argument is, therefore, questionable.
[9] For a lucid account of such scientific prejudices see, for example, STEPHEN JAY GOULD (1981) *The Mismeasure of Man* (New York, Norton).
[10] J. BENTHAM (1907) *An Introduction to the Principles of Morals and Legislation* (Oxford, Oxford University Press), p. 311.
[11] ATTFIELD, op. cit., p. 154.
[12] PAUL TAYLOR (1986) *Respect for Nature* (Princeton, Princeton University Press).
[13] For example, ATTFIELD, op. cit., having extended the boundary of moral considerability to those things capable of possessing interests then constructs a rational justification which, in effect, severely limits the degree of consideration we can actually give things to their degree of similarity to humans. See also the remarks on Taylor in this paper.
[14] TAYLOR, op. cit., p. 66.
[15] Ibid., p. 101.
[16] For a full explanation of Taylor's principles "for the fair resolution of conflicting claims" between humans and non-humans and the source of these quotations see op. cit., Ch. 6.
[17] Singer provides a clear example of the problems associated with humanist arguments when he refers to genetic arguments about racial differences: "Equality is a moral idea, not simply an assertion of fact. There is no logically compelling reason for assuming a factual difference in ability between two people justifies any difference in the amount of consideration we give to satisfying their needs and interests". PETER SINGER (1976) All Animals Are Equal, in: T. REGAN & P. SINGER (Eds) *Animal Rights and Human Obligations* (New Jersey, Prentice-Hall) p. 152. Singer must be arguing that since the boundaries of moral consideration based on racial differences are mistaken, using as they do, the wrong criterion of moral demarcation, we need no longer be interested in differences between individuals on these grounds. If moral status depends not upon intelligence, but upon sentience, the degree of intelligence someone possesses is superfluous. Singer just cannot mean what he says when he states that factual differences are immaterial to moral considerability. If sentience is the boundary delimiting moral standing then factual differences in ability to feel pleasure or pain will be crucial in moral deliberation. (All the racist need now do is focus attention on sensitivity rather than intelligence.) Like it or not, factual differences are central to Singer's and other humanists' arguments on the expanding circle.
[18] TAYLOR, op. cit., p. 61.
[19] J. BAIRD CALLICOTT (1985) Intrinsic Value, Quantum Theory, and Environmental Ethics, *Environmental Ethics*, 7, p. 274.
[20] Ibid., p. 275.

[21] A. A. LONG (1974) *Hellenistic Philosophy* (London, Duckworth). See Ch. 4, especially section 5.
[22] GILLES DELEUZE (1988) *Spinoza: Practical Philosophy* (San Francisco, City Light Books).
[23] For more advanced formulations of eco-holism see ARNE NAESS (1984) In Defense of the Deep Ecology Movement, *Environmental Ethics*, 6, pp. 265–270, and ANDREW BRENNAN (1988) *Thinking About Nature* (London, Routledge).
[24] J. E. LOVELOCK (1979) *Gaia, A New Look At The Earth* (Oxford, Oxford University Press).
[25] J. E. LOVELOCK (1988) *The Ages Of Gaia* (Oxford, Oxford University Press), p. 225.
[26] Ibid., p. 223.
[27] ALDOUS HUXLEY (1932) Wordsworth In The Tropics, in: *Rotunda* (London, Chatto & Windus). Stephen Clark has pointed out that the 'unknown modes of being' Huxley quotes from the Prelude (l. 393) is, in fact, followed by a reference to mountains as "... huge and mighty forms, that do not live like living men...".
[28] Ibid., p. 883.
[29] RUDYARD KIPLING (1931) *The Second Jungle Book* (London, Macmillan).
[30] IRIS MURDOCH (1970) makes the point that, "More naturally, as well as more properly, we take a self-forgetful pleasure in the sheer alien pointless independent existence of animals, birds, stones and trees". *The Sovereignty of Good* (London, Routledge & Kegan Paul), p. 85.
[31] STEPHEN CLARK (1983) in: R. ELLIOT & A. GARE (Eds) *Environmental Philosophy* (Milton Keynes, Open University Press), p. 187.
[32] DAVID HUME (1978) in: L. A. SELBY-BIGGE (Ed.) *A Treatise on Human Nature* (Oxford, Oxford University Press), p. 415.

[22]

THE BIG LIE:
HUMAN RESTORATION OF NATURE

Eric Katz

The trail of the human serpent is thus over everything.
—William James, *Pragmatism*

I

I begin with an empirical point, based on my own random observations: the idea that humanity can restore or repair the natural environment has begun to play an important part in decisions regarding environmental policy. We are urged to plant trees to reverse the "greenhouse effect." Real estate developers are obligated to restore previously damaged acreage in exchange for building permits.[1] The U.S. National Park Service spends $33 million to "rehabilitate" 39,000 acres of the Redwood Creek watershed.[2] And the U.S. Forest Service is criticized for its "plantation" mentality: it is harvesting trees from old-growth forests rather than "redesigning" forests according to the sustainable principles of nature. "Restoration forestry is the only true forestry," claims an environmentally-conscious former employee of the Bureau of Land Management.[3]

These policies present the message that humanity should repair the damage that human intervention has caused the natural environment. The message is an optimistic one, for it implies that we recognize the harm we have caused in the natural environment and that we possess the means and will to correct these harms. These policies also make us feel good; the prospect of restoration relieves the guilt that we feel about the destruction of nature. The wounds we have inflicted on the natural world are not permanent; nature can be made "whole" again. Our natural resource base and foundation for survival can be saved by the appropriate policies of restoration, regeneration, and redesign.

It is also apparent that these ideas are not restricted to policymakers, environmentalists, or the general public—they have begun to pervade the normative principles of philosophers concerned with developing an adequate environmental ethic. Paul Taylor uses a concept of "restitutive justice" both as one of the basic rules of duty in his biocentric ethic and as a "priority principle" to resolve competing claims.[4] The basic idea of this rule is that human violators of nature will in some way repair or compensate injured natural entities and systems. Peter Wenz also endorses a principle of restitution as being essential to an adequate theory of environmental ethics; he then attacks Taylor's theory for not presenting a coherent principle.[5] The idea that humanity is morally responsible for reconstructing natural areas and entities—species, communities, ecosystems—thus becomes a central concern of an applied environmental ethic.

In this paper I question the environmentalists' concern for the restoration of nature and argue against the optimistic view that humanity has the obligation and ability to repair or reconstruct damaged natural systems. This conception of environmental policy and environmental ethics is based on a misperception of natural reality and a misguided understanding of the human place in the natural environment. On a simple level, it is the same kind of "technological fix" that has engendered the environmental crisis. Human science and technology will fix, repair, and improve natural processes. On a deeper level, it is an expression of an anthropocentric world view, in which human interests shape and redesign a comfortable natural reality. A "restored" nature is an artifact created to meet human satisfactions and interests. Thus, on the most fundamental level, it is an unrecognized manifestation of the insidious dream of the human domination of nature. Once and for all, humanity will demonstrate its mastery of nature by "restoring" and repairing the degraded ecosystems of the biosphere. Cloaked in an environmental consciousness, human power will reign supreme.

II

It has been eight years since Robert Elliot published his sharp and accurate criticism of "the restoration thesis."[6] In an article entitled "Faking Nature,"

Elliot examined the moral objections to the practical environmental policy of restoring damaged natural systems, locations, landscapes. For the sake of argument, Elliot assumed that the restoration of a damaged area could be recreated perfectly, so that the area would appear in its original condition after the restoration was completed. He then argued that the perfect copy of the natural area would be of less value than the original, for the newly restored natural area would be analogous to an art forgery. Two points seem crucial to Elliot's argument. First, the value of objects can be explained "in terms of their origins, in terms of the kinds of processes that brought them into being."[7] We value an art work in part because of the fact that a particular artist, a human individual, created the work at a precise moment in historical time. Similarly, we value a natural area because of its "special kind of continuity with the past." But to understand the art work or the natural area in their historical contexts we require a special kind of insight or knowledge. Thus, the second crucial point of Elliot's argument is the co-existence of "understanding and evaluation." The art expert brings to the analysis and evaluation of a work of art a full range of information about the artist, the period, the intentions of the work, and so on. In a similar way, the evaluation of a natural area is informed by a detailed knowledge of ecological processes, a knowledge that can be learned as easily as the history of art.[8] To value the restored landscape as much as the original is thus a kind of ignorance; we are being fooled by the superficial similarities to the natural area, just as the ignorant art "appreciator" is fooled by the appearance of the art forgery.

Although Elliot's argument has had a profound effect on my own thinking about environmental issues, I believed that the problem he uses as a starting point is purely theoretical, almost fanciful.[9] After all, who would possibly believe that a land developer or a strip mining company would actually restore a natural area to its original state? Elliot himself claims that "the restoration thesis" is generally used "as a way of undermining the arguments of conservationists."[10] Thus it is with concern that I discover that serious environmentalist thinkers, as noted above, have argued for a position similar to Elliot's "restoration thesis." The restoration of a damaged nature is seen not only as a practical option for environmental policy but also as a moral obligation for right-thinking environmentalists. If we are to continue human projects which (unfortunately) impinge on the natural environment (it is claimed), then we must repair the damage. In a few short years a "sea-change" has occurred: what Elliot attacked as both a physical impossibility and a moral mistake is now advocated as proper environmental policy. Am I alone in thinking that something has gone wrong here?

Perhaps not enough people have read Elliot's arguments; neither Taylor nor Wenz, the principal advocates of restitutive environmental justice, list this article in their notes or bibliographies. Perhaps we need to re-examine the idea of recreating a natural landscape; in what sense is this action analogous to an art forgery? Perhaps we need to push beyond Elliot's analysis, to use his arguments as a starting point for a deeper investigation into the fundamental errors of restoration policy.

III

My initial reaction to the possibility of restoration policy is almost entirely visceral: I am outraged by the idea that a technologically created "nature" will be passed off as reality. The human presumption that we are capable of this technological fix demonstrates (once again) the arrogance with which humanity surveys the natural world. Whatever the problem may be, there will be a technological, mechanical, or scientific solution. Human engineering will modify the secrets of natural processes and effect a satisfactory result. Chemical fertilizers will increase food production; pesticides will control disease-carrying insects; hydroelectric dams will harness the power of our rivers. The familiar list goes on and on.

The relationship between this technological mind-set and the environmental crisis has been amply demonstrated, and need not concern us here.[11] My interest is narrower. I want to focus on the creation of artifacts, for that is what technology does. The recreated natural environment that is the end result of a restoration project is nothing more than an artifact created for human use. The problem for an applied environmental ethic is the determination of the moral value of this artifact.

Recently, Michael Losonsky has pointed out how little we know about the nature, structure, and meaning of artifacts. "[C]ompared to the scientific study of nature, the scientific study of artifacts is in its infancy."[12] What is clear, of course, is that an artifact is not equivalent to a natural object; but the precise difference, or set of differences, is not readily apparent. Indeed, when we consider objects such as beaver dams, we are unsure if we are dealing with natural objects or artifacts. Fortunately, however, these kind of animal-created artifacts can be safely ignored in the present investigation. Nature restoration projects are obviously human. A human built dam is clearly artifactual.

The concepts of function and purpose are central to an understanding of artifacts. Losonsky rejects the Aristotelian view that artifacts (as distinguished from natural objects) have no inner nature or hidden essence that can be discovered. Artifacts have a "nature" that is partially comprised of three features: "internal structure, purpose, and manner of use." This nature, in turn, explains why artifacts "have predictable lifespans during which they undergo regular and predictable changes."[13] The structure, function, and use of the artifacts determine to some extent the changes which they undergo. Clocks would not develop in a manner which prevented the measurement of time.

Natural objects lack the kind of purpose and function found in artifacts. As Andrew Brennan has argued, natural entities have no "intrinsic functions," as he calls them, for they were not the result of design. They were not created for a particular purpose; they have no set manner of use. Although we often speak as if natural individuals (for example, predators) have roles to play in ecosystemic well-being (the maintenance of optimum population levels), this kind of talk is either metaphorical or fallacious. No one created or designed the mountain lion as a regulator of the deer population.[14]

This is the key point. Natural individuals were not designed for a purpose. They lack intrinsic functions, making them different from human-created artifacts. Artifacts, I claim, are essentially anthropocentric. They are created for human use, human purpose—they serve a function for human life. Their existence is centered on human life. It would be impossible to imagine an artifact not designed to meet a human purpose. Without a foreseen use the object would not be created. This is completely different from the way natural entities and species evolve to fill ecological niches in the biosphere.

The doctrine of anthropocentrism is thus an essential element in understanding the meaning of artifacts. This conceptual relationship is not generally problematic, for most artifacts are human creations designed for use in human social and cultural contexts. But once we begin to redesign natural systems and processes, once we begin to create restored natural environments, we impose our anthropocentric purposes on areas that exist outside human society. We will construct so-called natural objects on the model of human desires, interests, and satisfactions. Depending on the adequacy of our technology, these restored and redesigned natural areas will appear more or less natural, but they will never be natural—they will be anthropocentrically designed human artifacts.

A disturbing example of this conceptual problem applied to environmental policy can be found in Chris Maser's *The Redesigned Forest*. Maser is a former research scientist for the United States Department of Interior Bureau of Land Management. His book attests to his deeply felt commitment to the policy of "sustainable" forestry, as opposed to the short term expediency of present day forestry practices. Maser argues for a forestry policy that "restores" the forest as it harvests it; we must be true foresters and not "plantation" managers.

Nonetheless, Maser's plans for "redesigning" forests reveal several problems about the concepts and values implicit in restoration policy. First, Maser consistently compares the human design of forests with Nature's design. The entire first chapter is a series of short sections comparing the two "designs." In the "Introduction," he writes, "[W]e are redesigning our forests from Nature's blueprint to humanity's blueprint."[15] But Nature, of course, does not have a blueprint, nor a design. As a zoologist, Maser knows this; but his metaphorical talk is dangerous. It implies that we can discover the plan, the methods, the processes of nature, and mold them to our purposes.

Maser himself often writes as if he accepts that implication. The second problem with his argument is the comparison of nature to a mechanism that we do not fully understand. The crucial error we make in simplifying forest ecology—turning forests into plantations—is that we are assuming our design for the forest mechanism is better than nature's. "Forests are not automobiles in which we can tailor artificially substituted parts for original parts."[16] How true. But Maser's argument against this substitution is empirical: "A forest cannot be 'rebuilt' and remain the same forest, but we could probably rebuild a forest similar to the original if we knew how. No one has ever done it. . . . [W]e do not have a parts catalog, or a maintenance manual. . . ."[17] The implication is that if we did have a catalog and manual, if nature were known as well as artifactual machines, then the restoration of forests would be morally and practically acceptable. This conclusion serves as Maser's chief argument for the preservation of old-growth and other unmanaged forests: "We have to maintain some original, unmanaged old-growth forest, mature forest, and young-growth forest as parts catalog, maintenance manual, and service department from which to learn to practice restoration forestry."[18] Is the forest-as-parts-catalogue a better guiding metaphor than the forest-as-plantation?

This mechanistic conception of nature underlies, or explains, the third problem with Maser's argument. His goal for restoration forestry, his purpose in criticizing the short-term plantation mentality, is irredeemably anthropocentric. The problem with present-day forestry practices is that they are "exclusive of all other human values except production of fast-grown wood fiber."[19] It is the elimination of other human values and interests that concerns Maser. "We need to learn to see the forest as the factory that produces raw materials. . . ." to meet our "common goal[:] . . . a sustainable forest for a sustainable industry for a sustainable environment for a sustainable human population."[20] Restoration forestry is necessary because it is the best method for achieving the human goods which we extract from nature. Our goal is to build a better "factory-forest," using the complex knowledge of forest ecology.

What is disturbing about Maser's position is that it comes from an environmentalist. Unlike Elliot's theoretical opponents of conservation, who wished to subvert the environmentalist position with the "restoration thesis," Maser advocates the human design of forests as a method of environmental protection and conservation for human use. His conclusion shows us the danger of using anthropocentric and mechanistic models of thought in the formulation of environmental policy. These models leave us with forests that are "factories" for the production of human commodities, spare-parts catalogs for the maintenance of the machine.

I began this section with a report of my visceral reaction to the technological re-creation of natural environments. This reaction has now been explained and analyzed. Nature restoration projects are the creations of human technologies, and as such, are artifacts. But artifacts are essentially the constructs of an anthropocentric world view. They are designed by humans for humans to satisfy human interests and needs. Artifactual restored nature is thus fundamentally different from natural objects and systems which exist without human design. It is not surprising, then, that we view restored nature with a value different from the original.

IV

To this point, my analysis has supported the argument and conclusions of Elliot's criticism of "the restoration thesis." But further reflection on the nature of artifacts, and the comparison of forests to well run machines, makes me doubt the central analogy which serves as the foundation of his case. Can we compare an undisturbed natural environment to a work of art? Should we?

As noted in Section II, Elliot uses the art/nature analogy to make two fundamental points about the process of evaluation: (1) the importance of a continuous causal history; and (2) the use of knowledge about this causal history to make appropriate judgments. A work of art or a natural entity which lacks a continuous causal history, as understood by the expert in the field, would be judged inferior. If the object is "passed off" as an original, with its causal history intact, then we would judge it to be a forgery or an instance of "faked" nature.

I do not deny that this is a powerful analogy. It demonstrates the crucial importance of causal history in the analysis of value. But the analogy should not be pushed too far, for the comparison suggests that we possess an understanding of art forgery that is now simply being applied to natural objects. I doubt that our understanding of art forgery is adequate for this task. L. B. Cebik argues that an analysis of forgery involves basic ontological questions about the meaning of art. Cebik claims that it is mistake to focus exclusively on questions of value when analyzing art forgeries, for the practice of forgery raises fundamental issues about the status of art itself.[21]

According to Cebik, an analysis of forgeries demonstrates that our understanding of art is dominated by a limiting paradigm—"production by individuals." We focus almost exclusively on the individual identity of the artist as the determining factor in assessing authenticity. "Nowhere ... is there room for paradigmatic art being fluid, unfinished, evolving, and continuous in its creation." Cebik has in mind a dynamic, communally based art, an ever-changing neighborhood mural or music passed on for generations.[22] Another example would be classical ballet, a performance of which is a unique dynamic movement, different from every other performance of the same ballet.

These suggestions about a different paradigm of art show clearly, I think, what is wrong with the art/nature analogy as a useful analytical tool. Natural entities and systems are much more akin to the fluid evolving art of Cebik's alternative model than they are to the static, finished, individual artworks of the dominant paradigm. It is thus an error to use criteria of forgery and authenticity that derive from an individualistic, static conception of art for an evaluation of natural entities and systems. Natural entities and systems are nothing like static, finished objects of art. They are fluid, evolving systems which completely transcend the category of artist or creator. The perceived disvalue in restored natural objects does not derive from a misunderstanding over the identity of the creator of the objects. It derives instead from the misplaced category of "creator"—for natural objects do not have creators or designers as human artworks do. Once we realize that the natural entity we are viewing has been "restored" by a human artisan it ceases to be a natural object. It is not a forgery; it is an artifact.

We thus return to artifacts, and their essential anthropocentric nature. We cannot (and should not) think of natural objects as artifacts, for this imposes a human purpose or design on their very essence. As artifacts, they are evaluated by their success in meeting human interests and needs, not by their own intrinsic being. Using the art/nature analogy of forgery reinforces the impression that natural objects are similar to artifacts—artworks—and that they can be evaluated using the same anthropocentric criteria. Natural entities have to be evaluated on their own terms, not as artworks, machines, factories, or any other human-created artifact.

V

But what are the terms appropriate for the evaluation of natural objects? What criteria should be used? To answer this question we need to do more than differentiate natural objects from artifacts; we need to examine the essence or nature of natural objects. What does it mean to say that an entity is natural (and hence, not an artifact)? Is there a distinguishing mark or characteristic that determines the descriptive judgment? What makes an object natural, and why is the standard not met through the restoration process?

The simple answer to this question—a response I basically support—is that the natural is defined as being independent of the actions of humanity. Thus, Taylor advocates a principle of non-interference as a primary moral duty in his ethic of respect for nature. "[W]e put aside our personal likes and our human interests.... Our respect for nature means that we acknowledge the sufficiency of the natural world to sustain its own proper order throughout the whole domain of life."[23] The processes of the natural world that are free of human interference are the most natural.

There are two obvious problems with this first simple answer. First, there is the empirical point that the human affect on the environment is, by now, fairly pervasive. No part of the natural world lies untouched by our pollution and technology. In a sense, then, nothing natural truly exists (anymore). Second, there is the logical point that humans themselves are naturally evolved beings, and so all human actions would be "natural," regardless of the amount of technology used or the interference on nonhuman nature. The creation of artifacts is a natural human activity, and thus the distinction between artifact and natural object begins to blur.

These problems in the relationship of humanity to nature are not new. Mill raised similar objections to the idea of "nature" as a moral norm over a hundred years ago, and I need not review his arguments.[24] The answer to these problems is twofold. First, we admit that the concepts of "natural" and "artifactual" are not absolutes; they exist along a spectrum, where various gradations of both concepts can be discerned. The human effect on the natural world is pervasive, but there are differences in human actions that make a descriptive difference. A toxic waste dump is different from a compost heap of organic material. To claim that both are equally non-natural would obscure important distinctions.

A second response is presented by Brennan.[25] Although a broad definition of "natural" denotes independence from human management or interference, a more useful notion (because it has implications for value theory and ethics) can be derived from the consideration of evolutionary adaptations. Our natural diet is the one we are adapted for, that is "in keeping with our nature." All human activity is not unnatural, only that activity which goes beyond our biological and evolutionary capacities. As an example, Brennan cites the procedure of "natural childbirth," that is, childbirth free of technological medical interventions. "Childbirth is an especially striking example of the wildness within us . . . where we can appreciate the natural at first hand. . . ." It is natural, free, and wild not because it is a nonhuman activity—after all, it is human childbirth—but because it is independent of a certain type of human activity, actions designed to control or to manipulate natural processes.

The "natural" then is a term we use to designate objects and processes that exist as far as possible from human manipulation and control. Natural entities are autonomous in ways that human-created artifacts are not; as Taylor writes, "to be free to pursue the realization of one's good according to the laws of one's nature."[26] When we thus judge natural objects, and evaluate them more highly than artifacts, we are focusing on the extent of their independence from human domination. In this sense, then, human actions can also be judged to be natural—these are the human actions that exist as evolutionary adaptations, free of the control and alteration of technological processes.

If these reflections on the meaning of "natural" are plausible, then it should be clear why the restoration process fails to meet the criteria of naturalness. The attempt to redesign, recreate, and restore natural areas and objects is a radical intervention in natural processes. Although there is an obvious spectrum of possible restoration and redesign projects which differ in their value—Maser's redesigned sustainable forest is better than a tree plantation—

all of these projects involve the manipulation and domination of natural areas. All of these projects involve the creation of artifactual natural realities, the imposition of anthropocentric interests on the processes and objects of nature. Nature is not permitted to be free, to pursue its own independent course of development.

The fundamental error is thus domination, the denial of freedom and autonomy. Anthropocentrism, the major concern of most environmental philosophers, is only one species of the more basic attack on the preeminent value of self-realization. From within the perspective of anthropocentrism, humanity believes it is justified in dominating and molding the non-human world to its own human purposes. But a policy of domination transcends the anthropocentric subversion of natural processes. A policy of domination subverts both nature and human existence; it denies both the cultural and natural realization of individual good, human and non-human. Liberation from all forms of domination is thus the chief goal of any ethical or political system.

It is difficult to awaken from the dream of domination. We are all impressed by the power and breadth of human technological achievements. Why is it not possible to extend this power further, until we control, manipulate, and dominate the entire natural universe? This is the illusion that the restoration of nature presents to us. But it is only an illusion. Once we dominate nature, once we restore and redesign nature for our own purposes, then we have destroyed nature—we have created an artifactual reality, in a sense, a false reality, which merely provides us the pleasant illusory appearance of the natural environment.

VI

As a concluding note, let me leave the realm of philosophical speculation and return to the world of practical environmental policy. Nothing I have said in this essay should be taken as an endorsement of actions that develop, exploit, or injure areas of the natural environment and leave them in a damaged state. I believe, for example, that Exxon should attempt to clean up and restore the Alaskan waterways and land that was harmed by its corporate negligence. The point of my argument here is that we must not misunderstand what we humans are doing when we attempt to restore or repair natural areas. We are not restoring nature; we are not making it whole and healthy again. Nature restoration is a compromise; it should not be a basic policy goal. It is a policy that makes the best of a bad situation; it cleans up our mess. We are putting a piece of furniture over the stain in the carpet, for it provides a better appearance. As a matter of policy, however, it would be much more significant to prevent the causes of the stains.

NOTES

1. In Islip Town, New York, real-estate developers have cited the New York State Department of Environmental Conservation policy of "no-net loss" in proposing the restoration of parts of their property to a natural state, in exchange for permission to develop. A report in *Newsday* discusses a controversial case: "In hopes of gaining town-board approval, Blankman has promised to return a three-quarter-mile dirt road on his property to its natural habitat. . . ." Katti Gray, "Wetlands in the Eye of a Storm," Islip Special, *Newsday*, April 22, 1990, pp. 1, 5.
2. *Garbage: The Practical Journal for the Environment*, May/June 1990, rear cover.
3. Chris Maser, *The Redesigned Forest* (San Pedro: R. & E. Miles, 1988), p. 173. It is also interesting to note that there now exists a dissident group within the U.S. Forest Service, called the Association of Forest Service Employees for Environmental Ethics (AFSEEE). They advocate a return to sustainable forestry.
4. Paul Taylor, *Respect for Nature: A Theory of Environmental Ethics* (Princeton: Princeton University Press, 1986), pp. 186-92, 304-06, and Chapters Four and Six generally.
5. Peter S. Wenz, *Environmental Justice* (Albany: State University of New York Press, 1988), pp. 287-91.
6. Robert Elliot, "Faking Nature," *Inquiry* 25: 81-93 (1982); reprinted in Donald VanDeVeer and Christine Pierce, eds., *People, Penguins, and Plastic Trees: Basic Issues in Environmental Ethics* (Belmont: Wadsworth, 1986), pp. 142-150.
7. Ibid., p. 86 (VanDeVeer and Pierce, p. 145).
8. Ibid., p. 91 (VanDeVeer and Pierce, p. 149).
9. Eric Katz, "Organism, Community, and the 'Substitution Problem,'" *Environmental Ethics* 7(1985): 253-55.
10. Elliot, p. 81 (VanDeVeer and Pierce, p. 142).
11. See, for example, Barry Commoner, *The Closing Circle* (New York: Knopf, 1971) and Arnold Pacey, *The Culture of Technology* (Cambridge: MIT Press, 1983).
12. Michael Losonsky, "The Nature of Artifacts," *Philosophy* 65: 88 (1990).
13. Ibid., p. 84.
14. Andrew Brennan, "The Moral Standing of Natural Objects," *Environmental Ethics* 6: 41-44 (1984).
15. Maser, *The Redesigned Forest*, p. xvii.
16. Ibid., pp. 176-77.
17. Ibid., pp. 88-89.
18. Ibid., p. 174.
19. Ibid., p. 94.
20. Ibid., pp. 148-49.
21. L.B. Cebik, "Forging Issues from Forged Art," *Southern Journal of Philosophy* 27: 331-46 (1989).
22. Ibid., p. 342.
23. Taylor, p. 177. The rule of noninterference is discussed on pp. 173-179.
24. J.S. Mill, "Nature," in *Three Essays on Religion* (London: 1874).
25. Andrew Brennan, *Thinking About Nature: An Investigation of Nature, Value, and Ecology* (Athens, GA: University of Georgia Press, 1988), pp. 88-91.
26. Taylor, p. 174.

[23]

The Restoration of Species and Natural Environments

Alastair S. Gunn*

My aims in this article are threefold. First, I evaluate attempts to drive a wedge between the human and the natural in order to show that destroyed natural environments and extinct species cannot be restored; next, I examine the analogy between aesthetic value and the value of natural environments; and finally, I suggest briefly a different set of analogies with such human associations as families and cultures. My tentative conclusion is that while the recreation of extinct species may be logically impossible, the restoration of natural environments raises only (formidable, no doubt) technical difficulties. Opponents of destructive developments which do *not* exterminate species, therefore, had better look elsewhere, rather than relying on the claim that restoration is logically impossible.

INTRODUCTION

Developments such as strip mining are often criticized on the grounds that they cause widespread and irreversible damage to natural environments, including loss of species, destruction of landforms, scarring of landscapes, and air and water pollution. At one time, such criticisms would either have been ignored or answered in a dismissive way: major damage will not occur, but (in case it does) that is the price of progress and anyway there's plenty more wilderness left.

Today, neither of these responses is very likely. In many countries today, including New Zealand, mining applicants are required to carry out at least a halfway honest survey of the natural resources of an area and to identify any damage their operations will cause. Environmental groups, increasingly with the support of farming and tourism interests, can be counted upon to do their own studies. Nor can environmental damage be just dismissed as the price of progress. Western countries do not need more development, and Third World countries need sustainable development rather than destruction.

In this changed climate of opinion, the developer's best line of defense is to deny that his or her operation will cause irreversible damage. The effects of

---

*Department of Philosophy, University of Waikato, Private Bag, Hamilton, New Zealand. Gunn's main teaching and research interest is applied ethics, especially environmental, engineering, and health care ethics. In addition, he has a strong interest in environmental education. He is Chair of the Waikato Area Health Board Committee on Ethics and a member of the New Zealand National Committee for Man and the Biosphere, and has represented New Zealand at several international UNESCO seminars. His book, *Environmental Ethics for Engineers*, co-authored with P. Aarne Vesilind of Duke University, was published by Lewis Publishers in 1986.

pollution will be only temporary. After the mining is finished, the land will be restored to its original physical condition, contoured and revegetated. Many animal species will recolonize naturally; others will be reintroduced from captured or captive bred sources. Perhaps species can be recreated by selective breeding or even recreated by genetic engineering. A variant is that the new environment, although admittedly different, will be an acceptable, equally (or more) valuable substitute for the destroyed original. In restoration proper, the damage is put right: there is no permanent loss. The value of the development outweighs the temporary loss of (for instance) recreational amenity or esthetic experience. Following Robert Elliot,[1] I call this defense the "restoration thesis."

Elliot argues against this thesis that even if restoration projects are successful, that is, even if an environment which has been mined, clear-cut, or otherwise destroyed is rebuilt to the point where no one can tell that it had ever been altered, the restoration will not be as valuable as the original, first, because the restoration lacks an unbroken history, and, second, because it is not natural. According to Elliot, the value of a forest, for example, is partly a function of its history, its continuity with past plant communities. A regenerated forest may also be valuable, but not as valuable as the original. In other words, Elliot is inviting us to use *continuity* and *natural origins* as criteria for valuing forests. If we wish to retain the value of natural systems, we ought to protect them from damage rather than allowing damage and then repairing them.

Elliot's argument is not intended to show that development is always or even usually wrong. Although the value of a natural area is a powerful reason for not destroying it, he recognizes that it cannot absolutely determine policy in every case, since there are other values which can also be promoted by development. Every environmental decision has an opportunity cost—that is, the price of a decision is the loss of the alternative uses to which a resource could have been put. If we clear-cut a forest, we lose the value of the forest as a wilderness. If we preserve the forest, we lose the value of the timber, employment possibilities, and so on. Elliot's argument is designed to show only that the opportunity costs are higher than some environmental engineers claim: that development followed by restoration results in a permanent, not merely a temporary loss of value.

Environmentalists are sometimes accused of ignoring the fact that the benefits and costs of environmental protection are not always fairly distributed. Species protection, for instance, is usually achieved by habitat protection, which means that planned developments do not take place, or take place less effectively. As a result, investors and landowners may lose opportunities and workers lose jobs, but the environmentalists lose nothing, while gaining the benefit of having the

---

[1] Robert Elliot, "Faking Nature," *Inquiry* 25 (1982): 81–93.

species preserved. In the western United States, wolves and coyotes are a cost to sheep farmers who may view them as an unmitigated nuisance. To members of the Sierra Club, most of whom do not have any sheep to lose, the predators are noble and beautiful and a necessary element in a balanced ecosystem. Even if it is true that "no food, no clothing, no shelter, no land, and certainly no luxury or technology is worth the irreplaceable loss of any species," it is not obvious that people who do not share that view of a threatened species should be forced to pay for it on behalf of those who do.[2]

Moreover, most species live in tropical Third World countries. The benefits of policies which destroy species and their habitat, including forest clearing for timber and agriculture, and the trade in wildlife products, are enjoyed mostly by the richer countries who import them. Nevertheless, moves to preserve tropical rain forests, and especially to set up national parks and to rescue endangered species, are often initiated by the developed world. Merely putting a stop to logging, setting up parks, and saving individual species benefits people like me, if only because of the existence value that I see in rain forests and their species. On their own, however, such measures do little to help poor countries to deal with such pressing problems as poverty, malnutrition, and overcrowding. Third World countries, moreover, often cannot afford to patrol national parks and borders to stop poaching and smuggling of theoretically protected wildlife.

To discuss environmental issues in terms of costs and benefits, however, strikes some as unduly shallow or merely reformist. No doubt, Elliot's position is not acceptable to the deep ecologist because it requires us merely to modify our existing policies rather than radically to rethink them, as deep ecologists demand. Moreover, deep ecologists view humans as part of nature, and therefore reject the distinction that Elliot makes between the human and the natural. Nonetheless, to oppose developments by reference to theory and without reference to costs and benefits is to forgo the opportunity to have an influence on actual environmental policy. For this reason, both environmentalists and developers have a stake in clarifying the status of the restoration thesis.

A CRITIQUE OF THE RESTORATION THESIS

The Importance of Origins

In general we do not object to the destruction of an object if it can really be replaced by something "just like it." A paradigm case of the replaceable is the mass produced item. If I lend you my digital watch or Mitsubishi car and you lose or damage it, you can certainly replace it with an identical and equally

---

[2] Editorial, *Oceans* 12 (September 1979): 8.

valuable watch or car.[3] At the other extreme, you certainly cannot replace my family if you were responsible for their loss, presumably, because I view my family and my relationship with them as unique.

Species and most natural areas seem to fall somewhere between these extremes. It is possible to take seriously both the view that a restored area or "recreated" species is really identical to a destroyed area or extinct species (and is therefore equally valuable), and the opposite view that it is not the same at all (and therefore not equally valuable). In contrast, there is something absurd about the idea of a replaceable family or an irreplaceable Mitsubishi. Claims to be able to undo damage or restore what was lost are of the utmost importance in debate about environmental policy, and whether genuine restoration is possible is a major source of controversy.

Elliot argues that the value of at least some things is a result of their being natural. Recognizing that "the distinction between the natural and the non-natural requires detailed working out," he asks us to accept that "for present purposes . . . 'natural' means something like 'unmodified by human activity'." Although Elliot does not claim "that what is natural is good and what is non-natural is not," he does argue that some things, in particular wilderness, are valued because they are (relatively, highly) natural; that "naturalness" is not replicable; and that therefore when "naturalness" is destroyed, value is irreversibly lost.[4] While he does not directly discuss species, he accepts that its evolutionary history is an essential condition of a species being what it is and, therefore, of it having the value that it has.[5]

Elliot does not claim that everything is either natural or non-natural, but recognizes that there are degrees of naturalness, and therefore of value. He gives an example of three situations in which John, a wilderness lover, may find himself. In the first, John is plugged into an experience machine which gives him non-veridical experiences of hiking through a wilderness. In the second, he is taken to a "simulated, plastic wilderness area" which he falsely believes to be real. In the third, he is taken to a regenerated forest on an area once devastated by strip mining, which he falsely believes to be primeval forest. In all three cases, Elliot claims, John is being "shortchanged," but not equally so:

---

[3] At least, identical in relevant respects. This does not mean having exactly the same scratches and dents in exactly the same places. It means that the replacement meets my criteria for valuing cars to the same extent as the original. Perhaps comparable is more accurate than identical. A car could be unique, of course. The original Darracq runabout used in the movie *Genevieve* sold for A$580,000 in August 1989 (*The Christchurch Press*, 21 August 1989). If the watch or the car were the gift of a loved one, it would have "sentimental value" and thus not be replaceable; however, that is not a property of the watch itself. Some natural areas are also irreplaceable if they have special cultural associations, for instance, sacred places.

[4] Elliot, "Faking Nature," 86–7.

[5] Elliot, personal communication.

In the same way that the plastic trees may be thought a (minimal) improvement on the experience machine, so too the real trees are an improvement on the plastic ones. In fact in the third situation there is incomparably more of value than in the second, but there could be more. The forest, although real, is not genuinely what John wants it to be. If it were not the product of contrivance he would value it more.[6]

Elliot is correct in asserting that we do attach importance to many things for reasons other than merely the properties they apparently have, including, sometimes, their history or origin. A particularly persuasive example he presents is one in which someone gives him a beautiful carved object. Elliot writes that he treasures and admires the object until he discovers

> that it is carved out of the bone of someone killed especially for that purpose. This discovery affects me deeply and I cease to value the object in the way that I once did. I regard it as in some sense sullied, spoilt by the facts of its origin. . . . The discovery about the object's origin changes the valuation made of it, since it reveals that the object is not of the kind that I value.[7]

With regard to the illusory, plastic, and regenerated forests, it is tempting to say that the reason why John is entitled to feel aggrieved is that he is being deceived, that this is the only respect in which he is being "shortchanged," and that if he were told in advance that the trees were part of an illusion, or plastic, or second growth, he would have no cause for complaint. However, the deception is not the only relevant feature of the three situations. If we imagine a bizarre situation in which John is deceived into believing that a primeval forest is really an illusion, or plastic, or second growth, it is surely reasonable to expect John to be relieved and delighted when the deception is exposed. Although he is not pleased to have been deceived, he is pleased that what he thought was second or third or fourth best is in fact first best. Of course, he resents being deceived and rightly so because deception is wrong on various grounds.[8] Nevertheless, the difference between Elliot's examples, in which John is deceived into thinking a fake forest is real, and mine, in which he is deceived into thinking a real forest is fake, shows that the difference in value between the fake and the real is not merely reducible to the wrongfulness of deception.[9]

---

[6] Elliot, "Faking Nature," 89.
[7] Ibid. pp. 85–86.
[8] For a thorough treatment of this topic, see Sissela Bok, *Lying* (New York: Vintage Books, 1979).
[9] Indeed, people may be sorry that a deception is unmasked, not because they feel aggrieved at having been deceived, but because they were happier when they held the false belief.

THE HUMAN AND THE NATURAL

Acceptance or rejection of the restoration thesis has implications for environmental rehabilitation in general. In addition to attempts to construct exact replicas, rehabilitation includes system conversion (for example, gravel pits turned into lakes) and the construction of analogies to natural features designed to reintroduce essential features of a damaged environment (fish ladders for migrating salmon or nest boxes for owls and parrots in clear-cut old-growth forest). The claim that the last two examples count as restoration at all depends on the fact that many species are adaptable and will accept changes in their environment—whether caused by humans or not—provided that certain features essential to their survival are maintained.

Implicit in the restoration thesis is the *refusal* to accept the sharp distinction between the "human" and the "natural." I have been unable to locate any published philosophical defense of the restoration thesis by strip miners, but environmental engineers and mining company personnel with whom I have discussed this topic appear to see the difference between the activities of humans and of other species as only a matter of degree. Some see humans as having a duty not to utilize our powers to destroy irresponsibly, and to try to make good the damage of past as well as current developments. One informant contrasts thoughtless strip-mining processes which turn forested hillsides into eroded deserts, such as parts of West Virginia, with such examples as gravel extraction in England. The pits left after the removal of the gravel fill up with water to become small lakes, which are carefully stabilized, contoured, landscaped, and planted with appropriate vegetation. They are then rapidly colonized by surplus populations of animal species which normally inhabit similar existing small lakes. The survival prospects of several rare and endangered species have been improved by this means. Soon, it all settles down and the human-made lakes become indistinguishable—aesthetically and ecologically—from the nonhuman-made ones.

Thomas Lovejoy makes a similar point in defense of a proposal to reintroduce the western subspecies of peregrine falcon, *Falco peregrinus pealei*, to the eastern United States in case the eastern subspecies, *Falco peregrinus anatum*, should become extinct:

> Surely it is better to have non-endemic peregrines in the eastern USA than to have no peregrines at all, and it can then be safely left to natural selection to work things out.[10]

---

[10] Thomas Lovejoy, "Tomorrow's Ark: By Invitation Only," in P. S. Olney, ed., *International Zoo Yearbook* (London: Royal Zoological Society, 1980), p. 183.

To the extent that opposition to the restoration thesis depends on the view that the products of human activity are not natural, this opposition seems incompatible with any concept of an environmental ethic in which humans are seen as part of nature. No doubt human behavior would have to be very different if we are to become "part of nature," but it seems arbitrary to define the possibility out of existence. The Shorter Oxford Dictionary lists seventeen distinct senses of *natural:* why should this one be the right one in this context? Just in order to have an extra stick with which to beat developers? Why not adopt a definition along different lines? Murie, for example, writes:

> My main concern . . . is to suggest a rapprochement with nature in which nature is respected but we people are not required to hate our presence on the planet. I define "natural environments" as any place where organisms exist in mutual relations that are free of direct management or other drastic human intervention; but presence or absence of people is not a criterion.[11]

Even if we accept Elliot's definition of *natural*, we cannot very well argue that all, and only, those areas and species whose evolution has been *completely unaffected by human actions* are "real" and are valuable as "natural" objects. Almost every area on Earth has been affected in some way by human activity. Deforestation and overgrazing have led to soil erosion, soil loss, and in extreme cases desertification. Traces of persistent pesticides and PCBs have been discovered in Arctic walruses and Antarctic penguins through the general dispersion of these substances. Most areas of the globe have experienced some human interference, thus tipping the survival balance in favor of adaptable or resilient species and of unusually adaptable or resilient individuals of other species, with unknown evolutionary consequences. Paleolithic exterminations of large fauna in the Americas and Europe left niches to be filled by other species. The American bison, *Bison bison* or buffalo, almost exterminated by European colonists, presumably owed its huge pre-European population to the extermination of its predators and larger competitors. Some species appear to have evolved along with humans, as parasites and scavengers—for example, the house or English sparrow, *Passer domesticus*. Human activity has diminished the range and numbers of many species but has provided increased habitat for others. Ignoring deliberate or accidental introductions, there are still many species which have increased due to land-use changes and other human induced factors—for example the European starling, *Sturnus vulgaris,* many gulls, and other scavengers. In various ways we have altered the genetic characteristics and their distribution within species, and therefore the course of evolution. But we do not

---

[11] M. Murie, "Evaluations of Natural Environments," in W. A. Thomas, ed., *Indicators of Environmental Quality* (New York: Plenum Press, 1972), pp. 43–51.

say that gulls, starlings, and bison do not *belong* in their expanded range, nor that the house sparrow is not a *real* species. Likewise, efforts to revive an endangered species, including captive breeding, if successful on their own terms do not produce anything inherently inferior to the original.

RECREATING EXTINCT SPECIES

EXTINCTION AND IDENTITY

The *recreation of extinct species* might seem to be another matter. To ask whether an extinct species could be recreated seems to mean something like this:

> Could it be that a species which existed at time T_1 and at a later time T_2 did not exist was caused to exist at a still later time T_3?

This formulation is neutral as to whether species are classes or, as some biologists and philosophers maintain, individuals.[12] If as I argue below, individuals cannot be recreated, and species are individuals, then their recreation is logically impossible. However, because in this paper I adopt the conventional view of species as classes, the question can be better formulated:

> Could it be that a species was a class with members at time T_1, and was a class without members (or null class) at a later time T_2, and was caused *to* be a class with members at a still later time T_3?

There are certainly many examples of classes of this sort. For instance, before the days of peacetime standing armies, "The British Army" was a class with members only in wartime. In times of peace, when the armies had all been demobilized, "The British Army" became a null class. Likewise, thanks to the temporary success of Henry VIII in disbanding the English religious houses, the class "English religious houses" has gone from being a class with members, to a null class, to a class with members.

Neither the later armies nor the later houses are necessarily identical with those which existed earlier, and certainly not numerically identical. But nor are the members of the class, brown kiwi, *Apteryx australis*, numerically identical with the birds which were members of the class fifty years ago. Nonetheless, they are

---

[12] See, e.g., Michael T. Ghiselin, "On Psychologism in the Logic of Taxonomic Controversies," *Systematic Zoology* 15 (1966): 207–15; "A Radical Solution to the Species Problem," *Systematic Zoology* 23 (1974): 536–44; "Biogeographical Units: More on Radical Solutions," *Systematic Zoology* 29 (1980): 80–86; "Can Aristotle Be Reconciled With Darwin?" *Systematic Zoology* 34 (1985): 457–60; David L. Hull, "Are Species Really Individuals?" *Systematic Zoology* 25 (1976): 174–91.

all members of the same class, unlike members of a different species such as the little spotted kiwi, *Apteryx oweni*.

Extinct species are not merely null classes. "Gryphon" and "basilisk" denote classes which have always been null: there never were gryphons or basilisks. To say that a species is extinct is to say something about its past as well as its present status—although there used to be moa, they no longer exist. It may be argued that *extinct* also says something about the future of a class—that once it becomes a null class, it can never come to have members again. It may even be claimed that this is what *extinct* means. If so, then the question, "Can extinct species be recreated?" is answered negatively by resort to what is sometimes called "definitional stop."[13]

Wildlife preservationists are always telling us, as a warning, that "Extinction is forever!" Perhaps this warning tells us no more about the world than "Bachelors are unmarried!" According to the Shorter Oxford Dictionary, the relevant meaning of *extinction* is: "Of a race etc: A coming to an end or dying out." This definition sounds permanent enough. An extinct species is one that has come to an end, has died out, is a permanently null class. As we use the word *extinct*, it seems, the recreation of an extinct species is a logical impossibility. It is for this reason that Robert Leo Smith refers to "this irrevocability of a species, this awesome finality of extinction."[14]

Because the ordinary language and scientific definitions of *extinction* are consistent, from a scientific perspective, it can also be said that the recreation of an extinct species is logically impossible. If two near identical organisms were to evolve at different times or in different places these biological events would be seen as convergent evolution, and the organisms would be classed as belonging

---

[13] This phrase was coined to denote a move made by defenders of a utilitarian account of the justification of punishment. S. I. Benn and R. S. Peters, in *Social Principles and the Democratic State* (London: Allen and Unwin, 1959) argued that punishment is justified for the good it does—particularly its deterrent effect. An objection is that the same effect could be produced by punishing an *innocent* person, thus avoiding the expense of catching a real criminal. Their reply was that doing so would be victimization, not punishment, which by definition is inflicted only on wrongdoers. Of course, this argument does not really meet the objection, which is that utilitarians are allegedly committed to harming the innocent in order to reduce crime, nevermind whether it is called punishment, or with John Rawls, "telishment." "Two Concepts of Rules," *The Philosophical Review* 64 (1955): 3–32.

[14] Robert Leo Smith, "Ecological Genesis of Endangered Species: The Philosophy of Preservation," *Annual Review of Ecology and Systematics* 7 (1976): 44. In fact, "contemporary biologists recognize levels of extinction, some less final than others," ranging from extinction of a population ("local extinction") to extinction of a whole species ("general extinction"). *Species extinction*, operationally, is "the death of the last individual that is referred to by a certain scientific binomial." Michael R. Soulé, "What Do We Really Know About Extinction?" in Christine Schonewald-Cox et al., *Genetics and Conservation* (Menlo Park, Calif.: Benjamin/Cummings, 1983) 111–24. "In the broadest sense, extinction occurs when populations cannot persist in the face of environmental change." Geerat J. Vermeij, "The Biology of Human-Caused Extinction," in Bryan G. Norton, ed., *The Preservation of Species* (Princeton: Princeton University Press, 1986), pp. 28–49.

to two different species because classification is supposed to exhibit evolutionary relationships, not mere morphological similarities: to provide explanations and not merely descriptions. For example, the scientific taxonomist recognizes the Australian red back spider and the American black widow as one species, *Lactrodactus mactans*, even though the red back was once seen as a separate species, *Lactrodactus hasseltii*; nevertheless, the similar New Zealand katipo is classed as a separate species, *Lactrodactus scelio*. If such similar products of *non*-human-assisted evolution are separate species, then surely so are the products of human genetic manipulation.

SELECTIVE BREEDING AND GENETIC ENGINEERING

The matter is not quite so simple, however. Common sense suggests that a species is effectively extinct when only one individual is left, at least in the case of sexually reproducing animals and dioecious plants. Careful selective breeding, however, might allow breeding from a single specimen with individuals of a closely related taxon. For instance, it was at one time proposed to try to save the dusky seaside sparrow, *Ammospiza maritima nigresens*, via selective breeding from five surviving individuals, all of whom were males, and existing female hybrids between *Ammospiza maritima nigresens* and Scott's seaside sparrow, *Ammospiza maritima penninsulae*.[15] Current programs are attempting to recreate, via selective breeding, extinct species such as the zebra-like quagga, *Equus quagga*. Perhaps genetic engineering could one day permit the production of individual organisms from the cells of museum specimens or from small pieces of tissue.[16] It is even conceivable that information on a species could be stored electronically, for later recreation.

Where should the line be drawn? Certainly we are in no doubt about the continued existence of species such as annual plants, whose entire population dies off each year. The seeds, which are all that survive the winter, are genes, not plants. Similarly, it is possible to imagine a species which survives the winter only in the form of unfertilized ova and sperm, relying on vector organisms or

---

[15] Tom J. Cade, "Hybridization and Gene Exchange Among Birds," in Schonewald-Cox, *Genetics and Conservation*, pp. 288–309.

[16] Thomas J. Foose believes that "reproductive technology [which] could help preserve diversity as stored germplasm . . . may be generally available for endangered species within 200 years." "Riders of the Last Ark: The Role of Captive Breeding in Conservation Strategies," in Les Kaufman and Kenneth Malloy, eds., *The Last Extinction* (Cambridge, Mass.: MIT Press, 1986), p. 145. In contrast, David Ehrenfeld believes that the claims of genetic engineers are the product of "technical euphoria in full swing. 'You want your species back? We'll recreate them,' the technophiles promise, ignoring all the accumulated knowledge of biology, physics and information theory simultaneously. Recreating from scratch even one species would be a feat which would make the stuffing of animals into Noah's ark simple by comparison. . . . [G]enetic engineering as a conservation tool is simply not worthy of consideration as a serious option." "Life in the Next Milennium: Who Will Be Left in the Earth's Community'" in Kaufmann and Malloy, *Last Extinction*, p. 175.

other environmental factors to arrange fertilization.[17] In such a case, even though no individual member of the species exists, not even in embryonic form, we would surely not want to say that a new species evolved each spring. Does any physical DNA need to exist? Does it make any difference whether the genetic information on a species is stored in DNA or on a floppy disk?

The difficulty in deciding where, if anywhere, to draw the line stems from the fact that our concept of identity was developed in the context of objects rather than biochemically or electronically coded information. A necessary condition of continuing identity as ordinarily understood is spatiotemporal continuity. If we think of a species' existence as consisting of (or requiring) the existence of actual organisms, we will draw the line at the death of the last individual. If we insist only on the continuity of at least some physical material, we may accept the tissue culture. If we see a species' existence solely in terms of genetic information, all the examples I have described will count as the continued existence of the same species. For what it is worth, intuition seems to exclude the floppy-disk case: there is something odd about the idea that the extinction of a species could occur at the moment the disk file is erased. (Perhaps, though, this conclusion is simply a limitation of my pre-electronic world view.)

RECREATING INDIVIDUALS

The situation may be clearer if we remember that the (so far unrealized) technology that would recreate species would create individual organisms which would be genetically identical to the individuals whose DNA was used. What would we say if an individual just like Jeremy Bentham was developed from cells from his embalmed body? Perhaps we would decide that *being genetically identical to Bentham* is not the same as *being Bentham*. After all, monozygotic twins are genetically identical, but are different individuals. Indeed, modern agriculture and forestry produce millions of genetically identical individual plants. The Bentham who emerged from the laboratories of the future would certainly be a different person, with a different history of experiences and a different role in life. An account of the world of Bentham (1748–1832) would certainly be very different from an account of the world of Bentham (2048–2132). A recreated species, likewise, could have a very different ecological role from the original, especially if there were a lengthy gap between its extinction and reintroduction, because it would not have evolved in the same situation. Indeed, it would not have evolved at all.[18]

---

[17] I owe this example and the next one to David Lewis.
[18] This does not mean that there is never a good reason for attempting to recreate a species. If a key species such as a major pollinator becomes extinct, a whole ecosystem might be threatened. A successful project to recreate the species (e.g., by cloning from a museum specimen) might enable us to save the ecosystem, provided that it could be reintroduced very quickly. However, the introduction of an exotic but closely related species might have the same effect; if so, this option would be equally effective in protecting the ecosystem, especially if it could be done immediately.

AESTHETICS AND ENVIRONMENTAL ETHICS

Writers on environmental ethics draw on aesthetics in two ways. First, the value of natural environments is sometimes said to *be* at least partly aesthetic. Second, analogies and parallels are drawn between art and nature. Some biologists as well as philosophers have enthusiastically supported the first view. Paul and Anne Ehrlich see insects as "an immense aesthetic resource" whose beauty may "often outshine the Mona Lisa"[19] and they quote with approval the view of Claude Levi-Strauss that "Any species of bug . . . is 'an irreplaceable marvel, equal to the works of art which we religiously preserve in museums.' "[20] David Ehrenfeld claims that "the disappearance of any species represents a great esthetic loss for the entire world. It can perhaps be compared to the destruction of a great work of art by a famous painter or sculptor."[21]

Because insects are not works of art, of course, for my purposes it is the alleged analogies between art and nature that are important. No doubt the Grand Canyon and the Mona Lisa are beautiful, but so, I am told, are the solution to the four-color problem, the structure of the DNA molecule, Concorde, aboriginal kinship systems, and Joe Louis' knockout punch. Perhaps they all produce identical goosebumps, but the significant difference between works of art and of nature is that the former are products of contrivance and design while the latter are not. Indeed, the fallacy of the theological argument from design lies precisely in the fact that (*pace* William Paley) watches do bear the marks of design while eyes do not.

Elliot uses the analogy of fakes and original works of art to illustrate or explain the difference he perceives between natural and restored environments. In the case of fake paintings, "There is a difference, and it is one which affects my perception, and consequent valuation, of the painting. The difference of course lies in the painting's genesis."[22] One's appreciation of a painting is not merely a recognition of its aesthetic qualities, which may be identical in copy and original.[23] We value the works of old masters just because they are the work of old masters—because Vermeer himself painted the picture in his studio, in a particular historical context, perhaps for a patron or client who wanted just that picture and no other. We stand in awe before famous works of art, but not before copies.

---

[19] Paul and Anne Ehrlich, *Extinction* (New York: Random House, 1981), p. 38.
[20] Ibid., p. 3.
[21] Ehrenfeld, "Non-Resources," p. 654, quoting A. F. Coimbra-Filho et al., "Vanishing Gold: Last Chance for Brazil's Lion Tamarins," *Animal Kingdom*, December 1975, pp. 20–26.
[22] Elliot, "Faking Nature," p. 85.
[23] According to Colin Radford, aesthetic value does include being an original rather than a fake. "Fakes," *Mind* 87 (1978): 66–76. The important point for my purposes, however, is the extent to which the value (not necessarily the aesthetic value) of a painting is a function of its being an original.

ORIGINALS, REPLICAS, FAKES, AND RESTORATION

Before discussing Elliot's argument, some conceptual distinctions need to be drawn.[24] First, a *replica* (or *copy*) is something that, if successful, is exactly like that of which it is a replica—or, at least, effectively indistinguishable. In order that A be a replica of B, B must already exist or have existed earlier. If we are fooled, we believe that A is B, though the purpose of making a replica need not be to fool anyone. Replicas of the *Venus de Milo* abound, but no one is meant to believe that any of them is the real, the original *Venus de Milo*.

Second, a *fake* (or *forgery*) is both a broader and a narrower category. It is narrower because to call something a fake is always to impute a dishonest motive to the faker, an intent to deceive. It is broader because it includes but is not limited to replicas. A replica of Van Gogh's *Sunflowers* is a fake, but so also is a painting done in the style of Van Gogh which is not a replica of an actual Van Gogh but which is intended to deceive people into thinking that it is a real, newly discovered Van Gogh.

Finally, a *restoration* is an original which has been damaged or partially destroyed in some way and has now been brought back to its original appearance. Paintings are restored by removal of grime and darkened varnish, sometimes by transferring the paint to a new canvas. Buildings are restored by replacing rotten timber, repiling, reroofing, and the like. There is typically no intent to deceive; on the contrary, the restoration of decaying artifacts is a source of pride. It is evidence of our respect for our cultural heritage. Unlike replicas and fakes, however, restorations cannot exist side by side with the unrestored originals. Restoration is, therefore, potentially controversial in a way that replication and fakery are not.

What we admire in human products—whether in the visual arts, music, theatre, furniture construction, landscape gardening, architecture, or engineering—is *creativity*. In contrast, the DNA molecule, the Ehrlichs' insects, and the Grand Canyon do not exhibit creativity. Nor does a replica of a famous painting. A fake which is not a replica exhibits a certain amount of creativity, since the faker has to conceive and design the painting before painting it. Van Meegeren at least had to think of the kind of subject Vermeer might have chosen and the elements and colors that he might have used, which is exactly what Vermeer himself might have done if he had a commission to execute and was feeling low on inspiration.[25] Still, the difference is that Vermeer developed his own style of

---

[24] This section deals only in a cursory way with what is a major topic in itself. For a detailed discussion, see Dennis Dutton, ed., *The Forger's Art* (Berkeley: University of California Press, 1983); also the journal *Restoration and Management Notes*.

[25] Radford, "Fakes," thinks that what is missing in Van Meegeren's bogus Vermeer *The Supper at Emmaus* is the deep religious conviction that it would have expressed, had it been an original, whereas Van Meegeren painted it as an expression of deep contempt for pretentious art critics. However, we do not know what Vermeer's feelings were when he was painting. Like Dickens, his mind may frequently have been on the rent money.

painting rather than merely painting in the style of another painter. Whether or not what I have called creativity is an *aesthetic* value, it is certainly part of what we admire in a work of art, but not in a natural scene, species, or ecosystem.

One of the criteria for valuing paintings, then, is that they are originals. Other things being equal, an original is significantly more valuable than a replica. However, originality is neither a sufficient nor a necessary condition of value. My own paintings, few in number, are original in the sense that they are not the result of an effort to replicate or fake another painting. They are, of course, quite valueless. On the other hand, the faked "Vermeers" painted by Van Meegeren *are* valuable, more so than many originals. If we discovered that a fresco long attributed to Giotto was in fact the work of Michelangelo, we would doubtless value it just as highly—even if we discovered that Michelangelo had painted it with deliberate intent to deceive, i.e., that it was a fake. However, we would do so because we value any painting by Michelangelo very highly. The situation would be quite different if we discovered that the "Giotto" was in fact the work of a nineteenth-century art student.

Originals, replicas, and fakes can all coexist, and a world which contains examples of all three is not thereby any worse (or better) than a world which contains only originals. To replicate or fake is not to *do* anything to the original; nothing is destroyed or damaged in the process. Restorations, however, cannot coexist with unrestored originals, for a restoration of A *is* A, restored. In the art world, as I understand it, restoration is controversial partly because of disputes about the technical skill with which it is sometimes done, and because a certain amount of guesswork may be needed to decide exactly what a painting looked like in its original state. Moreover, a restoration is no longer the work of just the original painter but also in part the work of the restorer.

Aging and other changes may be seen as a part of the life of a work of art. A piece of music is not intended by the composer to be altered, presumably, but arrangements and scoring are in fact major changes, and music written for one instrument may be rewritten for another. Moreover, the aesthetic experiences of performers and audiences vary according to the interpretation of the piece, the standard and mode of the performance, and the context. Musical composition, then, is not merely the production of static pieces. Antique furniture, outdoor sculpture, stone buildings, and oil paintings are also "dynamic": they are not *meant* to look as if they have just been finished. The life of a building may even be said to include its eventual decay. Some English and Welsh castles have been restored, but others have been allowed to fall into ruins. These are regarded as picturesque and Romantic, and any proposal to rebuild them would undoubtedly cause an outcry. In Rome, some historic monuments are continually being rebuilt, while others are left in ruins. Another possibility is to "freeze" a ruin in its present state, as has been done with some historic places in the United States. This is preservation rather than restoration.

Restoration of buildings may be justified on various grounds. First, buildings are designed to be used. To enable continued and safe use, routine maintenance is necessary. This is as true of Salisbury Cathedral as it is of an ordinary house. Restoration simply does in a short period a great deal of work which should have been done in smaller stages over the years. Second, while it is true that castles often decline gracefully, most neglected buildings become charmless, rat infested eyesores. Third, buildings are frequently not created all at one time: the older and larger a building is, the more additions and "improvements" usually have been made to it.

At some point, no doubt, a building is determined to be "just right," and proposals for further additions (but not restoration) are seen as aesthetic vandalism. Moreover, at some point we might say that the process of rebuilding has gone so far that none of the original building is left. Is it then the *same* building? This question sounds more like a call for a decision rather than a discovery. My inclination is to say that it is the same building, restored, if the main structural features are substantially intact (even if they are in need of considerable repair) at the time when restoration begins; however, when a castle is built on and includes parts of ruins, it is at best a replica. Perhaps such questions turn on what we want from a building. If we want veridical experiences of the achievements of the past, we might be more purist about what counts as a restoration than if we want only a location for horror movies.

Buildings may be a special case. Certainly we do not want people to start improving and adding to the great paintings of the past. Nevertheless, if we do accept the desirability of limited, careful restoration, and if we accept that a restored painting is at least as valuable as it was before restoration, it follows that there is nothing wrong with an original painting being restored by someone who is not the original artist. A restored Vermeer is still a real, an original Vermeer, unlike a replica or fake, which never was a Vermeer. The identity of a Vermeer includes its history. A Vermeer painting *means* a painting produced by Vermeer, not just a painting which looks as if it might have been painted by him. The number of Vermeer paintings is limited to those that exist today, including those yet to be discovered, and excluding those erroneously attributed to Vermeer. If some are destroyed, there will be fewer than there are at present, and if all are destroyed, there will never be any more. The class "Vermeer paintings" will become irreversibly null.

It is important to note that while fakes and replicas are not real Vermeers, they are still real paintings; indeed, they may really have the qualities of a "typical" Vermeer.[26] A completely rebuilt castle is certainly a real building, though it is

---

[26] A perfect hologram of a Vermeer would not be a real painting, and presumably would not really have the Vermeer like qualities it appeared to have; it would be neither a fake nor a replica.

not a real medieval castle, perhaps not even a real castle. "Sleeping Beauty's Castle" at Disneyland is not even a real building—reflecting the difference that Elliot sees between plastic "trees" and the replanted ones. The plastic objects are not real trees: they are not trees at all.

RESTORING NATURE

When an area of forest is clear-cut and replanted, it is unlikely that an exact replica is expected or intended. First, there is the sheer impossibility of putting everything back in place exactly as it was before the loggers moved in, and of replacing physically identical rocks and streams and genetically identical flora and fauna. More importantly, an ecosystem such as a forest is very complex and constantly changing. It is probably impossible to gather simultaneously all the ecological data about a forest; by the time the last piece of information is in, the situation will have changed. Even if a perfect ecological blueprint were possible, the new forest at time T_2 would not be a replica of the original forest *as it would have been* at T_2 had it not been clear-cut. At best, it would be a replica of the forest at T_1, before it was clear-cut. Had it not been logged, it would have continued to change in essentially unpredictable ways. Perhaps the area is prone to fires caused by lightning, to flooding, erosion, slips, earthquake, or plagues of temporarily devastating insects: we cannot know which, if any, of these events would have occurred, or what their effects might have been.

Exact replication, then, is neither possible nor desirable. The replanted forest cannot be said to lack the value of the original just because it is only a replica, for it is *not* a replica; and in any case, unlike a replica, it cannot exist at the same time as the original forest. On the contrary, the replanting is possible only because of the destruction of the original. For the same reason, it is not a fake. It is, in fact, a real forest, if done properly. For Elliot, however, it is not a restoration in my sense, because he believes that it is not the same forest.

The case of some New Zealand offshore islands might be counterexamples to Elliot's claim: they have had introduced species removed and native flora and fauna reintroduced. The intention was to undo the damage caused by humans, in the sense of creating something close to what we think would have turned out had humans not caused that damage. This project seems little different, in principle, from restoring a damaged painting. It is not exactly the same as the original in every respect, of course, but then it would have changed in some ways without our interference.

Replanting a clear-cut forest, however, does not seem to be very much like restoring a painting. A clear-cut forest is no longer a forest at all. If a painting were *destroyed*, we would not say that it could be restored but only replicated or faked. To repeat, the replica Vermeer—unlike the restored Vermeer—is not a Vermeer, because it is not a painting produced by Vermeer. The restoration *is* a

Vermeer, because he painted it, even though someone else put something into it later. The restored painting is Vermeer's work: it bears the mark of his creativity, and he brought it into existence. The island's original ecology, similarly, evolved without the assistance of the Department of Conservation, and the role of the latter is like that of the painting restorer. Nevertheless, the forest replanter does not restore what is already there in damaged form. He or she may very well be effectively starting from scratch, using bulldozers to produce suitable contours, trucking in topsoil and fertilizer, diverting streams, digging ponds, planting trees, and releasing captive bred animals. The replanted forest is not only different from the original in that it has some different properties. It is a different forest, and however similar it may be, it is not the same, the original forest.

NATURAL AND HUMAN-CAUSED EXTINCTIONS

This argument, it may be said, ignores the fact that destruction and devastation occur on an enormous scale without humans. The eruption of Mount St. Helen's flattened thousands of hectares of forest, polluted streams, and covered a vast area with ash and lava. The successive ice ages destroyed everything in their path: mature coastal forest in New Zealand's South Westland, for instance, now stands on areas which were covered in ice sheets some thousands of years ago—ice which destroyed the earlier forests. Because nature produces destruction on a scale totally beyond human powers, Elliot's argument does not seem to depend on the *scale* of the destruction. Indeed, destruction and change are "natural" in the sense that they occur without human intervention. Thus, Elliot's objection to the restoration thesis must depend instead on a sharp distinction between what is caused by humans and what is natural.

It is important to note the limitations of the analogies between human and nonhuman-caused disasters. For instance, we cannot argue that because ninety percent of the species that have ever existed died out before humans evolved, current exterminations are acceptable, for natural systems may not recover from the current pounding they are receiving at the hands of humans. Ecosystem resilience—the ability of natural systems to adapt to change—is enormous globally (though very limited in some areas), but certainly not infinite. The biosphere recovered from whatever wiped out the dinosaurs and most other species at the end of the Cretaceous Era—a collision with asteroids, perhaps[27]— but it could have been different. The atmosphere could have been torn away, leaving Earth an irreversibly lifeless planet. Life recovers from devastating events such as volcanic eruptions because they are infrequent and because devastated areas are like uninhabited islands, waiting to be colonized from

---

[27] Stephen Jay Gould, "The Belt of an Asteroid," *Natural History* 89 (June 1980): 26–33.

surrounding areas rich with fauna and flora. Increasingly, humans are creating biologically rich islands—protected areas—surrounded by degraded areas which may approach ecological deserts. The technical or ecological limitations on restoration in an increasingly stressed biosphere may be enough to defeat proposals to restore large areas that have been ecologically devastated by developments. Moreover, industrial societies may be chemically unique in the life of the Earth. That the biosphere can survive collisions with asteroids does not imply that it can survive saturation by toxic chemicals and other wastes which are not found in nature.

MORAL RESPONSIBILITY

The really significant ethical difference, though, may be a matter of moral agency and responsibility. The landscapes of the American Southwest are "wrecked,"[28] but Krakatoa was not. To be wrecked is to be destroyed by a wrecker: in the case of the Southwest, the owners of the cattle and sheep that overgrazed the land. In contrast, eruptions merely happen. It makes a difference whether the moa were exterminated by the Māori or just died out due to climate change. The former is a cause for regret, not merely that there are no more moa, but that they were exterminated. In a different world view, in which natural processes are supposed to be caused by nonhuman agents, the distinction may not be so significant. But we are not animists or medieval Christians, and we are, uniquely, *responsible* for environmental destruction. We therefore have an obligation to avoid causing it—and, I suggest, to restore what we can of the destroyed areas of the past.

Elliot is right to insist that there is a value difference between human and nonhuman caused environmental change. Nevertheless, the difference lies in the fact that we are morally responsible for what we produce, not in an inherent and necessary qualitative difference between environments which have been influenced by humans and environments which have not.

CODA: EXTINCT CULTURES

Not all environmental ethicists try to explain the value of natural areas in terms of some other value; some may consider it to be reductionist on the grounds that environmental value is *sui generis*. However, it does seem reasonable at least to attempt to explain environmental value in terms of or by analogy with other, familiar values. This is the point of attempts to show that natural environments

---

[28] Aldo Leopold, "The Land Ethic," *A Sand County Almanac* (New York: Ballantine Books, 1966), p. 242.

are valuable aesthetically, or for recreation, as playground, cathedral, or gymnasium, as a treasure house of resources, a laboratory for scientific study, or an inspiration for artists. . . .[29] After all, these are all familiar criteria of value that many of us accept, that can successfully be appealed to, and that are part of our culture. It may be that the highest order or most general level of value is what is deeply embedded in flourishing cultures,[30] and there may be an analogy between the restoration of cultures and of natural environments.

Many forms of human association come into existence, go through (often predictable) stages of development, flourish, are subject to internal stresses and external threats, and either die out or recover. Clubs, nations, traditions, and cultures become threatened, endangered, or moribund, but they may revive—or *be* revived with the help of outside agencies or individuals.

Poland was partitioned three times: I do not know whether to say that it continued to exist in between reunifications, but Poles (and their enemies) are in no doubt that the same entity was involved in each partition and reunification. Whether or not the current State of Israel is the same Israel that existed 2000 years ago may be controversial, and depends on the criteria of identity that one is willing to adopt, but even its most bitter enemies could not deny that plausible criteria could be adopted on which it would be the same state.

In the case of a culture, its members may be reduced to a handful of old people; yet, such cultures can be revived, and, to repeat, it makes no difference to the identity and integrity of the culture whether the revival is partly at the hands of sympathetic outsiders. This kind of revival has happened, and is happening, in New Zealand, in Wales, and in North America. It is even happening in Lithuania, Latvia, Estonia, and the other Soviet occupied states of Eastern Europe. If a stressed culture can be revived, so can a stressed environment, or an endangered species.

Because these cases are like restorations of works of art, particularly what I call "dynamic" works such as buildings, which are expected to change over time, the revival of a culture can be seen as analogous to the restoration of a building. A restored culture is not exactly "the same" as it would have been when it formerly flourished, or as it would have been had it not been damaged. For example, Māori lifestyle now includes European introductions such as a cash economy, vehicles, farm animals, two languages, and a written literature; for many it also includes Christianity. Nevertheless, all cultures except the most isolated are in part the product of interactions. Although Navajo jewelry includes material obtained by trade with other tribes, and to some extent its distinctive

---

[29] These and other approaches to valuing wilderness are discussed by Warwick Fox, *Approaching Deep Ecology: A Response to Richard Sylvan's Critique of Deep Ecology* (Hobart: Centre for Environmental Studies, University of Tasmania, 1986).

[30] This is the view that I, in fact, hold, but for reasons of space I do not attempt to defend it here.

forms and excellence were influenced by contact with Anglo-American entrepreneurs, no one denies that it is a *real* Navaho product.

The situation is different, however, when the *origins* of a human association are part of its essence. Anyone can learn to speak the Māori language, but not anyone can be a Māori; a dead language can be revived, but not a dead tribe. A family, race, or tribe is linked by genealogy. Certainly, kinship can be acquired via marriage and adoption, but only if *bona fide* members exist. Once the last member of the House of York or the Etruscans died, the tribe or family ceased to exist forever, just like the Huia, the Tasmanian tiger, and the passenger pigeon.

THE WILDERNESS IDEA REVISITED: THE SUSTAINABLE DEVELOPMENT ALTERNATIVE

J. Baird Callicott
University of Wisconsin—Stevens Point

Abstract. At the beginning of the 20th century, led by John Muir and Gifford Pinchot, respectively, American conservation thought was divided into two camps: the preservation of "wilderness" versus the wise use of "natural resources." By mid-century, Aldo Leopold placed wilderness preservation in a broader, third philosophy of conservation that emphasized a harmony between human economic activities and ecosystem health. As the 20th century gives way to the 21st, conservation *via* wilderness preservation is less viable, especially on a global scale, than conservation *via* sustainable development. The received wilderness idea, further, is conceptually flawed: it is dualistic, ethnocentric, and static. Large wildlife refugia connected by wild corridors should be integrated, instead, into patterns of land use governed by considerations of ecological, as well as economic and technical, feasibility. There exist historic and contemporary examples of human economic activities that even enhance ecosystem health. Universal restoration of native ungulates to western rangelands for sustainable harvesting might achieve economic and conservation desiderata simultaneously.

> It seems to me that sanctuaries are akin to monasticism in the dark ages. The world was so wicked it was better to have islands of decency than none at all. Hence decent citizens retired to monasteries and convents. Once established these islands became an alibi for lack of private reform. People said: 'We pay the bills for all this virtue. Let goodness stay where it belongs and not pester practical folks who have to run the world.' ... The more monasteries or sanctuaries the grimmer the incongruity between inside and outside —Aldo Leopold (1942, p. 263).

INTRODUCTION

In making selections for the new collection of Aldo Leopold's essays that we prepared for the University of Wisconsin Press, Susan L. Flader and I sometimes disagreed. Because of his close association with the wilderness movement in America, I was for including virtually all of Leopold's papers on wilderness. Flader thought that those from the 1920s were too similar to one another, albeit rhetorically varied, to include more than a few. She also thought, more generally, that the importance of wilderness to Leopold had been exaggerated and that we should not abet unwittingly such a misperception. In Flader's well-informed opinion, Leopold was more interested, or at least more fully engaged, in managing humanly inhabited and used land—not only or even principally for enhanced resource production, but for a flourishing wild flora and fauna—than in campaigning for wilderness set asides. His work, in other words, had a different focus than that of his contemporaries, Robert Marshall, Howard Zahnizer, and Olaus Murie.

Flader, however, was anxious to include Leopold's earliest discussion of the relationship of wilderness to conservation, "The Popular Wilderness Fallacy," first published in 1918 in a long defunct periodical called *Outer's Book—Recreation*. I argued for suppressing it, since it seemed so anomalous. The fallacy that Leopold explores in this essay is that we need wilderness to have game. On the contrary, Leopold argues, game penury is not an inevitable result of the march of empire any more than timber famine is. The eradication of "varmints," he insists, following the reduction of wilderness, will make more game available to increased numbers of hunters; fire suppression will save game and its food from destruction; flowages behind dams will replace drained and seasonally dried up marshes for waterfowl; and so on. In short, everything will be coming up roses (or, in this case, deer and ducks) after the complete conquest of the continent by the forces of civilization. Indeed, in "The Popular Wilderness Fallacy" Leopold (1991a, p. 50) even writes, "Nature was actually improved upon by civilization"—at least as far as game goes. Leopold's early persecution of predators is well known and was acknowledged personally and apologetically in *Sand County*'s famous essay, "Thinking Like a Mountain." But that Leopold was also at first a Philistine on the wilderness issue—a cause, by 1918, well publicized and vigorously

---

J. Baird Callicott is a professor of philosophy and natural resources at the University of Wisconsin-Stevens Point, Stevens Point, WI 54481. He is author of *In Defense of the Land Ethic* and, with Susan L. Flader, editor of *The River of the Mother of God and Other Essays by Aldo Leopold*. In 1971, Dr. Callicott taught the first course in environmental ethics. Since that time, he has continued to contribute to the literature of this rapidly growing interdisciplinary field.

A response to this paper by Holmes Rolston III will be published in volume 13(4) of the journal. A short reply by J. Baird Callicott will follow Rolston's paper.

championed by the late John Muir—seemed to me a sleeping dog that we might just as well let lie.

I now no longer think that "The Popular Wilderness Fallacy" is as aberrant as at first it seemed. Rather, I think that Leopold envisioned throughout his career an ideal of human unity and harmony with nature, rather than a trade off between human economic activities and environmental quality. One might even go so far as to say that he did not abandon altogether his 1918 proposition respecting the improvement of nature by civilization, though by no stretch of the imagination did he retain it in its original sanguine and *laissez faire* form. One can imagine the ecologically chastened Leopold suggesting not that nature actually was improved upon by civilization, but that a mature and humane civilization actually might improve upon it.

On the other hand, Leopold eventually came to see and expressly to formulate the important and absolutely vital role that wild refugia had to play in biological conservation. I certainly hope that my remarks here will not be construed to deny or undermine the importance and necessity of wild lands in that regard. But the dialectical history of American conservation philosophy has fostered a more recent popular wilderness fallacy. That fallacy has two closely connected formulations. The first is that the New World was, when Christopher Columbus stumbled upon it, in a totally "wilderness condition"—as Nash (1967) famously characterizes it. The second is that any human alteration of pristine nature degrades it, and therefore biological conservation is served best by wilderness preservation; that is, the best way to conserve nature is to protect it from human inhabitation and utilization.

Here I briefly review the history of American conservation philosophy and the role that the wilderness idea has played in it. Then I argue that wilderness preservation, as a conservation stratagem, needs to be . . . not replaced, certainly, since nature reserves now fill and will continue to fill a vital niche in a more broadly conceived struggle to conserve biodiversity, but . . . refined, and augmented by a complementary approach. Having said what I am revisiting critically, let me set out as clearly and explicitly as I can a few caveats and qualifications.

First, I am as ardent an advocate of those patches of the planet called "wilderness areas" as any other environmentalist. My discomfort is with an idea, the received concept of wilderness, not with the ecosystems, so called.

Second, to suggest that something is amiss with the concept of wilderness is to suggest, at the same time, that something is amiss with its antithesis, the concept of civilization. That is the point of the epigram inaugurating this article. Implicit in the most passionate pleas for wilderness preservation is a complacency about what passes for civilization. If all that we can feature is the present adolescent state of civilization and its mechanical motif continuing indefinitely into the future, then naturally the only way we can conceive of conserving nature is to protect bits of it from destructive development. A harmony-of-man-and-nature conservation philosophy such as Leopold espoused implies re-envisioning civilization as well as critically revisiting the wilderness idea.

Therefore, third, I do not advocate what is euphemistically called "multiple use" for all landscapes. I do not suggest we attempt to mix strip-mining, clear-cutting, stock-grazing, four-wheeling, downhill-skiing, and motor-camping with biological conservation. By suggesting that we try to shift the burden of conservation from wilderness preservation to sustainable development, I mean to suggest that we try to think up economic strategies that are compatible with ecosystem health and that are limited strictly by ecological exigencies. There is precious little designated wilderness as things stand. Such areas serve the cause of biological conservation most importantly as refugia for species not tolerant of or tolerated by people. Personally, I would like to see more wild lands designated as wilderness with this purpose in view, but given a global human population approaching six billion persons, the greatest part of the best land will be put to economic use whether we conservationists like it or not. We conservationists, however, may hope realistically that in the future, ecological, as well as technological, feasibility may be taken into account in designing new and redesigning old ways of human living with the land.

THE WILDERNESS IDEA IN HISTORIC AMERICAN CONSERVATION PHILOSOPHY

Ralph Waldo Emerson and Henry David Thoreau were the first notable American thinkers to insist, a century and a half ago, that wild nature might serve "higher" human spiritual values, as well as supply raw materials for meeting our more mundane physical needs. Nature can be a temple, Emerson (1989) enthused, in which to draw near to and commune with God. Too much civilized refinement, Thoreau argued, can over-ripen the human spirit; just as too little can coarsen it. "In wildness," he wrote, "is the preservation of the world" (Thoreau, 1962, p. 185).

Building on the nature philosophies of Emerson and Thoreau, John Muir (1901) spearheaded a national, morally-charged campaign for public appreciation and preservation of wilderness. People's going to forest groves, mountain scenery, and meandering streams for religious transcendence, aesthetic contemplation, and healing rest and relaxation puts these resources to a higher and better use, in Muir's opinion, than did the lumber jacks, miners, shepherds, and cowboys who went to the same places in pursuit of the Almighty Dollar, and who were inspired only by the Main Chance.

Critics today, as formerly, may find an undemocratic and un-American presumption lurking in the Romantic-Transcendental conservation philosophies of Emerson, Thoreau, and Muir. To suggest that some of the human satisfactions that nature affords are morally superior to others may reflect only aristocratic biases and class privilege. Let me hasten to say that personally, I agree with Muir *et al.* Birdwatching, for example, is, in my opinion, morally superior to dirtbiking. But there is a contingent of powerful and influential professionals who do not agree. An axiom of neoclassical economics is that all human preferences concerning "resource" use are morally equal and should be weighed one against the other in the marketplace (Randall, 1988).

At the turn of the century, Gifford Pinchot, a younger contemporary of John Muir, formulated a novel conservation philosophy that reflected the general tenets of the Progressive era in American history. Notoriously, the country's vast biological capital had been plundered and squandered for the benefit, not of all its citizens, but for the profit of a few. Pinchot (1947, pp. 325-326) crystalized a populist, democratic conservation ethic in a credo—"the greatest good of the greatest number for the longest time"—that echoed John Stuart Mill's famous Utilitarian maxim, "the greatest happiness for the greatest number." He bluntly reduced Emerson's "Nature" (with a capital "N") to "natural resources." Indeed, Pinchot (1947, p. 326) insisted, "There are just two things on this material earth—people and natural resources." He even equated conservation with the systematic exploitation of natural resources. "The first great fact about conservation," Pinchot (1947, p. xix) noted, "is that it stands for development"—with the proviso that resource development be scientific and thus efficient. For those who might take the term "conservation" at face value and suppose that it meant saving natural resources for future use, Pinchot (1947, p. xix) was quick to point out their error: "There has been a fundamental misconception," he wrote, "that conservation means nothing but the husbanding of resources for future generations. There could be no more serious mistake." And it was none other than Pinchot (1947, p. 263) who first characterized the Muirian contingent of nature lovers as aiming to "lock up" resources in the national parks and other wilderness reserves.

The infamous schism in the traditional American conservation movement thus was rent. Muir and Pinchot, once friends and allies, quarreled, and each followed his separate path (Nash, 1967). Pinchot appropriated the term "conservation" for his utilitarian philosophy of scientific resource development, and Muir and his exponents came to be called "preservationists."

The third giant in 20th century American conservation philosophy is, of course, Aldo Leopold. At the Yale Forest School, founded with the help of the Pinchot family fortune, Leopold was steeped in what Hays (1959) called "the gospel of efficiency"—the scientific exploitation of natural resources, for the satisfaction of the broadest possible spectrum of human interests, over the longest time. Moreover, for fifteen years Leopold worked for the Forest Service, whose first chief was Pinchot himself. Leopold's ultimately successful struggle for a system of wilderness reserves in the national forests was molded consciously to the doctrine of highest use,

and his new science of game management essentially amounted to the direct transference of the principles of forestry from a standing crop of large plants to a standing crop of large animals (Leopold, 1918; 1921). However, Leopold gradually came to the conclusion that Pinchot's utilitarian conservation philosophy was inadequate, because it was not well informed by the new kid on the scientific block, ecology (Flader, 1974; Meine, 1988). As Leopold (1939a, p. 727) put it:

> Ecology is a new fusion point for all the sciences....
> The emergence of ecology has placed the economic biologist in a peculiar dilemma: with one hand he points out the accumulated findings of his search for utility, or lack of utility, in this or that species; with the other he lifts the veil from a biota so complex, so conditioned by interwoven cooperations and competitions, that no man can say where utility begins or ends.

Conservation, Leopold came to realize, must aim at something larger and more comprehensive than a maximum sustained flow of desirable products (like lumber and game) and experiences (like sport hunting and fishing, wilderness travel, and solitude) garnered from an impassive nature. It must take care to ensure the continued function of ecological processes and the integrity of ecosystems. For it is upon them, ultimately, that human resources and human well-being depend, for the present generation as well as for those to come. Indeed, Leopold quietly transformed the concept of conservation from its pre-ecological to its present deep ecological sense— from conservation understood as the wise use of natural resources to conservation understood as the maintenance of biological diversity and ecological health.

The word "preserve" in the summary moral maxim of Leopold's (1949, pp. 224-225) famous land ethic ("A thing is right when it tends to preserve the integrity, stability, and beauty of the biotic community. It is wrong when it tends otherwise.") is unfortunate, because it seems to ally Leopold with the Preservationists in the familiar Preservation *vs.* Conservation feud. We tend to think of Leopold as having begun his career in the Conservationist camp and then gradually to have come over, armed with new ecological arguments, to the Preservationist camp. Leopold appears to be a mid-20th century conservation prophet emerging from the woods wearing the hat of Gifford Pinchot and speaking with the voice of John Muir. His historical association with the wilderness movement cements this impression.

While still with the Forest Service, Leopold had campaigned hard to preserve a few relics of the American frontier in which he and like-minded sportsmen might play at being pioneers (Leopold, 1925a; 1925b; 1925c; 1925d; Allin, 1987). After becoming a professor of wildlife management, he suggested that the designated wilderness areas he had helped to create might serve threatened species as biotic refugia (Leopold, 1936). In the last decade of his life, he suggested that representative undeveloped biomes might serve science as a "base datum of land health" (Leopold, 1941).

While I would be the first to agree that designated and *de facto* wilderness areas are important as "land laboratories" and vitally important as biotic refugia, Leopold's unfortunate—and unintended—legacy for the American conservation policy debate has been to intensify the familiar alternative: either efficiently exploit the remaining and dwindling wild lands or lock them up and preserve them forever as wildernesses. But a review of his unpublished papers and published, but long-forgotten, articles (now conveniently collected in the new book of his essays) confirms Flader's opinion that Leopold was concerned primarily, in theory as well as on the ground, with integrating an optimal mix of wildness with human habitation and economic utilization of land.

As *Sand County*'s upshot essay, "Wilderness," shows, Leopold was committed to wilderness preservation no less fully at the end of his career than at the beginning. However, his mature vision went beyond the either develop and necessarily destroy or lock up and preserve dilemma of American conservation as he inherited it from the generation of Pinchot and Muir. Wild sanctuaries, for Leopold, were a component of a much broader and subtler conservation philosophy, a conservation philosophy that he regarded as more developed and mature than simply saving wild remnants. As he put it, "This impulse to save wild remnants is always, I think, the forerunner of the more important and complex task of mixing a degree of wildness with utility" (Leopold, 1991b, p. 227). In a typescript composed shortly after a four-month trip to Germany in 1935, Leopold (1991b, pp. 226-227) wrote,

> To an American conservationist, one of the most insistent impressions received from travel in Germany is the lack of wildness in the German landscape. Forests are there.... Game is there.... Streams and lakes are there.... But yet, to the critical eye there is something lacking.... I did not hope to find in Germany anything resembling the great "wilderness areas" which we dream about and talk about, and sometimes briefly set aside, in our National Forests and Parks.... I speak rather of a certain quality [—wildness—] which should be, but is not found in the ordinary landscape of producing forests and inhabited farms.

In a more fully developed essay entitled "The Farmer as a Conservationist," Leopold (1939b) regales his reader with a rustic idyll in which the wild and domesticated floral and faunal denizens of a Wisconsin farmscape are feathered into one another to create a harmonious whole. In addition to cash and the usual supply of vegetables, chicken, beef, pork, lumber, and fuel wood, Leopold's envisioned farmstead affords its farm family venison, quail and other small game, and a variety of fruit and nuts from its wood lot, wetlands, and fallow fields; and its pond and stream yield pan-fish and trout. It also affords intangibles—songbirds, wild flowers, the hoot of owls, the bugle of cranes, and intellectual adventures aplenty in natural history. To obtain this bounty, the farm family must do more than permanently set aside acreage, fence woodlots, and leave wetlands undrained. They must sow food and cover patches, plant trees, stock the stream and pond, and generally thoughtfully conceive and skillfully execute scores of other modifications, large and small, of the biota that they inhabit.

Further, Leopold (1939b, p. 294) explicitly states the preservationist heresy that human economic activity may not only co-exist with healthy ecosystems, but also actually enhance them: "When land does well for its owner, and the owner does well by his land; when both end up better by reason of their partnership, we have conservation. When one or the other grows poorer, we do not."

Like Pinchot, Leopold (1949, p. 207) attempted to distill his philosophy of conservation into a quotable definition. Indeed, it is often quoted, but little analyzed or appreciated: "Conservation is a state of harmony between men and land." This definition represents a genuine third alternative to Pinchot's brazenly anthropocentric, utilitarian definition of conservation as efficient exploitation of "resources" and Muir's anti-anthropocentric definition of conservation as saving innocent "Nature" from inherently destructive human economic development.

Can we generalize Leopold's vision of an ecologically well-integrated family farm to an ecologically well-integrated technological society? Can we reconcile and integrate human economic activities with biological conservation? Can we achieve "win-win" rather than "zero-sum" solutions to development-environment conflicts? Can we design "sustainable economies," rather than zone the planet into ever-expanding sectors of conventional, destructive development and ever-shrinking wilderness sanctuaries? Can we succeed as a global technological society in enriching the environment as we enrich ourselves?

A THIRD WORLD CRITIQUE OF CONSERVATION VIA WILDERNESS PRESERVATION

I think we can. More to the point, I think we have to. The pressure of growing human numbers and rapid development, especially in the Third World, bodes ill for a global conservation strategy focused primarily on "wilderness" preservation and the establishment of nature sanctuaries. Such a strategy represents a holding action at best and a losing proposition at last.

A sobering "Third World critique" of conservation on a global scale *via* wilderness preservation has been advanced recently in the pages of *Environmental Ethics*. Therein, Gua (1989a, p. 79) points out that we Americans:

possess a vast, beautiful, and sparsely populated continent.... America[n's thus] can simultaneously enjoy the material benefits of an expanding economy and the aesthetic benefits of unspoilt nature. The poles of 'wilderness' and 'civilization' mutually coexist in an internally coherent whole....

Exporting the American pattern of conventional industrial development counterbalanced by conservation of nature *via* wilderness preservation to a "long settled, densely populated country" like India, Gua (1989a, p. 75) explains, results in a "direct transfer of resources from the poor to the rich"—because the poor are evicted from their homelands to create nature reserves and only wealthy foreigners and the in-country elite that benefit from industrial development can afford to enjoy the wilderness experience or just the knowledge that a bit of pristine nature is being protected.

Gua's critique notwithstanding, the United Nations biosphere reserve concept genuinely de-Americanizes the wilderness idea, in that it specifically requires planners to take account of and to integrate local peoples culturally and economically into reserve designs (von Droste, 1988). However, faced with the harsh realities of the coming century, the wilderness idea —even become the biosphere reserve concept—is, by itself, too little too late, and it is too defensive to save the planet and all of us, its people, from ecological collapse. We need to integrate wildlife sanctuaries into a broader philosophy of conservation that generalizes Leopold's vision of a mutually beneficial and mutually enhancing integration of the human economy with the economy of nature.

Here let me be both clear and emphatic. I am not suggesting that we open the remaining wild remnants to development, but that we begin to reconceive economic development in the light of ecology. Human economic activities should at least be compatible with the ecological health of the environments in which they occur, and, ideally, they should enhance it. In Leopold's agrarian idyll of conservation, not only does the ecological farm family actively manage its wild lands, but it also reforms its farming practices on the fields so that the two, the wild culture and the domestic culture, might coexist better. "Clean farming" was a frequent target of Leopold's pointed criticisms and wry wit; he may have been the first environmentalist clearly to recognize and lament the "industrialization" of agriculture during the 20th century and the conversion of the classic, relatively benign and sustainable farm, into an environmentally-destructive and unsustainable "food-factory" (Leopold, 1945).

Echoing the Leopold epigraph of this essay, Gua further charges that conservation primarily *via* wilderness preservation is an American subterfuge. We who enjoy the benefits of modern industrial civilization with all its environmental costs can salve our consciences by pointing to the few odds and ends of arid, rough, scenic, or remote country that we have set aside, first for our own recreational, aesthetic, and spiritual needs, and second as ecologic refugia—little places where our fellow denizens of planet Earth can live and blossom. In so doing, we can avoid facing up to the fact that the ways and means of industrial civilization lie at the root of the current global environmental crisis. To Gua (1989a, p. 79), the new voice for Third World environmental ethics, conservation *via* wilderness preservation "runs parallel to the consumer society without seriously questioning its ecological and sociopolitical basis."

Without such questioning, the call of wilderness advocates for the recrudescence of big wilderness is quixotic. Wilderness is one pole of a dualism. To want more of it is to oppose the forces of conventional industrial development, certainly, but not to challenge the conventional conception of economic development. And since it pits—as a one-or-the-other-but-not-both choice—human economic interests against the interests of nature, it has little political appeal, and thus little chance of success.

In the Third World, on the other hand, industrial development often has resulted in human tragedy as well as environmental disaster. The green revolution, for example, has dispossessed small land holders and actually increased chronic hunger as prices fall, costs increase, and crops are exported to First World consumers (Wright, 1984). Industrial forestry similarly has disrupted traditional patterns of sustainable forest use, as well as played havoc with forest ecosystems (Gua, 1989b). By contrast, according to Gua (1989a, p. 81), Third World peasants have far more basic needs.

> Their demands on the environment are far less intense, and they can draw on a reservoir of cooperative social institutions and local ecological knowledge in managing the 'commons'—forests, grasslands, and waters—on a sustainable basis.

Concerned to articulate a distinctive Third World environmental philosophy, Gua (1989a) may have conceded too readily that conservation principally *via* wilderness preservation is even practicable in the United States. The violent confrontations of the Redwood Summer of 1990 in California and the ongoing, increasingly acrimonious Old Growth/Spotted Owl impasse in the Pacific Northwest make me wonder if Gua (1989a, p. 79) were not perpetuating uncritically a Third World myth when he describes North America as a "vast," superabundant, and "sparsely populated continent."

The late 20th-century crescendo of cut/graze/plow/pave or preserve conflicts here and abroad lead me to undertake a generalization of Gua's "Third World critique" of the wilderness idea. I wish to strike deeper than he and suggest that the popular wilderness idea is as inherently flawed as its counterpart, the conventional development idea. Most conservation biologists today recognize the paradoxical necessity of managing (and hence artificializing?) wilderness areas in order for them to continue to play their vital part in biological conservation (Reed, 1990). Just as paradoxically, refugia for species that require lots of living space may be compromised by wilderness recreation for which purpose such areas originally were set aside pursuant to the popular concept of wildernesses as areas where man is a visitor who does not remain (Reed and Merigliano, 1990).

A THREE POINT CRITIQUE OF THE RECEIVED CONCEPT OF WILDERNESS

Upon close scrutiny, the simple, popular wilderness idea dissolves before one's gaze. First, the concept perpetuates the pre-Darwinian Western metaphysical dichotomy between "man" and nature, albeit with an opposite spin. (Fully aware that it is gender-biased, I use the term "man" both deliberately and apologetically to refer globally and collectively to the species *Homo sapiens*, because no other term carries the same connotation and flavor, a connotation and flavor that I wish to evoke in the course of this critique, including its decided sexism.) In fact, one of the principal psycho-spiritual benefits of wilderness experience is said to be contact with the radical "other," and wilderness preservation the letting be of the nonhuman other in its full otherness (Birch, 1990).

Second, the wilderness idea is woefully ethnocentric. It ignores the historic presence and effects on practically all the world's ecosystems of aboriginal peoples.

Third, it ignores the fourth dimension of nature, time. In a recent discussion, Cordell and Reed (1990, p. 31) say flatly, "Preservation implies cessation of change." In ecosystems, however, change is as natural as it is inevitable (Botkin, 1990); consequently, trying to preserve in perpetuity—trying to "freeze-frame"—the ecological status quo ante is as unnatural as it is impossible. A more sophisticated and refined concept of wilderness preservation among contemporary conservationists aims, rather, to perpetuate the integrity of evolutionary and ecological processes, instead of existing "natural" structures (Parsons *et al.*, 1986). Cordell and Reed (1990, pp. 30-31), in fact, understand wilderness preservation not as an effort to halt change, but to slow "accelerating rates of change" and to preserve the "dynamic operation of natural processes... fire, drought, disease, predation, and geological change." However, even such a dynamic, process-sensitive notion of wilderness preservation is incomplete if it ignores the role that *Homo sapiens* has played historically practically everywhere and if it would deny *Homo sapiens* the opportunity to reestablish a positive symbiotic relationship with other species and a positive role in the unfolding of evolutionary processes.

The Wilderness Act of 1964 beautifully reflects the conventional understanding of wilderness. It reads: "A wilderness, in contrast with those areas where man and his works dominate the landscape, is hereby recognized as an area where the earth and its community of life are untrammeled by man, where man himself is a visitor who does not remain" (Nash 1967, p. 5).

This definition assumes, indeed it enshrines, a bifurcation of man and nature. That the man-nature dichotomy insidiously infects even our well-intentioned and noble efforts to limit our own grasp should not be surprising. A major theme both in Western philosophy, going back to the ancient Greeks, and in Western religion, going back to the ancient Hebrews, is how man is unique and set apart from the rest of nature.

In the Judeo-Christian religious tradition, man alone among all the other creatures is created in the image of God. In the Greco-Roman philosophical tradition, among all the other animals, man is uniquely rational. Subsequently, philosophers as different from one another as Thomas Aquinas, the perennial philosopher of the Catholic church, and René Descartes, the father of modern philosophy, variously synthesized these two strands of Western thought, Thomas in the Middle Ages and Descartes at the dawn of the Scientific Revolution. The classical Western segregation of man from nature thus became ever more ingrained, a veritable cachet of Western ideology, both religious and scientific. Moreover, all the wonderful works of man—from the pyramids of Egypt to the Gothic cathedrals of France, to say nothing of all the marvels of modern technology—seemed to confirm the radical metaphysical rift between us and the brute creation.

Now that man's technological dominion over the Earth is virtually absolute, and its community of life will survive only if we permit, a rising chorus of voices is crying, if not in the wilderness, at least for it—for man to show a little mercy, to allow a little untrammeled, unconquered nature to exist here and there (Oelschlaeger, 1991). Until recently, man seemed the up-and-coming hero armed with Promethean science in the struggle with Titanic nature. Now, as the 20th century winds to a close, victorious man seems to be a tyrant, his conquest a spoils, and nature the victim. For many ardent wilderness advocates, the roles of hero and villain are reversed (Birch, 1990), but the underlying dichotomy goes unchallenged.

Since Darwin's *Origin of Species* and *Descent of Man*, however, we have known that man is a part of nature. We are only a species among species, one among twenty or thirty million natural kinds. The natural works of other species, everyone seems to agree, can help as well as harm the biotic communities of which they are a part. Pursuing their own economic interests, bees assist the reproduction of flowering plants and thus perform an invaluable community service. Hundreds of other similar examples—from nitrogen-fixing bacteria to scavenging turkey vultures—could be cited. Elephants, on the other hand, pursuing their own economic interests, can be very destructive members of their biotic communities. So can deer. Hundreds of similar examples—from cow birds to kudzu—could be cited.

ALTERNATIVE IDEAS OF WILDERNESS

If man is a natural, a wild, an evolving species not essentially different in this respect from all the others, as Snyder (1990) reminds us, then the works of man, however precocious, are as natural as those of beavers, or termites, or any of the other species that dramatically modify their habitats. And if entirely natural, then the works of man, like those of bees and beavers, in principle may be, even if now they usually are not, beneficial—judged by the same objective ecological norms—to the biotic communities which we inhabit.

In one important (and relevant) respect, we are different from other species. The pollinating services performed by bees and the decomposition of dead wood and soil treatment provided by termites are instinctive behaviors. The migration routes of birds, the hunting techniques of predators, and many other animal behaviors that are learned might be regarded as "cultural," or at least "protocultural," rather than strictly hereditary or "instinctive." However, the cultural component in human behavior is so greatly developed as to have become more a difference of kind than of degree. To suggest that the works of man are not natural is not to suggest that they are supernatural or preternatural, but that they are products of culture, not instinct. Still, the cultural works of man are evolutionary phenomena no less than are other massive structures created by living things like, say, coral reefs. They are, one and all, natural in that sense of the word (Lemons, 1987). Therefore, it is logically possible that they may be well attuned and symbiotically integrated with other contemporaneous evolutionary phenomena, with coral reefs and tropical forests, as well as the opposite.

Precisely because the works of man are largely cultural, they are capable of being reformed rapidly. Other animals cannot change what they do in and to their biotic communities, at least not very rapidly, and perhaps not ever consciously and deliberately. We can, since our economic behaviors are determined more by our cultures than by our genes. Whether to trammel nature totally—to "develop" every acre—or here and there to forbear is not the only question man should put to himself. How to work our works in ways that are at once humanly, socially, and ecologically benign rather than malignant—that's the more important, and problematic, question.

Now on to my second point, that the popular wilderness idea is ethnocentric. More than anyone else, Nash (1967) has molded the popular idea of wilderness in the contemporary American mind. He acknowledges, but skates rapidly over, American Indian complaints that the very concept of wilderness is a racist idea, and he expresses no doubt that the first European settlers of North America encountered a "wilderness condition" (Nash, 1967). In the recent (and excellent) Wilderness Idea film by Hott and Garey (1989), Nash is even more emphatic. He says that the pilgrims literally stepped off the Mayflower into a wilderness of continental dimensions.

Upon the eve of European landfall, most of temperate North America was not, *pace* Nash, in a wilderness condition—not undominated by the works of man—unless one is prepared to ignore the existence of its aboriginal inhabitants and their works or to insinuate that they were not "man," *i.e.*, not fully human human beings. In 1492, Antarctica was the only true wilderness land mass on the planet. (And by now, even a good bit of it has come under the iron heel of industrial man.)

Until rather recently, it was possible for environmental historians to minimize the ecological importance of the original human inhabitants of the New World, because the decimating effects of Old World diseases had not been taken into account in estimating their Pre-Columbian numbers. Kroeber (1939) calculated a hemispheric total of eight-and-a-half million souls at contact and placed the population of the lands now comprised of the 48 (geographically) United States, Canada, Greenland, and Alaska at fewer than one million. Dobyns (1966), adding the impact of Old World diseases on the immunologically-innocent New World aborigines to his demographic equations, proposed to increase Kroeber's estimates by, roughly, a factor of ten. Tragically, only one Indian in ten seems to have survived the epidemics of small- and chicken-pox, diphtheria, measles, scarlet fever, and other infections that swept through North America from east to west during the 15th and 16th centuries—often passed from Indian to Indian before the leading edge of the pale-face tide arrived. As Witthoft (1965, p. 28) described this inverse decimation, "Great epidemics and pandemics of these diseases are believed to have destroyed whole communities, depopulated whole regions, and vastly decreased the native population everywhere in the yet unexplored interior of the continent."

In the spring of 1492, North America (not including Mexico) may not have been densely populated, but it was largely inhabited by some ten million people. Nor were the Indians passive denizens of the continent's forests, prairies, swamps, deserts, and tundra, simply taking—like foraging Mandrill baboons—what usable plants and animals they happened to stumble upon. Most of temperate North America was managed actively by its aboriginal human inhabitants. In addition to domesticating and cultivating an extraordinarily wide range of food and medicine plants, native North Americans managed the continent's forest and savannah communities, principally with fire (Heizer, 1955; Pyne, 1982). Their pyrotechnology helped to determine the mix of species and reset succession in the various plant associations in which they lived (Day, 1953; Martin, 1973; Lewis, 1973). The European immigrants, in fact, found a man-made landscape, but they thought it was a wilderness because it didn't look like the man-made landscape that they had left behind (Cronon, 1983; Merchant, 1989).

It is important to note, however, that the same kind of country that now is designated or *de facto* wilderness in the United States also was frequented and utilized less by the American aborigines (Barrett, 1980; Devall, n.d.). The two per cent of the 48 (geographically) United States presently devoted to wilderness is mostly in high, rough, or arid lands—and often all three. A good argument, therefore, could be made that an expansion of the present system of wilderness reserves to mirror ancient human land use patterns on this continent is consistent with long-established Nearctic evolutionary and ecological regimes.

The incredible abundance of wildlife encountered in the western hemisphere by the first European intruders was not, however, a concomitant of a universal wilderness condition, that is, not due to the absence of inherently destructive *Homo sapiens* in significant numbers everywhere. Rather, the biological wealth of North America on the eve of European landfall is more attributable to the bioregional management programs of the indigenous human population than to low numbers (Hughes, 1983). Further, the ubiquity of grizzly bears throughout the west and big cats and wolves throughout the continent indicates a mutual tolerance of these species with *Homo sapiens americana* that was, apparently, disrupted when *Homo sapiens europi* began to persecute them as varmints (Martin, 1978).

I now take up my third point: Wilderness preservation, as the popular conservation alternative to destructive land use and development, suggests that, untrammeled by man, a wilderness will remain "stable," in a steady state. However, nature is inherently dynamic; it is constantly changing and ultimately evolving. Today, most of the pitifully small fenced-off patches of designated wilderness areas of temperate North America lack major components of their Holocene ecological complement—notably their large predators. Not only that, a fence, or the policy equivalent thereof, will not exclude all the exotic species that have accompanied or followed the migration of *Homo sapiens europi* to North America. Designated wilderness areas, paradoxically, must be restored and managed actively if they are to remain fit habitat for native species, but the necessity of means raises a question of ends, of values. Is maintaining "vignettes of primitive America" the most important and defensible goal of biological conservation? (Gordon *et al.*, 1989). (Here, again, let me be clear. I am excluding from consideration biologically-destructive economic desiderata such as hydroelectric impoundment.) Since we must manage nature actively and invasively to preserve the ecological status quo ante, the possibility of managing nature for more direct, less incidental conservation goals arises (Westman, 1990).

The whole notion of preserved wilderness areas—even were we to restore the wolf and the grizzly to representative spots in their original ranges, simulate the effects, where they existed, of *Homo sapiens americana*'s hunting and burning, and maintain a constant vigil against invasion by exotics—defies the fourth dimension of nature, time. In the course of time, ecological succession is reset continually by one or another natural disturbance. Paleo-ecological studies reveal, moreover, that species composition within successional seres—the structure, in other words, of biotic communities—

has changed over time (Thompson, 1988). Fluctuations in climate drive migrations of glaciers, forests, deserts, and grasslands (Botkin, 1990). Exotic species, with or without human help, invade new environments, and in the course of time, become naturalized citizens (Westman, 1990). Indeed, the concepts of exotic/native species are "relative, scale dependent (temporally and spatially), and about as ambiguous as any in our conservation lexicon" (Noss, 1990a, p. 242). Are the feral mustangs roaming the American west, for example, natives or exotics? Soulé (1990, pp. 234-235) speculatively envisions lions, cheetahs, camels, elephants, saiga antelope, yaks, and spectacled bears joining horses in the name of "the restoration to the Nearctic of the great paleomammalian megafauna" which disappeared from this continent "only moments ago in evolutionary time."

The wilderness idea has not only made conservation convenient—if Gua (1989a) is correct to argue that it has served American conservationists as a subterfuge, allowing us to enjoy the benefits of industrial development and over consumption, while salving our consciences by setting aside a few undeveloped remnants for nature—but it aslo has made conservation philosophy simple and easy. If we conceive of wilderness as a static benchmark of pristine nature in reference to which all human modifications may be judged to be more or less degradations, then we can duck the hard intellectual job of specifying criteria for land health in four-dimensional, inherently dynamic landscapes long inhabited by *Homo sapiens* as well as by other species. The idea of healthy land maintaining itself is more sensitive to the dynamic quality of ecosystems than is the conventional idea of preserving vignettes of primitive America. Moreover, if the concept of land health replaces the popular, conventional idea of wilderness as a standard of conservation, then we might begin to envision ways of creatively reintegrating man and nature.

Conservation biologists just now are coming to grips with the problem of setting out objective criteria of ecological health in dynamic, long-humanized landscapes. Recent efforts to do so have been made by Costanza (1991), Westman (1991), and Ulanowicz (1991). While insisting upon the naturalness of change, the importance of rate and scale of change for land health cannot be overemphasized. Cordell and Reed (1990, pp. 30-31) suggest that "'bad' change" is the result of "accelerating rates of environmental change" and Botkin (1990) again and again warns that, while change per se always has characterized the living earth, the current changes imposed upon nature by global industrial civilization are unprecedentedly rapid and radical and therefore, albeit natural, not normal.

CONSERVATION VIA SUSTAINABLE DEVELOPMENT

The new idea in conservation today is called "sustainable development" (Brundtland *et al.*, 1987), a term that can mean different things to different people. Under the essentially

ALTERNATIVE IDEAS OF WILDERNESS

economic interpretation of the Brundtland Report (1987), it means little more than what it says: initiation of human economic activity that can be sustained indefinitely, quite irrespective of whether such development is ecologically salubrious. Worse still, some economists would denominate a development path "sustainable," even if it leaves subsequent generations a depauperate natural environment, but sufficient technological know-how and investment capital to invent and manufacture an ersatz world (Passell, 1990). Following Dasmann (1988), I would like to mean by "sustainable development" initiation of human economic activity that is limited by ecological exigencies; economic activity that does not compromise ecological integrity seriously; and, ideally, economic activity that positively enhances ecosystem health. However, is sustainable development, so understood, possible? The surest proof of possibility is actuality. Here are some actual examples of mutually sustaining and enhancing human-nature symbioses.

The Desert Smells Like Rain, by ethnobotanist Nabhan (1982), is about present-day Papago dry farmers in the desert Southwest. From time immemorial, two oases some thirty miles apart, A'al Waipia and Ki:towak, had been inhabited by Papago. The former lies in the United States, in the Organ Pipe Cactus National Monument, and the latter in Mexico. The United States government designated A'al Waipia a bird sanctuary and stopped all cultivation there in 1957. Over in Mexico, Ki:towak still is being farmed in traditional style by a group of Indians. Nabhan reports visiting the two oases, accompanied by ornithologists, on back-to-back days, three times during one year. At the A'al Waipia bird sanctuary, they counted thirty-two species of birds; at the Ki:towak settlement, they counted sixty-five. A resident of Ki:towak explained this irony: "When the people live and work in a place, and plant their seeds and water their trees, the birds go live with them. They like those places. There's plenty to eat and that's when we are friends to them" (Nabhan, 1982, p. 98).

Ehrenfeld (1989, p. 9) concludes from this "parable of conservation" that "the presence of people may enhance the species richness of an area rather than exert the effect that is more familiar to us." Here, of course, we must be cautious. Species richness is not the only measure of ecosystem health. The quality, so to speak, of species is also important. In general, Nabhan (1982) suggests, the whole desert ecosystem in which they live, not just the Ki:towak oasis, is as adapted to and dependent upon the Papago as they are upon it. Their little charco fields, built to catch and hold the runoff from ephemeral desert rains, are home to a wide variety of coevolved uncultivated plants (some of which the Papago eat) and unfenced animals ("field meat" as the Papago think of them). Undoubtedly, the desert ecosystem has been enriched rather than impoverished by millennia of Papago habitation and exploitation.

Gomez-Pompa and Kaus (1988), on the basis of the higher incidence of fruit-bearing trees in the remnants of rainforest in southern Mexico, suggest that what appear to the untutored eye to be pristine patches of wilderness, rich in animal as well as plant life, actually are surviving fragments of an extensive lowland Maya permaculture.

Ecologists working in the Amazon have come to similar conclusions about the vast South American rainforest. Posey has studied the methods of living in the Amazon rainforest without destroying it devised by the Kayapo Indians (Stevens, 1990). The Kayapo fish, hunt, gather, and cultivate swiddens. In sharp contrast to the displaced Euro-Brazilian peasants who are entering the region, the Kayapo, through a complex cycle of planting, manage to cultivate a forest clearing for nearly ten years, instead of merely three or four. But after a decade of cultivation, neither is a Kayapo plot simply abandoned. Instead, the Kayapo manage the regeneration of the forest by planting useful native species—first, fast-growing short-lived early succession plants like banana, and later, long-lived canopy trees like Brazil nut trees and coconut and oil palms. Thus, their fallows become permanent resource patches from which they obtain fruit, nuts, medicines, thatch, and other materials in perpetuity.

I cite these indigenous New World examples of human-nature symbiosis not to suggest that we give the hemisphere back to the Indians or that we all go native and attempt somehow to recreate American Indian culture in the late 20th century. The three hundred million of us contemporary denizens of North America could not return to the lifestyles of the ten million Pre-Columbian inhabitants even if we wanted to, without exceeding carrying capacity. I simply wish to point out, rather, that the past affords paradigms aplenty of an active, transformative, managerial relationship of people to nature in which both the human and nonhuman parties to the relationship benefited. The human-nature relationship is an ongoing, evolving one. We can, I am confident, work out our own, postmodern, technologically-sophisticated, scientifically-informed, sustainable civilization, just as in times past the Minoans in the Mediterranean, the vernacular agriculturalists of Western Europe, and the Incas in the Andes worked out theirs.

The symbiotic win-win philosophy of conservation gradually is replacing the bifurcated zero-sum approach as the 20th century gives way to the 21st. For example, one of the most promising conservation stratagems in the Amazon rainforest today is the designation not of nature reserves from which people are excluded to protect the forest and its wildlife, but of so-called "extractive reserves" (Hecht and Cockburn, 1989). An extractive reserve is an area where traditional patterns of human-nature symbiosis—such as those evolved by the Amazonian Indians and more recently by the rubber tappers—are protected from loggers, cattle ranchers, miners, and hydroelectric engineers.

Peters *et al.* (1989) report that the nuts, fruits, oils, latex, fiber, and medicines annually harvested from a representative hectare of standing Amazon rainforest in Peru is of greater economic value than the saw logs and pulpwood stripped from a similar hectare—greater even than if, following clear-cutting and slash-burning, the land is, in addition, converted either to a forest monoculture or to a cattle pasture. From a painstaking econometric study, they conclude "without question, the sustainable exploitation of non-wood forest resources represents the most immediately profitable method for integrating the use and conservation of the Amazonian forests" (p. 656).

Is it possible to export this Third World approach to conservation to the First World? Here is a modest proposal that I think would have a greater chance of realization than Noss's (1991) suggestion that 50 per cent of the lower 48 states be allowed to return to a wilderness condition, a condition which as I have pointed out here has not existed in 50 per cent of what is now the lower 48 states for the 100 centuries since the original discovery of the Americas by eastward migrating pedestrian *Homo sapiens*. It would have a greater chance of realization, because it offers a forward-looking, sustainable economic alternative to the present pattern of destructive exploitation, and because it is not implicitly misanthropic.

First, I agree with Noss (1990b) that we need big wilderness to help conserve biodiversity, but is 50 per cent a reasonable amount? A more practicable alternative might be to start with presently designated wilderness areas and, guided by information on aboriginal settlement and use, expand their political boundaries to coincide with their ecological boundaries. Such expanded wild reserves might be connected where possible by wild corridors, permitting the migration of animals and gene flow between populations, as Noss (1990b) also suggests. These core regions and interconnective corridors would serve as sanctuaries for the large carnivores and other species, like the spotted owl, that need old growth or interior habitat to thrive. One-third of the lower 48 is publicly-owned land. These public lands are managed mostly by the USDA Forest Service, the Bureau of Land Management, and the National Park Service, but only by the latter in anything resembling the public interest. In designing a utopia we can stipulate that public agencies function as ideally they ought. Even in the real world, there are some winds of discontent blowing in the Forest Service and some stirrings of reform (DeBonis, 1989; USDA Forest Service, 1989). I shall assume that it is theoretically possible for the Forest Service to manage the national forests for ecological integrity and function as well as for a sustainable supply of forest products.

For the western rangelands, however, I propose a complete overhaul of current management philosophy. There is little economic and no ecological justification for grazing domestic animals like cattle and sheep on the western ranges (Conaway, 1987). Wuerthner (1991, p. 30) reports:

> Grazing [is] permitted on 89 percent of Bureau of Land Management lands and 69 percent of Forest Service lands.... Altogether, more than 265 million acres of federal lands...are leased under federal grazing programs. Despite the huge acreage involved, the Department of Agriculture estimates that these federal lands provide forage for less than two percent of the cattle and sheep produced annually in the United States. This trifling production comes at considerable ecological cost.

Adding insult to injury, the public treasury actually subsidizes this pathetic livestock industry on these public lands: grazing fees are significantly lower on public than on private lands, and taxpayers pay for a variety of "improvements," most of which, like water impoundment and predator control, are ecologically benighted (Wuerthner, 1991). So what is the justification for this "cowboy welfare" paid for by the public, not only in dollars but in a degraded natural heritage? A very few citizens—ranchers—benefit. According to Watkins (1991), the larger justification is essentially historic and mythic. Europeans brought their domestic stock with them and set up shop in the west, and a popular romance grew out of that phenomenon. The present western livestock industry, such as it is, is a relic of that period, and because the animals ranched are not suited to the climate, it will, sooner or later, inevitably wither away.

Why not sooner, before more damage is done to the soils, waters, plants, and animals? It's just like the old growth controversy in the Pacific Northwest. If we stop the cutting of primary forests, jobs will be lost. The same jobs will be lost after the ten to twenty years it will take to log out the old growth, but then the forest will be gone too. So why not bite the economic bullet now rather than later, while we still have significant old growth left?

The remaining old growth forests are, in my view, good candidates for biosphere reserve-style conservation, with the local economy confined to the periphery of the core sanctuaries and geared to ecotourism and other nonconsumptive uses of the "resource." The vast rangeland of the west, both public and private, might be exploited economically and conserved quite differently. According to Wuerthner (1991, p., 29), "The current condition of our rangelands is as much a factor of using the wrong animal in the wrong place as it is of lax management." Cattle are inherently destructive, especially in arid country.

So why not ranch the right animals in the right places? Could we take a page out of the approach to conservation advocated by Myers (1981) for wild African fauna and crop the indigenous ungulates of the American west? If I have expressed doubts here about the wilderness condition of most places in the New World before the European conquest, I certainly have no doubt that most New World ecosystems were then in a condition of robust health. The vast Pre-Columbian herds of bison and antelope, and large populations of mule deer, white-tailed deer, and elk did not denude stream banks and gully slopes like the cattle and sheep that have succeeded them. They also supported a sizable population of carnivorous primates, as well as four-legged predators. A return of these species in comparable numbers might support a new and very different western ranching system and significantly contribute to the diet of a much larger contemporary human population of North America that draws on many other food resources than could the hunting/gathering/dry farming indigenes.

How can we get from here to this range utopia? I don't know. Very real political and economic obstacles, to say nothing of the sheer inertia of habit and tradition, stand in the way. Nonetheless, before we can figure out how to get from here to anywhere, we have to have a vision to guide us. Utopias may be unattainable in reality, but they are not impractical. They help move us off dead center, and falling short of an ideal is still movement in the right direction.

So far, this much can be said: The idea has been tried successfully in Africa despite dismal infrastructural support, resulting in significant spoilage losses (Casebeer, 1978). With the capital and technical resources available in North America, the efficient processing and distribution of game meat should be well within reach, and demand for game meat has increased in recent years, in part because of its low fat content (White, 1987). Wild game meat is also "organic," a quality in both vegetables and meats commanding a premium price in many upscale retail outlets. (Personally, I might add, I find the conventional choice of animal foods boring. While the surf side of a surf and turf entree can be quite varied—anything from sea scallops to thresher shark—the turf fare is dismally limited to beef, pork, mutton, chicken, and turkey.)

Some ranchers, moreover, already are shifting from raising cattle to raising deer and selling permission to hunt on their lands to sportspersons, because it is more profitable (White, 1987). The concept I am suggesting here is not a universalization of the private hunting preserve. I am, rather, suggesting the total elimination of livestock from the western ranges, ripping out unnecessary roads, a massive restoration of native wildlife populations, and a conversion of cowboys to airborne, satellite-guided game managers and professional market hunters. The trend toward game ranching might be a foot in the door for a serious discussion about the biological, economic, political, legal, and social possibility of establishing an ecological and economic regime on the western ranges that is continuous with, and an extension of, those that prevailed before the advent of the cow, the horseback, and later, the pick-up truck cowboy.

Can we generalize this particular example of the sustainable development alternative to either conventional development, like intensive stock grazing, or wilderness designation and restoration (where none existed before)? Can we envision and work to create an eminently livable, systemic, postindustrial technological society well adapted to, and at peace and in harmony with, its organic environment? If illiterate, unscientific peoples can do it, can't a civilized, technological society also live, not merely in peaceful coexistence, but in benevolent symbiosis with nature? Is our current industrial civilization the only one imaginable? Aren't there more appropriate, alternative technologies? Can't we be good citizens of the biotic community, like the birds and the bees, drawing an honest living from nature and giving back as much or more than we take?

ACKNOWLEDGEMENTS

I thank Eugene C. Hargrove for providing an opportunity to broach the central ideas contained in this paper to the faculty of environmental ethics at the University of Georgia's Institute of Ecology. I have benefited from discussion of the role of wilderness in Aldo Leopold's philosophy of conservation with Susan L. Flader, Curt Meine, and Bryan Norton. John Lemons and an anonymous referee for *The Environmental Professional* generously and critically commented on an earlier draft of this paper and rescued it from more errors of fact and doctrine than those that remain.

REFERENCES

Allin, C. 1987. The Leopold Legacy and the American Wilderness. In *Aldo Leopold: The Man and His Legacy*, T. Tanner, ed. Soil Conservation Society of America, Ankeny, IA, pp. 25-38.

Barrett, S.W. 1980. Indian Fires in the Presettlement Forests of Western Montana. In *Proceedings of the Fire History Workshop*, Tucson, AZ. USDA Forest Service General Technical Report RM-81. Fort Collins, CO.

Birch, T. 1990. The Incarceration of Wilderness: Wilderness Areas as Prisons. *Environmental Ethics* 12: 3-26.

Botkin, D.B. 1990. *Discordant Harmonies: A New Ecology for the Twenty-First Century*. Oxford University Press, New York.

Brundtland, G.H., Chair, World Commission on Environment and Development. 1987. *Our Common Future*. Oxford University Press, New York.

Casebeer, R.L. 1978. Coordinating Range and Wildlife Management in Kenya. *Journal of Forestry* 76: 374-375

Conaway, J. 1987. *The Kingdom in the Country*. Houghton Mifflin, Boston.

Cordell, H.K., and P.C. Reed. 1990. Untrammeled by Man: Preserving Diversity through Wilderness. In *Preparing to Manage Wilderness in the 21st Century: Proceedings of the Conference*. Southeastern Forest Experiment Station, U.S. Department of Agriculture, Forest Service, Asheville, NC, pp. 30-33.

Costanza, R. 1991. Toward an Operational Definition of Ecosystem Health. AAAS annual meeting, Washington, DC. February 15.

Cronon, W. 1983. *Changes in the Land: Indians, Colonists, and the Ecology of New England*. Hill and Wang, New York.

Dasmann, R.F. 1988. Conservation, Land Use, and Sustainable Development. In *For the Conservation of Earth*, V. Martin, ed. Fulcrum, Inc., Golden, CO, pp. 68-70.

Day, G.M. 1953. The Indian as an Ecological Factor in Northeastern Forests. *Ecology* 34: 329-346.

DeBonis, J. 1989. Speaking Out: A Letter to the Chief of the US Forest Service. *Inner Voice* 1: 1, 3.

Dobyns, H.F. 1966. Estimating Aboriginal American Population: An Appraisal of Techniques with a New Hemispheric Estimate. *Current Anthropology* 7: 395-412.

Devall, B. n. d. Personal communication. Professor of Sociology, California State University, Humbolt.

Dysart, III, B.C., and M. Clawson. 1988. *Managing Public Lands in the Public Interest*. Praeger, New York.

Ehrenfeld, D. 1989. Life in the Next Millennium: Who Will Be Left in the Earth's Community? *Orion Nature Quarterly* 8 (spring): 4-13.

Emerson, R.W. 1989. *Nature*. Beacon Press, Boston.

Flader, S.L. 1974. *Thinking Like a Mountain: Aldo Leopold and the Evolution of an Ecological Attitude toward Deer, Wolves, and Forests*. University of Missouri Press, Columbia.

Gomez-Pompa, A., and A. Kaus. 1988. Conservation by Traditional Cultures in the Tropics. In *For the Conservation of the Earth*, V. Martin, ed. Fulcrum Inc., Golden, CO, pp. 183-194.

Gordon, J.C., Chair, Commission on Research and Resource Management Policy. 1989. *National Parks: From Vignettes to a Global View*. National Parks and Conservation Association, Washington, DC.

Gua, R. 1989a. Radical American Environmentalism: A Third World Critique. *Environmental Ethics* 11: 71-83.

Gua, R. 1989b. *The Unquiet Woods: Ecological Change and Peasant Resistance in the Himalaya*. University of California Press, Berkeley, CA.

Hays, S.P. 1959. *Conservation and the Gospel of Efficiency: The Progressive Conservation Movement*. Harvard University Press, Boston.

Hecht, S., and A. Cockburn. 1989. *The Fate of the Forest: Developers, Destroyers, and Defenders of the Amazon*. Verso, New York.

Heizer, R.F. 1955. Primitive Man as an Ecologic Factor. Kroeber Anthropological Society Papers, No. 13. Berkeley, CA.

Hott, L., and D. Garey. 1989. The Wilderness Idea: John Muir, Gifford Pinchot, and the First Great Battle for Wilderness. Florentine Films, Haydenville, MA.

Hughes, J.D. 1983. *American Indian Ecology*. Texas Western Press, El Paso.

Kroeber, A.L. 1939. *Cultural and Natural Areas of Native North America*. University of California Press, Berkeley.

Lemons, J. 1987. United States' National Park Management: Values, Policy, and Possible Hints for Others. *Environmental Conservation* 14: 329-340, 328.

Leopold, A. 1918. Forestry and Game Conservation. *Journal of Forestry* 19: 404-411.

Leopold, A. 1921. Wilderness and Its Place in Forest Recreation Policy. *Journal of Forestry* 19: 718-721.

Leopold, A. 1925a. Conserving the Covered Wagon. *Sunset Magazine* 54 (March): 21, 56.

Leopold, A. 1925b. The Last Stand of the Wilderness. *American Forests and Forest Life* 31: 599-604.

Leopold, A. 1925c. Wilderness as a Form of Land Use. *Journal of Land and Public Utility Economics* 1: 398-404.

Leopold, A. 1925d. A Plea for Wilderness Hunting Grounds. *Outdoor Life* 56: 348-350.

Leopold, A. 1936. Threatened Species: A Proposal to the Wildlife Conference for an Inventory of the Needs of Near-Extinct Birds and Animals. *American Forests* 42: 116-119.

Leopold, A. 1939a. A Biotic View of Land. *Journal of Forestry* 37: 727-730.

Leopold, A. 1939b. The Farmer as a Conservationist. *American Forests* 45: 294-299, 316, 323.

Leopold, A. 1941. Wilderness as a Land Laboratory. *The Living Wilderness* 6 (July): 3

Leopold, A. 1942. Land Use and Democracy. *Audubon* 44 (Sept./Oct.): 259-265.

Leopold, A. 1945. The Outlook for Farm Wildlife. *Transactions of the 10th North American Wildlife Conference*: 165-168.

Leopold, A. 1949. *A Sand County Almanac: And Sketches Here and There*. Oxford University Press, New York.

Leopold, A. 1991a. The Popular Wilderness Fallacy. In *The River of the Mother of God and Other Essays by Aldo Leopold*, S.L. Flader and J.B. Callicott, eds. University of Wisconsin Press, Madison, pp. 49-52.

Leopold, A. 1991b. Wilderness. In *The River of the Mother of God and Other Essays by Aldo Leopold*, S.L. Flader and J. B. Callicott, eds. University of Wisconsin Press, Madison, pp. 226-229.

Lewis, H.T. 1973. *Patterns of Indian Burning in California: Ecology and Ethnohistory*. Ballena Press Anthropological Papers, No. 1.

Martin, C. 1973. Fire and Forest Structure in the Aboriginal Eastern Forest. *Indian Historian* 6: 38-42, 54.

Martin, C. 1978. *Keepers of the Game: Indian-Animal Relationships and Fur Trade*. University of California Press, Berkeley.

Meine, C. 1988. *Aldo Leopold: His Life and Work*. University of Wisconsin Press, Madison.

Merchant, C. 1989. *Ecological Revolutions: Nature, Gender and Science in New England*. University of North Carolina Press, Chapel Hill.

Myers, N. 1981. A Farewell to Africa. *International Wildlife* 11 (Nov./Dec.): 36-46.

Muir, J. 1901. *Our National Parks*. Houghton Mifflin, Boston.

Nabhan, G.P. 1982. *The Desert Smells Like Rain: A Naturalist in Papago Country*. North Point Press, San Francisco.

Nash, R. 1967. *Wilderness and the American Mind*. Yale University Press, New Haven, CT.

Noss, R.F. 1990a. Can We Maintain Our Biological and Ecological Integrity? *Conservation Biology* 4: 241-243.

Noss, R.F. 1990b. What Can Wilderness Do for Biodiversity? In *Preparing to Manage Wilderness in the 21st Century: Proceedings of the Conference*, P. Reed, ed. Southeastern Forest Experiment Station, U.S. Department of Agriculture, Forest Service, Asheville, NC, pp . 49-61.

Noss, R.F. 1991. Sustainability and Wilderness. *Conservation Biology* 5: 120-122.

Oelschlaeger, M. 1991. *The Idea of Wilderness: From Prehistory to the Age of Ecology*. Yale University Press, New Haven, CT.

Parsons, D.J., D.M. Graber, J.K. Age, and J.W.V. Wagtendonk. 1986. Natural Fire Management in National Parks. *Environmental Management* 10: 21-24.

Passell, P. 1990. Rebel Economists Add Ecological Costs to Price of Progress. *New York Times* (Nov. 27): B5-B6

Peters, C.M., A.H. Gentry, and R.O. Mendelsohn. 1989. Valuation of an Amazonian Rainforest. *Nature* 339: 656-657.

Pinchot, G. 1947. *Breaking New Ground*. Harcourt, Brace, and Co., New York.

Pyne, S.J. 1982. *Fire in America, A Cultural History of Wildland and Rural Fire*. Princeton University Press, Princeton, NJ.

Randall, A. 1988. What Mainstream Economists Have to Say about the Value of Biodiversity. In *Biodiversity*, E.O. Wilson, ed. National Academy Press, Washington, DC.

Reed, P., ed. 1990. *Preparing to Manage Wilderness in the 21st Century: Proceedings of the Conference*. Southeastern Forest Experiment Station, U.S. Department of Agriculture, Forest Service, Asheville, NC.

Reed, P., and L. Merigliano. 1990. Managing for Compatibility Between Recreational and Nonrecreational Wilderness Purposes. In *Preparing to Manage Wilderness in the 21st Century: Proceedings of the Conference*, P. Reed, ed. Southeastern Forest Experiment Station, U.S. Department of Agriculture, Forest Service, Asheville, NC, pp. 95-107.

Snyder, G. 1990. *The Practice of the Wild*. North Point Press, San Francisco.

Soulé, M.E. 1990. The Onslaught of Alien Species, and other Challenges in the Coming Decades. *Conservation Biology* 4: 233-239.

Stevens, W.K. 1990. Research in 'Virgin' Amazon Uncovers Complex Farming. *New York Times* (April 3): B5-B6

Thompson, T.S. 1988. Vegetation Dynamics in the Western United States: Modes of Response to Climate Fluctuations. In *Vegetation History*, B. Huntly and T. Webb, eds. Kluwer Academic Publishers, Boston.

Thoreau, H.D. 1962. *Excursions*. Corinth Books, New York.

Ulanowicz, R. 1991. Ecosystem Health in Terms of Trophic Flow Networks. AAAS annual meeting, Washington, DC. February 15.

USDA Forest Service. 1989. New Perspectives: An Ecological Path for Managing Forests. Pacific Northwest Research Station, Portland, OR, and Pacific Southwest Research Station, Redding, CA.

von Droste, B. 1988. The Role of Biosphere Reserves at a Time of Increasing Globalization. In *For the Conservation of the Earth*, Vance Martin, ed. Fulcrum, Inc., Golden, CO, pp. 89-93.

Watkins, T.H. 1991. Aspects of Grass. *Wilderness* 54 (Spring): 27.

Westman, W.E. 1990. Park Management of Exotic Species: Problems and Issues. *Conservation Biology* 4: 251-260.

Westman, W.E. 1991. Restoration Projects: Measuring Their Performance. *Environmental Professional* 13(3): 207-215.

White, R.J. 1987. *Big Game Ranching in the United States*. Wild Sheep and Goat International Publishing Co., Mesilla, NM.

Witthoft, J. 1965. *Indian Prehistory of Pennsylvania*. Pennsylvania Historical and Museum Commission, Harrisburg.

Wright, A. 1984. Innocents Abroad: American Agricultural Research in Mexico. In *Meeting the Expectations of the Land: Essays in Sustainable Agriculture and Stewardship*, W. Jackson, W. Berry, and B. Colman, eds. North Point Press, San Francisco, pp. 135-151.

Wuerthner, G. 1991. How the West Was Eaten. *Wilderness* 54 (Spring): 28-37.

The Wilderness Idea Reaffirmed

Holmes Rolston, III
Colorado State University

Abstract. The concept of wilderness is coherent and vital for the protection of intrinsic natural values. Wild nature differs from human culture in radical ways. To suppose that humans can improve spontaneous wild nature deliberately is a contradiction in terms. Kinds of biodiversity can be protected by wilderness designation that are protected doubtfully by rural indigenous peoples. American landscapes, though sometimes affected by the aboriginal inhabitants, were not modified so dramatically or irreversibly as to make wilderness designation impossible. Though unavailable to aboriginal peoples, the wilderness ideal is critical today.

The need for sustainable development on agricultural lands does not prejudice the need for wilderness, nor does wilderness designation lead to complacency about sustainable development. Baird Callicott's nondiscriminating account of humans as entirely natural is a metaphysical confusion. Coupled with his anthropogenic value theory, the outcome operationally will be inadequate respect for intrinsic natural values.

INTRODUCTION

Revisiting the wilderness, Callicott (1991) is a doubtful guide; indeed he has gotten himself lost. That is a pity, because he is on the right track about sustainable development and I readily endorse his positive arguments for developing a culture more harmonious with nature. But these give no cause for being negative about wilderness.

The wilderness concept, we are told, is "inherently flawed," triply so. It metaphysically and unscientifically dichotomizes man and nature. It is ethnocentric, because it does not realize that practically all the world's ecosystems were modified by aboriginal peoples. It is static, ignoring change through time. In the flawed idea and ideal, wilderness respects wild communities where man is a visitor who does not remain. In the revisited idea(l), also Leopold's ideal, humans, themselves entirely natural, reside in and can and ought to improve wild nature.

HUMAN CULTURE AND WILD NATURE

Wilderness valued without humans perpetuates a false dichotomy, Callicott maintains. Going back to Cartesian and Greek philosophy and Christian theology, such a contrast between humans and wild nature is a metaphysical confusion that leads us astray and also is unscientific. But this is not so. One hardly needs metaphysics or theology to realize that there are critical differences between wild nature and human culture. Humans now superimpose cultures on the wild

---

Holmes Rolston, III. is a professor of philosophy in the Department of Philosophy, Colorado State University, Fort Collins, CO 80523, and the author of *Environmental Ethics* and *Philosophy Gone Wild*.

This article is a response to J. Baird Callicott's article "The Wilderness Idea Revisited: The Sustainable Development Alternative," which appeared in *The Environmental Professional* 13: 235-247.

---

nature out of which they once emerged. There is nothing unscientific or nonDarwinian about the claim that innovations in human culture make it radically different from wild nature.

Information in wild nature travels intergenerationally on genes; information in culture travels neurally as persons are educated into transmissible cultures. (Some higher animals learn limited behaviors from parents and conspecifics, but animals do not from transmissible cultures.) In nature, the coping skills are coded on chromosomes. In culture, the skills are coded in craftsman's traditions, religious rituals, or technology manuals. Information acquired during an organism's lifetime is not transmitted genetically; the essence of culture is acquired information transmitted to the next generation. Information transfer in culture can be several orders of magnitude faster and overleap genetic lines. I have but two children; copies of my books and my former students number in the thousands. A human being develops typically in some one or a few of ten thousand cultures, each heritage historically conditioned, perpetuated by language, conventionally established, using symbols with locally effective meanings. Animals are what they are genetically, instinctively, environmentally, without any options at all. Humans have myriads of lifestyle options, evidenced by their cultures; and each human makes daily decisions that affect his or her character. Little or nothing in wild nature approaches this.

The novelty is not simply that humans are more versatile in their spontaneous natural environments. Deliberately rebuilt environments replace spontaneous wild ones. Humans can therefore inhabit environments altogether different from the African savannas in which they once evolved. They insulate themselves from environmental extremes by their rebuilt habitations, with central heat from fossil fuel or by importing fresh groceries from a thousand miles away. In that sense, animals have freedom within ecosystems, but humans have freedom from ecosystems. Animals are adapted to their niches; humans adapt their ecosystems to their needs. The determinants of animal and plant behavior, much less the determinants of climate or nutrient recycling, are never anthropological, political, economic, technological, scientific, philosophical, ethical, or religious. Natural selection pressures are relaxed in culture; humans help each other out compassionately with medicine, charity, affirmative action, or headstart programs.

Humans act using large numbers of tools and things made with tools, extrasomatic artifacts. In all but the most primitive cultures, humans teach each other how to make clothes, thresh wheat, make fires, bake bread. Animals do not hold elections and plan their environmental affairs; they do not make bulldozers to cut down tropical rainforests. They do not fund development projects through the World Bank or contribute to funds to save the whales. They do not teach their religion to their children. They do not write articles revisiting and reaffirming the idea of wilderness. They do not get confused about whether their actions are natural or argue about whether they can improve nature.

If there is any metaphysical confusion in this debate, we locate it in the claim that "man is a natural, a wild, an evolving species, not essentially different in this respect from all the others" (p. 241). Poets like Gary Snyder perhaps are entitled to poetic license. But philosophers are not, especially when analyzing the concept of wildness. They cannot say that "the works of man, however precocious, are as natural as those of beavers," being "entirely natural," and then, hardly taking a breath, say that "the cultural component in human behavior is so greatly developed as to have become more a difference of kind than of degree" (p. 241). If this were only poetic philosophy it might be harmless, but proposed as policy, environmental professionals who operate with such contradictory philosophy will fail tragically.

"Anthropogenic changes imposed upon ecosystems are as natural as any other" (Callicott, 1990). Not so. Wilderness advocates know better; they do not gloss over these differences. They appreciate and criticize human affairs, with insight into their radically different characters. Accordingly, they insist that there are intrinsic wild values that are not human values. These ought to be preserved for whatever they can contribute to human values, and also because they are valuable on their own, in and of themselves. Just because the human presence is so radically different, humans ought to draw back and let nature be. Humans can and should see outside their own sector, their species self-interest, and affirm nonanthropogenic, noncultural values. Only humans have conscience enough to do this. That is not confused metaphysical dichotomy; it is axiological truth. To think that human culture is nothing but natural system is not discriminating enough. It risks reductionism and primitivism.

These contrasts between nature and culture were not always as bold as they now are. Once upon a time, culture evolved out of nature. The early hunter-gatherers had transmissible cultures but, sometimes, were not much different in their ecological effects from the wild predators and omnivores among whom they moved. In such cases, this was as much through lack of power to do otherwise as from conscious decision. A few such aboriginal peoples may remain.

But we Americans do not and cannot live in such a twilight society. Any society that we envision must be scientifically sophisticated, technologically advanced, globally oriented, as well as (we hope) just and charitable, caring for universal human rights and for biospheric values. This society will try to fit itself in intelligently with the ecosystemic processes on which it is superposed. It will, we plead, respect wildness. But none of these decisions shaping society are the processes of wild nature. There is no inherent flaw in our logic when we are discriminating about these radical discontinuities between culture and nature. The dichotomy charge is a half-truth, and, taken for the whole, becomes an untruth.

HUMANS IMPROVING WILD NATURE

Might a mature, humane civilization improve wild nature? Callicott thinks that it is a "fallacy" to think that "the best way to conserve nature is to protect it from human habitation and utilization" (p. 236). But, continuing the analysis, surely the fallacy is to think that a nature allegedly improved by humans is anymore real nature at all. The values intrinsic to wilderness cannot, on pain of both logical and empirical contradiction, be "improved" by deliberate human management, because deliberation is the antithesis of wildness. That is the sense in which civilization is the "antithesis" (p. 236) of wilderness, but there is nothing "amiss" in seeing an essential difference here. Animals take nature ready to hand, adapted to it by natural selection, fitted into their niches; humans rebuild their world through artifact and heritage, agriculture and culture, political and religious decisions.

On the meaning of "natural" at issue here, that of nature proceeding by evolutionary and ecological processes, any deliberated human agency, however well intended, is intention nevertheless and interrupts these spontaneous processes and is inevitably artificial, unnatural. (There is another meaning of "natural" by which even deliberated human actions break no laws of nature. Everything, better or worse, is natural in this sense, unless there is the supernatural.) The architectures of nature and of culture are different, and when culture seeks to improve nature, the management intent spoils the wildness. Wilderness management, in that sense, is a contradiction in terms—whatever may be added by way of management of humans who visit the wilderness, or of restorative practices, or monitoring, or other activities that environmental professionals must sometimes consider. A scientifically managed wilderness is conceptually as impossible as wildlife in a zoo.

To recommend that *Homo sapiens* "reestablish a positive symbiotic relationship with other species and a positive role in the unfolding of evolutionary processes" (p. 240) is, so far as wilderness preservation is involved, not just bad advice, it is impossible advice. The cultural processes by their very "nature" interrupt the evolutionary process; there is no symbiosis, there is antithesis. Culture is a post-evolutionary phase of our planetary history; it must be superposed on the nature it presupposes. To recommend, however, that we should build sustainable cultures that fit in with the continuing ecological processes is a first principle of intelligent action, and no wilderness advocate thinks otherwise.

If there are inherent conceptual flaws dogging this debate, we have located another: Callicott's allegedly "improved" nature. In such modified nature, the different historical genesis brings a radical change in value type. Every wilderness enthusiast knows the difference between a pine plantation in the Southeast and an old-growth grove in the Pacific Northwest. Even if the "improvement" is more or less harmonious with the ecosystem, it is fundamentally of a different order. Asian ring-tailed pheasants are rather well naturalized on the contemporary Iowa landscape. But they are there by human introduction, and they remain because farmers plow the fields, plant corn, and leave shelter in the fencerows. They are really as much like pets as like native wild species, because they are not really on their own.

BIODIVERSITY AND WILDERNESS

As an example of his recommended symbiosis where human culture enriches natural systems, Callicott cites a study (Nabhan et al., 1982) of two nearby communities, Quitovac (= Ki:towak) in Mexico, where sixty-five bird species were found, and Quitobaquito Springs (or A'al Waipia) in Organ Pipe National Monument, with only thirty-two species. His conclusion is that biodiversity is greater in such rural communities than in wild natural systems.

But this is an unusual case; the locale is desert, where water is the limiting factor. If you artificially water the desert, some things will come in that could not live there before. Similarly, if you heat up the tundra, where cold is the limiting factor. We will not be surprised if there are more birds around feeders offering food, water, and shelter than elsewhere. But bird feeders actually may not be increasing biodiversity. We will have to look more closely at what is meant by biodiversity and what is going on in the two communities.

A species count, uninterpreted, doesn't tell us much. In more sophisticated analyses, ecologists use up to a dozen and a half indices of diversity (Magurran, 1988; Pielou, 1975). These include within habitat diversity (alpha diversity), between habitat diversity (beta diversity) and regional diversity (gamma diversity). They include diversity of processes and heterogeneity of fauna and flora, and on and on. If all you do is count species, there are more animal species in the Denver zoo than in the rest of Colorado. Never mind that the processes of nature are entirely gone. Callicott knows that and wants ecosystem health as well as diversity.

Whether there is ecosystem health at Quitovac is less clear. Callicott thinks so, but Nabhan et al. (1982) are more circumspect. Though the bird species count was always higher at Quitovac, by a heterogeneity diversity index the avifauna at Quitovac has no advantage over Organ Pipe. (This asks what proportions of the birds are of what species, such as grackles, doves, English sparrows, pigeons.) They also find that Quitovac is "not nearly as diverse in mammals," that ever-present dogs, horses, and cattle, limit the presence of wild animals. Deer and javelina drink and browse frequently at Organ Pipe, seldom at Quitovac. Even rodents are more abundant at Organ Pipe.

They also found more plant diversity at Quitovac, one hundred and thirty-nine species there against eighty at Organ Pipe. It is hardly surprising that if you add some irrigated cultivated fields and orchards, new plant species will appear, and some insects will follow, and birds in turn follow the seeds and insects. They also note that seventeen of these plant species were planted intentionally and that of the fifty-nine species in fields and orchards, many were adventitious species, weeds of disturbed sites. Many were the Old World waifs that, like dandelions, have tagged along after civilization willy-nilly. Is this being offered as a wise symbiosis of nature and culture? Is that enhanced richness in biodiversity?

A species count, offered as evidence of biodiversity without further ado, assumes that if we have the species, we have what we want conserved. But we may have the parts, even extra, artificial parts, but no longer the composition of the former whole. Maybe Quitovac is about as much "ersatz world" (p. 243) as idyllic, humanized ecosystem health with optimized biodiversity. Even a new whole would not have the integrity of the once wild ecosystem. We can and ought to have rural nature, and we will be glad to have rural nature with a high bird count. But we can have a rural nature with a high species count and not have anywhere on the landscape the radical values of wild, pristine nature. That loss would not be compensated for by the stepped-up species count in agriculturally disturbed lands. In wilderness, we value the interactions as a fundamental component of biodiversity.

The predation pressures, for instance, are never the same on agricultural lands as they are on wildlands. Agriculture means an increase of disturbed soil, with most of these disturbances different in kind from those in wild nature. Different kinds of things grow in such soil, more r-selected species, fewer k-selected ones. Underground, the fungi and soil bacteria are different, so the decomposition regime is different, and that results in differences above ground. The energy flow and the nutrient cycling is different. It is often the case that the highest number of species are found in intermediately disturbed environments, but that considers species counts and alpha diversity alone. If all the environments are kept intermediately disturbed, we lose beta and gamma diversity. Indeed over the landscape as a whole, we lose even species counts, since in disturbed environments the sensitive species go extinct. We are not likely to retain the large carnivores.

Both these oases are water magnets for migrant birds. Quitovac, with its cultivated fields and orchards, draws more migrants into close proximity. Muddy shorelines attract some waders less frequent at Quitobaquito. All this tells us little about whether these migrating birds are safe in their wintering or breeding grounds. In fact, Central American agricultural development, destroying winter grounds, threatens many bird species. Quitovac may draw some breeding species that cannot survive at Quitobaquito or in the unwatered desert. But there are no bird species flourishing in Quitovac that were not flourishing already in their native habitats elsewhere. It is hard to think that much important bird conservation is going on there.

Quitobaquito Springs, far from being depauperate in birds, is one of the best-known sites for observing birds in that region of Arizona, and birders go there from all over the United States to see the migrants and to find the desert species. The oasis is but a small area. Organ Pipe Monument is designated to preserve many other kinds of habitats. Enlarging to consider beta and gamma diversity, the official Monument checklist contains 277 bird species, of which 63 are known to breed in various habitats there, and five more believed to breed as well (Groschupf et al., 1987). Only three are nonnative. Even if the diversity at Quitovac is greater, the diversity preserved by having both a rural area and a wilderness is higher than if we had two rural areas.

Also, whatever the possibilities, we do not want to forget the probabilities, which are that this (allegedly) idyllic picture will be upset by development pressures. Quitovac had been used for centuries, steadily but not intensively. When the comparisons here were made, only two or three dozen persons were using the area. The study concentrates on only a five hectare site, and the natives had only used ten percent of this for cultivated fields and orchards. Before the study could be completed, 125 hectares there were bulldozed to be used for intensive agriculture, including most of the study area, with disastrous results (Nabhan et al., 1982). There may be fewer species at Organ Pipe, but such a disaster is not likely to occur, owing to its sanctuary designation.

WILDERNESS AND CHANGE

Another alleged flaw in the concept of wilderness is that its advocates do not know the fourth dimension, time. That is a strange charge; my experience has been just the opposite. In wilderness, the day changes from dawn to dusk, the seasons pass, plants grow, animals are born, grow up, and age. Rivers flow, winds blow, even the rocks erode; change is pervasive. Indeed, wilderness is that environment in which one is most likely to experience geological time. Try a raft trip through the Grand Canyon.

On the scale of deep time, some processes continue on and on, so that the perennial givens—wind and rain, soil and photosynthesis, life and death and life renewed—can seem almost forever. Species survive for millions of years; individuals are ephemeral. Life persists in the midst of its perpetual perishing. Mountains are reliably there generation after generation. The water cycles back, always moving. In wilderness, time mixes with eternity; that is one reason we value it so highly.

Callicott writes as if wilderness advocates had studied ecology and never heard of evolution. But they know that evolution is the control of development by ecology, and what they value is precisely natural history. They do not object to natural changes. They may not even object to artificial changes in rural landscapes. But, since they know the difference between nature and culture, they know that cultural changes may be quite out of kilter with natural changes. Leopold uses the word "stability" when he is writing in the time frame of land-use planning. On that scale, nature typically does have a reliable stability, and farmers do well to figure in the perennial givens.

In an evolutionary time frame, Leopold knows that relative stability mixes with change. "Paleontology offers abundant evidence that wilderness maintained itself for immensely long periods; that its component species were rarely lost, and neither did they get out of hand; that weather and water built soil as fast or faster than it was carried away." That is why "wilderness ... assumes unexpected importance as a laboratory for the study of land health" (Leopold, 1968, p. 196). Wilderness is the original sustainable development.

With natural processes, "protect" is perhaps a better word than either "preserve" or "conserve." Wilderness advocates do not seek to prevent natural change. There is nothing illusory, however, about appreciating today in wilderness processes that have a primeval character. There, the natural processes of 1992 do not differ much from those of 1492, half a millennium earlier. We may enjoy that perennial character, constancy in change, in contrast with the rapid pace of cultural changes, seldom as dramatic as those on the American landscape of the last few centuries.

A management program in the U. S. Forest Service seeks to evaluate the "limits of acceptable change." This emphasis worries about the rapid pace of cultural change as this contrasts with the natural pace on landscapes. Cordell and Reed (1990) are trying to decide the limits of acceptable humanly-introduced changes, artificial changes, since these are of such radically different kind and pace that they disrupt the processes of wilderness. They do not oppose natural changes. At this point, we have an example of how and why environmental professionals will make disastrous decisions, if confused by what is and is not natural. Callicott warns them that they do have to worry about "accelerating rates of environmental change" (p. 242). No one can begin to understand these rates of changes if the changes are thought of as being introduced by a species that is "entirely natural."

When we designate a desert wilderness in Nevada, there really isn't any problem deciding that mustangs are feral animals in contrast with desert bighorns, which are indigenous. There might have been ancestral horses in Paleolithic times in the American West, but they went extinct naturally. The present mustangs came from animals that the Europeans brought over in ships, originally from the plains of Siberia. Bighorns are what they are where they are by natural selection. Mustangs are not so. There is nothing conceptually problematic about that—unless one has never gotten clearly in mind the difference between nature and culture in the first place.

ABORIGINAL PEOPLES AND WILDERNESS

What of the argument that we cannot have any wilderness, because there is none to be had? This is a much stronger claim than that there is no real wilderness left on the American landscape after the European cultural invasion. Even the aboriginals had already extinguished wilderness. Now we have a somewhat different account of the human presence from that earlier advocated. The claim is no longer that the Indians were just another wild species, "entirely natural," but that they actively managed the landscape, so dramatically altering it that there was no wilderness even when Columbus arrived in 1492. It is ethnocentric to think otherwise. This is because we Caucasians exaggerate our own power to modify the landscape and diminish their power. This is a judgment based on prejudice, not on facts.

How much did the American Indians modify the landscape? That is an empirical question in anthropology and ecology. We do not disagree that where there was Indian culture, this altered the locales in which they resided, so that these locales were not wilderness in the pure sense. In that respect, Indian culture is not different in kind from the white man's culture. What we need to know is the degree. Had the Indians, when the white man arrived, already transformed the pre-Indian wilderness beyond the range of its spontaneous self-restoration?

Callicott concedes (pp. 241-242), rightly, that most of what has been presently designated as wilderness was infrequently used by the aborigines, since it is high, rough, or arid. We have no reason to think that in such areas the aboriginal modifications are irreversible. Were the more temperate regions modified so extensively and irreversibly that so little naturalness remains as to make wilderness designation an illusion? Callicott has "no doubt that most New World ecosystems were in robust health" (p. 244). That suggests that they were not past self-regeneration.

The American Indians on forested lands had little agriculture; what agriculture they had tended to reset succession, and, when agriculture ceases, the subsequent forest regeneration will not be particularly unnatural. The Indian technology for larger landscape modification was bow and arrow, spear, and fire. The only one that extensively modifies landscapes is fire. Fire is—we have learned well by now—also quite natural. Fire suppression is unnatural, but no one argues that the Indians used that as a management tool, nor did they have much capacity for fire suppression. The argument is that they deliberately set fires. Does this make their fires radically different from natural fires? It does in terms of the source of ignition; the one is a result of environmental policy deliberation, the other of a lightning bolt.

But every student of fire behavior knows that on the scale of regional forest ecosystems, the source of ignition is not a particularly critical factor. The question is whether the forest is ready to burn, whether there is sufficient ground fuel to sustain the fire, whether the trees are diseased, how much duff there is, and so on. If conditions are not right, it will be difficult to get the fire going and it will burn out soon. If conditions are right, a human can start a regional fire this year. If some human does not, lightning will start it next year, or the year after that. On a typical summer day, the states of Arizona and New Mexico are each hit by several thousand bolts of lightning, mostly in the higher, forested regions. Doubtless the Indians started some fires too, but it is hard to think that their fires so dramatically and irreversibly altered the natural fire regime in the Southwest that meaningful wilderness designation is impossible today.

We do not want to be ethnocentric, but neither do we want to be naïve about the technological prowess of the American Indian cultures. They had no motors, indeed no wheels, no domestic animals, no horses (before the Spanish came), no beasts of burden. The Indians had a hard time getting so simple a thing as hot water. They had to heat stones and drop them in skins or tightly woven baskets. They lived on the landscape with foot and muscle, and in that sense, though they had complex cultures, they had culture with very reduced alterative power. Even in European cultures, in recent centuries the power of civilization to redo the world has accelerated logarithmically.

In Third World nations, perhaps areas that seem "natural" now are often the result of millennia of human modifications through fire, hunting, shifting cultivation, and selective planting and removal of species. This will have to be examined on a case-by-case basis, and we cannot prejudge the answers. We do not know yet how intensively the vast Brazilian rainforests were managed and whether no wilderness designation there is ecologically practicable, even if we desired it. Nor do wilderness enthusiasts advocate that such peoples be removed to accomplish this, were it is possible. What is protested is modern forms of development. Extractive reserves may be an answer, but extractive reserves for latex sold in world markets and manufactured into rubber products can hardly be considered aboriginal wisdom.

Sometimes we will have to make do with what wildness remains in the nooks and crannies of civilization. Meanwhile, where wilderness designation is possible and where

THE WILDERNESS IDEA REAFFIRMED

there is an exploding population, what should we do? No one objects to trying to direct that explosion into more harmonious forms of human-nature encounter. But constraining an explosion takes some strong measures. One of these ought to be the designation of wilderness.

Perhaps the American Indians did not have enough contrast between their culture and the nature that surrounded them to produce the wilderness idea. It was not an idea that, within their limited power to remake nature, could occur to them. If you have only foot, muscle, bows, arrows, and fire, you do not think much about wilderness conservation. But we, in the twentieth century, do have the wilderness idea; it has crystallized with the possibility, indeed the impending threat, of destroying the last acre of primeval wilderness. It also has crystallized with our deepening scientific knowledge of how wild nature operates, of DNA, genes, and natural selection, and how dramatically different in kind, pace, and power the processes of culture can be. The Indians knew little of this; they lived still in an animistic, enchanted world.

And we need the wilderness idea desperately. When you have bulldozers that already have blacktopped more acreage than remains pristine, you can and ought to begin to think about wilderness. Such an idea, when it comes, is primitive in one sense: it preserves primeval nature, as much as it can. But it is morally advanced in another sense: it sees the intrinsic value of nature, apart from humans.

Ought implies can; the Indians could not, so they never thought much about the ought. We in the twentieth century can, and we must think about the ought. When we designate wilderness, we are not lapsing into some romantic atavism, reactionary and nostalgic to escape culture. We are breaking through culture to discover, nonanthropocentrically, that fauna and flora can count in their own right (an idea that Indians also might have shared). We realize that ecosystems sometimes can be so respected that humans only visit and do not remain (an idea that the Indians did not need or achieve). A "can" has appeared that has generated a new "ought."

Even some modern American Indians concur. In western Montana, the Salish and Kootenai tribes have set aside 93,000 acres of their reservation as the Mission Mountains Tribal Wilderness; in addition, they have designated the South Fork of the Jocko Primitive Area. In both areas, the Indian too is "a visitor who does not remain"; they want these areas "to be affected primarily by the forces of nature with the imprint of man's work substantially unnoticeable" (*Tribal Wilderness Ordinance*, 1982). Indeed, in deference to the grizzly bears, in the summer season, the Indians do not permit any humans at all to visit 10,000 acres that are prime grizzly habitat. In both areas, they can claim even more restrictive environmental regulation on what people can do there than in the white man's wilderness. What, when, and how they hunt is an example.

Not a word of the above discussion disparages aboriginal Indian culture. To the contrary, that they survived with the bare skills they had is a credit to their endurance, courage, resolution, and wisdom. A wilderness enthusiast, if he or she has spent much time in the woods armed with only muscle and a few belongings in a backpack, is in an excellent position to appreciate the aboriginal skills.

SUSTAINABLE DEVELOPMENT AND WILDERNESS

Finding out how to remake civilization so that nature is conserved in the midst of sustainable development is indeed a more difficult and important task than saving wild remnants. Little wilderness can be safe unless the sustainable development problem is solved also. I can only endorse Callicott's desire to conserve nature in the midst of human culture. "Human economic activities should at least be compatible with the ecological health of the environments in which they occur" (p. 239). No party to the debate contests that. But this does not mean that wilderness ought not to be saved for what it is in itself.

"The farmer as a conservationist" is quite a good thing, and Leopold does well to hope that "land does well for its owner, and the owner does well by his land"; perhaps where a farmer begins, as did Leopold, with lands long abused, "both end up better by reason of their partnership." In that context, "conservation is a state of harmony between men and land" (p. 238). But none of that asks whether there also should be wilderness. Leopold tells us what he thinks about that after his trip to Germany. There was "something lacking.... I did not hope to find in Germany anything resembling the great 'wilderness areas' which we dream about and talk about." That was too much to hope; he could dream that only in America. But he did hope to find "a certain quality [—wildness—] which should be, but is not found" in the rural landscape, and, alas, not even that was there (Leopold, quoted by Callicott, p. 238). "In Europe, where wilderness now has retreated to the Carpathians and Siberia, every thinking conservationist bemoans its loss" (Leopold, [1949] 1968, p. 200). That loss would not be restored if every farmer were a restoration ecologist. All that Leopold says about sustainable development is true, but there is no implication that wilderness cannot or ought not to be saved. Affirming sustainable development is not to deny wilderness.

MONASTIC WILDERNESS AND CIVILIZED COMPLACENCY

Nor is affirming wilderness to deny sustainable development. Callicott alleges, "Implicit in the most passionate pleas for wilderness preservation is a complacency about what passes for civilization" (p. 236). Not so. I cannot name a single wilderness advocate who cherishes wilderness "as an alibi for the lack of private reform," any who "salve their consciences" by pointing to "the few odds and ends" of wilderness and thus "avoid facing up to the fact that the ways and means of industrial civilization lie at the root of the current global environmental crisis" (p. 239). The charge is flamboyant; the content runs hollow. Wilderness advocates want wilderness and they also want, passionately, to "re-envision civilization" (p. 236) so that it is in harmony with the nature that humans do modify and inhabit. There is no tension between these ideas in Leopold, nor in any of the other passionate advocates of wilderness that Callicott cites, nor in any with whom I am familiar.

The contrast of monastic sanctuaries with the wicked everyday world risks a flawed analogy. Unless we are careful, we will make a category mistake, because both monastery and lay world are in the domain of culture, while wilderness is a radically different domain. Monastery sets an ideal unattainable in the real civil world (if we must think of it that way), but both worlds are human, both moral. We are judging human behavior in both places, concerned with how far it can be godly. By contrast, the wilderness world is neither moral nor human; the values protected there are of a different order. We are judging evolutionary achievements and ecological stability, integrity, beauty—not censuring or praising human behavior.

Confusion about nature and culture is getting us into trouble again. We are only going to get confused if we think that the issue of whether there should be monasteries is conceptually parallel to the issue of whether there should be wilderness. The conservation of value in the one is by the cultural transmission of a social heritage, including a moral and religious heritage, to which the monastery was devoted. The conservation of value in the other is genetic, in genes subject to natural selection for survival value and adapted fit. There is something godly in the wilderness too, or at least a creativity that is religiously valuable, but the contrast between the righteous and the wicked is not helpful here. The sanctuary we want is a world untrammelled by man, a world left to its own autonomous creativity, not an island of saintliness in the midst of sinners.

We do not want the whole Earth without civilization, for we believe that humans belong on Earth; Earth is not whole without humans and their civilization, without the political animal building his *polis* (Socrates), without peoples inheriting their promised lands (as the Hebrews envisioned). Civilization is a broken affair, and in the long struggle to make and keep life human, moral, even godly, perhaps there should be islands, sanctuaries, of moral goodness within a civilization often sordid enough. But that is a different issue from whether, when we build our civilizations for better or worse, we also want to protect where and as we can those nonhuman values in wild nature that preceded and yet surround us. An Earth civilized on every acre would not be whole either, for a whole domain of value—wild spontaneous nature—would have vanished from this majestic home planet.

INTRINSIC WILDERNESS VALUES

I fear that we are seeing in Callicott's revisiting wilderness the outplay of a philosophy that does not think, fundamentally, that nature is of value in itself. Such a philosophy, though it may protest to the contrary, really cannot value nature for itself. All value in nature is by human projection; it is anthropogenic, generated by humans, though sometimes not anthropocentric, centered on humans. Callicott has made it clear that all so-called intrinsic value in nature is "grounded in human feelings" and "projected" onto the natural object that "excites" the value. "Intrinsic value ultimately depends upon human valuers." "Value depends upon human sentiments" (Callicott, 1984, p. 305).

He explains, "The source of all value is human consciousness, but it by no means follows that the locus of all value is consciousness itself.... An intrinsically valuable thing on this reading is valuable for its own sake, for itself, but it is not valuable in itself, i.e., completely independently of any consciousness, since no value can in principle...be altogether independent of a valuing consciousness.... Value is, as it were, projected onto natural objects or events by the subjective feelings of observers. If all consciousness were annihilated at a stroke, there would be no good and evil, no beauty and ugliness, no right and wrong; only impassive phenomena would remain." This, Callicott says, is a "truncated sense" of value where "'intrinsic value' retains only half its traditional meaning" (Callicott, 1986, pp. 142-43, p. 156, and p. 143).

Talk about dichotomies! Only humans produce value; wild nature is valueless without humans. All it has without humans is the potential to be evaluated by humans, who, if and when they appear, may incline, sometimes, to value nature in noninstrumental ways. "Nonhuman species...may not be valuable in themselves, but they may certainly be valued for themselves.... Value is, to be sure,. humanly conferred, but not necessary homocentric" (Callicott, 1986, p. 160). The language of valuing nature for itself may be used, but it is misleading; value is always and only relational, with humans one of the relata. Nature in itself (a wilderness, for example) is without value. There is no genesis of wild value by nature on its own. Such a philosophy can value nature only in association with human habitation. But that—not some elitist wilderness conservation for spiritual meditation—is the view that many of us want to reject as "aristocratic bias and class privilege" (p. 237).

Sustainable development is, let's face it, irremediably anthropocentric. That is what we must have most places, and humans too have their worthy values. But must we have it everywhere? Must we have more of it and less wilderness? Maybe the value theory here is where the arrogance lies, not in some alleged ethnocentrism or misunderstood doctrine of the dominion of man.

A truncated value theory is giving us a truncated account of biodiversity. Callicott hardly wants wildernesses as "sanctuaries," only as "refugia" (p. 236). A refugia is a seedbed from which other areas get restocked. That is one good reason for wilderness conservation, but we do not want wilderness simply as a place from which the game on our rural lands can be restocked, or even, if we have a more ample vision of wildlife recreation, from which the wildlife that yet persists on the domesticated landscape can be resupplied steadily. Wildernesses are not hatcheries for rural or urban wildlife. Nor are they just "laboratories" (p. 238) for baseline data for sound scientific management. Nor are they raw materials on which we can work our symbiotic enhancements. Nor are they places that can excite us into projecting truncated values onto them. Some of these are sometimes good reasons for conserving wilderness. Leopold sums them up as "the cultural value of wilderness" (Leopold [1949] 1968, p. 200). But they are not the best reasons.

LEOPOLD AND WILDERNESS

Leopold pleads in the "Upshot," in his last book in the penultimate essay, entitled "Wilderness": "Wilderness was an adversary to the pioneer. But to the laborer in repose, able for the moment to cast a philosophical eye on his world, that same raw stuff is something to be loved and cherished, because it gives definition and meaning to his life" (Leopold, [1949] 1968, p. 188). He does not mean that wilderness is only a resource for personal development, though it is that. He means that we never know who we are or where we are until we know and respect our wild origins and our wild neighbors on this home planet. We never get our values straight until we value wilderness appropriately. The definition of the human kinds of values is incomplete until we have this larger vision of natural values.

Concluding his appeal for "raw wilderness" (p. 201), Leopold turns to the "Land Ethic," "The land ethic simply enlarges the boundaries of the community to include soils, waters, plants, and animals, or collectively: the land.... A land ethic of course cannot prevent the alteration, management, and use of these 'resources,' but it does affirm their right to continued existence, and, at least in spots, their continued existence in a natural state." We may certainly assert that the founder of the Wilderness Society believed that wilderness conservation is essential in this right to continued existence in a natural state.

"I am asserting that those who love the wilderness should not be wholly deprived of it, that while the reduction of wilderness has been a good thing, its extermination would be a very bad one, and that the conservation of wilderness is the most urgent and difficult of all the tasks that confront us" (Leopold, quoted in Meine, 1988, p. 245). We must take it as anomalous (else it would be amusing or even tragic) to see Leopold's principal philosophical interpreter, himself a foremost environmental philosopher who elsewhere has said many wise things, now trying to revisit the wilderness idea and de-emphasize it in Leopold.

Just before Leopold plunges into his passionate plea for the land ethic, he calls for "wilderness-minded men scattered through all the conservation bureaus." "A militant minority of wilderness-minded citizens must be on watch throughout the nation, and available for action in a pinch" (Leopold [1949], 1968, p. 200). Alas! His trumpet call is replaced by an uncertain sound. Robert Marshall saluted Leopold as "The Commanding General of the Wilderness Battle" (cited in Meine, 1988, p. 248). How dismayed he would be by this dissension within his ranks.

On Earth, man is not a visitor who does not remain; this is our home planet and we belong here. Leopold speaks of man as both "plain citizen" and as "king." Humans too have an ecology, and we are permitted interference with, and rearrangement of, nature's spontaneous course; otherwise there is no culture. When we do this there ought to be some rational showing that the alteration is enriching, that natural values are sacrificed for greater cultural ones. We ought to make such development sustainable. But there are, and should be, places on Earth where the nonhuman community of life is untrammeled by man, where we only visit and spontaneous nature remains. If Callicott has his way, revisiting wilderness, there soon will be less and less wilderness to visit at all.

REFERENCES

Callicott, J.B. 1984. Non-anthropocentric Value Theory and Environmental Ethics. *American Philosophical Quarterly* 21: 299-309.

Callicott, J.B. 1986. On the Intrinsic Value of Nonhuman Species. In *The Preservation of Species*, B.G. Norton, ed. Princeton University Press, Princeton, NJ.

Callicott, J.B. 1990. Standards of Conservation: Then and Now. *Conservation Biology* 4: 229-232.

Callicott, J.B. 1991. The Wilderness Idea Revisited: The Sustainable Development Alternative. *The Environmental Professional* 13: 235-247.

Cordell, H.K., and P.C. Reed 1990. Untrammeled by Man: Preserving Diversity through Wilderness. In *Preparing to Manage Wilderness in the 21st Century: Proceedings of the Conference*. Southeastern Forest Experiment Station, U.S. Department of Agriculture, Forest Service, Asheville, NC, pp. 30-33.

Groschupf, K., .B.T. Brown, and R.R. Johnson. 1987. *A Checklist of the Birds of Organ Pipe Cactus National Monument*. Southwest Parks and Monument Association, Tucson, AZ.

Leopold, A. [1949] 1968. *A Sand County Almanac*. Oxford University Press, New York.

Leopold, A.S., S.A. Cain, C.M. Cottam, I.N. Gabrielson, and T.L. Kimball. 1963. Wildlife Management in the National Parks, Report of the Advisory Board on Wildlife Management. U.S. Government Printing Office, Washington, DC, March 4.

Meine, C. 1988. *Aldo Leopold: His Life and Work*. University of Wisconsin Press, Madison.

Magurran, A.E. 1988. *Ecological Diversity and its Measurement*. Princeton University Press, Princeton, NJ.

Nabhan, G.P., A.M. Rea, K.L. Reichhardt, E. Mellink, and C.F. Hutchinson. 1982. Papago Influences on Habitat and Biotic Diversity: Quitovac Oasis Ethnoecology. *Journal of Ethnobiology* 2: 124-143.

Pielou, E.C. 1975. *Ecological Diversity*. John Wiley, New York.

Tribal Wilderness Ordinance of the Governing Body of the Confederated Salish and Kootenai Tribes. 1982.

THAT GOOD OLD-TIME WILDERNESS RELIGION

J. Baird Callicott
University of Wisconsin-Stevens Point

In this paper, I respond to Rolston's (1991) critique of my recent article on wilderness (Callicott, 1991). I knew that critically revisiting the wilderness idea would provoke a reaction about as pleasant as poking a stick into a hornet's nest. My friend and colleague Holmes Rolston is the first wasp out of the hive to sting me. I fear a swarm of others will follow.

So let me reiterate a vital point that Rolston has distorted in his first sentence and his last. I find fault with the wilderness idea, not the places so called. On p. 236 I write, "I am as ardent an advocate of those patches of the planet called 'wilderness areas' as any other environmentalist.... Such areas serve the cause of biological conservation most importantly as refugia for species not tolerant of or tolerated by people." My use of the term "refugia" throughout may have been imprudent since the designation of "game refuges" was advocated by turn-of-the-century sportspersons—for purely selfish reasons. I advocate, passionately, the provision of habitat so that large predators and those species that require old growth forest, extensive grasslands, wetlands, and so on, may survive and thrive—for their inherent value. Indeed, the more wildlife sanctuaries (a term Rolston prefers and one I also use [p. 239]), the better—as I am on record as insisting (p. 244).

Rolston's clinic on sonoran avifauna conservation is a much appreciated elaboration of my caution that "the quality, so to speak, of species is also important" (p. 243).

Rolston attributes my unorthodox reflections on the wilderness idea to theories about the value of nature published elsewhere. As these were taken out of context, it is necessary for me to add an explanation. Most philosophers "cry 'Heresy' on everyone whose sympathies reach a single hair's breadth beyond the boundary epidermis of our own species,"

J. Baird Callicott is a professor of philosophy and natural resources at the University of Wisconsin-Stevens Point, Stevens Point, WI 54481. He is author of *In Defense of the Land Ethic* and, with Susan L. Flader, editor of *The River of the Mother of God and Other Essays by Aldo Leopold*. In 1971, Dr. Callicott taught the first course in environmental ethics. Since that time, he has continued to contribute to the literature of this rapidly growing interdisciplinary field.

to quote Muir (1916, p. 139). The standard view in both science and philosophy is that value is consciousness-dependent. Assuming the standard view, I argue (in the places Rolston cites) that our sympathies may reach well beyond the boundary epidermis of our own species—to all the others, to ecosystems, even to the biosphere as a whole. Because my argument was with the anthropocentrists, my language may have suggested that I do not think that other conscious beings also value things. But I do in fact think that they may. Value is not only anthropogenic; it is perhaps "vertebragenic."

If it is fair for Rolston to diagnose my dissidence within the ranks of Bob Marshall's wilderness militia by reference to what I have elsewhere written, he cannot cry foul if I diagnose his orthodoxy in similar fashion. However much value Rolston may find in nature, he firmly holds that "Morality appears in humans alone and is not, and never has been, present on the natural scene"—experimental evidence of thoughtful altruism among rhesus monkeys, and anecdotal evidence of voluntary celibacy among wolves, notwithstanding (Masserman et al., 1964; Mowat, 1963; Rolston, 1988, p. 38). Rolston is a former Presbyterian minister and remains committed to Christian ideology, a major tenet of which is that man is a case apart from (the rest of) nature, as his book *Science and Religion* amply testifies (Rolston, 1987). Here culture serves Rolston as a secular surrogate for the image of God cleanly segregating man from nature. The crux (no pun intended) of the debate between us concerns the relation of culture to nature. I follow Darwin in thinking that human culture is continuous with primate and mammalian proto-culture and that, no matter how hypertrophic it may lately have become, contemporary human civilization remains embedded in nature. To claim, as Rolston does, that no cultural modification of nature can be ecologically (not anthropocentrically) beneficial, measured by objective criteria, is merely question-begging dogmatism.

Rolston also epitomizes the faithful who remain committed to a parallel dichotomy—between civilization and savagery—that relegates aboriginal peoples to the status of two-legged wildlife. Here, fortunately, I can defend two points at once: that the difference between native species and exotics is temporally scale-dependent and that the environmental impact of Pre-Columbian American peoples was significant. Are feral mustangs in the American west native or exotic?, I asked (p. 242). Exotic, Rolston answers: "ancestral horses...in the American West went extinct naturally." Naturally for sure, as I understand naturally. But was a contributing agent to this natural extinction cultural *Homo sapiens*? *Equus* evolved in North America and roamed here up to and shortly after Siberian big game hunters wandered across the Bering land bridge into the western hemisphere (Simpson, 1956). Two species of elephant and 30 other genera of hemispherically extinct large mammals were also here when the spearmen appeared (Wright et al., 1965). Did atlatl launched missiles armed with Clovis warheads conspire with global warming to push these animals over the brink? If, as Martin (1973) has argued, the sudden extinction of the American megafauna is attributable to a newly arrived predator overkilling prey totally unprepared for projectile-hurling carnivorous apes, then the environmental impact of the Siberian pioneers, measured in terms of biodiversity loss, greatly exceeds that of their technologically superior European counterparts some 100 centuries later. What would the New World biota have been like in 1492 if Columbus had really been its discoverer? Very different, I would suppose; much the same, Rolston seems to think.

After the spasm of (anthropogenic?) extinctions in the western hemisphere at the beginning of the Holocene, Martin (1973, p. 972) speculates that "Major cultural changes would begin." This is the lesson that I wish to draw from American prehistory: Eastward migrating pedestrian *Homo sapiens* found a genuinely pristine and virgin wilderness. There can be little doubt about that. Though controversial, some archaeologists are inclined to believe that these first pioneers raped it. If so, subsequent generations of Indians apparently learned to live symbiotically with the depauperate natural heritage that their immigrant ancestors bequeathed them. They could because cultural adaptation is more rapid and versatile than genetic adaptation, as Rolston so eloquently and forcefully argues. Perhaps we postmodern Euro-, Afro-, and Asian-Americans can do likewise. Our contemporary overpopulation and technological sophistication may preclude that possibility, in which case all we can do about conservation is defensively protect relatively undisturbed patches of nature—which I'm all for—and occasionally visit them to salve our spirits while living in a world of wounds. But a new generation of post-industrial technologies may make it possible for us to pursue many of our economic activities without compromising ecosystem health. If so, we may envision an alternative, and complementary, approach to the conservation of biodiversity as I tried to do in "The Wilderness Idea Revisited" (Callicott, 1991).

REFERENCES

Callicott, J.B. 1991. The Wilderness Idea Revisited: The Sustainable Development Alternative. *The Environmental Professional* 13: 235-247.
Martin, P. S. 1973. The Discovery of America, *Science* 179: 969-974.
Masserman, J. H., S. Wechkin, and W. Terris. 1964. Altruistic Behavior in Rhesus Monkeys, *American Journal of Psychiatry* 121: 584-585.
Mowat, F. *Never Cry Wolf*. Little, Brown. Boston.
Muir, J. 1916. *A Thousand Mile Walk to the Gulf*. Houghton-Mifflin, Boston.
Rolston, H. 1987. *Science and Religion: A Critical Survey*. Random House, New York.
Rolston, H. 1988. *Environmental Ethics: Duties to and Values in the Natural World*. Temple University Press, Philadelphia.
Rolston, H. 1991. The Wilderness Idea Reaffirmed. *The Environmental Professional* 13: 370-377.
Simpson, G. G. 1956. *Horses: The Story of the Horse Family in the Modern World through Sixty Million Years of Evolution*. Scribner's, New York.
Wright, H. E. and D. G. Frey, eds., 1965. *The Quaternary of the United States*. Princeton University Press, Princeton, New Jersey.

Part VI
Policy, Dilemmas and Pluralism

Some Problems with Environmental Economics

Mark Sagoff*

In this essay I criticize the contigent valuation method in resource economics and the concepts of utility and efficiency upon which it is based. I consider an example of this method and argue that it cannot—as it pretends—substitute for public education and political deliberation.

I. INTRODUCTION

In a paper appearing recently in this journal, Steven Edwards points out, correctly, that "[m]ore and more, economists use willingness-to-pay surveys to elicit valuations of the natural environment."[1] Economists have devised surveys, questionnaires, and other experimental instruments to determine how much individuals are willing to pay, for example, to preserve natural environments ("preservation" or "existence" value), to maintain the option of using natural environments ("option" value); and to leave the environment unspoiled for future generations ("bequest" value). This "contingent valuation method," Edwards notes, "is being used increasingly to assess the economic value of recreation, scenic beauty, air quality, water quality, species preservation, and bequests to future generations" (p. 80). Edwards defends the contingent valuation method against its critics "as an instrument for elucidating nonmarket and nonconsumptive values derived from the natural environment" (p. 74).

Edwards offers three important arguments to back up his "defense of environmental economics in general and the contingent valuation method of valuing natural resources in particular" (p. 74). First, he describes "the utility-theoretic foundation of the method" (p. 80), which asserts a relation between willingness to pay and personal utility. Edwards writes: "Maximum willingness-to-pay is *the change in income that holds personal utility constant*" (p. 76). This suggests that

---

* Center for Philosophy and Public Policy, University of Maryland, College Park, MD 20742. Sagoff is a research scholar at the Center for Philosophy and Public Policy. His research has been supported by grants from the National Science Foundation to the Center for Philosophy and Public Policy. The views expressed, however, are those of the author and not necessarily of any governmental agency. This paper draws on arguments Sagoff presents in a book, *The Economy of the Earth: Philosophy, Law, and the Environment,* soon to be published by Cambridge University Press.

[1] Steven Edwards, "In Defense of Environmental Economics," *Environmental Ethics* 9 (1987): 74–85. Quotation at p. 75. Subsequent page references in the text are to this essay.

willingness-to-pay measures or at least varies with something valuable, namely, personal utility.

Edwards' second point begins with an important assumption. He assumes that the sort of economic theory he defends applies as a useful analysis of traditional markets. These are markets in which buyers and sellers transfer property voluntarily at prices they agree upon. Edwards argues that "[l]imiting economics to the analysis of traditional markets is arbitrary" (p. 77). He notes that people derive personal utility from publicly owned assets, and they may be willing to pay for them, even though the absence of exclusive property rights in these assets prevents traditional markets from setting prices for them.

Third, Edwards defends the contingent valuation method (CVM) against the difficulty posed by the "frequency of protest responses . . . of up to fifty percent . . . " (pp. 80-81). He attributes this observed resistance or noncooperation to a kind of hypothetical and potential bias with which economists are familiar (p. 81).

In this paper, I reply to Edwards' presentation, and, in particular, to these three arguments. I argue, first, that the relation between willingness-to-pay and personal utility, as it occurs in Edwards' paper and environmental economics generally, is an analytic, not an empirical or contingent one. Willingness-to-pay does not measure, reflect, or vary with happiness, pleasure, contentment, or personal utility in any substantive sense, that is, any sense that allows us to understand why personal utility is valuable. The relation Edwards asserts is entirely stipulative: the term *personal utility* is simply a stand-in for "willingness-to-pay" and has no independent meaning or normative significance.

Second, I argue that the kind of economic analysis Edwards defends has no basis in or application to traditional markets in which buyers and sellers must agree upon prices. His willingness-to-pay approach, on the contrary, envisions a kind of universal bidding game in which resources go to those willing to pay the most for them, even if the owners refuse to sell their rights. Thus, Edwards is correct, but only in a trivial way, when he says that it is arbitrary to limit welfare economic analysis to traditional markets. It is actually "contingent" markets that constitute the paradigm for the kind of allocation-to-the-highest-bidder that Edwards favors. This kind of allocation has little relevance to traditional markets, which do not typically allocate resources to the highest bidder, since willingness to sell is just as important as willingness to pay.

Finally, I argue that Edwards is mistaken in his characterization of protest bids and other kinds of resistance by respondents to contingent valuation surveys. Apparently, Edwards attributes this sort of noncooperation to "strategic" bidding behavior, e.g., to bluffing. I argue, on the contrary, that respondents may enter protest bids because they believe the contingent valuation method conflicts with representative democracy, political deliberation, and the rule of law. In other words, respondents may believe that the property rights in question are not for sale—a concept entirely familiar in traditional markets, but unknown in Edwards' sort of analysis.

In this essay, I argue for each of these views in turn. I then apply these criticisms to a prominent example of the contingent valuation method.

II. EDWARDS' DEFENSE OF ENVIRONMENTAL ECONOMICS

Edwards speaks of "income" and "personal utility" as if these concepts, as he employs them, were logically distinct. He writes, for example (p. 76):

> Clearly if economic man's income increases, so will personal utility, since more money is available to increase the amounts of things that provide personal satisfaction. Similarly, if income increases, utility will decrease.

If we assume that these sentences express an empirical judgment—roughly, that money buys happiness—then we must conclude that they are false. There is no evidence (and Edwards cites none) to show that people become happier, more satisfied, or better off in some substantive sense (after basic needs are met) when their income, and, therefore, the amount they can and will pay for things, increases. Rather, the extensive empirical evidence that exists runs strongly in the opposite direction.[2]

Economists have long followed ordinary wisdom in acknowledging that we become happier—our welfare increases in a substantive sense—not insofar as we satisfy our preferences on a willing-to-pay basis, but insofar as we improve those desires or overcome or outgrow them. A. O. Hirschman, for example, has argued that "acts of consumption . . . which are undertaken because they are expected to yield satisfaction, also yield disappointment and dissatisfaction."[3] And Frank Knight, following Mill and other classical utilitarians, observes that the education, not necessarily the satisfaction, of desire leads to happiness. "The chief thing which the common sense individual actually wants," Knight says, "is

---

[2] For a discussion of relevant surveys, see Nicholas Rescher, *Welfare: The Social Issues in Philosophical Perspective* (Pittsburgh: University of Pittsburgh Press, 1972), esp. chap. 3, and *Unpopular Essays on Technological Progress* (Pittsburgh: University of Pittsburgh Press, 1980), chap. 1. See also A. Campbell, P. E. Converse, and W. Rodgers, *The Quality of American Life: Perceptions, Evaluations, and Satisfactions* (New York: Russell Sage Foundation, 1976). Welfare in a substantive sense does increase with the satisfaction of basic needs, of course, but this is an argument for justice or equality, not necessarily for efficiency, in the allocation of resources.

[3] Albert O. Hirschman, *Shifting Involvements: Private Interest and Public Action* (Princeton: Princeton University Press, 1982), p. 10.

not the satisfaction of the wants he has, but more, and better wants. . . . [T]rue achievement is the refinement and elevation of the plane of desire, the refinement of taste."[4]

We may conclude, then, that Edwards does not mean to argue that economic value (in other words, willingness-to-pay) varies with or is determined by personal utility in an empirical or contingent sense, for this thesis flies in the face of ordinary wisdom and is obviously false.[5] Rather, Edwards might mean to point out, correctly, that resource economists often define "personal utility" as that which willingness-to-pay measures—in other words, he might mean that the concepts are equivalent. To contend that policies that satisfy preferences ranked by willingness-to-pay to that extent maximize personal utility, then, is to defend an empty and uninformative tautology. It is not to provide a "utility-theoretic foundation" for contingent market valuation.

Edwards writes that cost-benefit analysis is a special case of utilitarianism (p. 74), but this is not so, since welfare economics and the techniques of cost-benefit analysis it employs have no relation to substantive conceptions of the good, such as pleasure or happiness, of the kind that utilitarianism values. Welfare economists earlier in this century, imbued with a positivistic philosophy of science popular at the time, took pains to divorce their theory from any such substantive conception of utility, like happiness, since it could not be quantified, and insisted instead on defining utility in relation to measureable quantities, such as willingness-to-pay, even if these have no normative significance and no basis, therefore, in utilitarianism. Thus, sophisticated advocates of welfare economics, like Richard Posner, point out that the efficiency criterion is independent of utilitarian ethical theory. "The most important thing to bear in mind about the concept of value [in the economist's sense]," he writes, "is that it is based on what people are willing to pay for something rather than the happiness they would derive from having it."[6]

---

[4] F. H. Knight, *The Ethics of Competition and Other Essays* (New York: Harper and Brothers, 1935), pp. 22–23. Compare John Stuart Mill, "What Utilitarianism Is," in *The Utilitarians* (Garden City, N.Y.: Doubleday, 1961), p. 410, where he argues that it is better to be Socrates dissatisfied than a fool satisfied. As far as I know, no economist has offered any evidence to show that people are only willing to pay, in general, for what makes them better off in a substantive sense (consider tobacco). Economists who have studied this question have concluded, on the contrary, that willingness-to-pay has no non-stipulative or non-definitional relationship with personal welfare. For arguments to this effect, see: Tibor Scitivsky, *The Joyless Economy* (Oxford: Oxford University Press, 1976); Richard Easterlin, "Does Money Buy Happiness?" *Public Interest* 30 (1973): 3–10; and Fred Hirsch, *Social Limits to Growth* (Cambridge, Mass.: Harvard University Press, 1976).

[5] I have argued at length in other places that by satisfying consumer preferences (in the sense of "meeting" or "filling" them) we do not as a rule produce satisfaction (in the sense of contentment or happiness). The assertion that the satisfaction of preferences produces consumer or any other kind of satisfaction is either false, tautological, or merely a bad pun. See, for example, "Values and Preferences," *Ethics* 96 (1986): 301–16.

[6] Richard Posner, *The Economics of Justice* (Cambridge, Mass.: Harvard University Press, 1981), p. 60.

Let me now turn to Edwards' second point (p. 77), namely, that it is arbitrary to limit economic analysis to traditional markets. While this is true, it leads us to ask whether the kind of economic analysis that is relevant to the "contingent" valuation of resources applies to traditional markets in the first place. I believe, on the contrary, that it is arbitrary to *limit* it because it is arbitrary to *apply* this sort of analysis to traditional markets.

The big difference between traditional markets, in which property rights are well defined, and "contingent" markets, in which resources go, in principle, to the highest bidder, is this. In traditional markets people can and do refuse to sell their property even to those who will pay the highest price for it; rather than sell out to a trespasser at the highest price, an individual is likely to try to enjoin the trespass. Thus, if your neighbor starts operating a stamping mill, drowning you in noise, you may proceed against him, and a court will enjoin the nuisance, especially if the zoning ordinances are on your side. The rights in question here are not for sale to the highest bidder; the owner may defend them in court. The rights to exclude and not to transfer are the most common incidents of ownership; they cannot be overridden, even by theorists who believe, for some reason, that all resources should be put up for auction to the highest bidder.

It is a commonplace that traditional markets will not allocate resources to those willing to pay the most for them, because of market "failures" and for other reasons, one of which is that many owners will refuse on principle to "sell out," e.g., to polluters.[7] If resources are to be allocated efficiently, an agency of the government must transfer them to the highest bidders whether the original owners (including the public that owns "common" resources) consent to that transfer or not.

The economic analysis that Edwards defends applies to an auction in which every item is determined beforehand to be for sale to the highest bidder. Edwards may believe that "publicly" owned resources are essentially "unowned" and should be auctioned off; hence, economic analysis applies to them at least as well as it does to privately owned resources. An enormous structure of public law, like the Clean Air and Clean Water Acts, however, establishes that the public knows it owns environmental resources and has decided not to market them even

---

[7] Many commentators recognize that cost-benefit analysis is an instrument of centralized government planning and that efficiency can be achieved only in an authoritarian system that substitutes bureaucratic control for free markets. The central authority, to justify its authoritarian allocation, need only say it is correcting a market failure. As Duncan Kennedy writes, it is rare that an analyst "lacks a handy externality to justify a particular . . . measure" (Duncan Kennedy, "Cost-Benefit Analysis of Entitlement Problems: A Critique," *Standard Law Review* 33 (1980): 419. I am urging an additional reason for the same conclusion, namely, that in traditional markets owners for ethical, cultural, or other reasons can refuse to sell, while in the hypothetical markets envisioned by economic analysis, resources essentially are auctioned off to the highest bidder. The point I am making here has nothing the do with the gap between "bid" and "asked" prices for property rights, which economists recognize. My point is that environmental law deals with resources that people in fact refuse to sell at all—and this includes publicly as well as privately owned resources.

to the highest bidder. The public has decided to prohibit or enjoin certain takings, as it were, rather than simply to accept them for a price. Likewise in traditional markets, an owner may refuse to sell to polluters rights to person and property even for a profit.

Let me put this point in technical terms. In traditional markets, property rights are backed by property rules. When such rules apply, victims of a trespass (for example, a nuisance such as noise pollution) can get injunctive relief in court. In the sort of "market" that the welfare economist envisions, however, property rights are backed at most by liability rules. When property rights are backed by liability rules, victims must endure pollution and other violations of those rights. There is no exclusivity. Instead a court awards the victim of a nuisance damages in the amount that it appraises the relevant property rights to be worth. The victim, in other words, must transfer the relevant rights to polluters willing to pay the price that an agency of the state, usually a court, determines those rights are worth.[8]

As two legal scholars explain, "[A]n entitlement is protected by a property rule to the extent that someone who wishes to remove the entitlement from its holder must buy it from him in a voluntary transaction in which the value of the entitlement is agreed upon by the seller."[9] In the hypothetical "markets" that welfare economists imagine, in contrast, anyone can remove an entitlement from anyone else as long as he is willing to pay the highest price the entitlement might fetch in an auction. This is the amount a court might say the "damage" is objectively worth. "Whenever someone may destroy the initial entitlement if he is willing to pay an objectively determined value for it, an entitlement is protected by a liability rule."[10]

To see this distinction, imagine you are a farmer and your neighbor, a rancher, sends his sheep to graze on your corn. In a traditional market—one which backs up property rights with property rules—you can get an injunction to compel your neighbor to stop the trespass. This is true because traditional markets recognize the traditional incidents of property, including the right to exclude.

In the sort of "market" welfare economists envision, injunctive relief of this sort is unavailable; you have to bargain with your neighbor to reach an "efficient" allocation of the resource. If bargaining breaks down, a court determines what your corn is "objectively" worth and awards you damages in that amount. What

---

[8] A Kaldor-Hicks or "potential" Pareto improvement criterion for efficiency would permit the pollution as long as the polluter could compensate his victims at this "objective" price; compensation need not in fact be paid.

[9] Guido Calabresi and A. Douglas Melamed, "Property Rules, Liability Rules, and Inalienability: One View of the Cathedral," *Harvard Law Review* 85 (1972): 1092. I have presented the argument I make here more extensively in "The Principles of Pollution Control Law," *Minnesota Law Review* 71 (1986): 19–95, esp. pp. 46–55.

[10] Calabresi and Melamed, "Property Rules," p. 1092.

matters is how much each of you is willing to pay; ownership determines at most the direction in which payment is made.

If a victim of some market "externality," such as pollution, refuses to sell the relevant rights and seeks injunctive relief instead, this signifies, to the economic theorist, that the property owner is "uncooperative;" he or she is trying to "gouge" the tortfeasor by holding out for a higher price. The economic theorist, by interpreting the actions of property holders in this way, easily overrides their right not to sell, and has the state impose an "efficient" bargain, e.g., by awarding damages, instead.[11]

The nature, meaning, and extent of property rights depend entirely on the legal regime that backs them up. A legal regime that takes property rights seriously, and thus allows people (including the public as a whole) to refuse to sell out to polluters, is consistent with traditional markets in which actual consent is necessary for a transfer to take place. A regime that takes property rights seriously, then, will not allocate resources efficiently, i.e., to those willing to pay the most for them, because people—and the public—may prefer not to sell at any price.[12]

The kind of economic analysis Edwards proposes, in contrast, eliminates exclusivity in principle by allocating property rights to those willing to pay the most for them, even if the original owners, including the public that owns many resources, go on record (as the public has done in a variety of statutes) to insist that they do not wish to market, but rather to regulate the use of those resources. This kind of analysis thus applies to abstract bidding games among individuals who are abstractions created by the analysis itself, not to traditional or actual markets. It is perverse for Edwards to suggest that this sort of analysis has anything to do with property rights. It is little more than a studied refusal to acknowledge the ideals of exclusivity and consent that we associate with those rights.

---

[11] This is the reason that "libertarianism rejects in principle the use of cost-benefit analysis as a basis to justify pollution." Libertarians recognize that in traditional markets, where property rules are backed by property rights, "processes of production which involve pollution, so long as the harmful imposition upon others occurs, without the consent of the victims, . . . may not be carried out." Tibor R. Machan, "Pollution and Political Theory," in *Earthbound*, ed. Tom Regan (New York: Random House, 1984), p. 98. A society that takes property rights and consent seriously, such as ours, will then at least enact environmental laws that seek to minimize and eventually eliminate pollution. A planned or centralized economy, in contrast, may permit and may even require pollution and any other transfer of property rights, without the consent of the initial owners, as long as the transfer is efficient or the benefits exceed the costs.

[12] Prosser observes that "the great majority of nuisance suits have been in equity and concerned primarily with the prevention of future damage." William Prosser, *Handbook on the Law of Torts*, 4th ed. (St. Paul: West Publishing Company, 1971), par. 87, p. 576. Environmental groups that routinely sue polluters seek to stop the pollution. Although they may assert economic injury to establish standing, they seek injunctive relief.

Put more generally, Edwards' position illustrates an unwillingness to take liberty and consent seriously. Edwards acknowledges that many respondents refuse to cooperate with contingent valuation surveys, e.g., they refuse to stipulate amounts at which they are willing to buy or sell environmental goods and resources. These refusals are noteworthy because the famous Milgram experiments have demonstrated that in social science research settings people are so cowed by authority that they will do anything, even torture and murder, when asked to do so.[13] It seems that the only kinds of experiments that respondents reject in large numbers are contingent valuation surveys conducted by resource economists. Why is this?

I imagine the primary reason is that respondents believe that environmental policy—for example, the degree of pollution permitted in national parks—involves ethical, cultural, and aesthetic questions over which society must deliberate on the merits, and that this has nothing to do with pricing the satisfaction of preferences at the margin. Respondents may know that a representative democracy possesses excellent processes of public discussion and debate for settling issues fraught with moral and political significance; they may also be aware that there are statutes that regulate and refuse to market the use of many publicly owned environmental resources. Accordingly, these respondents may resist "backdoor" cost-benefit analyses that defy in principle the letter and the spirit of the legislation. They may also reject these surveys because they see through the circular definitions of utility on which the survey methodology is based.

Edwards, however, interprets this rejection differently. He ascribes it to sources of bias with which economists "are quite familiar" (p. 81). I assume Edwards refers to "strategic" bias, that is, the strategic misrepresentation of one's "bid" or "asked" price in order to influence the overall result to which one's own contribution would otherwise be small. Once refusals are construed as misrepresentations in this way, they can be excluded from the analysis. In this way, resource economists adjust to the "cognitive dissonance" that they encounter when citizens reject their surveys.

In doing so, however, resource economists refuse to acknowledge the possibility that citizens may believe that environmental resources should be allocated on normative, political, and cultural grounds, which citizens can understand, rather than on an efficiency principle, which may appeal to no one but the economists who invented it. By not recognizing refusals to sell for what they are, analysts show that they have no interest in property rights, in exclusivity, or in consent. They are concerned only to allocate resources in a way that maximizes some technical notion that they apparently understand, e.g., potential Pareto improvement, Kaldor-Hicks efficiency, consumer and factorial surplus, or some other

---

[13] Stanley Milgram, *Obedience to Authority* (New York: Harper and Row, 1974).

arcane and academic concept that they learned as graduate students. When citizens refuse to see the value and legitimacy of these notions, these resource economists attribute this rejection to strategic bias and unwillingness to bargain in good faith.

In order to weigh Edwards' defense against my criticisms, the reader may wish to consider an example of a contingent valuation of an environmental resource. In the following sections, I describe one such experimental survey and I consider the extent to which it provides a legitimate basis for social choice.

III. TANGIBLE AND INTANGIBLE VALUES

In 1975, the Environmental Protection Agency, enforcing the "Prevention of Significant Deterioration" (PSD) requirement of the Clean Air Act, directed states to amend their implementation plans to protect air quality in areas where it exceeds national health and safety minimums. The regulation, in other words, intends to keep clean air clean beyond health and safety requirements. While these PSD requirements may appeal to us on aesthetic and on ethical grounds, they may, nevertheless, impede economic growth and development: for example, they may conflict with plans to locate a network of power plants in the southwest, where coal and clean air are abundant. What to do? How do we enforce idealistic regulations when they blink at important economic facts?

In a recent article, "An Experiment on the Economic Value of Visibility," three economists from the University of Wyoming have tackled this problem. The authors attempt to interpret and to evaluate PSD requirements in economic terms. "Aesthetics," the authors say, "will play a major role. The PSD requirements amount to formal governmental admission that aesthetics, at least as embodied in atmospheric visibility, is a 'good' that might have a positive value."[14]

These writers point out that economists "generally have shied away from attempting to quantify aesthetic phenomena because they are usually defined as intangible."[15] This is correct. By and large, economists who engage in cost-benefit analysis assign prices only to goods and services of the sort that are typically traded in markets and thus that can easily be priced. These economists generally list other values as "intangibles" to bring them to the attention of the political authority. The "intangible" values involved in environmental, health, and safety policy may often be more important than "tangible" ones, of course,

---

[14] R. Rowe, R. D'Arge, and D. Brookshire, "An Experiment on the Economic Value of Visibility," *Journal of Environmental Economics and Management* (1980): 1. For a similar study and useful bibliography, see John Balling and John Falk, "Development of Visual Preference for Natural Environments," *Environmental Behavior* 14 (1982): 5–28.

[15] Rowe et al., "An Experiment," p. 2.

since they include social, moral, aesthetic, and cultural goals that have carried the day before Congress—hence, the PSD requirements of the Clean Air Act.

This is not to say that Congress gives regulators, for example, the Administrator of the EPA, a great deal of discretion in weighing "intangible" benefits against those markets price. On the contrary, legislation may instruct the Administrator to preserve and protect environmental quality without regard to the effect on consumer markets. The Clean Air Act, to continue the example, requires that "economic growth will occur in a manner consistent with the preservation of existing clean air resources."[16] The law sets the prevention of significant deterioration of air quality, then, as a normative constraint on economic growth and development, at least in certain areas, e.g., those in and around national parks.

The Wyoming economists, however, apparently interpret the law differently. Evidently, they believe that the law does not require that the Administrator protect air quality in and around national parks against the "intrusions" of industrial civilization. Instead, they apparently read the law as instructing the Administrator to strike an economically efficient balance between the costs and benefits likely to result, for example, from protecting visibility or from providing more electric power. They suggest that, if a way could be found to give "intangible" values an accurate "shadow" or surrogate market price, then this "balancing" might take place within the framework of cost-benefit analysis. Accordingly, the authors write:

> The perspective that aesthetic phenomena are unquantifiable employing economic analysis may be unduly pessimistic. Beauty, or aesthetic phenomena, given that some physical measure is available which is perceivable with human senses, should be measurable in economic terms. Further, PSD regulations indirectly necessitate quantification. How then, has the economist responded to the intangible which must of necessity become tangible?[17]

To solve this problem, the authors showed a variety of people photographs of scenes in the southwest. In some of these photographs the air quality was better or at least the visibility was greater than in others. The authors asked the participants how much they would be willing to pay on their monthly utility bills to preserve the visibility depicted in one photograph rather than to switch to that shown in the next. The economists attempted in this way to establish a surrogate market in which "intangible" aesthetic values could be priced.

Many statutes other than the Clean Air Act set strong normative constraints on commercial exploitation of the environment. The Endangered Species Act, for

---

[16] The Clean Air Act as amended August 1977 (Public Law 95-11), sec. 160(3).
[17] Rowe et al., "An Experiment," p. 2.

example, expresses this aspect of the national conscience.[18] It requires all federal agencies to "insure that actions authorized, funded, or carried out by them do not jeopardize the continued existence of such endangered species."[19] As Chief Justice Burger wrote for the Supreme Court: "One would be hard pressed to find a statutory provision whose terms were any plainer than those . . . of the Endangered Species Act. . . . The language admits of no exception."[20]

Because the plain language as well as the judicial interpretation of the Endangered Species Act explicitly prohibit an interest-balancing or cost-benefit test, the statute has worked rather well. Developers by and large have found mitigating strategies to protect species that their projects might otherwise eradicate. Conflicts have given way so quickly to deliberation and negotiation on a case-by-case basis, indeed, that a special Endangered Species Committee, set up to grant exemptions, has only met twice,[21] and very few cases have been litigated under the act.[22]

Some analysts, however, would take a different approach to endangered species policy. "The existence of such statutes as the Endangered Species Act," Judith Bentkover writes, "provides evidence that man values preservation of species and ecological diversity, although the art of converting those values into economic terms is relatively undeveloped."[23] Bentkover observes that economists have developed theoretical means for quantifying "intangible" benefits of this kind, but that "the application of these methodologies is fraught with difficulties."[24] In recent years, economists have dealt extensively with the methodological difficulties involved in assessing environmental benefits.[25] They have attempted, in this way, to place endangered species and other forms of environmental policy on a rational and scientific basis.

The Wyoming researchers followed the economic literature in supposing that environmental preservation may have value in various ways to which consumer

---

[18] The Sixth Circuit Court in *TVA v. Hill* described the statute as an expression of the "public conscience." 549 F.2d 1064, 1074 (6th Cir. 1976), *aff'd*, 437 U.S. 153 (1978).

[19] 16 U.S.C. Section 1536 (1976).

[20] 437 U.S. 153, 173 (1978).

[21] The 1978 amendments to the Endangered Species Act created a high-level Endangered Species Committee to deal in a juridical way with "irresolvable conflicts." The committee unanimously voted to deny an exemption in the Tellico Dam case. It permitted the Grayrocks reservoir to continue after conditions were met to mitigate its effect on the habitat of the whooping crane.

[22] I have discussed the legislative and judicial history of the Endangered Species Act in "On the Preservation of Species," *Columbia Journal of Environmental Law* 7 (1980): 33–67.

[23] Judith Bentkover, "The Role of Benefits Assessment in Public Policy Development," in Judith Bentkover, Vincent Covello, and Jeryl Mumpower, *Benefits Assessments: The State of the Art* (Boston: D. Reidel, 1986), p. 10.

[24] Ibid., p. 11.

[25] For major surveys, discussions, and bibliographies, see Bentkover et al., *Benefits Assessments*, and George L. Peterson and Alan Randall, eds., *Valuation of Resource Benefits* (Boulder, Colo.: Westview, 1984).

markets may not adequately respond. First, it may have recreational use value, and this may be estimated by reference to entrance fees, travel costs, and the like.[26] Second, a preserved environment may have "existence" value either because it gives a person an option to use the resource or because it provides him or her the ideological satisfaction of merely knowing it is there.[27] Hence, these economists write:

> Individuals and households who may never visit the Grand Canyon may still value visibility there simply because they wish to preserve a natural treasure. Individuals also may wish to know that the Grand Canyon retains its relatively pristine air quality even on days when they are not visiting the park. Concern about preserving air quality at the Grand Canyon may be just as intense in New York or in Chicago as in nearby states and communities.[28]

To speak more generally, the Wyoming economists treated visibility, e.g., in southwestern national parks, as a pure public good, that is, a good any person can enjoy without thereby lessening the amount that may be enjoyed by others.[29] They were concerned, in part, with determining the "preservation" or "existence" value of visibility, i.e., "the value assigned to the existence of a certain level of visibility aesthetics at a site even though one does not *ever* intend to participate in activity at the site."[30] The economists wrote another questionnaire to determine how much *users* of parks would pay in additional entrance fees to protect air quality or visibility. In this way, they devised bidding games or "contingent" or hypothetical markets in which to estimate, on a willingness-to-pay basis, the value of an amenity resource.

The attempt to "price" aesthetic or "existence" values, however it might serve to buttress arguments for environmental protection, invites a variety of objections. I have suggested one: the law directs the Administrator to keep clean air clean "in national parks, national wilderness areas, national monuments, national seashores, and other areas of special national or regional natural, recreational, scenic, or historic value." The PSD requirements as they stand (of course, they

---

[26] For a sample of this literature, see Marion Clawson and Jack Knetsch, *Economics of Outdoor Recreation* (Baltimore: Resources for the Future, 1966).

[27] For discussion of "existence value," see J. V. Krutilla, "Conservation Reconsidered," *American Economic Review* 57 (1967): 777–86. For discussion of "option value," see Burton A. Weisbrod, "Collective-Consumption Services of Individual Consumption Goods," *Quarterly Journal of Economics* 78 (1964): 471–77.

[28] William Schulze et al., "The Economic Benefits of Preserving Visibility in the National Parklands of the Southwest," *Natural Resources Journal* 23 (1983): 149–73; quotation at p. 154.

[29] For a definition of public goods, see Paul Samuelson, *Economics*, 10th ed. (New York: McGraw-Hill, 1976), pp. 159–60.

[30] Robert D. Rowe and Lauraine G. Chesnut, *The Value of Visibility: Economic Theory and Applications for Air Pollution Control* (Cambridge, Mass.: Abt Associates, 1982), p. 10.

could be amended by Congress) do not just set goals but also make rules; they establish air quality as a normative constraint on economic development. Thus, the law does not indirectly or directly necessitate quantification of aesthetic "benefits." It does not ask us to make the intangible tangible. The Wyoming economists might lobby Congress to have the law changed, but until it is changed, it does not permit, much less require, a cost-benefit or "balancing" test.[31]

The Wyoming economists might plausibly reply that we cannot always take laws at their face value. Environmental legislation, in particular, sets lofty, noble, and aspirational national goals that we may not fully achieve without bringing the economy to a screeching halt. Any project that causes air pollution in the southwest, for example, could arguably affect air quality in a national park. Yet no one would insist, therefore, on forbidding all polluting activities in that area. The ideal of a perfectly unpolluted environment—like the ideal of a completely risk-free workplace—is a chimera. At some point, the Administrator of the EPA (and of the other regulatory agencies) has to recognize not only the law of the land, but also the law of diminishing returns. The question will then arise: how much safety, purity, or whatever are we willing to pay for? How much clean air—as opposed to other goods and services—is enough?

This reply makes an important point that everyone—even those who interpret environmental laws as strong normative constraints on economic development—must concede. We must acknowledge, however idealistic we may be, that clean air, workplace safety, etc., have a price, and that at some point the additional amount we may buy may be grossly disproportionate to the goods and services we must forgo in order to pay for it.[32] It hardly seems reasonable to ask industry to pay hundreds of millions of dollars, for example, to provide a tiny or insignificant improvement in workplace safety; yet, it is surely appropriate to require companies to pay even large sums to prevent significant risks. But how to determine what is appropriate from an ethical point of view? What counts as a "significant" risk or a "significant" deterioration of air quality? When should we apply the law of diminishing returns?

---

[31] Congress intended clean and safe air—not allocatory efficiency—to be the goal of the Clean Air Act. Thus, the statute precludes the kind of cost-benefit balancing envisioned by the Wyoming economists. See *American Textile Manufacturers Institute* v. *Donovan*, 452 U.S. 490, 510 (1981), which states: "When Congress has intended that an agency engage in cost-benefit analysis, it has clearly indicated such intent on the face of the statute." The D.C. Circuit, in permitting EPA to consider costs in regulating vinyl chloride emissions, distinguishes taking costs into account from cost-benefit analysis. See *NRDC* v. *USEPA*, 824 F. 2d 1146, 1160–1161, note 6 (D.C. Cir. 1987).

[32] Thus, it seems to be EPA policy that, in the presence of scientific uncertainty concerning risk, the cost of regulations should not be "grossly disproportionate" to health benefits. See the *Vinyl Chloride* case cited above.

IV. TWO APPROACHES TO RATIONALITY

In order to evaluate the economic approach to enviromental policy, it is important to distinguish two senses in which social policy decisions might be described as "rational" and as "scientific."[33] In one sense, a decision is "rational" if it uses mathematical criteria and methodologies, laid down in advance, to infer policy recommendations from independent or exogenous preferences in the client society. This approach conforms to a philosophy of science that stresses notions like "value neutrality," "replicable experiments," and "correspondence to an independent reality."

A decision or policy might be described as "rational" or as "scientific" in another sense if it is based on good reasons—reasons that are open and yet stand up to criticism. The words *rational* and *scientific* in this second sense, as Richard Rorty writes, means something more like "sane" and "reasonable" than "methodical." The term *rationality*, on this second approach, "names a set of moral virtues: tolerance, respect for the opinions of others, willingness to listen, reliance on persuasion rather than force."[34]

In accordance with the first approach, economists study data that have to do with prices and with the preferences consumers reveal or express in actual and surrogate markets. Economic analysts can then answer questions like "how safe, clean, etc., is safe or clean enough?" in terms of data about preferences—data that therefore represent independent variables or exogenous states of the world. In doing so, they are able to balance the benefits of environmental protection, measured in this way, against the opportunity costs of economic development.

The second approach, in contrast, uses a juridical or deliberative model to weigh various normative constraints, established by statute, against these opportunity costs. In accordance with this approach, public officials must not only recognize both the legal and ethical force of these constraints, but at the same time take account of technical, economic, and other realities, since no one can pursue a goal without adjusting to the obstacles that stand in the way of achieving it. Because there is no methodology for making this sort of judgment, public officials have only statutory language, judicial interpretation of that language, their general knowledge and experience, and the virtues of inquiry to rely upon. This is the reason why statutes generally require that officials respond to views presented at public hearings that they set policies that are reasonable and feasible, and that they create a record of their deliberations which can be reviewed by the courts.

---

[33] I have explained this distinction in more detail in "Where Ickes Went Right or Reason and Rationality in Environmental Law," *Ecology Law Quarterly* 14 (1987): 265–323.

[34] Richard Rorty, "Science as Solidarity," unpublished manuscript, 1984, p. 3.

The problem-solving approach of the Endangered Species Act, which sets up a committee to mitigate conflicts, illustrates this ethical and juridical approach to social regulation. This approach ties the rationality of the policy-making process to virtues, particularly, the virtues of deliberation, for example, intellectual honesty, civility, willingness to see a problem in a larger context, and openness of mind. It does not require decisions to conform to criteria, methodologies, or guidelines laid down in advance. Rather, it depends upon an open process in which decision makers respond on the record to the merits of arguments and proposals.

Economists, in measuring the value of "unpriced" social and environmental benefits, approach situations in a way that brings these two conceptions of science and rationality into serious conflict. This conflict becomes apparent when analysts must decide how much information to present to subjects and how much discussion, deliberation, and education to allow as part of a survey experiment. An analysis or assessment can be "scientific" in the sense of "gathering data on exogenous variables" only if it allows no discussion, education, or deliberation to take place. An approach which is "scientific" and "neutral" in this way, however, cannot be "scientific" in the sense of being "reasonable," "civilized," or "intelligent."

V. THE WYOMING EXPERIMENT

The Wyoming economists attempted to make their analyses "scientific" by basing them on quantitative methodologies and on independently existing data that can be verified through replicable experiments. Accordingly, these economists sought to develop quanitified methodologies to identify exogenous preferences as data and to aggregate them in a way that permits the calculation of a social decision. In other words, they tried to make intangible values tangible. Can this be done?

Anyone trying to deal with these intangibles has to answer a lot of questions. How valuable is atmospheric visibility in parks, wilderness areas, and so on? How important is it for us to be able to stand on a mountaintop in Yosemite and contemplate an "integral vista" free of power plants, hotels, highways, or other signs of industrial civilization? What are the expressive or symbolic values of nature untouched by man and how much are these worth to us? How draconian should prohibitions on development be in order to keep the wilderness experience pristine?

Let us stick to the example of visibility in and near national parks. In measuring the value of visibility, we need to know, first, how a loss of atmospheric clarity or quality is caused. A mist or fog hanging on the mountains, for example, can be very beautiful, perhaps more beautiful than a clear view, as the Japanese show us in their paintings. A mist or fog, then, need not impair aesthetic value. Even a volcano which distributes ash over hundreds of miles may be viewed as an aesthetic marvel; people will come from as far just to see it. If soot and precipitates from a power plant impede visibility, however, the resulting loss of air quality, even if indistinguishable from that caused by a volcano, has a completely different meaning. We no longer think of it as natural or compare it with aspects of nature and its beauty; we may perceive it, rather, as an assault on nature and as destructive of its integrity.

The Wyoming economists faced something of a dilemma when they designed their experiment: they had to decide whether or not to explain to the participants how the visibility would be lost in the vistas presented in the photographs. If they let the participants assume that the cause would be natural, e.g., an approaching storm, then they might elicit a preference for *less* visibility, since oncoming storms in deserts can be considered beautiful. If the experimenters identified the cause as the belching smokestacks of Humongous Megawatt, a coal-fired utility, however, the respondents might not reveal aesthetic but political preferences. They might express opinions, for example, about the inadvisability of increasing supply as opposed to decreasing demand for energy through conservation. They might even offer legal arguments based on the PSD provisions of the Clean Air Act.

In fact, this is what happened. The Wyoming team (appropriately, I believe) informed the respondents that the visibility would be obscured by pollution from a power plant. They described the amount of energy (in kilowatt hours) to be produced, the location of the facilities, the levels of emission of various pollutants, and so on. I do not know whether the economists gave the subjects of the experiment information about the PSD requirements of the Clean Air Act; the respondents, however, may have had that information. The economists asked the subjects, first, how much they would pay (the "equivalent" or "ES" measure of consumer surplus) to prevent the deterioration of the visibility caused by the power plant. They then asked for "compensating" or "CS" values, which is to say, the amounts that the respondents would accept to allow the power plant to emit that much pollution.

When the respondents were asked how much they would demand in compensation (the "CS" or "WTA" value) to permit the loss of visibility shown in the photographs, at least half of them used the question as an occasion to express a political opinion. The Wyoming experimenters report:

> The CS values . . . put the liability for maintaining visibility with the power companies and presupposes [sic] that the power companies will attempt to buy off consumers rather than cleanse the air. If respondents reject this concept of "being bought off to permit pollution" they might increase their compensation. Strategically, respondents may give large or infinite valuations as an indication that this concept is unacceptable. This is partially supported in that slightly over one-half of

the sample required infinite compensation or refused to cooperate with the CS portion of the survey instrument.[35]

The experimenters found even in their own experiment that a majority of a sample of citizens rejected a cost-benefit or "consumer surplus" approach to trade-offs between health, safety, or environmental quality and economic growth, an approach which also seems to be precluded by the Clean Air Act, the Occupational Safety and Health Act, and by other legislation.[36] Attempting to make their approach practicable, if not legal, they ended up in an awkward position: they asked citizens participating in the experiment to accept the concept of trading dollars for pollution "rights," a concept that many citizens reject,[37] and most of the subjects responded by entering protest bids or by refusing to cooperate with the experiment.

VI. THE PROBLEM OF INFORMATION

In an excellent paper on "Information Disclosure and Endangered Species Evaluation," a group of economists from Hawaii describe bidding games and surrogate markets, i.e., the contingent valuation method (CVM), they used to determine citizen willingness to pay to preserve endangered species. The Hawaii group observed that WTP values are deeply influenced by the information subjects receive in the survey or experiment. These authors write:

> . . . willingness to pay (WTP) to preserve a particular animal is significantly influenced by information provided about the animal's physical and behavioral characteristics, and about its endangered status. While this proposition may appear obvious, it bears important implications for the proper type and amount of information disclosed in preservation valuation studies.[38]

In the Hawaii experiment, subjects were asked how much they were willing to contribute to a fund for preserving humpback whales, an endangered species. Then an experimental group saw *The Singing Whale*, a Jean Cousteau film describing the humpback and the threats to its survival. A control group viewed a film unrelated to whales, *The Sixty Minute Spot: The Making of a Television Commercial*. All subjects were then asked to reevaluate or reconsider their bids.

---

[35] Rowe et al., "An Experiment," p. 9.
[36] *American Textile Manufacturers* v. *Donovan (Cotton Dust)*, 452 U.S. 490 (1981).
[37] For discussion, see Steven Kelman, *What Price Incentives: Economists and the Environment* (Boston: Auburn House, 1981).
[38] Karl Samples, John Dixon, and Marcia Gowen, "Information Disclosure and Endangered Species Evaluation," *Land Economics* 62 (1986): 306–12; quotation at 306.

After seeing the films, one-third of the experimental group and one-fifth of the control group increased their bids. The authors note that this "lends support to the view that preferences are learned during the interview process, even in the absence of new relevant information."[39]

The Hawaii economists point out that relevant information can influence preservation bids in many ways. An individual is likely to decrease his bid, for example, if he learns that the population of a particular species is so large that it will survive or so small that it will go extinct no matter how much he and others contribute. A reasonable individual, in other words, is likely to apply some principle of triage to deal with the number and characteristics of endangered species. The economists conclude that "information disclosure can influence perceived marginal efficiency investment in a preservation fund, and thereby result in changes in an individual's budget allocation strategy."[40]

The Hawaii experimenters recognized the importance of their results for the contingent valuation of preservation, amenity, and other benefits of environmental protection. They identified a methodological question about the extent to which respondents are given information or otherwise allowed to educate themselves, discuss, or deliberate over the issues. Should valuation be based on the immediate, untutored, *ex ante* preferences of the respondents or should valuation refer to their informed or educated judgment instead?

One alternative, the authors note, "is to accept the state of the respondents' ignorance about the resource as given, and provide only enough information about the resource to create a realistic market situation." This alternative has the advantage of keeping the response exogenous to or independent of the experiment. It has the disadvantage (as we saw in the Wyoming experiment), however, "that respondents may not readily accept operating in a hypothetical market situation with unknown payoffs and opportunity costs."[41]

At the other extreme, "the analyst could provide vast amounts of information to respondents about the resource being valued, along with complete information about its substitutes and complements." The respondents might discuss, in the visibility case, for example, various alternatives to constructing power plants near national parks, e.g., the possibility of energy conservation. They might try to size up or define the problem in terms which allow a different sort of solution. This kind of approach, the Hawaii economists point out, "could change the preference mappings of respondents and therefore make individual values endogenous to the valuation process."[42]

How should we choose between these alternatives? Should we accept the first

---

[39] Ibid., p. 310.
[40] Ibid., p. 311.
[41] Ibid., p. 312.
[42] Ibid.

alternative, insisting that the valuation of environmental benefits be "rational" in the sense of being methodical, derived from exogenous variables, and determined by criteria laid down in advance? Should we prefer the second alternative, emphasizing the virtues of deliberation rather than the methods of derivation, and hence a conception of "rationality" which is less akin to "methodological" than to "civilized," "reasonable," and "sane"?

An analogy may help us answer this question. Let us suppose that a person has been called to perform jury duty. The judge informs each juror that a Mr. Smith has been accused of robbing a liquor store. Then the judge asks asks each juror separately whether Mr. Smith is guilty. If the judge is methodologically sophisticated, indeed, he or she may ask how much each juror is willing to pay for the preferred verdict. The judge may then report the verdict in terms of the mean, the average, or some statistical transformation of the weighted average of the jurors' preferences.

If you were a juror, how would you respond to the judge? You might complain that the methodology is flawed—the judge should use the average rather than the mean bid to set the sentence. The judge may point, however, to a large literature which investigates all the ins and outs of the statistical methodologies—perhaps the software—used by the court. He or she may reply, moreover, that the verdict rests entirely on *ex ante* preferences which remain completely independent or exogenous to the decision-making process.

You might, on the other hand, ask the judge to let the jury hear the case—the evidence for and against—and to deliberate to reach a consensus in good faith. The judge could rule this out on the grounds that the verdict would then be biased by the means of obtaining it. What is more, he could point out that no quantified methodology exists for reaching a verdict through deliberation on the evidence. To be scientific, so the judge might reason, the verdict must be derived from exogenous variables by quantified criteria laid down in advance. Jurors might be permitted to make use of any hearsay evidence that they may have picked up beforehand from the newspapers. No further inquiry, however, may bias or prejudice preference.

If you were faced with this situation, what would you think? You would think that the judge is *crazy*. You would probably refuse to cooperate with this sort of "valuation." You might protest, for example, or just vote to acquit Mr. Smith.

Economists often confront this kind of resistance to their surveys. Their subjects may reject cost-benefit balancing as an inappropriate and illegal framework for making social policy. Two resource economists observe:

> Bidding questions for changes in air quality are not always well received by respondents due to rejection of the hypothetical scenario, rejection of the implied property rights or liability rules presented in a situation, or rejection for moral and ethical reasons. . . . Rejection and protest bids have varied from 20 percent to 50

percent for specific applications of the bidding technique. *In these cases, respondents' true values remain unknown and unaccounted for.*[43]

I contend that just the reverse is true: it is only in this way—by lodging a protest—that respondents can begin to make their values known. These respondents may not perceive themselves as bundles of exogenous preferences, but rather as thinking beings capable of reaching informed judgments in the context of public inquiry and deliberation. They may regard themselves as a jury who might reach a considered judgment after discussion of all relevant views and information, including the relevant statutes. The contingent valuation method (CVM), however, insofar as it tries to make respondents express preferences rather than form judgments, denies their status both as thinking and political beings.[44] This is possibly the major reason that respondents so often enter protest bids or otherwise resist this sort of experiment.

---

[43] Rowe and Chestnut, *The Value of Visibility,* pp. 80–81 (citations omitted, italics added). These authors cite three studies that encountered a 50 percent protest or rejection rate.

[44] I have argued elsewhere that resource economists commit a "category mistake" by asking of objective beliefs and judgments a question that is appropriate only to subjective preferences and wants. See "Economic Theory and Environmental Law," *Michigan Law Review* 79 (1981): 1393–419, esp. pp. 1410–18.

[28]

Ethical Dilemmas and Radioactive Waste: A Survey of the Issues

Kristin Shrader-Frechette*

The accidents at Three Mile Island and Chernobyl have slowed the development of commercial nuclear fission in most industrialized countries, although nuclear proponents are trying to develop smaller, allegedly "fail-safe" reactors. Regardless of whether or not they succeed, we will face the problem of radioactive wastes for the next million years. After a brief, "revisionist" history of the radwaste problem, I survey some of the major epistemological and ethical difficulties with storing nuclear wastes and outline four ethical dilemmas common to many technological and environmental controversies. I suggest two solutions to these ethical dilemmas and show why they are also economical and realistic proposals.

INTRODUCTION

Egyptians have been unable to protect the tombs of the Pharoahs for less than *four* thousand years, and some of them were looted within centuries. Italians have been unable to protect Renaissance art treasures for less than *one* thousand years.[1] Yet we in this generation have to be able to protect nuclear wastes, for *hundreds* of thousands to millions of years, for a period longer than all recorded history, because of their long-lived radionuclides (like plutonium 239, carbon 14, and iodine 129).[2]

There are no plausible inductive arguments based on data for radwaste storage for *one* century, much less data for the minimum storage period of *thousands* of centuries. What data we do have is not comforting. Regulatory problems and safety questions have repeatedly forced the Energy Department to delay planning and opening facilities for nuclear waste. In the past, hundreds of people have

---

*Department of Philosophy, University of South Florida, Tampa, FL 33620. Shrader-Frechette's professional work is in the area of philosophy of science, technology assessment/risk assessment, and environmental ethics.

[1] P. Z. Grossman and E. S. Cassedy, "Cost-Benefit Analysis of Nuclear Waste Disposal," *Science, Technology, and Human Values* 10, no. 4 (Fall 1985): 49; hereafter cited as "Cost-Benefit Analysis."

[2] David Hawkins, *Considerations of Environmental Protection Criteria for Radioactive Waste* (Washington, D.C.: U.S. Environmental Protection Agency, 1978), p. 1.

died in radwaste accidents. The worst occurred when twenty-two square miles in Soviet Kasli were rendered uninhabitable by high-level radwaste that went critical three decades ago.[3] At the premier U.S. high-level facility, in Hanford, Washington, over 500,000 gallons of high-level waste have leaked into the soil, the Columbia River, and the Pacific Ocean.[4]

At the nuclear waste facility containing more plutonium than any other commercial site in the world, Maxey Flats, experts were wrong by six orders of magnitude when they predicted how fast the stored plutonium would migrate. They said that it would take 24,000 years for the plutonium to travel one-half inch on-site. It went two miles off-site in ten years. Current plans for future U.S. storage of high-level radioactive waste require the steel canisters to resist corrosion for as little as 300 years. Nevertheless, the U.S. Department of Energy admits that the waste will remain dangerous for longer than 10,000 years. Government experts agree that "there is no doubt that the repository will leak over the course of the next 10,000 years."[5]

The U.S. government has extrapolated, on the basis of past leaks at its nuclear waste facilities, and has said that future leaks should occur at a rate of two to three per year. Using U.S. government-estimated exposure levels (580 person rem) at each radwaste site, each existing facility could cause approximately 12 cancers and 116 genetic deaths per century,[6] and ultimately, tens of thousands of cancers per storage site.

---

[3] See, for example, J. Raloff, "Nuclear Waste Still Homeless," *Science News* 136, no. 3 (15 July 1989): 47; R. Monastersky, "More Questions Plague Nuclear Waste Dump," *Science News* 135, no. 25 (24 June 1989): 389; R. Monastersky, "Opening Delayed for Nuclear Waste Site," *Science News* 134, no. 13 (24 September 1988): 199; Zhores Medvedev, *Nuclear Disaster in the Urals*, trans. George Sanders (New York: Norton, 1979).

[4] U.S. Energy Research and Development Administration, *Final Environmental Statement: Waste Management Operations, Hanford Reservation, Richland, Washington*, vol. 1 (ERDA-1538) (Springfield, Virginia: National Technical Information Service, 1975), p. x-28; hereafter cited as "ERDA-1538."

[5] See James Neel, "Low-Level Radioactive Waste Disposal," statement in U.S. Congress before a subcommittee of the Committee on Government Operations, House of Representatives, 94th Congress, Second Session, 23 February, 12 March, and 6 April 1976 (Washington, D.C.: U.S. Government Printing Office, 1976), p. 258. See also U.S. Geological Survey, vertical file, "Maxey Flats: Publicity" (Louisville, Kentucky: Water Resources Division, U.S. Division of the Interior, n.d.); hereafter cited as U.S.G.S.-P. (The Louisville office of the U.S.G.S. is responsible for monitoring the Maxey Flats radioactive facility.) Finally see A. Weiss and P. Columbo, *Evaluation of Isotope Migration—Land Burial*, NUREG/CR-1289 BNL-NUREG-51143 (Washington D.C.: U.S. Nuclear Regulatory Commission, 1980), p. 5. For future U.S. radwaste storage and requirements, see R. Monastersky, "The 10,000-Year Test," *Science News* 133, no. 9 (27 February 1988): 139–41. See also G. Hart, "Address to the Forum", U.S. EPA, *Proceedings of a Public Forum on Environmental Protection Criteria for Radioactive Wastes* (ORP/CSD-78-2) (Washington, D.C.: U.S. Government Printing Office, May 1978), p. 6.

[6] U.S. ERDA, "ERDA-1538," vol. 1, pp. x-74. See chap. 2, notes 13 and 16; vol. 1, pp. ii, 1-57. U.S. Atomic Energy Commission, *Comparative Risk-Cost-Benefit Study of Alternative Sources of Electrical Energy* (WASH-1224) (Washington, D.C.: U.S. Government Printing Office, December 1974), pp. 3–83. See also I. Amato, "Dangerous Dirt: An Eye on DOE," *Science News* 130, no. 14 (4 October 1986): 221; hereafter cited as "DOE."

Perhaps because they are worried about human error and about scientists' claims to store waste safely in perpetuity, virtually no one wants it in his or her backyard. Over the past two years, Congress has been besieged with more than thirty bills proposing to delay, abandon, or change the repository program established under the 1982 U.S. Nuclear Waste Policy Act.[7]

A REVISIONIST HISTORY OF COMMERCIAL NUCLEAR FISSION

Standard accounts of nuclear history are often misleading regarding the source and the magnitude of radwaste. Some persons have claimed that most nuclear waste arises (1) from military activities and important hospital uses of nuclear medicine and (2) because a number of utilities were eager to provide inexpensive electricity. Both these myths are untrue.

High-level radwaste is, for the most part, spent fuel rods from fission reactors and residues from fuel reprocessing.[8] Less than one percent of high-level radwaste is from medical activities.[9] Moreover, at least in the U.S., approximately half of the high-level radwaste now needing storage is from commercial nuclear fission, not military activities.[10] The commercial half of the waste, primarily spent fuel, is expected to rise dramatically by 1995, to eleven times the metric tons that now need to be stored, while the military waste will increase very slowly and remain close to current levels. By 1995, most high-level and low-level radwaste will be from commercial reactors, not military activities, and certainly not medical processes.[11]

Nor do we have the problem of nuclear waste because industry was eager to generate electricity, and fission was an economical means of doing so. At the beginning of the atomic era, industry was reluctant, both on economic and on safety grounds, to use fission to generate electricity. Worried about safety, every major U.S. corporation with nuclear interests refused to generate nuclear electricity unless some indemnity legislation was passed to protect them in the

---

[7] J. Raloff and I. Peterson, "Trouble With EPA's Radwaste Rules," *Science News* 132, no. 5 (1 August 1987): 73.

[8] J. P. Murray, J. J. Harrington, and R. Wilson, "Chemical and Nuclear Waste Disposal," *The Cato Journal* 2, no. 2 (Fall 1982): 569; hereafter cited as "Waste Disposal."

[9] Sierra Club, *Low-Level Nuclear Waste: Options for Storage* (Buffalo: Sierra Club Radioactive Waste Campaign, 1984), p. 2.

[10] J. M. Deutch and the Interagency Review Group on Nuclear Waste Management, *Report to the President* (T1D-2817) (Springfield, Virginia: National Technical Information Service, October 1978); hereafter cited as *Report*.

[11] D. MacLean, "Introduction," to D. Bodde and T. Cochran, "Conflicting Views on a Neutrality Criterion for Radioactive Waste Management," University of Maryland, College Park, Center for Philosophy and Public Policy, 23 February 1981, p. 3. See also IRG, *Report*, pp. D-11; D-12; D-14, and D-19.

event of a major accident.[12] Commercial nuclear fission began, and was pursued, only because government hoped to justify continuing military expenditures in nuclear-related areas and to obtain weapons-grade plutonium.[13] Moreover, at least in the U.S., nuclear fission began only because government provided more than $100 billion in subsidies (for research, development, waste storage, and insurance) to the nuclear industry. It also gave the utilities a liability limit (the Price-Anderson Act) that protects licensees from ninety-nine percent of all claims in the event of a catastrophic nuclear accident.[14]

Twenty years after commercial fission reactors began operating, in 1976, the *Wall Street Journal* proclaimed them an economic disaster. Nuclear electricity has proved so costly that year-2000 projections for commercial fission reactors are now approximately one-eighth of what they were in the mid-seventies. No new reactors have been ordered in the U.S. since 1974.[15]

The few U.S. nuclear manufacturers that are still in business have survived by selling reactors to other nations, often developing countries. Yet, as I argue below, many of the commercial reactors going to these nations may not be in their best interests. Indeed, commercial nuclear fission may be the current version of infant formula. In the last two decades, U.S. and multinational corporations made great profits by exporting infant formula to developing nations, but they were able to do so only by coercive sales tactics and by misleading foreign consumers.

Some diplomats also have charged that developing nations are seeking fission-generated electricity as a subterfuge for obtaining weapons capability,[16] through the plutonium byproduct. India exploded its first nuclear bomb, for example, by using plutonium from a reactor exported by Canada. Whether or not this is the reason for the survival of nuclear fission, it is not obviously a safe, inexpensive way to boil water and run a turbine.

RADWASTE STORAGE: TECHNICAL, EPISTEMOLOGICAL, AND ETHICAL PROBLEMS

With this revisionist nuclear history behind us, let's look at some of the main technical problems posed by radwaste, especially high-level radwaste. Each year, each 1000-megawatt reactor discharges about 25.4 metric tons of high-

---

[12] W. S. Caldwell et al., 'The "Extraordinary Nuclear Occurrence" Threshold and Uncompensated Injury Under the Price-Anderson Act,' *Rutgers-Camden Law Journal* 6, no. 2 (Fall 1974): 379. See also K. Shrader-Frechette, *Nuclear Power and Public Policy* (Boston: D. Reidel, 1983), pp. 10–11; hereafter cited as *NPPP*.

[13] Cited by Sheldon Novick in a taped interview with Carl Walske in *The Electric War* (San Francisco: Sierra Club Books, 1976), pp. 32–33. See also Shrader-Frechette, *NPPP*, pp. 8–9.

[14] See Shrader-Frechette, *NPPP*, pp. 75–81.

[15] Christopher Flavin, *Nuclear Power: The Market Test* (Washington, D.C.: Worldwatch Institute, 1983), p. 33.

[16] A. Lovins and L. Lovins, *Energy/War* (San Francisco: Friends of the Earth, 1980), chaps. 2–3.

level waste, spent fuel.[17] For 300 commercial reactors, worldwide, the annual high-level radwaste would be 7,620 metric tons per year. Compare this to the fact that ten micrograms of plutonium is almost certain to induce cancer, and that several grams of plutonium, dispersed in a ventilation system, are enough to cause thousands of deaths.[18] Moreover, each of the 7,620 metric tons of high-level waste produced annually has the potential to cause hundreds of millions of cancers for at least the first 300 years of storage, and then tens of millions of cancers for the next million years.[19]

These cancers could be prevented, of course, with perfect isolation of the wastes for a million years. But the U.S. Environmental Protection Agency (EPA) has warned that we cannot count on institutional safeguards for the waste beyond one hundred years.[20] Moreover, there are geological and hydrological problems with all forms of storage.[21] The famous U.S. Interagency Review Group on nuclear-waste management reported that the scientific feasibility of dry storage in geologic repositories, deep in salt beds, or hard rock, "remains to be established."[22] As a result, granite storage sites in Sweden have been vetoed as unsafe, Kansas salt beds have been rejected because they are riddled with holes, and the first model U.S. repository for high-level radwaste will not be ready before 2008 or later.[23]

In the absence of proof that we can successfully store radwaste, doing so requires a great gamble—that our descendants will not breach the repositories through war, terrorism, or drilling for minerals; that water and heat will not combine to create nuclear reactors in underground waste, as already happened in the USSR; and a gamble that ice sheets and geological folding will not uncover the wastes. Nuclear proponents who have ignored the waste problem have been like contractors who built houses without toilets, and then alleged that constructing the toilets would be easy.[24] Since no country has a permanent high-level disposal facility,[25] perhaps the task is not so easy as has been alleged.

Most of the epistemological difficulties with radwaste arise from the fact that secure storage cannot be guaranteed. Hence, regardless of the technology used, anyone who favors a particular method of radwaste management must use some

---

[17] E. Winchester, "Nuclear Wastes," *Sierra*, July-August 1979.
[18] Grossman and Cassedy, "Cost Benefit Analysis," p. 48.
[19] Murray, Harrington, and Wilson, "Waste Disposal," p. 586.
[20] Hawkins, *Considerations*, pp. 27–29.
[21] Winchester, "Nuclear Wastes." See also R. Schneider and J. Trask, *U.S. Geological Survey Research in Radioactive Waste Disposal—Fiscal Year 1982*, Water Resources Investigations Report 84-4205 (Reston, Va.: U.S. Geological Survey, 1984), p. 38.
[22] Winchester, "Nuclear Wastes."
[23] Mark Crawford, "DOE, States Reheat Nuclear Waste Debate," *Science* 230, no. 4722 (11 October 1985): 151.
[24] Flavin, *Nuclear Power: The Market Test*, p. 31.
[25] Cynthia Polluck, *Decommissioning: Nuclear Power's Missing Link* (Washington, D.C.: Worldwatch Institute, 1986), p. 13.

form of the fallacy of the appeal to ignorance. Namely, "I know of no way in which containment could be breached; therefore, containment will probably not be breached." In the past, we were wrong to use an appeal to ignorance when we dumped radwaste into the sea, when we treated mastitis with radiation, when we used X-rays to determine shoe fit, and when we subjected U.S. soldiers to nuclear-test fallout during peacetime. We may likewise be wrong to use an argument from ignorance to justify storage of radwastes.

Another epistemological problem is how we can completely isolate the radwaste from the biosphere and yet monitor it to insure that containment has not been breached. Complete isolation appears to preclude adequate monitoring, and adequate monitoring appears to preclude isolation. Also, how can we guarantee the so-called "neutrality criterion," that the levels of risk to which future generations will be subjected, because of the waste, will be no greater than those of present persons?[26] Because of the absence of good inductive evidence, any suggestion that this criterion can be met amounts to an argument from ignorance.[27]

Most of the ethical problems associated with radwaste focus on the issue of equity. Kasperson places them in three main groups: locus problems, legacy problems, and labor-laity problems. The locus issues have to do with where and how to site radwaste facilities. The labor-laity problems focus on whether to maximize the safety of the public or that of radwaste workers, since both cannot be accomplished at once. The legacy problems concern exporting radwaste risks to future generations.[28]

The key question raised by legacy concerns is whether one can justify intergenerational inequity by mortgaging the future, by imposing our debts of radwaste on subsequent generations. If we saddle our descendants with our medical and financial debts of waste, then taxpayers in later centuries could be forced to pay an annual tab for radwaste storage between $3.8 and $1.9 million per reactor per year.[29] This expenditure is obviously questionable because future generations ought not be saddled with other persons' debts. Also, there is little public funding by taxpayers for decentralized energy technologies, like solar, which are less likely to burden the future.[30]

A second legacy question is what sort of criteria might justify environmentally irreversible damage to the environment, like that caused by deep-well storage of high-level radwaste. Radwaste management schemes which are irreversible

---

[26] See note 11.
[27] Polluck, *Decommissioning*, p. 15.
[28] R. E. Kasperson, *Equity Issues in Radioactive Waste Management* (Cambridge, Mass.: Oelgelschlager, Gunn, and Hain, 1983).
[29] Shrader-Frechette, *NPPP*, pp. 57–58.
[30] J. Berger, *Nuclear Power* (Palo Alto, Calif.: Ramparts, 1976), p. 150.

theoretically impose fewer management burdens on later generations, but they also preempt future choices about how to deal with the waste. Schemes which are reversible allow for greater choices for future generations, but they also impose greater management burdens.

Perhaps the most important legacy question concerns the contribution of radwaste production and storage to the "plutonium economy" which is necessary for building nuclear weapons.[31] Still other legacy questions have to do with the use of social discount rates. Any alleged economies or safety claims associated with high-level radwaste storage are in large part questionable because of their dependence on a particular discount rate. Using a discount rate amounts to discounting future costs (of radwaste storage, like radiation-related deaths or injuries) at some rate of x percent per year. Thus, at a discount rate of ten percent, effects on people's welfare twenty years from now count only for one-tenth of what effects on people's welfare count for now. With a discount rate of five percent, effects next year count for 1000 times more than effects 200 years from now. Or, more graphically, with a discount rate of five percent, a billion deaths in 400 years counts the same as one death next year. Yet, it is not obvious that the moral importance of future events, like the death of a person, declines at some x percent per year.[32] Without discounting, however, it would be impossible to justify the dangers and costs of storing radwaste for centuries.

Imposing nuclear waste on future generations might also be questionable from a practical point of view. Uranium will not be available much beyond the year 2000, even to supply existing fission reactors. Hence, after having generated tons of highly toxic waste, nuclear energy (without the breeder reactor and without fusion) will not have provided a long-term technological fix for our energy problems. From a practical point of view, it is not clear that the *temporary* benefits of nuclear fission are worth the *permanent* costs of radwaste.[33]

In addition to the legacy issues, there are locus or siting problems associated with managing radioactive waste. One of the key difficulties here is vesting, allowing a company to obtain a return on its initial capital investment in a

---

[31] See K. Shrader-Frechette, "Nuclear Arms and Nuclear Power: Philosophical Connections," in M. Fox and L. Groarke, eds., *Nuclear War: Philosophical Perspectives* (New York: Peter Lang Publishers, 1985); S. Cohen, *Arms and Judgment* (Boulder: Westview, 1989); and K. Kipnis and D. Meyers, eds., *Political Realism and International Morality* (Boulder: Westview, 1987).

[32] Derek Parfit, "Energy Policy and the Further Future," University of Maryland, College Park, Center for Philosophy and Public Policy, 23 February 1981, pp. 1–19, especially p. 1; H. S. Burness, "Risk: Accounting for the Future," *Natural Resources Journal* 21, no. 4 (October 1981): 723–34; Grossman and Cassedy, "Cost Benefit Analysis," p. 51.

[33] See P. L. Joskow, "Commercial Impossibility, the Uranium Market, and the Westinghouse Case," *Journal of Legal Studies* 6, no. 1 (January 1977): 119–76.

radwaste site. Are there ethical grounds for limiting property rights, even when such limitations fly in the face of current vesting doctrine?[34]

Still other, and even more far-reaching, locus questions arise because of the emergent field of land ethics.[35] How does one justify siting any land for radwaste storage? This is a use which amounts to exercising the most extreme form of property rights, since it preempts both present and future choices about land use at that site. Another important siting or locus issue is geographical equity, especially since it is a foregone conclusion that radwaste depositories will be located in rural areas, away from major population centers.[36] Is it fair to impose a risk on a person just because she lives in a rural community rather than a large city? Likewise, is it ethical for one geographical subset of persons to receive the benefits of nuclear-generated electricity, while a much larger set of persons bears the costs? Despite the fact that utilities contribute ($1 for every 1000-kilowatt hours of electricity that they generate) toward a radwaste management fund,[37] these contributions cover only a small fraction of storage costs. The bulk of storage expenditures comes from the $100 billion in nuclear subsidies already spent by U.S. taxpayers, part of which is for nuclear waste storage.[38] In the past decade, for example, government subsidies to the nuclear industry, in the form of write-off for capital invested in plants not completed, was $4 billion in the U.S. alone.[39]

Once taxpayer subsidies for costs like waste storage and decommissioning are included in the calculations, nuclear power can be shown to be more expensive than every other energy alternative.[40] The only way to make it viable is to remove it entirely from the discipline of the market and to increase taxpayer subsidies.[41] Already this removal from the market has resulted in taxpayers and members of future generations picking up the tab for billions of dollars of waste storage that should be borne by the nuclear industry and its beneficiaries.

---

[34] See Mark Sagoff, "Property Rights and Environmental Law," *Philosophy and Public Policy* 8, no. 2 (Spring 1988): 9–12.

[35] See previous note. See also L. C. Becker, *Property Rights* (Boston: Routledge and Kegan Paul, 1977).

[36] P. J. Leahy, U.S. Senator, Vermont, "The Socioeconomic Effects of a Nuclear Waste Storage Site on Rural Areas and Small Communities," statement before the Subcommittee on Rural Development, Committee on Agriculture, Nutrition, and Forestry, U.S. Senate, Washington, D.C., 26 August 1980 (Washington, D.C.: U.S. Government Printing Office, 1980), p. 2.

[37] Flavin, *Nuclear Power: The Market Test*, p. 31.

[38] Berger, *Nuclear Power*, pp. 94–97, 106–112, 144–47. See also J. Gofman and A. Tamplin, *Poisoned Power* (Emmaus, Pa.: Rodale Press, 1971), pp. 177, 199; J. Primack and F. Von Hippel, "Nuclear Reactor Safety," *Bulletin of the Atomic Scientists* 30, no. 8 (October 1974): 7; E. Muchnicki, "The Proper Role of the Public in Nuclear Power Plant Licensing Decisions," *Atomic Energy Law Journal* 15, no. 2 (Spring 1973): 45.

[39] Flavin, *Nuclear Power: The Market Test*, p. 41.

[40] Ibid., esp. pp. 1–33. See also Saunders Miller, *The Economics of Nuclear and Coal Power* (New York: Praeger, 1976), p. 105. Finally, see Shrader-Frechette, *NPPP*, p. 123.

[41] Flavin, *Nuclear Power: The Market Test*, p. 42.

PHILOSOPHICAL DILEMMAS

Rather than discuss each of these ethical problems, a task beyond the scope of a short paper, I now examine four classical philosophical dilemmas posed by radwaste storage. I call them the consent dilemma, the federalism dilemma, the threshold dilemma, and the contributor's dilemma. The consent dilemma is that siting radwaste facilities and employing waste management workers requires the consent of those put at risk; yet those most able to give free, informed consent are usually unwilling to do so, and those least able to validly consent are often willing to give alleged consent.

To see how the consent dilemma arises, consider a typical case. When West Valley, New York was proposed as a storage site for low-level radioactive waste, townspeople were eager to obtain the economic benefits they believed the facility would bring. City leaders predicted a "boomtown," and this prediction was the basis of community acceptance of the site. Although the boomtown never occured, the facility paid twenty percent of the town and county taxes. From a superficial perspective, it might look as if West Valley citizens gave free, informed consent to the waste site. But consider what sorts of communities would be most likely to consent to a radwaste facility nearby. Probably not those whose residents had high incomes, job security, and high levels of education. Communities full of people with these characteristics would know enough to be wary of the risk, and they probably would already have excellent jobs. Hence, they would not need to take any risks in order to better themselves economically.[42]

Or consider the case of workers at another radwaste facility. A British study of 35,000 living workers from the Hanford high-level radwaste facility showed increased chromosomal damage in workers exposed to less than one-half of the annual allowable exposure of five rems of radiation.[43] Workers at the facility were justified in having these risks imposed on them, according to classical economic theory, because of an alleged compensating wage differential. According to the theory behind the alleged differential, the riskier the occupation, the higher the wage required to compensate the worker for bearing the risk, all things being equal.[44]

According to classical ethical theory, imposition of these higher workplace

---

[42] See R. E. Kasperson, "The Socioeconomic Effect of a Nuclear Waste Storage Site on Rural Areas and Small Communities," Hearing before the Subcommittee on Rural Development of the Committee on Agriculture, Nutrition, and Forestry, U.S. Senate, 96th Congress, Second Session, 26 August 1980 (Washington, D.C.: U.S. Government Printing Office, 1980), pp. 61–62. For substantiation of the claim about those that consent to locating a risky facility nearby, see note 45.
[43] See Amato, "DOE"; see also Dr. Alice Stewart's study cited in Winchester, "Nuclear Wastes."
[44] Flavin, *Nuclear Power: The Market Test*, esp. pp. 40–42.

risks is legitimate apparently only after the worker consents, with knowledge of the risks involved, to perform the work for the agreed-upon wage. The dilemma arises once one considers who is most likely to give legitimate informed consent. It is a person who is well educated and adequately informed about the risk, especially its long-term and probabilistic effects. It is a person who is not forced under dire financial constraints to take a job which he knows is likely to harm him. Yet, sociological data reveals that, as education and income rise, persons are less willing to take risky jobs, and those who do so are primarily those who are poorly educated or financially strapped. Sociological data also shows that the alleged compensating wage differential does not operate for poor, unskilled, minority, or non-unionized workers. Yet these are precisely the persons most likely to work at risky jobs like storing radwaste.[45] As a result, the very set of persons *least* able to give free, informed consent to workplace radwaste risks are precisely those who most *often* work in risky jobs and are alleged to have given consent.

If this observation about worker radwaste risk and community acceptance of a radwaste site is accurate, then medical experimentation may have something to teach us about risk assessment. We know that the promise of early release for a prisoner who consents to risky medical experimentation provides a highly coercive context which could jeopardize his legitimate consent. Likewise, high wages for a desperate worker who consents to take a risky job provides a highly coercive context which could jeopardize his legitimate consent. We must, therefore, either admit that our classical ethical theory of free, informed consent is wrong or we must question whether those closely affected by radwaste risk genuinely consent to it.

Consider now a second classic dilemma. Liberty and grass-roots self-determination require local control of whether a radwaste facility is sited in a particular area. Yet, equality of consideration for people in all locales and the minimization of overall risk require federal control. Do we say that the local community can veto a radwaste site, even though that site may be the best in the country and may provide for the most equal protection of all people? Or do we say that the national government can impose a radwaste site on a local community, even though the imposition is at odds with their free and self-determined choice?[46] How do we resolve this dilemma? Noted law professor, R. B. Stewart, claims that it is impossible to maximize both liberty and environmental integrity, or both

---

[45] K. Shrader-Frechette, *Risk Analysis and Scientific Method* (Boston: D. Reidel, 1985), pp. 107–12; hereafter cited as *RASM*. K. Shrader-Frechette, *Science Policy, Ethics, and Economic Methodology* (Boston: D. Reidel, 1985), p. 137; hereafter cited as *SP*.
[46] Shrader-Frechette, *SP*, p. 214.

liberty and equality.[47] The dilemma cannot be resolved in the sense of maximizing all three values.[48]

A third problem, the threshold dilemma, is based on the fact that society must declare some threshold below which risk is declared to be negligible or minimal, so far as its acceptability is concerned. Typically this threshold level is set at what would cause less than a 10^{-6} increase in one's average annual probability of fatality.[49]

The reasoning behind setting such a level is that a zero-risk society is impossible, and some standard needs to be set, especially in order to determine pollution-control expenditures. Choosing the 10^{-6} standard also appears reasonable, both because society must attempt to reduce larger risks first, and because 10^{-6} is the natural hazards death rate.[50]

The dilemma arises because no threshold standard is able to provide *equal* protection from harms like radwaste to all citizens. Any such standard guarantees merely that an *average* annual probability of fatality is associated with some hazard. Because this 10^{-6} threshold seems acceptable on the average, however, does not mean that it is acceptable to each individual. Blacks, for example, face higher risks from air pollution than do whites, even though they share the same "average" exposure.[51] Likewise, those around radwaste facilities are exposed to radiation levels for average persons, not for them as individuals with unique needs.

Most civil rights, however, are not accorded on the basis of the *average* needs of persons, but on the basis of *individual* characteristics. For example, we do not accord constitutionally guaranteed civil rights to public education on the basis of

---

[47] Ibid., p. 215.

[48] Ibid., p. 214. See R. B. Stewart, "Pyramids of Sacrifice? Problems of Federalism in Mandating State Implementation of National Environmental Policy," in F. A. Strom, ed., *Land Use and Environment Law Review—1978* (New York: Clark Boardman, 1978), pp. 162–63. See also R. B. Stewart, "Paradoxes of Liberty, Integrity, and Fraternity: the Collective Nature of Environmental Quality and Judicial Review of Administrative Action," *Environmental Law* 7, no. 3 (Spring 1977): 472. Finally, see M. Markovich, "The Relationship between Equality and Local Autonomy" in W. Feinberg, ed., *Equality and Social Policy* (Urbana: University of Illinois Press, 1978), p. 96.

[49] See Shrader-Frechette, *RASM*, sec. 3.3.2 of chap. 2. See C. Zracket, "Opening Remarks," in Mitre Corporation, *Symposium/Workshop . . . Risk Assessment and Governmental Decision Making* (McLean, Va.: Mitre Corporation, 1979), p. 3 (The Mitre publication will hereafter be cited as *Symposium/Workshop*). See C. Starr, "Benefit-Cost Studies in Socio-technical Systems," in Committee on Public Engineering Policy, ed., *Perspectives on Benefit-Risk Decision Making* (Washington, D.C.: National Academy of Engineering, 1972); D. Okrent and C. Whipple, *Approach to Societal Risk Acceptance Criteria and Risk Management*, PB-271 264 (Washington D.C.: U.S. Department of Commerce, 1977); A. P. Hull, "Discussion," in Mitre, *Symposium/Workshop*, pp. 171–72; and N. Rescher, *Risk: A Philosophical Introduction* (Washington D.C.: University Press of America, 1983), pp. 35–40.

[50] C. Starr and C. Whipple, "Risks of Risk Decisions," *Science* 208 (6 June 1980): 1114–19. See also C. Starr, *Current Issues in Energy* (New York: Pergamon, 1979), p. 15.

[51] Shrader-Frechette, *SP*, p. 134. See also Kasperson, "Socioeconomic Effect," pp. 62–63.

average characteristics of students. If we did, then retarded children or gifted children would have rights only to education for children at the average level. Instead, we say that according "equal" civil rights to education means according "comparable education," given one's aptitudes and needs. That is why the state can provide special schools for both the retarded and the gifted.

This example from the field of education raises an interesting question for radwaste risks and indeed for all risks requiring a threshold to be set: if civil rights to education are accorded on the basis of individual, not average, characteristics, then why are civil rights to equal protection from risks not accorded on the basis of individual, rather than average, characteristics? Why is a 10^{-6} average threshold accepted for everyone, without compensation, when adopting it poses risks higher than 10^{-6} for the elderly, for children, for persons with previous exposures to radiation, for those with allergies, for persons who must lead sedentary lives, and for the poor? Which should we choose: average protection or equal protection, efficiency in regulation or equality of protection?

A fourth difficulty is what I call the contributor's dilemma. It arises out of the fact that citizens are subject to numerous small risks, e.g., to certain carcinogens, each of which is allegedly acceptable; yet, together such exposures are clearly unacceptable. In the case of radioactive waste, selected groups of citizens, such as those living near a storage facility or a radwaste transport route, are exposed to radiation. Each of these small exposures is alleged to be acceptable because it is below the threshold at which some statistically significant increase in harm occurs; yet together these exposures can cause serious damage.

Statistically speaking, twenty-five to thirty-three percent of us are going to die from cancers. The U.S. Office of Technology Assessment says that ninety percent of these cancers are environmentally induced and hence theoretically preventable.[52] Many of the cancers are obviously caused by the aggregation of numerous exposures to carcinogens, like radiation, no one exposure of which is alone alleged to be harmful. The contributor's dilemma is especially problematic in cases involving the devising of ethical regulations for small risks, like radiation. Risk assessors who condone sub-threshold risks, but who condemn the deaths caused by the aggregate of these sub-threshold risks, are something like the bandits who eat the tribesmen's lunches in the famous story of Jonathan Glover:

> Suppose a village contains 100 unarmed tribesmen eating their lunch. One hundred armed bandits descend on the village and each bandit at gun-point takes one tribesman's lunch and eats it. The bandits then go off, each having done a

---

[52] J. C. Lashof, et al., Health and Life Sciences Division of the U.S. Office of Technology Assessment (OTA), *Assessment of Technologies for Determining Cancer Risks from the Environment* (Washington D.C.: U.S. Office of Technology Assessment, 1981), pp. 3, 6.

discriminable amount of harm to a single tribesman. Next week, the bandits are tempted to do the same thing again, but are troubled by new-found doubts about the morality of such a raid. Their doubts are put to rest by one of their number [a government assessor]. . . . They then raid the village, tie up the tribesmen, and look at their lunches. As expected, each bowl of food contains one hundred baked beans. . . . Instead of each bandit eating a single plateful as last week, each [of the one hundred bandits] takes one bean from each plate. They leave after eating all of the beans, pleased to have done no harm, as each has done no more than sub-threshold harm to each person.[53]

The obvious question raised by this example is how a risk assessor can say that sub-threshold radwaste exposures are harmless, as the data allegedly indicates, and yet that the additivity, or contribution, of these doses causes great harm. It appears that risk regulators need to amend their theory regarding synergistic or additive risks like cancer.

TWO SOLUTIONS BY ETHICS: CONSENT AND COMPENSATION

Although there is inadequate space here to defend the sorts of solutions that would help resolve these four dilemmas, it is possible to outline briefly the arguments which, when developed, would provide such a defense. One solution to the consent dilemma and to the federalism dilemma is to apply the same strict standards of informed consent, already well-known in doctor-patient or experimenter-patient relationships, to the radwaste siting issue.

We all probably believe that it would be unethical for a doctor to impose a risky treatment on a patient without her free, informed consent. Yet most of us need to develop our moral sensibilities so as to see radwaste workers or radwaste siting in the same light. Until or unless a risk imposition receives the consent of those who are its potential victims, it cannot be justified. This means that those who wish to impose societal risks on others need to do whatever is necessary to compensate them to a degree adequate to obtain their free, informed consent. If no compensation is adequate to obtain free, informed consent, then it is questionable whether the risk imposition is justified.

Whenever the federalism and consent dilemmas have arisen in radwaste siting issues, these have often been resolved by means of increasing the compensation to the risk takers likely to be affected by the waste site. In some cases in which it was not possible to meet the compensatory demands of a community proposed as a radwaste site, the facility was not located. In the cases in which compensation was able to be negotiated, however, often this was sufficient to insure community consent. Typically industry-supplied compensation in radwaste cases includes

---

[53] Quoted by Derek Parfit, *Reasons and Persons* (Oxford: Clarendon Press, 1984), p. 511.

tax breaks or funding community projects or schools. According to social scientists (studying waste siting), however, one form of compensation dominates all negotiation. The one factor almost always essential to achieving local consent is giving citizens/workers funding to control health and safety monitoring at the facility. By forcing the waste managers to pay for outside monitors, citizens and workers are freed from relying on company monitoring. Once they have greater control of their safety, they appear ready to give informed consent to the risk.[54]

Compensation also provides one solution to the threshold dilemma and to the contributor's dilemma, both of which raise equity issues. By providing full compensating benefits for all unavoidable radwaste risks, industry can offset the inequities generated by some of the locus and legacy problems.[55] Admittedly, compensation of future generations who bear the brunt of the legacy problems might be difficult. Obtaining free, informed consent and guaranteeing compensation in such cases requires that the consent of future persons be obtained by means of representatives acting in their best interests. It also requires that those who store radwaste actually set up a fund for compensation of future persons possibly harmed by the waste. To the degree that we cannot guarantee that persons a million years from now will have equal opportunity to protect themselves from our radwaste hazards, then to that same degree we cannot justify generating radwaste or imposing it upon others.[56] The rationale for requiring full compensation and consent, even regarding future persons, is that any technology ought to "pay its own way." If nuclear power cannot "pay its own way," especially in regard to waste storage, it will reinforce all the old income distributions and inequities in which the poor and the uneducated bear the social costs of contemporary technology.

Also, if someone can impose a bodily risk of harm on another without his consent, then there is no right to life. If someone can profit by imposing a threat of physical harm on another without compensating him, then there is no right to due process. These considerations suggest that, if we refrain from requiring genuine informed consent and from guaranteeing complete compensation for radwaste risks, then radwaste will put more at risk than our health and that of future generations. It will put at risk our most basic rights to justice and equity.

---

[54] E. Peele, "Innovative Process and Inventive Solutions," unpublished manuscript, 1986, p. 6 (available from Peele at the Energy Division, Oak Ridge National Laboratory, Oak Ridge, Tennessee 37831); hereafter cited as Peele, "Innovative Process." See also E. Peele, "Mitigating Community Impacts of Energy Development," *Nuclear Technology* 44 (1979): 132–40, esp. pp. 133–36.

[55] W. A. O'Connor, "Incentives for the Construction of Low-Level Nuclear Waste Facilities," in *Low Level Waste*, Final Report of the National Governors Association Task Force on Low-Level Radioactive Waste Disposal, Washington D.C., 1980. E. Peele, "Siting Strategies for an Age of Distrust," unpublished essay (available from Peele at the address listed in note 54); finally, see Peele, "Innovative Process."

[56] Bodde and Cochran, "Radioactive Waste Management," p. 7.

THESE ETHICAL SOLUTIONS MAKE ECONOMIC SENSE

Providing equity, informed consent, and full compensation regarding the risks associated with managing radwaste requires that we forgo use of nuclear power and the generation of additional radwastes. Avoiding additional nuclear plants, however, is also an economically sound decision, for a variety of reasons. Virtually all commercial nuclear construction in developed nations has come to a halt, largely because fission generation of electricity is uneconomical and likely unsafe. France is the only developed country with an ongoing nuclear program, allegedly the most successful in the world.

A closer look at the French situation, however, reveals that it is questionable. For one thing, the French use a breeder reactor, not the fission technology employed in other energy programs around the world. Most nations have decided against the breeder because it generates inordinate amounts of radioactive waste—materials for nuclear weapons—and hence creates a "plutonium economy." Moreover, the apparent reasons for the success of the French nuclear program—the fact that it is centralized and government owned—are artificial and not transferable to other nations. The French utility is protected from market forces, protected from environmentalists' criticisms, and protected from public participation in decision making. Even with these benefits, the French utility now has a debt of $30 billion, due to its commitment to nuclear power. In 1982 this debt had accumulated to $152 billion, although part of it was forgiven by the government. In light of the French deficit, developing nations, many of which are on the edge of insolvency, will be hard pressed to pay the enormous capital costs for nuclear reactors.[57] Developing nations, in particular, will probably be hard pressed to pay the costs of decommissioning and waste storage associated with their approximately thirty nuclear reactors. Some experts claim that waste storage costs will add an estimated five to ten percent to the total cost of nuclear power.[58] U.S. industry experts maintain that the decommissioning of a reactor could amount to twenty-four percent of the original construction costs. Such an estimate translates to more than $500 million for recently built facilities. If French nuclear-industry experts are correct, then decommissioning currently would run at least forty percent of initial construction costs of the reactor.[59]

Apart from paying for decommissioning the reactor and managing the waste, most nations (and especially developing ones) will likely have a difficult time paying for nuclear power at all. Since the mid-seventies, nuclear-plant construction costs have doubled every four years. More than twenty-five percent of these

---

[57] Polluck, *Decommissioning*, p. 27. Flavin, *Nuclear Power: The Market Test*, pp. 45–47.
[58] Flavin, *Nuclear Power: The Market Test*, p. 31.
[59] Polluck, *Decommissioning*, p. 27.

costs are for financing.[60] Part of the reason for these increases is that, because of additional safety measures needed in the wake of accidents, the amount of concrete, piping, and cable used in the average plant has more than doubled, while labor requirements have more than tripled.[61] No one in any country is going to build a nuclear plant without knowing its full costs, including decommissioning and waste storage.

A particular obstacle to nuclear plants in developing nations is the small size of electricity grids in most of them. If a single power plant provides more than fifteen percent of the grid's capacity, the whole system will "crash" if that plant is shut down. Yet only four or five developing countries have grids large enough for a conventional 1000 megawatt reactor. This problem could be addressed by plants smaller than the 1000 megawatt ones, but the per-kilowatt construction cost for a nuclear plant of 200 megawatts is more than twice that for a 1000 megawatt one, and even the larger ones are not cost-effective compared to other energy alternatives. Because of the capital intensity of nuclear power, it seems unattractive for debt-strapped developing nations.[62]

Right now small hydropower and cogeneration plants are all much cheaper than nuclear fission.[63] Hydropower is particularly attractive because, although North America and Europe have developed sixty percent and thirty-six percent of their hydropower potential, respectively, Asia has used just nine percent, Latin America eight percent, and Africa only five percent. In China, for example, 76,000 small hydropower plants supply almost 10,000 megawatts of power in rural areas. By the year 2000, cogeneration can account for ten times that amount in some countries.[64]

Biomass is also an inexpensive alternative for rural electrification. The greatest use of biomass residues is found in the relatively treeless plains of Northern India, Bangladesh, and China, where crop residues and dung provide as much as ninety percent of household energy in many villages and a considerable proportion in urban areas.[65] Likewise, in inner Mongolia, 2,000 small wind turbines are used for lighting, running television, electrifying corral fences, and projecting movies.[66]

Compared to such alternatives, nuclear power is also a questionable energy

---

[60] Flavin, *Nuclear Power: The Market Test*, p. 15.
[61] Ibid., p. 26.
[62] Ibid., p. 52.
[63] Ibid., pp. 54–59.
[64] But much of this includes reliance on natural gas and some dependence on coal. See Christopher Flavin and Alan Durning, *Building on Success: The Age of Energy Efficiency* (Washington, D.C.: Worldwatch Institute, March 1988), p. 35.
[65] Jessica Tuchman Mathews, *World Resources 1986* (New York: Basic Books, 1986), p. 111.
[66] Kosta Tsipis, "Nuclear Power and Energy Needs of the Third World Economies," *Church and Society*, Report and Background Papers, Meeting of the Working Group, World Council of Churches, Glion, Switzerland, September 1987, pp. 227, 229.

source because it creates fewer jobs and requires more dependence upon foreign companies and governments than almost any other investment a developing nation can make. Also, nuclear energy is a target for military and terrorist abuse in a politically unstable region. Besides, it is likely to serve only the minority that uses electricity. It bypasses the majority who rely on fuelwood and charcoal. All these points suggest that investment in rural electrification, using small-scale renewables, is a better way for developing nations to go.[67]

CONCLUSION

Apart from the ethical and economical reasons for avoiding the generation of radioactive waste, there might be a number of general philosophical considerations that suggest that the commercial, and not only the military, nuclear path is wrong. On this view, how we deal with the radwaste problem might be a barometer for the collective sanity and morality of our society. A thousand years ago, the world's finest architectural and engineering talents were mobilized to build cathedrals. It is ironic that comparable talents and even more skills are today dedicated to devising foolproof nuclear garbage dumps.[68]

---

[67] Flavin, *Nuclear Power: The Market Test*, pp. 54–55.
[68] Winchester, "Nuclear Wastes."

[29]

Sustainability, Human Welfare, and Ecosystem Health

BRYAN NORTON

School of Public Policy
Georgia Institute of Technology
Atlanta, GA 33032, USA

ABSTRACT: Two types of sustainability definitions are contrasted. 'Social scientific' definitions, such as that of the Brundtland Commission, treat sustainability as a relationship between present and future welfare of persons. These definitions differ from 'ecological' ones which explicitly require protection of ecological processes as a condition on sustainability. 'Scientific contextualism' does not follow mainstream economists in their efforts to express all effects as interchangeable units of individual welfare; it rather strives to express sensitivity to different types and scales of impacts that present activities can exert on the future. We can therefore express the moral obligation to act sustainably as an obligation to protect the natural processes that form the context of human life and culture, emphasizing those large biotic and abiotic systems essential to human life, health, and flourishing culture. Ecosystems, which are understood as dynamic, self-organizing systems humans have evolved within, must remain 'healthy' if humans are to thrive. The ecological approach to sustainability therefore sets the protection of dynamic, creative systems in nature as its primary goal.

KEYWORDS: Sustainability, ecological management, obligations to future, welfare, intergenerational equity, irreversibility.

The goal of 'sustainability' has emerged as a rallying cry for a broad spectrum of advocates of both environmentalism and rational development. The term sustainability was first popularized in the field of resource use, and it initially had a fairly precise application in phrases such as 'maximum sustainable yield', which represents the highest level of exploitation consistent with maintaining a steady flow of resources from a forest or fishery. Today, however, the term is used much more broadly to include, for example, levels of pollution and degradation of natural systems that are consistent with maintaining current levels of use and enjoyment of those systems. In the context 'sustainable development', it must be used in the broader sense, and hence it is in this broader sense that the term has become a shibboleth of mainstream environmentalists.

It is no doubt useful, in policy discussions, to have a term like 'sustainability', which, like 'conservation' in days of old, can stand as a label for the many activities of environmentalists. The danger is that the term, like 'conservation' before it, will become a cliché.[1] Nobody opposes it because nobody knows exactly what it entails. To avoid this trap it will be necessary for environmentalists, with the help of scientists and philosophers, to develop, explain and justify a theory of environmental practice that gives form and specificity to the goal of sustainability. In particular, what is needed is a set of principles, derivable from a plausible core idea of sustainability, but sufficiently specific to provide significant guidance in day-to-day decisions and in policy choices affecting the environment.

As a first step in giving form to the definition, it is useful to note that the term implies sustainable *use*, so it would appear to exclude severely moralistic approaches, such as positions of extreme deep ecologists who argue that the natural world ought not to be considered 'resources' for human use at all.[2] At the other extreme, advocates of unlimited economic growth, who argue that it is wrong to place any constraints on the ability of the free market to generate goods and services in response to consumer demands, would reject the implication that environmental concerns justify any constraints on the use of nature.[3] Between these extreme positions, however – and I think it is safe to say that these extreme positions have very few advocates – lie the vast majority of environmentalists,[4] who believe that use of the environment is morally acceptable, but that this use is constrained by obligations not to misuse the environment in unsustainable ways.

PART I: SUSTAINABILITY AND HUMAN WELFARE

Today, the most often-cited definition of sustainability is that of the Brundtland Commission's report, Our *Common Future*: "Sustainable development is development that meets the needs of the present without compromising the ability of future generations to meet their own needs."[5] The Commission followed this definition with a formulation of the "two key concepts" of their definition: "the concept of 'needs,' in particular the essential needs of the world's poor, to which overriding priority should be given", and "the idea of limitations imposed by the state of technology and social organization on the environment's ability to meet present and future needs".[6]

Since the exact meaning of sustainability will depend upon the specification of the 'limitations' mentioned in the second concept, it is notable that the Brundtland definition states these as determined essentially by "the state of technology and social organization". Sustainability is therefore defined as an intertemporal relationship between *human needs* and *human productive ca-*

pacities, as a relationship between human welfare at different stages of human development. While the environment is mentioned, it appears as a passive element in the equation – needs are human-determined, and limitations are seen as human limitations. The environment does not impose any non-negotiable limits on sustainable use, independent of limitations on the abilities of humans to control it. Any limitation on use of the environment may in principle be overcome by some new breakthrough in technology and social organization. Our obligation, on this view, is to balance present fulfilment of needs against the ability of future generations to fulfil their needs.

The Brundtland definition, then, can stand as characteristic of one broad approach to sustainability, which I will call the 'social scientific' approach, both because it is popular among social scientists, such as demographers and economists, and because it focuses most empirical attention on human demands and on characteristics of technical and social innovation.[7]

While the Brundtland definition was intended as a relatively 'neutral' definition, attractive to a broad range of environmentalists and developmentalists, we can now see that it may not be. The implication that there can be no insuperable shortages in resources precludes, by the very definition of sustainability, limitations imposed by characteristics of the environment itself: characteristics that might limit its ability to produce consumable goods or absorb human wastes. On the Brundtland approach, projections of economic and social growth can be calculated without accounting for the *scale* of human activities.

Intuitively, this implication that nature sets no natural limits on economic uses is implausible; it implies that no human activity will, in principle, be precluded by shortages of resources. This implication seems to contradict the obvious fact that the stocks of any given resource are finite, and that some of them, such as copper ore, are quite limited.[8] The denial of natural limits does not challenge this fact directly, however. It recognizes that stocks of non-renewable resources will decline and the price of raw resources will rise; the key to maintaining this position rests on a high degree of confidence in the intersubstitutability of resources. The finitude of copper does not cause a limit on economic growth because, as the price rises, a substitute resource will replace it. Similarly, as the cost of disposing of pollutants and wastes increase, entrepreneurs will be stimulated to develop alternative means of recycling and disposal. I am suggesting, then, that social scientific definitions of sustainability presuppose a very strong principle of intersubstitutability of resources, indeed, a Principle of Infinite Intersubstitutability (PII). This principle is inherent in the definition of sustainability as a simple balance of 'human welfare' across time. Environmentalists, I submit, will question PII. They should, therefore, be wary of attempts to define sustainability simply as a matter of human technology and welfare.[9]

It can be argued that the assumption of PII is intimately tied to the unidimensional value analysis of the mainstream economic paradigm. One will find PII plausible only if one assumes the interchangeability of labour, resources and capital, and that all value can be represented as prices in markets. Interchangeability is essential to the central idea of mainstream economics; that all choices can be understood incrementally, as consumer choices at the margin.[10] If sustainability is to be a simple problem of balancing welfare across generations, then human welfare must be understood incrementally and interchangeably as it is in mainstream economics. Provided we leave our descendants *richer* than we are, according to this analysis, we cannot have done wrong; the future can simply trade its wealth for amenities, substitutes for lost resources, or a pollution-free environment. In an incremental system of value in which all values are interchangeable and all resources have, with requisite capital, adequate substitutes, environmental constraints need be given no special pre-emptive status.

To recognize limits inherent in nature itself would be to introduce discontinuities into the analysis. If over-consumption of passenger pigeons were analysed in 1900, according to the mainstream economic paradigm, profits resulting from over-exploitation could have been deemed 'beneficial' to the future as capital capable of generating new sources of protein. If, however, one insisted that passenger pigeons represented an irreplaceable resource, one would have argued that continued consumption of squab, even as the stocks plummeted toward extinction, represented an unrecompensable harm perpetrated by one generation on subsequent ones.

It is tempting to set out to show that the economic paradigm, despite its unquestioned advantage of simplicity (in that it can represent all values on a single scale of welfare), is too simplistic to deal with questions of intergenerational equity. In particular, it could be argued that the incrementalist model of mainstream economics (which seems to be presupposed in the Brundtland definition) is ill-suited to deal with policy problems in which incremental choices can have irreversible effects that will have impacts over very long periods of time.[11] Space will not permit such an argument here. Instead, an alternative conception of intertemporal welfare will be proposed and explained. This conception, 'scientific contextualism', does not flatten out all decisions into interchangeable units of individual welfare, but instead, retains a sensitivity toward different types and scales of impacts that the present can exert on the future.

PART II: A CLASSIFICATION OF RISKS

The flattening-out approach to judging intergenerational impacts, measuring intertemporal welfare according to a single scale of present valuations, usually dollars, ignores apparently important differences in the types of impacts the present can have on the future. A broad and inclusive conception of sustainability must gauge the ability of the future to deal with pollution and waste as well as with declining stocks of resources.[12] For the sake of a convenient terminology, and because it seems reasonable to treat some present activities as creating a 'risk' of future shortages of resources and sinks for waste products, let us propose an intuitive scale for classifying types of risks that the present may impose on the future, as in Figure 1.

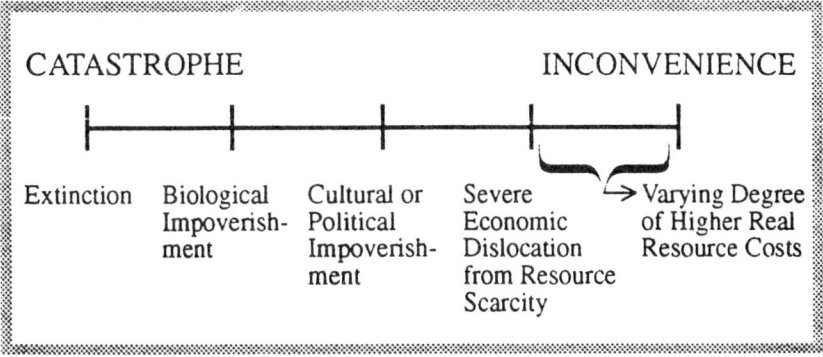

FIGURE 1.
Typology of Risk Severities

This scale recognizes the apparent difference between activities such as burning fossil fuel in great quantities, which may include considerable risk that the planet will become uninhabitable by future humans at one extreme; and less cataclysmic results, such as filling all available waste dumps, which might force future generations to give up disposable diapers and return to the old-fashioned practice of washing diapers. Because one of the apparent weaknesses of the incrementalist model is that it does not deal well with irreversibilities such as species extinctions, we can remedy this weakness by introducing a scale of comparative reversibility of present decisions. If some decisions we make today are easily reversible, then capital or know-how may be a reasonable substitute for some forms of environmental protection. Conversely, major cataclysms would be irreversible. Therefore, we can plot our intuitive scale of types of future risks against a scale of reversibility, creating a decision grid as represented in Figure 2.

BRYAN NORTON

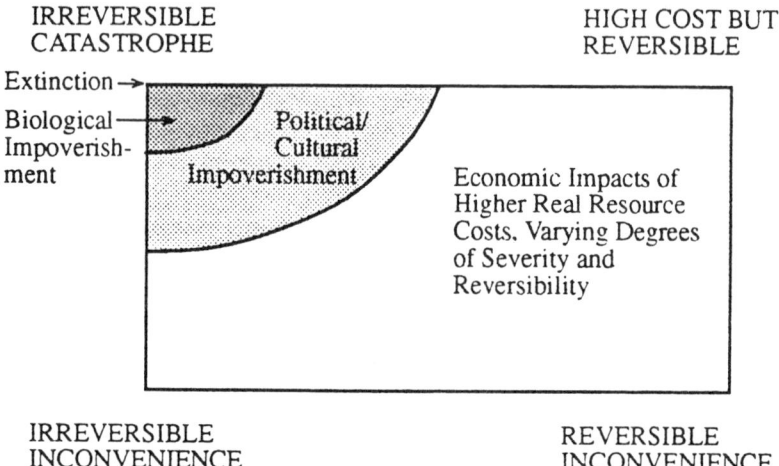

FIGURE 2.
Risk Typology: Severity and Reversibility

The decision space here defined will include, in the far east, the southeast, and the far south portion of the space, decisions that we will consider simple trade-offs. If the negative impacts of our activities on the future result only in easily reversible changes, or if the impacts result only in minor inconveniences, we can figure that we have been 'fair' to the future provided we make available capital and techniques sufficient to reverse or counterbalance those effects. Decisions that have possibly cataclysmic impacts or irreversible consequences, on the other hand, will not be open to trade-offs. Decisions that fall in the northwest portion of the decision space will be governed by non-negotiable constraints. A slightly less constraining decision rule, the Safe Minimum Standard of conservation (SMS), might be applied in this 'red' area. This rule advises: Protect the resource (natural process, species, etc.) provided the costs are bearable.[13]

I am hypothesizing, then, that our obligations will be least negotiable in the NW corner of the decision space and that these obligations will decline along all vectors away from that corner. Figure 2 also represents the above-mentioned insight that the degree to which one believes in intersubstitutability of resources will determine the relative size of the decision space governed by non-negotiable constraints. In the limiting case of a belief in PII, the decision space governed by non-negotiable constraints will be null – all risks are recompensable with adequate capital and technological development. As one's faith in intersubstitutability decreases, the northwestern space will expand, representing more and more decisions as governed by non-negotiable constraints.

We can refer to approaches to sustainability that recognize some decisions with future impacts as governed by non-negotiable constraints and some decisions governed by trade-offs as 'hybrid theories' or as 'two-tier systems'.[14] They recognize at least two measures of value that cannot be aggregated together – one set of obligations may trump another. Two-tier approaches therefore differ from the single-tier systems of micro-economics and of other utilitarian approaches, which see only interchangeable units of welfare as the measure of sustainability, by recognizing some constraints that cannot be traded off. The two-tiered approach eschews simple aggregation in favour of lexically ordered rules.

The moral status of these non-negotiable obligations in the NW corner is, of course, open to much debate. Some of my colleagues in environmental ethics would insist that these obligations be formulated as protecting values 'intrinsic' or 'inherent' in nature, positing values independent of humans.[15] Others would posit basic rights of future persons, which would trump mere consumptive interests of present persons.[16] Another understanding of non-negotiable obligations, the one to be explained here, is morally anthropocentric, but based in a holistic conception of the natural systems on which humans depend. If it turns out that nonhuman rights or rights of the not-yet-existent can be specified later, they could be added on, further strengthening the already strong obligations involved in contextualism. The point I wish to stress here is that the logic of two-tiered, hybrid systems is similar, and this sets them apart from the single scale of values approach of many social scientists.

PART III: SCIENTIFIC CONTEXTUALISM

Consider an approach to sustainability that recognizes obligations of the present generation to future generations, but views these obligations holistically – as not reducible to individual satisfactions or preferences.[17] These obligations are of the type that would be suggested by Edmund Burke's understanding of a society as "a partnership not only between those who are living, but between those who are living, those who are dead and those who are to be born",[18] and these obligations would be based on a belief that the continuance of the human species is a good thing.

Having posited such a value and cross-generational obligations not defeasible into individual satisfactions, let me now argue that, if we have such obligations – and I think we do – we can now posit an alternative approach to understanding sustainability. This approach recognizes that there are non-negotiable obligations regarding our use of resources (the NW corner of the decision space is not empty), and that those obligations can be understood as the obligations the present has to perpetuate the conditions necessary for the continuation of the human species and of its culture. The exact nature of these obligations must be determined scientifically, as we understand the impacts human activities have on their larger context. If, following Aldo Leopold's land ethic, we insist that this larger context can only be understood as a complex ecological system, sustainable activities are activities that do not destabilize the large-scale, dynamic, biotic and abiotic systems on which future generations will depend. Scientific contextualism applies a variety of moral rules, placing priority on different values in different situations. If plausible scientific models indicate that a realistic, if not necessarily probable, chain of processes could result in cataclysmic effects, we are in the 'red zone', and the SMS standard applies.

Admittedly, the information necessary to act sustainably in this sense would be very hard to obtain. But our concern here is mainly conceptual. Assume, for example, that models showing rapid and accelerating warming of the atmosphere in response to a build-up of greenhouse gases were strongly verified, and that these models showed increases, 50-100 years in the future, too rapid for civilization to survive. I believe that most people would say, once most scientific uncertainty was removed, that such a scenario would trigger non-negotiable constraints limiting current behaviour. Examples such as this are important, because they help to shift the burden of proof from those who would institute constraints on economic growth to those who would flirt with cataclysmic changes in the context of human adaptation. If there are clear examples in which non-negotiable constraints would exist, growth enthusiasts must show that their proposed activities violate no such constraints.

We know that undisturbed natural systems are able to maintain themselves across time, that they will keep their energy pathways open, and that they will maintain their productivity. Once a system is disturbed, effects cascade through the system; if those effects are of the sort that the system is used to, and can assimilate, it does so. If, on the other hand, the disturbance is so pervasive or so new that the system has no means by which to damp out its effects, the system crosses a threshold; and humans, as well as members of other species, who are adapted to systems with a given set of characteristics, may be unable to adapt to accelerating and cascading changes in their habitat. For example, humans with light skin pigmentation, a trait which evolved in a time of relatively small exposures to ultraviolet light in temperate climates, may be unable to adapt to an earth with less upper-atmosphere ozone.

In a contextual analysis, individual behaviours are not the main focus of environmental ethics – it is trends in those behaviours that determine whether they will have intergenerational impacts. For example, if one farmer cuts and clears his woodlot to plant wheat, this is not wrong, as long as his land is not on a highly erodible slope and provided his action is not copied by all of his neighbours. The moral status of this activity depends not only on the content, but also on the context, of the action. If Farmer Jones plants wheat and Farmer Smith lets his wheat field go fallow, no trend is instituted, and there is little likelihood that the action will trigger non-negotiable constraints. If most farmers follow

Jones, conditions ripe for a dust bowl or desertification may be created. Scale is crucial in determining when a red-zone decision is faced.

Expressed metaphorically, contextualism is organicism – the biota is a living system which has an internal, self-perpetuating organization – but organicism minus teleology. Contextualism need not posit a metaphysical value in the supraorganism, just as it need not posit independent value of wholes.[19] But contextualism does recognize the importance of protecting the processes sustaining self-organizing systems through time. For example, once Leopold fully understood the implications of a systems-oriented approach, he fell back upon the recommendation that we practice something akin to preventive medicine.[20]

Leopold's theory of sustainable management envisioned hierarchically organized systems, with human activities impacting them, not individually, but in larger trends.[21] Technology and population growth have given human cultures the ability to alter larger and normally slowly-changing systems of ecology, climate and atmosphere, and to initiate oscillations and fluctuations in these systems. Since we have evolved to live within systems that change slowly, such activities play Russian Roulette with the options of the future. Our generation could cause irreversible changes much too rapid for future generations to adapt to, either physiologically or culturally.

Scientific contextualism places a heavy burden on scientific models to help us determine which activities may have long-delayed, but potentially catastrophic, consequences. The contextualist paradigm of environmental management interprets the larger systems under impact from human activities in mainly ecological terms. An essential element of the contextualist approach to management will be a commitment (non-negotiable constraints) to protect the health and integrity of ecological systems. The contextualist paradigm is not, however, *simply* an ecological paradigm – as human impacts grow, biotic systems, and also atmospheric and climatological systems, are inexorably affected by the aggregated impacts of human economic and other activities.

While I agree with Leopold that these larger impacts should be understood through ecology, because humans are, after all, evolved animals who relate as living things both to the biota and to larger, abiotic systems, contextualism also recognizes the role of non-biological sciences in defining the limits of our impacts on our larger context. For that reason, I call the approach to sustainability sketched here 'scientific' contextualism.

PART IV: HEALTH, INTEGRITY, AND SUSTAINABILITY

The idea that there is an obligation to protect the health and integrity of ecological systems rests firmly upon the premise that natural systems are self-organizing in an important sense. This is a difficult concept, and it must be carefully explained and qualified – a task that can only be begun here.[22] Recognizing that natural systems change constantly, intertemporal stability is conceived as a scalar relation between human activities and their larger environmental context. Contextualism assumes that the self-organization of large systems is essential to future generations and that the ability of those large, self-organizing systems to assimilate human impacts is large, but not infinite. These systems provide the context within which we have evolved. Because they change more slowly than culture, these large systems set the 'stage' for human activities. They therefore give meaning to human culture. At the same time, the system is unquestionably dynamic. Stability only exists relative to differing scales of time; we might say that stability is a well-founded illusion.

Autonomous systems overcome entropy; autonomy is the characteristic of systems that allows self-organization. Given this operational definition of autonomy, we can define sustainability as follows. Sustainability is a relationship between dynamic human economic systems and larger, dynamic, but normally slower-changing ecological systems, such that: (a) human life can continue indefinitely; (b) human individuals can flourish; (c) human cultures can develop; but in which (d) effects of human activities remain within bounds so as not to destroy the health/integrity of the environmental context of human activities.

But how are we to define 'health' and 'integrity'? I doubt that one can understand a definition of sustainability without understanding the system of concepts and principles that surround it. Sustainability, when understood within the atomistic, incrementalist paradigm of welfare economics, reduces to a question of balancing interchangeable units of welfare across time. If, as we are hypothesizing here, there are non-negotiable constraints that mandate protection of large autonomous systems of nature, those constraints must be expressed in a richer and more complex paradigm.[23]

So the goal of specifying practical guidelines will require a 'paradigm' of ecological/contextual management, a set of concepts and principles that can guide attempts to protect and restore ecological systems. Let me begin by citing and agreeing with the definition of Faber, Manstetten, and Proops, who note that 'ecology' combines, etymologically, the Greek ideas of 'house' with their idea of 'logos', which they translate as concept/structure, and define ecology as "the science of the principles of the self-organization of nature".[24] Thus defined, it is part of the specification of the field of ecological management that its subject matter is self-organizing. This definition also ensures that ecology, not economics, will be the new 'fusion point' of the sciences, because economic activities are understood as one type of ecological activity, one that takes place within the economic system of one (however dominant) species.[25]

Our current task is to build upon this approach to management by providing more elements of the ecological paradigm of environmental management. To that end, I suggest five Axioms of ecological management:

(1) The Axiom of Dynamism: Nature is more profoundly a set of processes than a collection of objects; all is in flux.

(2) The Axiom of Relatedness: All processes are related to all other processes.

(3) The Axiom of Hierarchy: Processes are not related equally, but unfold in systems within systems, which differ mainly regarding the temporal and spatial scale on which they are organized.

(4) The Axiom of Creativity: The autonomous processes of nature are creative and represent the basis for all biologically based productivity. The vehicle of that creativity is energy flowing through systems which in turn find stable contexts in larger systems, which provide sufficient stability to allow self-organization within them, through repetition and duplication.

(5) The Axiom of Differential Fragility: Ecological systems, which form the context of all human activities, vary in the extent to which they can absorb and assimilate human-caused disruptions in their autonomous processes.

These Axioms function, in practice, in conjunction with a normative definition of ecosystem health/integrity. I begin by proposing a definition of integrity. An ecological system has maintained its integrity if it retains:

(a) the total diversity of the system, the sum total of the species and associations that have held sway, historically;[26] and

(b) the autonomous processes (systematic organization) that maintain that diversity, including, especially, the multiple layers of complexity through time.[27]

It is useful to have two, related concepts to describe ecosystem wellbeing, as Leopold noted in his comparison of succession after the plough in Kentucky and in the American Southwest.[28] In both cases, the integrity of the system was compromised – including loss of total diversity and invasion by exotics. The difference, Leopold noted, was that bluegrass represented a new stable point, capable of maintaining itself across time. This system lacked the integrity of systems like Rio Gavilan, that maintained their 'aboriginal health' (including their historical mix of species), but maintained a 'healthy' equilibrium nonetheless.

I am suggesting that we use the term 'integrity' as the stronger term – though it certainly can admit of degrees – while we use 'health' to designate the somewhat weaker concept that describes the Kentucky bluegrass system. Integrity, in other words, emphasizes both clause (a) and (b). System S1 maintains greater integrity than system S2 if S1 retains not only enough complexity to maintain autonomous functioning, but also maintains more of its original species, populations, micro-habitats, and processes of interaction. A system is healthy if it maintains its complexity and autonomy/self-organization.

Within this dynamic, contextualist paradigm we can understand the centrality of the goal of protecting biological complexity. Complexity is directly related to self-organization, and self-organization is an essential part of ecosystem health and integrity. And thus we understand the non-negotiable obligation to protect biodiversity: it is an obligation to future generations to protect the diversity and, even more important, the complexity of self-organizing systems. This obligation requires protection of complex processes of ecosystems.

CONCLUSION

I have sketched two broad approaches to understanding sustainability, recognizing that this keystone concept of modern environmentalism can only be given meaning as a part of a constellation of concepts and methods – a paradigm, as some would say. The social scientific approach, which sees sustainability as a relationship between levels of welfare in the present and the future, defines sustainability within an incrementalist paradigm that interprets all values in a common measure, such as dollars or present satisfactions. The advantage of this paradigm, and its proposed approach to sustainability, is that this approach expresses the sustainability relation in interchangeable units; judgements regarding intergenerational fairness can therefore be understood as a balance between commensurable values across time. This incrementalist approach, however, has the attendant disadvantage that it does not deal very well with discontinuities and irreversibilities – and those who worry about global environmental problems such as the greenhouse effect and loss of species diversity emphasize concerns of precisely those kinds.

I have therefore sketched an alternative framework for understanding sustainability, based on a two-tiered system of values, some of which are interchangeable and able to be traded off, and some of which are non-negotiable. Scientific contextualism relies on information and models from the natural sciences to determine which decisions carry significant risk of cataclysmic and irreversible results; and, hence, when non-negotiable moral constraints trump interchangeable measures of individual welfare. This approach *balances* short-term economic and long-term ecological concerns, but does not *reduce* them to a common metric. Environmental policy is constrained by *both* ecological and economic limits; economic concerns predominate when risks are not catastrophic or irreversible and when the areas affected are relatively small. Non-negotiable, intergenerational obligations predominate when decisions carry risk of irreversible or catastrophic change in those large-scale systems on which the human species depends.

NOTES

[1] See Caldwell, 1990, p. 177.

[2] It seems to me questionable that anyone consistently holds this position in its most extreme form, even though certain passages in the writings of deep ecologists seem to consider all use of nature a violation of its intrinsic value. For a fuller discussion of the policy implications of deep ecology, see Norton, 1991, chapter 12.

[3] See, for example, Kahn, 1982.

[4] Data from the late 1970s showed only eight per cent of environmentalists advocating all of the tenets of deep ecology (Mitchell, 1980). I am unaware of more recent data on this subject. As for the other end of the spectrum, I have argued in *Toward Unity Among Environmentalists* that advocacy of some market constraints to protect environmental values provides a minimal defining characteristic of all environmentalists.

[5] World Commission on Environment and Development, 1987, p. 43.

[6] Ibid.

[7] See Ehrlich and Ehrlich, 1986, pp.8-10.

[8] See Woodwell, 1985, for a useful, recent discussion of this difficult issue.

[9] See Daly and Cobb, 1989, pp. 72-6, for a more positive characterization of the Brundtland definition. Daly and Cobb describe the Brundtland definition as "vague", but artfully so, and believe that it will give rise to more specifically biological criteria.

[10] For a careful and detailed explanation of how the mainstream economic paradigm reduces environmental values to increments of willingness-to-pay, and how this paradigm consequently ignores problems of scale and magnitude of throughput, see Daly and Cobb, 1989.

[11] This argument is made, for example, by Kneese and Schulze (1985), who can be considered, in general, proponents of the mainstream micro-economic paradigm.

[12] Indeed, many environmental and resource analysts now believe that problems of waste disposal will prove far more intractable than will problems of resource availability. See Faber, Manstetten and Proops, 1990.

[13] Ciriacy-Wantrup, 1952.

[14] For a classic economic treatment, see Page, 1977. Also see Page, 1991, and Toman and Crosson, 1991.

[15] See, for example, Callicott, 1989, and Rolston, 1988.

[16] I have argued elsewhere that both of these approaches suffer serious conceptual difficulties and will not repeat those arguments here. See Norton, 1982a and 1982b.

[17] See Norton, 1989, for an explanation of a decision model that relies on general obligations to the future – obligations that are owed to no specifiable individuals.

[18] Burke, *Reflections on the Revolution in France*, pp. 93-4.

[19] See Ulanowicz, 1986, p. 25, for a concise explanation of how self-organization can be treated independently of teleology.

[20] See Leopold, 1939, and Hargrove, 1989.

[21] See Norton, 1990.

[22] See Prigogine and Stengers, 1984; Gleick, 1987, and Ulanowicz, 1986, for a comprehensive examination of this and related concepts.

[23] See Carpenter, 1990.

[24] See Faber, Manstetten, and Proops, 1990.

[25] See Leopold, 1939.

[26] Total diversity (Gamma diversity) is defined as a function of within-habitat diversity (Alpha diversity) and cross-habitat diversity (Beta diversity). It is the diversity characteristic of a landscape composed of many habitats and micro-habitats. See Norton, 1987, pp. 32-3.
[27] These axioms were introduced in Norton, 1991.
[28] Leopold, 1949, p. 206.

REFERENCES

Burke, E. *Reflections on the Revolution in France*. London, Dent, 1910 edition.
Caldwell, L.K. 1990 *Between Two Worlds*. Cambridge, Cambridge University Press.
Callicott, J.B. 1989 *In Defense of the Land Ethic*. Albany, State University of New York Press.
Carpenter, S.R. 1990 "Sustainability and Forms of Life", unpublished manuscript presented at the World Bank conference, The Ecological Economics of Sustainability, Washington, D.C., May 1990.
Ciriacy-Wantrup, S.V. 1952 *Resource Conservation*. Berkeley, University of California Press.
Daly, H. and Cobb, J. 1989 *For the Common Good*. Boston, Beacon Press.
Erlich, P. and Erlich, A. 1986 "Population and Development Misunderstood", *The Amicus Journal*, **8**: 8-10.
Faber, M.; Manstetten, R. and Proops, J. 1990 "Towards an Open Future: Ignorance, Novelty and Evolution", in *Ecosystem Health: New Goals for Environmental Management*, edited by R. Costanza, B. Norton and B. Haskell. Covelo, CA, Island Press.
Gleick, J. 1987 *Chaos*. New York, Penguin Books.
Hargrove, E.C. 1989 *Foundations of Environmental Ethics*. Englewood Cliffs, NJ, Prentice Hall.
Kahn, H. 1982 *The Coming Boom: Economic, Political, and Social*. New York, Simon and Schuster.
Kneese, A.V. and Schulze, W.D. 1985 "Ethics and Economics", in *Handbook of Natural Resources and Energy Economics*, edited by J.L. Sweeney. Amsterdam, North Holland.
Leopold, A. 1939 "A Biotic View of Land", *Journal of Forestry*, **37**: 727-30.
Leopold, A. 1949 *A Sand County Almanac*. New York, Oxford University Press.
Mitchell, R. 1980 "How 'Deep,' 'Soft' or 'Left'? Present Constituencies in the Environmental Movement for Certain World Views", *Natural Resources Journal* **20**: 352.
Norton, B.G. 1982a "Environmental Ethics and Nonhuman Rights", *Environmental Ethics* **4**: 17-36.
Norton, B.G. 1982b "Environmental Ethics and Rights of Future Generations", *Environmental Ethics* **4**: 319-37.
Norton, B.G. 1987 *Why Preserve Natural Variety?* Princeton, Princeton University Press.
Norton, B.G. 1989 "Intergenerational Equity and Environmental Decisions: A Model Based on Rawls' Veil of Ignorance", *Ecological Economics* **1**: 137-159.

Norton, B.G. 1990 "Context and Hierarchy in Aldo Leopold's Theory of Environmental Management", *Ecological Economics*, 2: 119-27.
Norton, B.G. 1991 *Toward Unity Among Environmentalists*. New York, Oxford University Press.
Page, T. 1977 *Conservation and Economic Efficiency*. Baltimore, The Johns Hopkins University Press.
Page, T. 1991 "Sustainability and the Problem of Evaluation", in *Economics and Sustainability: Balancing Trade-Offs and Imperatives*, edited by R. Costanza. New York, Columbia University Press.
Prigogine, I. and Stengers, I. 1984 *Order Out of Chaos*. New York, Bantam Books.
Rolston III, Holmes 1988 *Environmental Ethics*. Philadelphia, Temple University Press.
Toman, M. and Crosson, P. 1991 Resources for the Future Discussion Paper ENR 9105.
Ulanowicz, R.E. 1986 *Growth and Development: Ecosystem Phenomenology*. New York, Springer-Verlag.
Woodwell, G.M. 1985 "On the Limits of Nature", in *The Global Possible: Resources, Development, and the New Century*, edited by R. Repetto, pp. 47-66. New Haven, Yale University Press.
World Commission on Environment and Development 1987 *Our Common Future*. Oxford, Oxford University Press.

[30]

Moral Pluralism and the Course of Environmental Ethics

Christopher D. Stone*

Environmental ethics has reached a certain level of maturity; further significant advances require reexamining its status within the larger realm of moral philosophy. It could aim to extend to nonhumans one of the familiar sets of principles subject to appropriate modifications: or it could seek to break away and put forward its own paradigm or paradigms. Selecting the proper course requires as the most immediate mission exploring the formal requirements of an ethical system. In general, are there constraints against bringing our moral relations with different sorts of things under different rules of governance? In particular, how much independence can an environmental ethic (or ethics) aim to have?

INTRODUCTION

With this volume, *Environmental Ethics* concludes its first decade. It may be a good time to ask what the environmental ethics movement has to show for itself, where it is, and where it should be heading. Without doubt, and particularly in view of the short time span, the contributions assembled are impressive. Many (some might surmise all) of the basic issues have been clarified. Perhaps most valuable is the body of literature focusing attention on what I will call "the obstacles" (below).

Good work continues, to be sure, but I fear we have reached a plateau. The signs include a tendency to reiterate the well-worn "need" for an environmental ethics "whose time has come," and then to work over the increasingly familiar themes about the restricted reach of mainstream theories, et cetera. Part of the problem is that we have yet to establish a clearly defined sense of mission. Where does environmental ethics situate itself within the larger world of moral philosophy?

---

* School of Law, University of Southern California, Los Angeles, California, 90089-0071. Stone, the author of *Should Trees Have Standing?* has recently published *Earth and Other Ethics: The Case for Moral Pluralism* (New York: Harper & Row, 1987). His current research interest is in the development of institutional responses to global pollution. He serves on USC's Institutional Animal Care and Use Committee and is a member of the Editorial Advisory Board of *Environmental Ethics*. The author would like to thank Holmes Rolston, III and Martin Krieger for criticisms of particular aspects of the manuscript.

As an applied ethics is one response. But, if so, we still need to ask what such a status entails.[1] Does it mean we are to regard environmental ethics as applying certain invariant fundamental moral principles—"core principles," let us say—to deal with the peculiar properties of nature, the way mathematics's core principles (of algebra and topology) are said to be extended and refined by statistics and probability theory to suit them for application to their special "materials"?[2] If that is the commitment, then certain other questions follow. What are the invariant moral principles that environmental ethics, as an applied field, is applying? (In whose service do we place ourselves?) What leeway do the appliers have to supplement and deform the "purest" and most abstract propositions in the core when they bruise against the concrete riddles of the world?

An alternate, considerably larger ambition is to assemble forces under the banner of a new, independent ethic and proceed to mount an assault on the core itself with an aim either to overthrow and replace the reigning premises or to establish some sort of co-regency.

A third alternative is the most far-reaching. It would use the environmental ethics movement as the occasion to reexamine the metaethical assumptions that underlie all of moral philosophy.

It is my position that each of these missions has some validity, but the third must dominate attention now, for we have not yet made clear, neither to ourselves nor to others, what exactly are the aims and ground rules that govern the composition of an ethical viewpoint.

THE OBSTACLES

Certainly I am not going to presume to summarize the body of literature that has appeared to date. The writers of the past ten years have identified a cluster of obstacles that environmental ethicists face. Most of these are familiar to readers of *Environmental Ethics* and require, therefore, only a brief recapitulation here.

First is the question of putting the *objective* into coherent form. On this score the proponent of an environmental ethic is tempted to fall back upon negatives, to speak of what such an ethics is *not:* the aim is to inject into moral reasoning considerations that are not sheerly homocentric, that do not appeal solely or decisively to human preferences or utility. Here the first difficulty appears. Even if the environmentalist can persuade others that trees and trout *have value* (in some sense), only humans *do the valuing;* it is, after all, humankind, not trees or

---

[1] See J. Baird Callicott, "Non-Anthropocentric Value Theory and Environmental Ethics," *American Philosophical Quarterly* 21 (1984): 299–300.

[2] See Lynn Arthur Steen, "Mathematics Today," in *Mathematics Today,* ed. Steen (New York: Springer-Verlag, 1970), pp. 7–8. Note that in the model of mathematics Steen presents the flow of ideas and valuable information runs in two directions: the inventory of the most highly abstract ideas in the core are available for equipping application in the outer regions; in turn, the core is fueled with the new ideas that concrete application sends back from the field.

trout, that the environmentalist is seeking to persuade. Does this requirement to appeal to human consciousness and preferences land us in a contradiction, a sort of homocentrism after all?

Second is the related question of the *foundation*. Even if we can intelligibly express an environmental ethic's objective, on what rational basis can it possibly rest? We could conceive ourselves to be working within an applied field, and then figure out which dominant ethic to apply. Subordination to utilitarianism is unappealing because it is an alliance that values nature only so far as it is instrumental to human welfare. Union with the neo-Kantians is rebuffed, for while we are glad that they do not kick their dogs, the justification—duties to their own selves, not to the dogs—is unacceptable. The prevailing mood is to uncover some "good" that is not wholly instrumental either to human welfare or to human virtue, one that is somehow situated outside ourselves in nature. The challenges of identifying and legitimating such an intrinsic or inherent good are substantial, however, and increase the further we wander beyond intelligence or life as its foundation. The animal rights advocate has, at least, some of the goods of familiar moral theory to work with: a life that can be snuffed out, a plan that can be frustrated, a nerve that can transmit pain. The person who supports the moral considerateness of an inanimate object confronts the task of identifying some comparable basis, some "intrinsic worth" of something that cannot be killed, frustrated, or pained.

Third, what is being sought is not just a moral viewpoint that accounts for nature in principle. We need a moral viewpoint detailed and ingenious enough to maneuver us through the *ontological conundrums*. By reference to what principles is the moral and legal world to be carved up into those "things" that count and those that do not? This is a problem that can be approached as one of ethic's *boundaries:* that is, if self-consciousness is not the key to moral considerateness, nor sentience, nor life . . . how does one draw the line so that an argument favoring a lake does not apply with equal force to a lamp? The same sort of dilemma crops up in other forms: is the unit of our concern the individual ant, the anthill, the family, the species, or the ant's habitat?

Fourth, suppose that we can do the carving up correctly, that is, identify those objects toward which some prima facie moral regard is justified, e.g., perhaps a certain mountain. There will remain the question, even if moral obligations to a mountain are conceded to exist in principle, how they can be *discharged*. In familiar, interpersonal moralities, the discharge of duties toward another is connected with respect for the other's wants and welfare. But how does one "do right by" a mountain?

Fifth, there are the *distributional dilemmas*. It is not enough to carve up the world, establishing what is to be morally considerate. Nor is it enough to agree how that regard translates into prima facie good and bad acts. What are we to do in the case of conflicting indications? For example, one can imagine a life-respecting moral framework whose basic principle is "more life is better than less." One can imagine, too, support for the preservation of a singular, pristine desert. But then, how do we judge an irrigation project that offers to transform the desert into a habitat teeming with vegetation? In general terms, the problem is the familiar one of weighing: even if the continued existence of a species, or the state of a river is demonstrated to be a (noninstrumental) good, how strongly does that good withstand the moral force of other, competing goods?

While each of these questions is hard—the fact we are in the tenth volume of this journal says as much—we can take some heart from the fact they are, in kind, no more formidable than those with which the proponents of every moral theory have been vexed: how to establish the meaning and legitimacy of moral reasoning in general, to demonstrate that it is cogent and defensible to sacrifice evident ego-pleasures to further something else. Those who appreciate the difficulties of substantiating the human community as that "something else" cannot sniff at those who find some plausible candidacy in the biotic. That granted, the development of an ethic that gives good moral guidance for our conduct respecting nature is not a quantum leap more perplexing than the task of putting together (or discovering) an ethics for our conduct respecting persons.

THE METAETHICAL ASSUMPTIONS

The larger—in all events, prior—questions require further consideration of the implicit metaethical assumptions. What are environmental ethicists trying to achieve, and what are the standards for success? In other words, what, more exactly, is an ethics supposed to look like and do? To illustrate, for years environmental ethicists have been stimulated by Aldo Leopold's conviction that we should develop a "land ethic." But how much thought has been given to what such a project implies? Are the proponents of a land ethic committed to coming up with a capacious replacement for all existing ethics, one capable of mediating all moral questions touching man, beast and mountain, but by reference to a grander, more all-encompassing set of principles? Or can the land ethic be an ethic that governs man's relations with land alone, leaving intact other principles to govern actions touching humankind (and yet others for actions touching, say, lower animals, and so on)?

If we are implying that there are different ethics, then there are a host of questions to face. What is an ethical system, and what are its minimum requirements? Need its "proofs" be as irresistible as a geometry's? Is it required to provide for each moral dilemma that it recognizes as a dilemma one right, tightly defined answer? Or is it enough to identify several courses of action equally acceptable, perhaps identifying for elimination those that are wrong or unwelcome? How—by reference to what elements—can one ethic differ from another? What possibilities of conflicting judgments are introduced by multiple frameworks, and how are they to be resolved?

These are among the questions that, sooner or later, environmental ethicists will have to confront. Upon their answer hinges nothing less than the legitimacy of environmental ethics as a distinct enterprise.

MORAL MONISM

The environmental ethics movement has always known that if it is to succeed, it has to challenge the prevailing orthodoxy. But the orthodoxy it has targeted is only the more obvious one, the orthodoxy of morals: that man is the measure (and not merely the measurer) of all value. Certainly calling that gross presumption to question is a valid part of the program. But the orthodoxy we have to question first is that of metaethics—of how moral philosophy ought to be conducted, of the ground rules.

Note that I am not claiming that we lack controversy at the level of *morals*. There is no shortage of lively contention in the philosophy literature. But underneath it all there is a striking, if ordinarily only implicit agreement on the metaethical sense of mission. It is widely presumed, by implication when it is not made explicit, that the ethicist's task is to put forward and defend a single overarching principle (or coherent body of principles), such as utilitarianism's "greatest good for the greatest number" or Kant's categorical imperative, and to demonstrate how it (the one correct viewpoint) guides us through all moral dilemmas to the one right solution.

This attitude, which I call moral monism, implies that in defending, say, the preservation of a forest or the protection of a laboratory animal, we are expected to bring our argument under the same principles that dictate our obligations to kin or the just deserts of terrorists. It suggests that moral considerateness is a matter of either-or; that is, the single viewpoint is presumably built upon a single salient moral property, such as, typically, sentience, intelligence, being the subject of a conscious life, etc. Various entities (depending on whether they are blessed with the one salient property) are *either* morally relevant (each in the same way, according to the same rules) *or* utterly inconsiderate, out in the moral cold.[3]

---

[3] Consider the argument that a proponent of using animals in medical research throws up to the animal rights advocate: "If all forms of animal life . . . must be treated equally, and if therefore . . . the pains of a rodent count equally with the pains of a human, we are forced to conclude (1) that neither humans nor rodents possess rights, or (2) that rodents possess all the rights that humans possess." Carl Cohen, "The Case for the Use of Animals in Biomedical Research," *New England Journal of Medicine* 315 (1986): 865, 867. An alternative "pluralist" position would examine the possibility that a laboratory bred animal has rights, but not the same as humans. The rodent might have no "right" to life, but have a "right" to be free from suffering. This distinction could be operationalized by saying that the proponent of an experiment that took a laboratory animal's life painlessly would only have to show a clear likelihood of an advance of human welfare; animal suffering, however, would (alternatively) never be allowed, or allowed only when it could be shown that there was a very high probability that the experiment would result in the saving of human lives or the reduction of human suffering—never because it would alleviate mere inconveniences in human life, such as baggy eyelids.

Environmentalists, more than most philosophers, have at least an intuitive reason for supposing that this attitude is mistaken, for it is they whom the attitude is the first to bridle. Environmentalists wonder about the possible value in a river (or in preserving a river), but cannot rationalize those feelings in the familiar anthropocentric terms of pains and life-projects that they would apply to their own situations. By contrast, mainstream ethicists, concentrating on interpersonal relations, constrict their attention to a relatively narrow and uncontroversial band of morally salient qualities. Persons can speak for themselves, exercise moral choice, and—because they share a community—assert and waive many sorts of claims that are useful in governing their reciprocal relationships. Orthodox ethics has understandably tended to identify all ethics with this one set of morally salient properties: the paradigmatic moral problems have historically been interpersonal problems; the paradigmatic rules, person-regarding.

Thus, while vying camps have arisen within the orthodox tradition, none is ordinarily forced to account for the significance of properties that lie outside the common pool of human attributes. It is only when one starts to wonder about exotic clients, such as future generations, the dead, embryos, animals, the spatially remote, tribes, trees, robots, mountains, and art works, that the assumptions which unify ordinary morals are called into question. Need the rules that apply be in some sense, and at some level of generality, "the same" in all cases? The term *environmental ethics* suggests the possibility of a distinct moral regime for managing our way through environment-affecting conduct. But in what respects that regime is distinct from other regimes and how conflicts among the regimes are to be mediated are crucial matters that have not been generally and directly addressed.

In default of well-worked out answers, the prevailing strategy of those who represent nonhumans is one of extension: to force one of the familiar person-oriented frameworks outward and apply one of the familiar arguments to some nonhuman entity. But such arguments too often appear just that—forced. Utilitarianism's efforts to draw future generations under its mantle (a relatively easy extension, one would suppose) ties it in some awkward, if not paradoxical knots. Do we include, for example, those who might be born—obliging us to bring as many as possible of them into existence in order to aggregate more pleasures? Nor is it clear that utilitarianism, unqualified by a complex and ill-fitting rights appendage, can satisfy the concerns that drive the animal liberation movement.

The shortcomings of (let us call it) moral extensionism[4] are not peculiar to utilitarianism. Extensions of utilitarianism's principal contenders all require, in various ways and with various justifications, putting oneself in the place of another to test whether we can really wish the conduct under evaluation if we assume the other's position, role, and/or natural endowment. While such hy-

---

[4] The term was suggested to me by Holmes Rolston.

pothetical trading of places and comparable techniques of thought experiment are always problematical, they are most satisfactory when we are trading places with (or universalizing about) persons who share our culture, and whose interests, values, and tastes we can therefore presume with some confidence. But even that slender assurance is destined to erode the further we venture beyond the domain of the most familiar natural persons. With what conviction can we trade places with members of spatially and temporally remote cultures, or with our own descendants in some future century? And, of course, if we wish to explore our obligations in regard to the dead, trees, rocks, fetuses, artificial intelligence, species, or corporate bodies, trading places is essentially a blind alley. It is one thing to put oneself in the shoes of a stranger, perhaps even in the hooves of a horse—but quite another to put oneself in the banks of a river.

Certainly, the fact that orthodox moral philosophies, each with its own ordinary-person orientation, have difficulty accommodating various nonhumans is not, in itself, proof that the conventional moral schools are wrong, or have to be amended beyond recognition. One alternative, the position of an ardent adherent to one of the predominant schools, is that any unconventional moral client that it cannot account for, except perhaps in a certain limited way, cannot (save in that limited way) have any independent moral significance or standing.

But there is another response to the dilemma, one that is more challenging to the assumptions that dominate conventional moral thought. In accordance with this approach we need to ask several new questions. How imperialistic need a moral framework be? Need we accept as inevitable that there be one set of axioms or principles or paradigm cases for all morals—operable across all moral activities and all diverse entities? Are we constrained to come forward with a single coherent set of principles that will govern throughout, so that any ethic we champion has to absorb its contenders with a more general, abstract and plenary intellectual framework? My own view is that monism's ambitions, to unify all ethics within a single framework capable of yielding the one right answer to all our quandaries, are simply quixotic.

First, the monists's mission sits uneasily with the fact that morality involves not one, but several distinguishable *activities*—choosing among courses of conduct, praising and blaming actors, evaluating institutions, and so on. Is it self-evident that someone who is, say, utilitarian in his or her act evaluation is committed to utilitarianism in the grading of character?

Second, we have to account for the *variety of things* whose considerateness commands some intuitive appeal: normal persons in a common moral community, persons remote in time and space, embryos and fetuses, nations and nightingales, beautiful things and sacred things. Some of these things we wish to account for because of their high degree of intelligence (higher animals); with others, sentience seems the key (lower life); the moral standing of membership groups, such as nation-states, cultures, and species has to stand on some additional footing, since the group itself (the species, as distinct from the individual whale) manifests no intelligence and experiences no pain. Other entities are genetically human, either capable of experiencing pain (advanced fetuses) or nonsentient (early embryos), but lack, at the time of our dealings with them, full human capacities. Trying to force all these diverse entities into a single mold— the one big, sparsely principled comprehensive theory—forces us to disregard some of our moral intuitions, and to dilate our overworked person-wrought precepts into unhelpfully bland generalities. The commitment is not only chimerical; it imposes strictures on thought that stifle the emergence of more valid approaches to moral reasoning.

MORAL PLURALISM

The alternative conception toward which I have been inviting discussion, what I call *moral pluralism*,[5] takes exception to monism point by point. It refuses to presume that all ethical activities (evaluating acts, actors, social institutions, rules, states of affairs, etc.) are in all contexts (in normal interpersonal relations, across large spaces and many generations, between species) determined by the same features (intelligence, sentience, capacity for emotions, life) or even that they are subject, in each case, to the same overarching principles (utilitarianism, Kantianism, nonmaleficence, etc.). Pluralism invites us to conceive the intellectual activities of which morals consist as being partitioned into several distinct frameworks, each governed by its own appropriate principles.

Certainly, one would expect pain-regarding principles to emerge as pivotal in establishing obligations toward all those things that experience pain. Not pain alone, but preferences of some sort, e.g., the projection of a life plan, have to be accounted for in our relations with a second level of creature. Still richer threads (such as a sense of justice, and rights of a sort that can be consensually created, extinguished, traded, and waived) form the fabric of the moral tapestry that connects humans who share a common moral community. Other principles, perhaps invoking respect for life, for a natural unfolding, seem fit as a basis for

---

[5] Moral pluralism ought not to be confused with moral relativism, the view, roughly, that all morals are context-dependent. A pluralist can be agnostic with respect to the moral realist position that there are absolutely true answers to moral quandaries, as invariable across time, space, and communities as the value of pi. There may be "really right" and not just relatively right answers, but the way to find them is by reference not to one single principle, constellation of concepts, etc., but by reference to several distinct frameworks, each appropriate to its own domain of entities and/or moral activities (evaluating character, ranking options for conduct, etc.).

forming our relations with plants.⁶ Indeed, should we pursue this path, we would multiply subdivisions even within the interpersonal realm. The Kantians, emphasizing the place of nonwelfarist duties, make rightful ado about our not saving our child from drowning because it is "best on the whole." But this does not mean that classic utilitarianism is wrong. Maybe it is of only limited force in parsing out obligations among associates and kin. Utilitarianism strikes me as having considerable validity for legislation (an activity) affecting large numbers of largely unrelated persons (an entity set) who are therefore relatively unacquainted with each other's cardinal preferences.

That monism should have become so firmly established in morals is understandable (it echoes one God, one grand unified theory), but is hardly inevitable. Geometers have long relinquished the belief that Euclid's is the only geometry.

> This discovery led to the pluralization of mathematics (itself already a strangely plural noun); where we once had geometry, we now have geometries and, ultimately, algebras rather than algebra, and number systems rather than a number system.⁷

A comparable partitioning has taken place in the empirical and social sciences. The body politic is commonly viewed as being comprised of groups: groups of humans, each of which is made up of more groups, groups of cells, molecules, atoms, and subatomic particles, and/or waves. What happens at one level of description is undoubtedly a product, in some complex way, of what is occurring at another. Many, perhaps most scientists feel that "in principle" there is a single unifying body of law—the laws of nature—that at some level of simple generality hold throughout. If so, one may harbor the hope not only of abolishing all lingering pockets of ignorance and chaos, but of connecting phenomena on every plane with phenomena on another, of someday unifying, say, the laws that govern the movement of subatomic particles with those that govern social conduct. But we are far from it. What we actually work with, for all intents and

---

⁶ See Paul W. Taylor, *Respect for Nature* (Princeton: Princeton University Press, 1986); J. L. Arbor presents a coherent and persuasive plea for plants—coldly logical, however heartfelt—in "Animal Chauvinism, Plant-Regarding Ethics, and the Torture of Trees," *Australasian Journal of Philosophy* 64 (1986): 335.

⁷ Steen, "Mathematics Today," pp. 4–5. To pursue the mathematical model for a further moment, Godel and others have laid to rest the hope of ever producing a complete and consistent formal system powerful enough to prove or to refute every statement it can formulate. Although what happens in math is hardly a conclusive model of what should go on in morals, it does make one wonder how much of moral philosophy implicitly proceeds on the assumption that a morality not only has axioms (or even solider starting points), but that they are axioms more powerful than math's! And if that is not the assumption, what takes its place?

purposes, and to almost everyone's satisfaction, are separate bodies of law and knowledge.

The issue I am raising is this. If, as I maintain, ethics comprises several activities and if it has to deal with subject matters as diverse as persons, dolphins, cultural groups, and trees, why has ethics not pursued the same path as the sciences—or, rather, paths? That is, why not explore the possibility that ethics can also be partitioned?

Perhaps the analogy is simply too weak. However free science may be to partition, one might argue, ethics appears to be under peculiarly strong constraints to remain monistic. The argument might go like this. Alternative descriptions of how the world is (or might be) can peacefully coexist over a broad latitude without logical conflict,—e.g., in most contexts, one can indulge either in a particle or a wave version of light without chafing. And even where apparently irreconcilable conflict does erupt at one level (say, at the subatomic) the participants at other levels (those doing cellular biology) can ordinarily remain agnostic. By contrast, ethics (one is tempted to say) is not merely descriptive. It has as its ultimate aim choosing the right *action*. Unlike describing, in which subtly overlapping nuances of adjective and predicate are tolerable, acting seems to lend itself to, if not to demand, binary yes/no, right/wrong alternatives.

If this is the argument why morals require monism, it appears to me unpersuasive. There is, to begin with, the question of agenda: one wants from moral reasoning not merely the verdict, whether or not to do act *a*, but also what the choice set is: *a, b, c, . . .*? Moral thought is a service when it is populating and clarifying the range of morally creditable alternatives. Hence, attention to plural approaches would find justification if, by stimulating us to define and come at problems from different angles, it were to advance our grasp of alternatives.[8]

Perhaps most importantly, let us remind ourselves that actions are in the physical world; the evaluation of them is intellectual. Many persons (are these the "moralists"?) would probably be pleased if our moral reasoning had the power to map a unique, precise moral evaluation for each alternative action. It would give us much the same pleasure (tinged with a not entirely ingenuous surprise) that mathematicians derive from confirmation that the world "out there," while theoretically at liberty to go its own haphazard way, is conforming

---

[8] Note that this rationale for pluralism could be endorsed on heuristic grounds by a monist, even by a moral realist who presumed (as I do not) that all the candidates for truth *disclosed* by this many-angled attack on the problem will in the end be submitted to a single adjudicatory principle to decide which of them is *uniquely and truly right*. Compare the position Paul Feyerabend adopts with respect to the natural sciences, viz., that the history of sciences reveals an incompleteness and even inconsistency of each framework which should be regarded as routine and inevitable, and that a pluralism of theories and metaphysical viewpoints should be nourished as a means of advancing on the truth. Feyerabend, *Against Method* (London: Verso, 1978): 35–53.

in general to the elegant inventions of our intellects.[9] Why, when we set out to apply our best moral theories to the unruly world of human conduct should we confidently expect more—a more meticulous isomorphism, more freedom from inconsistency, more power of resolution?

Specifically, it may be a (not terribly interesting) truth that an act can be defined in such a way that we are left with no alternative but to do it or not—a feature of the world that makes monism superficially attractive. But even if so, it is a fact about the world that our best moral reasoning may just not be able to rise to or to map. The rightness and wrongness of some acts may lie beyond our power to deduce or otherwise discover. Key moral properties may not lend themselves to produce a transitive ordering across the choice set.

THE VARIABLES

If we are to explore bringing our relations with different sorts of things under different moral governances, then we face the question: by reference to what intellectual elements might governances vary domain to domain?

(a) *Grain of description*. Morals is concerned with comparing actions, characters, and states of affairs. To compare alternatives, as a logical first step, we have to settle upon the appropriate vocabulary of description. For example, in evaluating our impact on humans, we consistently adopt a grain of description that individuates organisms: each person counts equally. In evaluating other actions, there is often intuitive support for some other unit, e.g., the hive or the herd or the habitat. I am not claiming that these intuitions are self-validating, only that they, and their implications, merit sustained and systematic attention. Each vying grain of description is integral to a separate editorial viewpoint. Suppose that a bison naturally (of its own action) faces drowning in a river in a national park. Should we rescue it, or let "nature take its course"? One viewpoint emphasizes the individual animal; another (favored, apparently, by the park service)[10] consigns the individual animal to the background and emphasizes the larger unit, the park ecosystem. Another viewpoint emphasizes species. Each focus brings along its allied constellation of concepts. In invoking the finer grain, focusing upon the individual animal, we scan for such properties as the animal's capacity to feel pain, its intelligence, its understanding of the situation, and its suffering. None of these terms apply to the park. Instead, the ecosystem version brings out stability, resilience, uniqueness, and energy flow.

(b) *Mood*. What I mean by mood may best be illustrated by a contrast between morals and law. Law, like morals, often speaks in negative injunctions, i.e.,

---

[9] See E. P. Wigner, "The Unreasonable Effectiveness of Mathematics in the Natural Sciences," in Wigner, *Symmetries and Reflections* (Cambridge: M.I.T. Press, 1970).

[10] See Jim Robbins, "Do not Feed the Bears?" *Natural History*, January 1984, p. 12.

"Thou shalt not kill . . ." and "Thou shalt not park in the red zone. . . ." But the law always proceeds to specify, in each case, a sanction which expresses the relative severity of the offense, viz., ". . . or face the death penalty," ". . . or face a $12 fine." The result is a legal discussion endowed with fine-tuned nuances. By contrast, much of moral philosophy, inspirited with monism, is conducted at a level of abstraction at which every act is assumed to be either-or, either good or bad; there is either a duty to do *x* or a duty not to do *x;* a right to *y* or no right to *y*. Monist moral discourse, then, lacks the refinements of expression that enrich legal discourse. As long as monism reigns, significant distinctions between cases, distinctions marked by nuances of feeling and belief that moral reflection might investigate and amplify, lack a semantic foothold.

By contrast, pluralism welcomes diversified material out of which moral judgments can be fashioned, particularly as we cross from one domain to another. Moral regard for lakes may seem silly—or even unintelligible—if we are required to flesh it out by reference to the same rules, and express our judgments in the same mood, as those that apply to a person. But there are prospective middle grounds. Our lake-affecting actions might have to be judged in terms of distinct deontic operators understood to convey a relatively lenient mood, perhaps something like "that which is morally welcome" or that which will bring credit or discredit to our character.

(c) *Logical (formal) texture.* Every system of intellectual rules is girded on a number of properties that endow it with a distinct logical texture. These range from whether it is subject to closure (whether it is capable of yielding one unique solution for each question that can be opened within it) to its attitude on contradictions and inconsistencies. As to closure, the monist implicitly assumes that morals must be modelled on ordinary arithmetic. There is one and only one solution to $4 + 7$; so too there should be, for each dilemma of morals, one right answer. And monism rejects, too, any system of ethical postulates from which we could derive conflicting and contradictory prescriptions. After all, what would we think of a system of geometry from whose postulates we could derive both that two triangles were, and that they were not, congruent?

Pluralism is not so dogmatic—or perhaps one should just say not so "optimistic"—about the prospects of assimilating morals to (slightly idealized conceptions of) arithmetic or geometry. We simply may not be able to devise a single system of morals, operative throughout, that is subject to closure, and in which the laws of noncontradiction[11] and excluded middle[12] are in vigilant command.[13]

---

[11] The law of contradiction holds that it cannot be the case that both a proposition *p* and its negation *-p* are true.

[12] The law of excluded middle maintains that either a proposition *p* or its negation *-p* must be true; there is no middle possibility.

[13] See Freidrich Waismann, "Language Strata," in *Logic and Language,* ed. Anthony Flew (New York: Anchor Books, 1965), p. 237. The notion I present of multiple conceptual planes with systematically varying formal requirements owes much to Waismann's musings about "language strata."

RECONCILING THE DIFFERENCES

There are many problems with this pluralistic approach. Many of the stumbling blocks—those that I could identify by myself, or with a little help from my friends—are dealt with in *Earth and Other Ethics*.[14] It can be defended from the obvious charge that it must stumble into moral relativism of the rankest sort.[15] But it faces comparable problems that are not so easy to dismiss. It would appear that a pluralist, analyzing some choice situation in one framework (say, one that accounts for species in an appropriate way) may conclude that act a is right. The same person, analyzing the situation in another framework (one built, say, from a person-regarding viewpoint) concludes b. Are not such conflicts paralyzing? And do they not therefore render pluralism methodologically unacceptable?

To begin with, the fact that morals might admit of several allowable viewpoints does not mean that each and every dilemma will require several competing analyses. Assuming that remotely probable and minimal consequences can be ignored, some choices may be carried through solely within one framework. For example, whatever morality has to say about whether to uproot an individual plant could be provided, presumably, by the appropriate one-plant framework. No excursion into the agent's obligations to the plant's species, or to mankind, or to kin or whatever would be called for.

We can anticipate myriad other circumstances in which thorough analysis requires defining and processing the situation in each of several frameworks. But in some subset of those situations each of the various analyses will endorse the same action. We all know that vegetarianism, for example, can be supported both within a framework that posits the moral considerateness of animals and one that values humans alone, viz., that by eating animals the planet uses protein inefficiently, therefore reducing aggregate human welfare, even robbing badly undernourished persons of a minimally human existence. (What we do not know—and ought to examine—is why approaching such a question from several angles, a technique well-accepted in other areas,[16] should be indicted as an ignoble and impure way to go about doing philosophy).

There is a third set of cases in which more than one framework will appear appropriate, and the different frameworks, rather than mutually endorsing the same result, reinforce different, even inconsistent actions. The potential for conflicts is there—but no more so than in any moral system that deems the proper

---

[14] Christopher D. Stone, *Earth and Other Ethics* (New York: Harper & Row, 1987).
[15] See note 5 above.
[16] I do not mean only lawyers, who do this sort of thing unabashedly all of the time. As for the natural sciences, see Feyerabend, *Against Method*. In mathematics, Gorg Polya, *How to Solve It* (Princeton: Princeton University Press, 1957) is a classic exposition of how mathematicians may stalk a single problem with widely assorted techniques (indirect proofs, reductio ad absurdums, analogy), ultimately to be convinced of the truth of a solution by the dual standards of formal proof and intuition.

choice to be a function of several independent criteria: welfare maximization, duties to kin, respect for life, the values of community and friendship. How do we "combine," where rights analysis says one thing, utility analysis, another?

One possibility is to formulate a lexical ordering rule. For example, our obligations to neighbor-persons, as determined on a framework built on neo-Kantian principles, might claim priority up to the point where our neighbor-persons have reached a certain level of comfort and protection. But when that level has been reached, considerations of, say, species preservation as determined per another framework, or of future generations per another, would be brought into play.

One might claim, with partial justification, that in those circumstances in which we accepted mediation by reference to a master rule, we are reintroducing a sort of monism "after all." But even in these cases, it is an "after all" significant enough to keep pluralism from collapsing into monism. Under monism, a problem is defined appropriately for evaluation by the relevant standard, in such a way that all the "irrelevant" descriptions are left behind from the outset. The problem, so defined, is worked through to solution without further distraction. Under pluralism, a single situation, variously described, may produce several analyses and various conclusions. If a master rule is to be introduced, it is to be introduced only after the separate reasoning processes have gone their separate ways to yield a conflicting set of conclusions, a, b, c, d. The master rule is brought to bear on that set, none of whose members would necessarily have been constructed had the procedure been subjected to the monist stricture that a single standard, such as utilitarianism, had to be applied consistently and exclusively from the start.

Finally, and most troublesomely, there are quandaries for which each of our multiple analyses not only endorse inconsistent actions, but for which no lexical rule is available, and for which further intuitive reflection[17] reveals no further, best-of-all, alternative. We can imagine as a "worst case scenario" an outcome not merely of the form a is mandatory per one framework and b is mandatory per the other (and we cannot do both), but rather of the form a is mandatory and $-a$ is mandatory (a is impermissible). One must, and must not, pull the trigger. What then?

This much is clear: those two edicts, taken together, tell us (logically) nothing. We would say of the total system of beliefs that it had *disappointed us in the particular case*. We would have to agree, too, that if such out-and-out conflicts were in each and every case endemic to pluralist methodology, the whole system

---

[17] I mean by intuitive reflection a process of analysis that leads to a right-feeling judgment, but one for which, even after the conclusion, we cannot offer any proof, perhaps not even specify the premises.

we constructed, would have to be abandoned. But suppose that such outcomes, while possible, should prove exceptional. Then we could regard their occasional occurrences as a particularly poignant indication of the total system's indeterminacy.

This prospect illustrates one of the principal monist-pluralist dividing lines referred to earlier: How fatal is it to a system of moral rules if it fails to furnish a single unambiguous answer to each choice we recognize as morally significant? If we cannot devise a whale-regarding moral framework that gives us one confident right answer to every action affecting whales, do we have to withdraw whales from consideration (except as resources in a human-oriented framework) entirely? If our whale-regarding and our person-regarding edicts conflict, does one or the other or both of the systems responsible have to be dismantled?

As I have already indicated, such a standard, if to be applied with an even hand (and fin) throughout, would cramp the range of morals significantly. Better to come right out and consider the alternative: that we may have to abandon the ambition to find perfect consistency and the "one right answer" to every moral quandary, either because a single answer does not exist, or because our best analytical methods are not up to finding it.[18]

In some circumstances, if we can identify and eliminate the options that are morally unacceptable, we may have gone as far as moral thought can take us. It may be that the choices that remain are equally good or equally evil or equally perplexing.[19]

This does not mean that as a moral community we are relieved from striving for a higher, if ultimately imperfect consensus on progressively better answers.[20] Nor does it mean that, as regards the indeterminate set, one can be arbitrary—as though, from that point on, flipping a coin is as good as we can do. It is by the choices we affirm in this zone of ultimate uncertainty that we have our highest opportunity to exercise our freedoms and define our characters. Particularly as the range of moral considerateness is extended outward from those who are (in various ways) "near" us, people who take morals seriously, who are committed

---

[18] As Hilary Putnam puts it, "The question whether there is one objectively best morality or a number of objectively best moralities which, hopefully, agree on a good many principles or in a good many cases, is simply the question whether, given the desiderata . . . [of] the enterprise . . . will it turn out that these desiderata select a best morality or a group of moralities which have a significant measure of agreement on a number of significant questions." Hilary Putnam, *Meaning and the Moral Sciences* (Boston: Routledge & Kegan Paul, 1978), p. 84.

[19] See Leibniz's stumper: "It is certain that God sets greater store by a man than a lion; nevertheless it can hardly be said with certainty that God prefers a single man in all respects to the whole of lion-kind." *Theodicy*, trans. E. M. Hoggard (New Haven: Yale University Press, 1952), sec. 118.

[20] One might even expect this endeavor to take the form of integrating, or at least striving to integrate, originally independent "plural" frameworks into something grander and more unified—much as the theoretical physicist will continue to scout about for a grand unified field theory. But in the meantime, the practical and even playful work of significance will take place on humbler levels.

to giving good reasons, will come to irreconcilably conflicting judgments on many issues. But the main question now is this: what model of decision process provides the best prospect for constructing the best answers reason can furnish?

[31]

The Case against Moral Pluralism

J. Baird Callicott*

Despite Christopher Stone's recent argument on behalf of moral pluralism, the principal architects of environmental ethics remain committed to moral monism. Moral pluralism fails to specify what to do when two or more of its theories indicate inconsistent practical imperatives. More deeply, ethical theories are embedded in moral philosophies and moral pluralism requires us to shift between mutually inconsistent metaphysics of morals, most of which are no longer tenable in light of postmodern science. A *univocal* moral philosophy—traceable to David Hume's and Adam Smith's theory of moral sentiments, grounded in evolutionary biology by Charles Darwin, and latterly extended to the environment by Aldo Leopold—provides a unified, scientifically supported world view and portrait of human nature in which *multiple*, lexically ordered ethics are generated by multiple human, "mixed," and "biotic" community memberships.

WHY MORAL PLURALISM SHOULD ARISE ESPECIALLY IN CONNECTION WITH ENVIRONMENTAL ETHICS

It is not at all accidental—or, now that it is here, surprising—that moral pluralism would pop up in close association with *environmental* ethics. Fifteen years ago a few academic philosophers, I among them, went looking for a moral theory that would ethically enfranchise nonhuman natural entities and nature as a whole. We wanted to articulate, as Tom Regan so forcefully put it in the third volume of this journal, not an ethic for the *use* of the environment, a "management ethic," but an ethic *of* the environment.[1] Or, put another way, we wanted to develop what I then called a "direct," not an "indirect," "environmental ethic," or what Holmes Rolston, III, still earlier called a "primary," not a

---

*Department of Philosophy, University of Wisconsin-Stevens Point, Stevens Point, WI 54481. Callicott is author of *In Defense of the Land Ethic* (SUNY Press, 1989), editor of *Companion to A Sand County Almanac* (University of Wisconsin Press, 1987), and, with Roger T. Ames, co-editor of *Nature in Asian Traditions of Thought* (SUNY Press, 1989). An earlier version of this paper was presented to the Society for Philosophy and Technology meeting in conjunction with the sixty-third annual meeting of the Pacific Division of the American Philosophical Association in Berkeley, California, 23 March 1989. The author thanks George Sessions, Holmes Rolston, III, Christopher D. Stone, Eugene C. Hargrove, S. K. Lehman, and an anonymous referee for valuable critical comments.

[1] Tom Regan, "The Nature and Possibility of an Environmental Ethic," *Environmental Ethics* 3 (1982): 19–34.

"secondary," "ecological ethic"—an ethic, in any case, which situates the environment as the object, not merely the arena, of human moral concern.[2]

We wanted to bring the natural environment within the purview of ethics, to be sure, but we also wanted to keep human well-being and the human social fabric in sharp moral focus.[3] Between lay a spectrum of concerns—the welfare of future people, of domestic animals, and so on—neglected in traditional Western moral theories that many other philosophers also felt compelled, either by novel circumstances (modern technology) or by the dialectic of rapidly evolving moral sensibility (civil rights, followed by women's liberation, followed by universal human rights), to try to bring within the reach of ethical theory. By working with *one* ethical theory, chosen to accommodate our special concern for the environment, how could we also account for our traditional interpersonal responsibilities and social duties, accommodate all these intermediate new moral concerns to boot, and then order and mutually reconcile the whole spectrum of traditional and novel ethical domains?

Christopher Stone, one of the fathers of environmental ethics and an early architect of the extensionist enterprise, now claims, in *Earth and Other Ethics: The Case for Moral Pluralism*, that we cannot. The Earth and other ethical requirements simply stretch any given moral theory to the breaking point. One thus seems confronted with two choices: moral cynicism or moral pluralism. We can either give in to moral overload and theoretical burn out, or pick up the pieces, one by one, and work theoretically with each separately. I am not attracted to either alternative and propose a third in the last section of this discussion.

In a *précis* of his book, published in these pages, Stone laments that environmental ethics has "reached a plateau. The signs include a tendency to reiterate the well-worn 'need' for an environmental ethics 'whose time has come' and then work over the increasingly familiar themes about the restricted reach of mainstream theories, *et cetera*."[4] Stone claims to draw this conclusion from a survey of the first ten volumes of *Environmental Ethics* (which he modestly declines to "presume to summarize") and allied monographs that appeared during the same decade—1979–1989.[5] But both his book and the spinoff article give the lie to

---

[2] J. Baird Callicott, "Elements of an Environmental Ethic: Moral Considerability and the Biotic Community," *Environmental Ethics* 1 (1979): 71–81; and Holmes Rolston, III, "Is There an Ecological Ethic?," *Ethics* 85 (1975): 93–109.

[3] Dave Foreman, editor of *Earth First!*, and associated environmental activists have been accused of going over the the top and recommending that we substitute an environmental ethic for a human social ethic. See George Bradford, "How Deep is Deep Ecology? A Challenge to Radical Environmentalism" *The Fifth Estate* 22 (Fall 1987): 3–33.

[4] Christopher D. Stone, "Moral Pluralism and the Course of Environmental Ethics," *Environmental Ethics* 10 (1988): 139.

[5] Ibid., p. 140.

that claim. The "need" for an environmental ethic was in fact the burden of the 1970s generation of philosophical environmental literature.[6] Over the 1980s, thanks in large measure to the forum for exploration and critical discussion provided here by Eugene C. Hargrove, environmental philosophers have actually developed an impressive array of fairly well worked out theories of environmental ethics.[7] State-of-the-art environmental ethics also exhibits lateral theoretical diversity; in other words, as each theoretician attempts his or her own vertical integration of multiple moral spheres. It is partly our success in creating a wide variety of compelling, but distinct and mutually inconsistent, environmental ethical systems—however great their tendency may be to weaken when asked to cover all our moral concerns—that has resulted in an embarrassment of riches, ripe for pluralist plucking.

As the 1990s arrive and *Environmental Ethics* the journal settles into its second decade, there exist a fairly wide selection of nonanthropocentric ethical theories, each of which, proclaim its proponents, is superior to all the others. Though many distinct voices have been heard in the environmental ethics choir during the past decade, most are improvising on one or another familiar melody.

A neo-Kantian family of environmental ethics (united by conation as a criterion for moral considerability) seems to be attracting more converts as time goes on. Paul Taylor's biocentrism is the purest neoclassical exemplar of this type. But more baroque variations on the conation theme have been set out by Robin Attfield and Holmes Rolston.[8] Although Rolston frequently quotes Aldo

---

[6] In addition to Rolston's "Is There an Ecological Ethic?" notable seventies-generation papers "calling for" the development of primary or direct environmental ethics are Richard Routley (now Sylvan), "Is There a Need for a New, an Environmental Ethic?" in Bulgarian Organizing Committee. ed., *Proceedings of the Fifteenth World Congress of Philosophy* (Sophia: Sophia Press, 1973), pp. 205–10; and Kenneth E. Goodpaster, "From Egoism to Environmentalism," in K. E. Goodpaster and K. M. Sayre, eds., *Ethics and Problems of the 21st Century* (Notre Dame: University of Notre Dame Press, 1979), pp. 21–35.

[7] For a representative sample see Robin Attfield, *The Ethics of Environmental Concern* (New York: Columbia University Press, 1983); Paul W. Taylor, *Respect for Nature: A Theory of Environmental Ethics* (Princeton: Princeton University Press, 1986); Holmes Rolston, III, *Philosophy Gone Wild: Essays in Environmental Ethics* (Buffalo: Prometheus Books, 1986), and *Environmental Ethics: Duties to and Values in the Natural World* (Philadelphia: Temple University Press, 1988); Eugene C. Hargrove, *Foundations of Environmental Ethics* (Englewood Cliffs, N.J.: Prentice Hall, 1989); and J. Baird Callicott, *In Defense of the Land Ethic: Essays in Environmental Philosophy* (Albany: State University of New York Press, 1989) Many of these full-dress theories of environmental ethics were first broached in the pages of this journal.

[8] See Taylor, *Respect for Nature;* Robin Attfield is clearest about his position, I think, in "The Good of Trees," *Journal of Value Inquiry* 15 (1981): 35–54. Rolston's earlier work was theoretically promiscuous, in my opinion, but has, in *Environmental Ethics*, settled into a stable relationship with conativism. Contemporary conativism in environmental ethics is traceable to a loose remark made by Joel Feinberg in "The Rights of Animals and Unborn Generations," in William Blackstone, ed., *Philosophy and Environmental Crisis* (Athens: University of Georgia Press, 1974): "A mere thing . . . has no good of its own. The explanation of that fact, I suspect, consists in the fact that mere things have no conative life: no conscious wishes, desires, and hopes; or urges and

Leopold and shares certain temperamental affinities with the great American conservationist, the immediate intellectual ancestor of contemporary conativism is Albert Schweitzer's reverence-for-life ethic.[9]

A second family of environmental ethics (united by "a more tender and widely diffused" altruism, to quote Darwin, with intellectual roots in Hume) has sprung, in fact, from the Aldo Leopold land ethic. I have been the most vocal champion of this theoretical approach, but fellow travelers include Edward O. Wilson, William Godfrey-Smith, and Richard and Val Routley (now Sylvan and Plumwood, respectively).[10]

A third family—centered upon Self-realization (with a capital S), based upon the unity between self and world suggested by ecology—has been advocated by the more philosophical exponents of deep ecology.[11]

Now that we had a good feel for the lay of the theoretical land, I assumed—before Stone came along with his powerful and seductive case for moral pluralism—that we could begin to work toward the creation of an intellectual federation and try to put an end to the Balkanization of nonanthropocentric moral

---

impulses; or unconscious drives, aims, and goals; or latent tendencies, direction of growth, and natural fulfillments" (p. 49). The implication, clearly, is that a minimally conative life, absent conscious wishes, desires, hopes, urges, and impulses, but possessing latent tendencies, directions of growth, and natural fulfillments has a good of its own. Tom Regan toyed with the theoretical possibilities for environmental ethics, inadvertently provided by Feinberg, in "Feinberg on What Sorts of Beings Can Have Rights," *Southern Journal of Philosophy* 14 (1976): 485–98, but eventually gave up on conations as a sufficient condition for something having a good of its own in favor of conscious wishes, desires, hopes, urges, and impulses, summed up in his concept of "subject-of-a-life." Kenneth Goodpaster, next picked up on the idea in "On Being Morally Considerable," *Journal of Philosophy* 75 (1978): 306–25; however, after publishing one more paper in environmental philosophy, "From Egoism to Environmentalism" (in which conation plays no crucial role), Goodpaster defected from environmental ethics to business ethics and has not been heard from since. Paul Taylor brought Feinberg's offhanded remark to its full fruition—inventing, parallel to Regan's notion of a subject-of-a-life, the notion of a teleological-center-of-a-life—in "The Ethics of Respect for Nature," *Environmental Ethics* 3 (1981): 197–218 and then, more fully still, in his book, *Respect for Nature*.

[9] See Albert Schweitzer, *Philosophy and Civilization*, trans. John Naish (London: A. & C. Black, 1923). The intellectual affinities between Schweitzer and an earlier generation of environmental conativists is developed in J. Baird Callicott, "Non-anthropocentric Value Theory and Environmental Ethics," *American Philosophical Quarterly* 21 (1984): 299–309.

[10] See J. Baird Callicott, *In Defense of the Land Ethic*; Edward O. Wilson, *Biophilia* (Cambridge: Harvard University Press, 1984); William Godfrey-Smith "The Rights of Non-humans and Intrinsic Values," and Richard and Val Routley, "Human Chauvinism and Environmental Ethics," both in D. Mannison, M. McRobbie, and R. Routley, eds., *Environmental Philosophy*, Monograph Series no. 2 (Canberra: Department of Philosophy, Australian National University, 1980), pp. 30–47 and 96–189, respectively.

[11] See Arne Naess, "Self-Realization: An Ecological Approach to Being in the World," in John Seed, Joanna Macy, Pat Fleming, and Arne Naess, ed., *Thinking Like a Mountain: Towards a Council of All Beings* (Philadelphia: New Society Publishers, 1988); Warwick Fox, *Approaching Deep Ecology: A Response to Richard Sylvan's Critique of Deep Ecology*, Occasional Paper no. 20 (Hobart: University of Tasmania, 1986) and Freya Matthews, "Conservation and Self-Realization: A Deep Ecology Perspective," *Environmental Ethics* (1988): 347–55.

philosophy. Recently, I took a step in that direction. Back in 1981, in "Animal Liberation: A Triangular Affair," I contemptuously dismissed the moral enfranchisement of *individual* animals *qua* individuals, because the ecocentric ethic adumbrated by Leopold—that had at first inspired me, and that I was attempting rigorously to ground—enthroned the integrity, stability, and beauty of the biotic *community* as the ultimate measure of the rightness and wrongness of human actions. Although I now wince at its stridency when I reread "Triangular Affair," and wish that I were not so closely identified with this particular piece of work, that essay did serve to delineate sharply the theoretical differences between animal welfare ethics and one approach—ecocentrism, as it has come to be called—to a primary or direct environmental ethic. Personally, however, I am not unmoved by the pain and suffering of individual sentient animals and believe that we ought to extend them moral considerability, if not rights. Thus, I have recently tried to effect a reconciliation between animal welfare ethics and environmental ethics with a little palinode entitled "Animal Liberation and Environmental Ethics: Back Together Again," which I held out as an olive branch to our colleagues interested primarily in the study of ethics and individual animals.[12] Nevertheless, in that paper I didn't simply say, "Where animals are concerned, I'll go with Singer or Regan, if they meet me half way and agree to go with me (or Rolston) on species and ecosystems." Rather, I tried to find a coherent theory that would provide at once for the moral considerability of individual animals—differently "textured," incidentally, for domestic animals on the one hand and for wild animals on the other—*and* for species and ecosystems.[13]

In "Triangular Affair" I even argued that the worth of individual *human* beings must, *if* one acceded to a demand for ruthless consistency, be measured against Leopold's holistic *summum bonum*, and suggested that its degree of misanthropy might be the litmus test of whether a stance or policy is in agreement with the land ethic. I never actually endorsed such a position, since it is obnoxious and untenable, and I now no longer think that antihuman prescriptions can be deduced from the Leopold land ethic, as I have subsequently explained.[14] I certainly feel that we have duties and obligations to fellow humans (and to humanity as a whole) that supersede the land ethic, although I have by no means abandoned the land ethic. Before Stone, I just *assumed* that a complete environmen-

---

[12] J. Baird Callicott, "Animal Liberation and Environmental Ethics: Back Together Again," *Between the Species* 4 (1988): 163–69.

[13] *Texture* is one of the many quasi-technical terms in Stone's wonderfully rich and creative book. I hope that I have used it correctly here.

[14] See J. Baird Callicott, "The Search for an Environmental Ethic," in Tom Regan, ed., *Matters of Life and Death*, 2d ed. (New York: Random House, 1986); and "The Conceptual Foundations of the Land Ethic," in J. Baird Callicott, ed., *Companion to A Sand County Almanac: Interpretive and Critical Essays* (Madison: University of Wisconsin Press, 1987), pp. 186–217.

tal ethic would begin with a carefully chosen theory of interpersonal and social human ethics, and unite animal welfare and environmental ethics under the same theoretical umbrella. Although I still think that this is the appropriate way to proceed, Stone has offered an easy and appealing alternative that demands a thoughtful reply.

Moral pluralism, crudely characterized—I hope not crudely caricatured—invites us to adopt one theory to steer a course in our relations with friends and neighbors, another to define our obligations to fellow citizens, a third to clarify our duties to more distantly related people, a fourth to express the concern we feel for future generations, a fifth to govern our relationship with nonhuman animals, a sixth to bring plants within the purview of morals, a seventh to tell us how to treat the elemental environment, an eighth to cover species, ecosystems, and other environmental collectives, and perhaps a ninth to explain our obligations to the planet, Gaia, as a whole and organically unified living thing. Stone himself provides an illustration of a pluralist posture in the following image:

> The Moral Pluralist holds that a public representative, a senator, for example, might rightly embrace utilitarianism when it comes to legislating a rule for social conduct (say, in deciding what sort of toxic waste program to establish). Yet, this same representative need not be principally utilitarian, nor even a consequentialist of any style, in arranging his personal affairs among kin or friends, or deciding whether it is right to poke out the eyes of pigeons. And surely being committed to utilitarianism as a basis for choosing legislation does not entail judging a person's character solely by reference to whether, on balance, he advances the greatest good for the greatest number of persons.[15]

THOSE WHO ADVOCATE PLURALISM IN ENVIRONMENTAL ETHICS AND THOSE WHO DON'T

Because environmental ethics invites pluralist parsing, and because Christopher Stone has now put together such a frank, strong, and eminently reasonable case for moral pluralism, one wonders if a pluralistic turn can be detected in the earlier unconventional environmental ethics literature.

In "Back Together Again" I gave critical attention to Mary Anne Warren's plea for moral pluralism published in the early 1980s. Although Warren argues that animals, like human beings, have rights, she also argues that animals do not enjoy the same rights as human beings and, *pace* Regan, that the rights of animals do not equal human rights. Animal rights and human rights are grounded in different metaphysics of morals. Human beings have "strong" rights because

---

[15] Stone, *Earth and Other Ethics*, p. 118.

we are autonomous (à la Kant). Animals have "weaker rights" because they are sentient (à la Bentham). Because plant liberation (à la Paul Taylor) had not been vociferously championed or well-formulated when Warren was writing, she has nothing specifically to say about the moral entitlements of individual living things. Nevertheless, a holistic environmental ethic, Warren suggests, rests upon still other foundations—the instrumental value of "natural resources" to us and to future generations (à la Gifford Pinchot) and the "intrinsic value" we may intuitively find in species, "mountains, oceans, and the like" (à la Aldo Leopold).[16] Warren doesn't come right out and say that she is advocating moral pluralism, but that's what her eclectic program amounts to. In "Back Together Again" I (mistakenly) thought that it was enough simply to point out that Warren's approach was eclectic and pluralistic, in order to set it aside and get on with the serious business of searching for the Holy Grail of environmental ethics—the coherent, inclusive super-theory.

Although he does not actually so label his posture, moral pluralism is, nevertheless, also detectable in Eugene C. Hargrove's metaethical discussion of the role of rules in ethical decision making. In light of recently emerged "moral perceptions" respecting animals and the environment, writes Hargrove, "moral philosophers will have to abandon for the most part the search for a rational set of universal principles which moral agents can mechanically follow."[17] Hargrove, incidentally, anticipates Stone's comparison of moral theorizing to map making. Hargrove, in another conceit (that Stone seems not to have thought of but might find useful), suggests that we understand moral rules to be similar to nonconstitutive chess rules—not the rules of the game that inflexibly govern the movement of the pieces, but the ad hoc rules for effective play in various situations. According to Hargrove, in ethics as in chess,

> the body of rules has no ultimate unifying principle, the principles themselves are not logically related to one another (the omission of one or the addition of another in no way affects the group as whole), they are not organized in any meaningful hierarchy, . . . and there are innumerable cases which can be brought forward with regard to each of them in which following the proper rule leads to disaster in a board [or, analogously, in a real-life] situation.[18]

Almost simultaneously with Stone's extensive case for moral pluralism, Andrew Brennan and Peter Wenz have expressly advocated "ethical polymorphism"

---

[16] Mary Anne Warren, "The Rights of the Non-human World," in Robert Elliot and Aaran Gare, ed., *Environmental Philosophy: A Collection of Readings* (University Park: The Pennsylvania State University Press, 1983), pp. 130–31.
[17] Eugene C. Hargrove, "The Role of Rules in Ethical Decision Making," *Inquiry* 28 (1985): 30.
[18] Ibid., p. 22.

and a "pluralistic theory" of environmental ethics, respectively.[19] Brennan remarks that

> an ethic by which to live is not to be found by adopting one fundamental, substantive principle relative to which all our deliberations are to be resolved. Instead we are prey to numerous different kinds of consideration originating from different directions, many of them with a good claim to be ethical ones.[20]

Brennan goes on to rummage through a few of these different kinds of consideration, showering most of his attention on the competing claims of environmental individualism and environmental holism to which our ethical intuitions are prey. Wenz, for his part, after glossing the same (by now familiar) biocentric (Taylorian) and ecocentric (Callicottian) theories, as something of an afterthought at the end of an elaborate review of conventional theories of justice and an evaluation of the direction they each give for equitably dividing the environmental pie, feels less inclined to settle on one best approach than to float them all at once:[21]

> We found . . . that none of the above theories of justice was flexible enough to accommodate all of our considered views about how particular matters of environmental justice should be decided. . . . But because each theory and many of the principles contained in each theory seem reasonable when applied to certain kinds of cases, they should not be abandoned entirely. They should be modified and blended to form an all embracing pluralistic theory. A theory is pluralistic when it contains a variety of principles that cannot be reduced to or derived from a single master principle.[22]

G. E. Varner's review of *Earth and Other Ethics*, underscores, *correctly* I think, "the fundamental metaethical challenge" to environmental ethics posed by Stone.[23] In casting about for harbingers of pluralism, however, Varner, quite understandably, completely overlooks Warren's essay—one small, dated item in

---

[19] Andrew Brennan, *Thinking About Nature: An Investigation of Nature, Value, and Ecology* (Athens: The University of Georgia Press, 1988); Peter S. Wenz, *Environmental Justice* (Albany: State University of New York Press, 1988).

[20] Brennan, *Thinking About Nature*, p. 186.

[21] In the introduction to *Environmental Justice*, Wenz tells how as a boy he and his friend Billy used to divide a pizza. One cut it in half and the other chose first. Wenz often returns to this childhood image of distributive justice throughout his discussion. For Wenz, the environment is a big pizza. The ethical question is how to divide it equitably among human consumers. Rolston notices this feature of Wenz's discussion in his review of *Environmental Justice* in *Between the Species* 5 (1989): 147–53.

[22] Wenz, *Environmental Justice*, p. 310. Wenz is aware that the phrase "pluralistic theory" may seem an oxymoron to many philosophers.

[23] G. E. Varner's review of Christopher D. Stone, *Earth and Other Ethics*, in *Environmental Ethics* 10 (1983): 264.

a literature that is growing exponentially—and Hargrove's which is about ethics in general, not environmental and animal welfare ethics specifically. Wenz's and Brennan's very recent books, on the other hand, could not have come to his attention before his review of Stone was all but set in stone.

Yet, Varner goes on to find, *incorrectly* I think, intimations of pluralism in Paul Taylor's work and my own. He writes,

> There are hints—but still only hints—of pluralism emerging in recent work by Paul Taylor and Baird Callicott. In *Respect for Nature*, Taylor stresses that environmental ethics rests on very different foundations than human ethics, and that it accordingly embodies very different principles. In his recent "Search for an Environmental Ethic," Callicott similarly abandons the vitriolic rhetoric of his "Triangular Affair" article and stresses that Leopold characterizes the land ethic as an accretion that supplements, rather than replaces, previous ethics.[24]

How accurately my work may be characterized as pluralistic may be gathered from this essay. As to Paul Taylor's, his theory strikes me as about as clear a case of what Stone calls moral monism as any theory of environmental ethics could possibly be. While Taylor grants moral rights to human teleological centers of life and withholds them from nonhuman teleological centers of life, he insists that, their rights notwithstanding, the former are in no way morally superior to the latter and that *all* teleological centers of life are of *equal* inherent worth.[25]

Basically, Taylor slips conation into the slot reason fills in Kant's moral philosophy.[26] To be sure, Taylor, a neo-Kantian, cannot bring himself to completely renounce the classical Kantian emphasis on special respect for persons and sometimes speaks of "both systems of ethics"—respect for persons and respect for nature—as if he were juggling two independent principles. But Taylor's moral theory is monistic, as Wenz clearly recognizes. When he considers conflicts of interest, Taylor does not treat respect for persons and respect for nature as two mutually incommensurable systems of ethics. For Taylor, "the good of other species and the good of humans make claims that must equally be taken into consideration."[27] It's just that human beings, as self-conscious and morally autonomous beings, have certain goods that other teleological centers of life lack. Thus, while it may be contrary to a tree's good to cut it down, it is not contrary to a tree's good, as it is to a human being's, to be lied to or to be fenced in. Respect for persons (and associated human rights) takes into account the

---

[24] Ibid.
[25] Taylor, *Respect for Nature*, p. 260.
[26] See Bryan Norton's review of Paul Taylor, *Respect for Nature*, in *Environmental Ethics* 9 (1987): 261–67.
[27] Taylor, *Respect for Nature*, p. 259.

peculiar, but not better making, endowment of human beings in comparison with other forms of life.

Taylor does not, in other words, resort to other theories than his basic biocentrism in order to guarantee that we human beings (at least we genteel, culturally rich, but materially modest human beings) can go on living the lives to which we have grown accustomed. He tries to make things come out right—so that we can eat vegetables, build wooden houses, and generally get on with our human projects (at least our more refined, low-impact human projects)—by means of an elaborate set of hedges enabling us consumptively to use our fellow entelechies within the limits of his extremely broad egalitarian theory.[28] To show that Taylor's attempt to deal, under the auspices of a single theory, with all our considered moral sensibilities—from respect for human rights to the moral enfranchisement of individual plants—collapses under the load, Wenz employs, very effectively in my opinion, a technique of philosophical refutation delightfully satirized by Stone: he "volleys hypothetical quandaries onto" Taylor that his "principles cannot handle."[29]

Varner might somewhat more profitably have looked into Holmes Rolston's theory of environmental ethics for "hints" of pluralism. In *Environmental Ethics* the book (a recent consolidated and definitive statement of his revolutionary moral theory) Rolston devotes a chapter each to the intrinsic value of, and corresponding duties owed to, "higher animals," "organisms," "species," and "ecosystems." Along the way he finds intrinsic value in evolutionary *processes*—going all the way back to the Big Bang.

Because Rolston does not strive for unambiguous clarity and clean, crisp definitions, as Taylor does, it may seem that in each of these chapters he develops independent arguments—not all of which, even within each chapter, are univocal—for the value of each of these natural things. Nevertheless, I think that most of his arguments for intrinsic value in nature cluster around a central, pivotal notion—conation, once again. Organisms have, each of them, a *telos*, Rolston reminds us—unconscious drives and aims, or (at least) latent tendencies, directions of growth and natural fulfillments. His defense of their intrinsic value seems to settle upon the fact that organisms, even unconscious organisms, compete for nutrients and a place in the sun, and defend their own lives—in an astonishing variety of ingenious ways. They have, thus, a good of their own. Sentient organisms are aware of their strivings, and feel urges and impulses, some more keenly than others; and we human organisms are *self*-aware, conscious of ourselves as beings with wishes, desires, hopes, and goals. Unlike Taylor, Rolston assigns a value bonus to consciousness and a double bonus to self-consciousness which he adds on to the value base constituted by conation.

---

[28] Ibid., chap. 6.
[29] Wenz, *Environmental Justice*, chap. 13. The quotations are from Stone, *Earth and Other Ethics*, p. 117.

From this value epicenter, conation, Rolston hoes the row the other way. Each organism, Rolston says, in a characteristic argument by verbal trope, represents—i.e., "re-presents"—its species.[30] Each struggles, not only to survive and flourish, but also to reproduce. Each is a token of its type. Its type is indeed its *telos*, just as Aristotle would have it. Each strives to be "good-of-its-kind" and to "defend its own kind of good."[31] Hence, each kind—each species—is intrinsically valuable, even though species are holistic entities which are not conative *per se*. Ecosystems, similarly, are the matrices which give birth to the myriad intrinsically valuable kinds. Rolston says that ecosystems and evolutionary processes are "projective."[32] They do not themselves possess *teloi* (Rolston observes with strict consistency the ateleological conventions of evolutionary biology), but they have "projected," or thrown up, a good many ordered systems and organized entities. The quasi-conative character of ecosystems and evolutionary processes earns them a value dividend in Rolston's theory of environmental ethics.

Given that even Rolston is not really a pluralist after all, one begins to wonder why our best, most systematic, and thoroughgoing environmental philosophers cling to moral monism. Although one can find scattered outbreaks of pluralism in the literature, so far pluralism has not become epidemic. Stone, a lawyer, does make moral pluralism seem so reasonable, and its opposite a silly and parochial preoccupation; yet, Taylor and Rolston have mounted veritably epic efforts to save the philosophical integrities of their respective systems. Taylor will save his biocentrism at the cost of patent sophistries (which Wenz revels in exposing) and Rolston can save his only by resorting to ambiguity and courting equivocation. Why? Why don't we all just become merry moral pluralists? I take this to be the metaethical challenge that Varner says that Stone has thrown down.

MORAL PLURALISM'S ACHILLES HEEL: THE HARD CHOICE BETWEEN CONTRADICTORY INDICATIONS

Wenz, pluralist convert though he may have become, clearly articulates one reason to beware of its siren lure:

> Without a single master principle in the background, what is to be done . . . when one of the independent principles in the pluralistic theory requires a course of action different from and incompatible with the course of action required by one of the other independent principles . . . ? In this kind of situation, the theory yields either no recommended course of action or contradictory recommendations.[33]

---

[30] Rolston, *Environmental Ethics*, p. 143.
[31] Ibid., p. 101.
[32] Ibid., chap. 6.
[33] Wenz, *Environmental Justice*, p. 313.

Consistency is not just a shrine before which philosophers worship. There is a reason for wanting consistency, insured by organization around or derivation from a "master principle," among one's practical precepts. Attempting to *act* upon inconsistent or mutually contradictory ethical principles results in frustration of action altogether or in actions that are either incoherent or mutually canceling.

Stone, of course, has thought of this problem. It is worth noting, he points out, that a multiplicity of independent principles might just as well all converge on a single course of action. The practical necessity of such a plurally mandated course of action would be reinforced, rather than frustrated or negated. *Earth and Other Ethics* is wonderfully creative, not only in its advocacy of the idea of pluralistic ethics, but in working out methods for solving ethical conundrums drawing upon the resources of a variety of moral systems. Stone asks us to think of various maps of a single territory. One map might show human population distribution, another land-use patterns, a third the vegetation types, a fourth contours, and so on. If we regard a situation in which we must do something as the "territory" and various theories as the "maps" (or "planes," as Stone later calls them), we may overlay the "planes"/"maps" and see if they indicate a clear path of action. Why, he asks, should we expect several overlays to yield interference patterns more usually than sympathetic vibrations?

Still, what do we do when we put all our systems down, each "plane" layered over the "territory" (the actual or hypothetical moral quandary in which we find ourselves), and the indications are inconsistent or contradictory? The actual case of a bison trapped in a frozen Yellowstone river, a case that Stone fully develops in his book, serves as a good example. Animal welfare ethics indicate that we ought to try to save individual animals from unnecessary pain, suffering, and eventual death, while ecocentric environmental ethics indicate that we ought to let bison (and all the other nonhuman members of the biotic community) alone to struggle for their lives and live and pass their tested genes on to the next generation or die and become food for the carrion eaters. Stone tells us what happened to this particular bison on this particular occasion, and how the moral theories that various people held affected their actions and inactions, but he never tells us what these people ought to have done or not done for the animal.

One problem with moral pluralism, noticed by Hargrove—but that neither Wenz nor Stone directly address—is that it invites a kind of moral promiscuity. Hargrove notes a potential "fear that the open form in which decisions naturally and normally take place will allow unscrupulous or weak moral agents to waver and [choose] principles to their own immoral advantage."[34] With a variety of

---

[34] Hargrove, "The Role of Rules," p. 26.

theories at our disposal, each indicating different, inconsistent, or contradictory courses of action, we may be tempted to espouse the one that seems most convenient or self-serving in the circumstances.

If Stone can, in a charming and friendly way, tweak us philosophers about our foibles, then turn about is fair play. Lawyers are notoriously adversarial. They are trained to use scholarship and logic, not to seek the truth, or implement justice, but to represent a client or win a case—regardless of where the truth and the right lie. The overall structure of *Earth and Other Ethics* does not give one much comfort about the worry that moral pluralism might provide a sophisticated scoundrel with a bag of tricks to rationalize his or her convenience or self-interest—rather than a box of tools to work his or her way through the moral complexities of life in the twentieth and twenty-first centuries.

In the early 1970s, Stone began with a *desideratum*—how *legally* to enfranchise "non-persons," as he calls everything from ships and corporations to wild rivers and endangered species. Extending them (limited) legal rights, operationally defined, was his answer in "Should Trees Have Standing?."[35] Since then the courts have considerably liberalized standing, but "standing . . . does nothing but get you through the courthouse door."[36] The law can provide all sorts of legal fictions and devices for the legal considerateness of non-persons. That's no problem. But once environmentalists or animal liberationists get a hearing for their "wards," Stone asks, then how do they go about "justifying" legal accommodation and how can they prevail against competing interests?[37] This question leads Stone from the realm of law to the realm of ethics. In other words, like a lawyer, he begins with a spectrum of practical ends in view—leaving some resources and wilderness for the use and enjoyment of future generations, protecting animals from pointless or needless experimentation, saving species, etc.—and looks for a spectrum of persuasive theories, as means, to secure those ends. Then finally, the apparent hopelessness of ordering that very caboodle of practical ends and kit of theoretical means within a single comprehensive moral philosophy leads him on to plead for pluralism and to challenge the very impetus to univocal theory construction.

To the worry that moral pluralism invites moral promiscuity, had Stone expressly confronted it, he might have replied, more or less as Hargrove does, that all moral philosophy presupposes persons of good will. Pluralism may supply a scoundrel with another sort of rationalization for ducking his or her responsibilities, but moral philosophy generally—monistic no less than pluralistic—is underdeterminate and, in the hands of a skilled, but unscrupulous,

---

[35] Christopher D. Stone, "Should Trees Have Standing?—Towards Legal Rights for Natural Objects," *Southern California Law Review* 45 (1972): 450–501.
[36] Stone, *Earth and Other Ethics*, p. 10.
[37] Ibid., p. 42.

advocate can be made to justify all manner of action or inaction. (Didn't lesser Nazi war criminals at their Nüremberg trials hide behind Kant's noble notions of apodictic duty?) One might argue, by parity of reasoning, that the ethical lives of sincere persons of good will are proportionately enriched and empowered by moral pluralism, thus offsetting the invitation to the abuse that pluralism inadvertently affords persons less noble of character.

Granted that moral pluralists may be sincere persons of good will, how *do* they decide between the inconsistent or contradictory indications of their several theories? Stone, of course, has thought about this too. He suggests we bring to bear a "lexical" procedure for reaching a decision.

Lexical, in this context, is a euphemism for hierarchical ordering— prioritizing, if you will. Baldly put, what Stone suggests is this: we take our many moral "maps," "planes," and "frameworks" (our polyglot ethical systems or theories), lay them out over the "territory" (the problem, quandary, or conundrum) and, if they jibe, fine. If they don't, then we prioritize them.

But how? For pluralists there's no "master principle," no super-theory, by definition. His back against the wall, Stone frankly endorses appeal to what Regan calls "considered intuition," and to cultivated moral tastes, sensibilities, feelings, and "moral faculties."[38] (Wenz similarly, asks us to use unspecified "good judgment," while Hargrove, for his part, is simply willing to live with more of a mess than Stone.[39])

But what does that really tell us? When push comes to shove, how do we choose between theory A with its recommended course of action a, B with b, and C with c? According to Stone, "when we turn to the selection of planes—what things [people, animals, plants, species], as bundled in what governance [utility-maximizing, person-respecting], count?—we are removed to another jurisdiction. . . ."[40] The final resolution of the intractable dilemmas that will inevitably confront the moral pluralist, Stone tells us, lies in "selecting a version of the world."[41] Stone points in his final pages, in other words, to metaphysics. To "buy into" a "plane"—a set of ethically enfranchised entities (a "moral ontology") as Stone revealingly calls such a set, and a "governance" (how exactly those entities are endowed with rights or equal consideration of interests, respect, or whatever)—is to buy into a world view. When we are forced to choose between "planes," we make a metaphysical commitment as well as a moral choice.

But then the real work, the ethical *Grundlegung* should commence anew. We might think of our hard moral choices (in my opinion misleadingly mystified by

---

[38] Regan endorses "considered" or "reflective" intuition in *The Case for Animal Rights.* Taylor, the consummate neo-classical monist spurns them in *Respect for Nature,* on the grounds that we should always follow the dictates of principle, whether they conform to our intuitions or not.

[39] Wenz, *Environmental Justice,* p. 314; Eugene C. Hargrove, personal communication.

[40] Stone, *Earth and Other Ethics,* p. 256.

[41] Ibid.

Stone with his talk about intuition, literature, art, and humor) as actually revelatory of the deeper—but because deeper less fully conscious—structures of our thinking.[42] The task then is to call these organizing concepts up on the screen—to articulate in a more self-conscious and deliberate way the world view, the metaphysics, which has rationally (we hope), but subconsciously, arbitrated among the divers moral "maps." However, once we have carried through on that project, then the many "maps," and the whole apparatus of moral pluralism become otiose. Some of the "maps" will be seen to assume untenable versions of the world or versions of human nature and we may readily consign them to the historical rubbish heap—along with "maps" to ferret out and punish blasphemy and hunt and burn witches, which involve concepts like God's literal word and efficacious satanic rites (and verses), and other notions belonging to a version of the world which centuries of experience and critical thinking have, let's face it, invalidated.

Stone seems to think, however, that such metaphysical questions lie beyond philosophical competence: "It is just that these 'big' questions lie outside the province of academic and legal philosophy, which are more at home working *within* or talking *about* planes."[43] Here, finally, is the crux of what I think is wrong with moral pluralism. It severs ethical theory from moral philosophy, from the metaphysical foundations in which ethical theory is, whether we are conscious of it or not, grounded.

THE CASE FOR A UNIFIED MORAL PHILOSOPHY

Can ethics be so severed from moral philosophy, the metaphysical groundwork of ethical theories? The medieval world view seems to most latter-day moderns to be a quaint anachronism and its associated moral "maps" to be curious—when they are not sinister—relics of a bygone mentality. But the modern world view is itself rapidly becoming history. How, therefore, can we reject, out of hand, condemnations and death sentences for "blasphemy," witch hunts, and book burnings, and continue uncritically to ascribe, in a pluralistic spirit, to equally—though more recently—obsolete eighteenth-century moral theories like hedonic utilitarianism or "pure reason" Kantianism?

---

[42] Stone writes that "we are removed to another jurisdiction in which our minds operate less by appeals to consistency than provocations of irony and humor. Here the dynamic involves the demonstration of buried contradictions in our lives, rather than of inconsistency among our ideas. Emotion has a more legitimate rein (or reign); suppressed feeling and insight are released and mobilized. . . . [we] are less under the sway of the stuff we academics do than of literature, folk songs, war, art, landscape, and poetry. . . . Poetry and literature, obviously, are high forms of intellect; but rather than to derive 'truths,' they make them manifest" (*Earth and Other Ethics,* p. 256).

[43] Ibid.

Why, more pointedly, we may ask, has Stone borrowed his "maps" and "planes" from the moral philosophies of Bentham, Mill, Kant, and other modern philosophers, and from those of Schweitzer, Singer, Leopold, Taylor and other contemporary philosophers, but neglected to borrow with equal alacrity from medieval philosophers like St. Thomas, and St. Augustine, and ancient philosophers like Plato and Aristotle? Or why do we not find "maps" and "planes" featured in *Earth and Other Ethics* taken from the Old Testament—say the Mosaic Decalogue—or from the Koran? The answer is, I suggest, that Stone and most of us who are thoughtful enough to worry about the human treatment of whales, rain forests, and the ozone layer do not buy into divine revelation of moral commandments or the independent existence of the Good (with a capital *G*)—concepts that are among the various metaphysical background assumptions of Aquinas, Augustine, Plato, Aristotle, the late Ayatolla Ruholla Khomeini, and the Reverend Jerry Falwell.

Nevertheless, each of the modern ethical systems which Stone so deftly employs in his pluralistic ethical tool kit also comes wrapped in its metaphysical vestments. Consider the threadbare metaphysical cloth from which classical utilitarianism is cut. Utilitarianism assumes a radical individualism or rank social atomism completely at odds with the relational sense of self that is consistent with a more fully informed evolutionary and ecological understanding of terrestrial and human nature. Bentham could not have more clearly revealed his reductive assumptions when he wrote, "the community is a fictitious body composed of the individual persons who are considered as constituting as it were its members. The interest of the community then is what?—the sum of the interests of the several members who compose it."[44]

Bentham, the founder of utilitarianism, also invests intrinsic value and disvalue exclusively in psychological experiences—pleasure and pain, in all their protean forms. Such a psychologized understanding of good and evil is historically linked with, if it does not literally follow from, the radical Cartesian split, ubiquitous in modern philosophy, between subject and object, and the resulting alienation of the self from the "external world"—which includes our own bodies. "Sense data" and sensations are, from the prevailing modern point of view, the nearest—and hence the dearest—realities.

To employ Stone's several utilitarian "planes," therefore, involves buying into a vision of human (and animal) nature in which isolated egos (*subjects*-of-a-life) are imprisoned within alien mechanical objects (their bodies), and look fearfully out on a foreign "external world." The only "things" to which good and evil can attach, given such a world view, are, naturally, positive and negative private *subjective* psychological states.

To adopt Kant's moral theory is to buy into a vintage European Enlightenment

---

[44] Jeremy Bentham, *An Introduction to the Principles of Morals and Legislation* (Oxford: The Clarendon Press, 1823), chap. 1, sec. 4.

philosophy of human nature in which Reason (with a capital *R*) constitutes the essence of "man," somewhat in the way that the "image of God" constitutes the essence of "man" in the biblical conception of human nature. It is revealing, I think, to note that Kant himself never considers the "marginal cases," which animal welfare ethicists routinely volley onto contemporary Kantians. He never seems to notice, in other words, that, by his principles, subrational human beings such as infants, imbeciles, and the senile are also "mere things" (as he characterizes sentient animals lacking reason) and might be treated accordingly. That's not because Kant didn't think his theory through, but because he understood reason to be less an organic function or capacity than a kind of philosophically sanitized, Enlightenment equivalent of the *imago dei* inhabiting all human beings, quite irrespective of their functional rationality.

Now let's return once more to Stone's senator, a model moral pluralist, and consider a variation on Stone's description of her agile ethical experience. After "embracing utilitarianism" on a floor vote in the morning "(deciding what sort of toxic waste program to establish)," her staff reminds her that in the afternoon a vote will be taken on a forest service plan to "road" the Gila wilderness. Our senator has read *A Sand County Almanac* and added the land ethic to her moral repertoire. So, having given utilitarianism a turn in the morning, embracing the land ethic in the afternoon, she votes against the forest service. This means that over lunch she has blithely stepped out of the atomized, mechanical, and dualistic view of nature and human nature inspiring utilitarianism and into the organic, internally related, holistic view of nature and human nature animating the land ethic—a world view in which human beings are not privatized pleasure-loving egos, but integrated plain members and citizens of social and biotic communities. Then upon leaving the Capitol, she remembers that she has promised to help her son write a school essay that evening. So, tired though she may be with handling the public business all day long, during dinner she slips into a Kantian mode, and considers that a promise is a promise, and that (as Stone represents it at least) one should not use utilitarian cost-benefit analyses appropriate to funding toxic waste programs in deciding what to do in family affairs. Thus, buying into a Prussian view of ironclad categorical, apodictic imperatives, grounded in metaphysical Reason, she works past her bedtime editing her son's essay, "How I Spent My Summer Vacation."

Moral pluralism, in short, implies metaphysical musical chairs. I think, however, that we human beings deeply need and mightily strive for consistency, coherency, and closure in our personal and shared outlook on the world and on ourselves in relation to the world and to one another. Stone is skeptical that Truth, with a capital *T*, may be had in matters metaphysical.[45] I am no more

---

[45] Stone writes, "No simple formula about truth is available when we ascend to the level of selecting a version of the world for scientific purposes, that is in deciding whether and in what form to posit matter, space, energy, time, and their relationships" (*Earth and Other Ethics*, p. 256).

sanguine than he, but I do think that we can expect to generate comprehensive conceptual systems that fully embrace our ever-growing body of empirical knowledge, scientific theory, and self-discovery. Although in many ways *Earth and Other Ethics* is a much better book than *Should Trees Have Standing?* in the years between *Trees* and *Earth* Stone seems to have abandoned a project that I think it is more important now than ever to try to advance. In the earlier book he wrote,

> The time may be on hand when these sentiments, and the early stirrings of the law, can be coalesced into a radical new theory or myth—felt as well as intellectualized—of man's relationships to the rest of nature. I do not mean "myth" in the demeaning sense of the term, but in the sense in which, at different times in history, our social "facts" and relationships have been comprehended and integrated by reference to the "myths" that we are co-signers of a social contract, that the Pope is God's agent, and that all men are created equal.... What is needed is a [new] myth that can fit our growing body of knowledge of geophysics, biology, and the cosmos. In this vein, I do not think it too remote that we may come to regard the Earth, as some have suggested, as one organism, of which mankind is a functional part—the mind, perhaps: different from the rest of nature, but different from the rest of nature as a man's brain is from his lungs.[46]

THE DISTURBING CONNECTION BETWEEN MORAL PLURALISM AND DECONSTRUCTIVE POSTMODERNISM

Absent such a comprehensive model to focus and order our competing moral concerns, we are left with kaleidoscopic and random, albeit enriched, moral lives—individually. Collectively, socially, we are left with irreconcilable fractional disputes. Stone's happy-go-lucky moral pluralism, culturally generalized and interpreted, is allied with—if not equivalent to—deconstructive postmodernism. Absent a comprehensive and culturally shared new myth, we are left with plural points of view, perspectives, multiple outlooks—each of which has an equal claim on "truth."

The "postmodern turn" in environmental philosophy has recently been taken by Jim Cheney.[47] Joining other ecofeminists running out to bark at deep ecology,

---

[46] Christopher D. Stone, *Should Trees Have Standing?: Toward Legal Rights for Natural Objects* (Los Altos, Calif.: William Kaufmann, 1974), pp. 51–52. Notice the rhetorical allusion in the first sentence to Bentham's famous remark, quoted *ad nauseum* in the animal welfare literature: "The day may come when the rest of the animal creation may acquire those rights...." Stone was writing in the early 1970s before the phrase "animal liberation" had been coined by Peter Singer and before women's liberation had made a reference to a "man's brain" and "his lungs" appear gender-biased.

[47] Jim Cheney, "Postmodern Environmental Ethics: Ethics as Bioregional Narrative," *Environmental Ethics* 11 (1989): 117–34; and "The Neo-Stoicism of Radical Environmentalism," *Environmental Ethics* 11 (1989): 293–325.

Cheney snaps at deep ecology's "totalizing" vision—a vision not unlike Stone's regrettably abandoned one—which attempts to "colonize" other outlooks. By *totalizing* vision Cheney seems to mean one that is comprehensive. By *colonizing* outlook he seems to mean one that claims to be the best available. Cheney employs these neologisms repeatedly and they have the distinct ring of an argot or cant. This is the closest Cheney comes to defining them:

> One form that understanding can take is that of the construction of a totalizing theory designed to assimilate the other into a unifying conceptual framework. . . . One of the functions of a totalizing theory is to rationalize the "colonizing" of the other, the control of the other by means of control over naming.[48]

A metaphysical system which tries to embrace our ever-expanding human experience, to comprehend it, *and* to make sense of it is part of the problem, Cheney thinks, not the solution. In any case, comprehensive system building, Cheney also seems to think, is a decidedly modern preoccupation. With the "demise of modernism" there has occurred a "shattering into a world of difference, the postmodern world."[49]

Postmodernism is a term associated with two very different things. One we might call "constructive" postmodernism, following Frederick Ferré.[50] Modern natural philosophy, essentially classical mechanics, has been overturned by the new physics. Everything else modern—capitalism (and anti-capitalist Marxism) in economics, utilitarianism in ethics, the social contract theory in political philosophy, etc.—that has orbited modern natural philosophy has been left without a center or a foundation. We live in a time not unlike that of Plato or Descartes. The old order has passed, but the new one has not yet arrived. For Plato the old order was the Homeric-Hesiodic world; for Descartes, it was the Aristotelian-Thomistic cosmos; for us it is the Cartesian-Newtonian universe. In our time, contemporary academic and legal philosophers "working within and talking about planes," as Stone puts it, are like the poets and rhapsodes scorned by Plato, and the Schoolmen contemned by Descartes—an army of craftspersons working frantically to keep a crumbling old edifice (to employ Descartes' architectural metaphor) in some semblance of repair.

---

[48] Cheney, "Neo-Stoicism," p. 310. The neologism, *totalizing*, seems to be a hybrid intended to evoke the political ogre of totalitarianism in the vocabulary of a Valley Girl, as in, "I was *totally* totalized."

[49] Ibid. p. 302.

[50] See Frederick Ferré, "Toward a Postmodern Science and Technology," in David Ray Griffin, ed., *Spirituality and Society: Postmodern Visions* (Albany: State University of New York Press, 1988). David Ray Griffin has established a Center for a Postmodern World in Santa Barbara, California to further postmodernism of this "constructive" sort and edits a series for SUNY Press called "Constructive Postmodern Thought."

In the Ferréian sense of the term, *postmodernism* is a place marker. While we revolutionary philosophers turn to the task of razing the old structures and rebuilding a new metaphysics from the ground up—by distilling the abstract ideas out of quantum theory and ecology, as Descartes distilled the abstract ideas out of Copernican astronomy and Galilean mechanics—and while we remodel, accordingly, such satellite areas of philosophy as ethics and political theory, we call this interlude "postmodernism," since all we know for sure is that modernism is dead (though not gone). Because we can't be quite sure yet what modernity's successor will turn out to be be, we remain cautious and wait for "organicism" or "systems theory" or some such label to take hold.

The other sense of postmodernism, following Jacques Derrida and Richard Rorty, is exclusively deconstructive—and essentially nihilistic.[51] For the constructive postmodernists the fractured present is an existentially distressing moment, but a moment of great creative intellectual opportunity—exhilarating, exciting, and stimulating. Deconstructive postmodernists, on the other hand, are content to deconstruct the old texts and declare that there will be no new master narratives, no new *Organons, Meditations,* or *Principia* to set the course for generations to come. We don't just need a *new* metaphysics, they seem to think, we need to get off the metaphysical treadmill altogether; we don't just need to reorganize our world *view*—to respond to and accommodate fundamental changes in natural philosophy—we need to see (oops, realize, rather) that a "view," a "vision" of any sort is a modernist hang-up.

Cheney's references to deep ecology's metaphysical pretensions and vision-constructing aspirations drip with contempt. "Why would it occur to one," Cheney writes, "that alienation is to be overcome by, of all things, metaphysics, by the empathetic internalization of a highly abstract, humanly constructed *vision* of wholeness, connectedness, and health."[52] It seems that like Ronald Reagan, who consigned "liberal" to the dictionary of dirty words, Cheney wants us to smirk at the deep ecologists' innocent description of their "vision."

Who knows whether future postmodern philosophy will be constructive or deconstructive. If deconstructivism is the wave of the future, however, then we should not call it "postmodernism," but "post-Western civilizationism," since metaphysics and comprehensive intellectual system building, with natural philosophy at the core, go way back in the Western tradition, considerably further back than the modern period.

Although the impetus to deconstructive postmodernism is largely political, its

---

[51] See Jacques Derrida, *Positions,* trans. Alan Bass (Chicago: University of Chicago Press, 1981); and Richard Rorty, *Philosophy and the Mirror of Nature* (Princeton: Princeton University Press, 1979).

[52] Cheney, "Neo-Stoicism," p. 301; emphasis in original.

argumentative fulcrum is epistemological. We have given up on Truth (with a capital *T*). To mirror nature with the mind has been a common ambition of philosophers from Thales to Russell in the Western tradition. Past Western philosophers hoped to arrive at and guarantee the truth by inductive/empirical methods, or by deductive/rational methods, or by a judicious combination of the two. Through centuries of error, they kept the faith that we would eventually arrive at a conceptual model that would correspond, point for point, to Reality (with a capital *R*). That dream has become more elusive now than ever. Newton seemed at last to have grasped Reality by the tail and to have put a lock on Truth. But then along came Planck, Einstein, and finally Heisenberg. Uncertainty is now a cornerstone of foundational physics.

Grant skepticism the freest rein. Must we, therefore, accede to nihilism and relativism, as the deconstructive postmodernists seem to think? Not necessarily, I would argue. Though we may not hope to marry Truth to Reality, we may hope to find an intellectual construct that comprehends and systematizes more of our experience and does so more coherently than any other. That's exactly what Stone meant in *Trees*, I think, when he urged us not to make a final assault on Reality and Truth, but to seek a new "*myth*"—one "that can fit our growing body of knowledge."

But such an honest and reasonable compromise between old fashioned truth seeking and nihilism is politically suspect. According to Cheney, it represents a last ditch effort to "colonize" "other" points of view—though he wouldn't want to own the visual metaphor—for there is no one true view. Hence, everyone is entitled to his or her own myth, one that grows out of his or her own personal experience, however limited and uninformed it may be. We require a "politics of difference" in which truth (with a small *t*) is reached by "social negotiation," not "colonization."

To my knowledge none of the philosophical deep ecologists who are the targets of Cheney's attack has proposed conversion to Ecosophy T or Ecosophy S at sword point, as the colonizing Spanish spread Christianity; nor have they suggested that harmony with nature grows out of the barrel of a gun. It seems, rather, that they are guilty of assuming that people can agree about what is what—if they are willing to come up to speed on our growing body of knowledge and rationally think through how that knowledge may be integrated into a single coherent world view. They have put their best foot (or feet) forward with Ecosophy T and Ecosophy S and I think that they are willing to argue these ecosophies on their merits. Why must we resort to *negotiation* between intractable parties when hope of agreement remains? Why not listen to one another . . . and be open to *persuasion?*

Because, it seems clear from Cheney's paper, the ground rules for persuasion themselves are totalizing/colonizing. Our growing body of knowledge comes from science and "science has constructed itself in such a way that it has insulated itself from social negotiation."⁵³ Never mind that science is an international activity, the hallmark of which is the falsifiable hypothesis and the repeatable experiment. Cheney even considers it to be politically suspect to suggest making *reason* an arbiter of what is worth believing and what isn't—since reason may be a colonizing device of patriarchy.

Thus, in the new dark age of deconstructive *différence*, without even the minimum methodological agreements required for resolving differences of opinion by informed, reasoned argument, negotiation is our only recourse. Or is it? Mention of the Spanish Conquistadores' technique for securing consensus reminds us that a far more likely option for a *Realpolitik* of difference in a shattered and fragmented world is naked power—backed either by bullets or bucks. Why negotiate with someone with whom agreement is hopeless when you can have your own way—when the "other" can be bombed, terrorized, bought, pacified, or sweet-talked?

A SINGLE MORAL PHILOSOPHY UNITING MULTIPLE MORAL COMMUNITIES

The moral pluralists' inability clearly to articulate a criterion for choosing among several inconsistent courses of action, indicated by several incommensurable moral theories, is not itself a terribly serious problem. (Monistic theories—Kant's notoriously, with its conflicting *categorical* duties—sometimes run aground on the same sorts of intractable practical contradictions.) Rather, in my opinion, it is a symptom of a deeper, more distressing malaise—the disengagement of ethics from metaphysics and moral philosophy. Hargrove, for example, simply regards the recent emergence of environmental and animal welfare ethics to have proceeded from new "moral perceptions"—rather than from newly acquired cognitive lenses which have reorganized what our senses have perceived all along—as if suddenly we acquired keener senses for empirical moral properties.⁵⁴ The environment and animals have been around all along. That we now regard them as appropriate beneficiaries of ethics depends not on our recent perception of something in them that we didn't perceive before, but on our recent understanding of what and who they are in relation to what and who we are.

On the other hand, I completely agree with Stone, who is too kind to put it so flatly, that a monistic system like Paul Taylor's simply fails to integrate our many and genuinely diverse moral concerns. So how can we have our cake and eat it too? We must operate effectively within a multiplicity of moral spheres—family obligations, the duties associated with our professional lives, our public lives, our interspecies, and ecosystemic and biospheric relationships—each with

---

⁵³ Ibid., p. 308.
⁵⁴ Hargrove, "The Role of Rules."

its very different set of demands that often compete, one with another. At the same time we feel (or at least I feel) that we must maintain a coherent sense of self and world, a unified moral world view. Such unity enables us rationally to select among or balance out the contradictory or inconsistent demands made upon us when the multiple social circles in which we operate overlap and come into conflict. More importantly, a unified world view gives our lives purpose, direction, coherency, and sanity.

Stone's term *lexical*, which he uses when talking about setting priorities among the multiple moral spheres through which we move, not only connotes an alphabetical hierarchy, but also, ever so subtly, a hierarchy that has a principle in a home-base or root notion—for what "lexical (in contradistinction to *grammatical*) meaning" means is the base or root meaning of a word (in contradistinction to the multiple but clustered meanings of its grammatical forms and variations). In my view, the base or root moral concept—which may serve as a univocal "lexical" root, with a multiplicity of "grammatical" permutations—is what Aldo Leopold called "the community concept."[55] The version of the world or "myth" in which the community concept is embedded is provided by Charles Darwin's general evolutionary epic.

In *The Descent of Man* Darwin constructed a communitarian moral philosophy consistent with and embedded in the larger evolutionary world view outlined in *The Origin of Species*. It goes something like this: the proto-moral sentiments of affection and sympathy (upon which David Hume and Adam Smith erected their moral philosophies) were naturally selected in mammals as a device to ensure reproductive success. The mammal mother in whom these sentiments were strong more successfully reared her offspring to maturity. For those species in which larger and more complex social organization led to even greater reproductive success, the filial affections and sympathies spilled over to other family members—fathers, siblings, grandparents and grandchildren, uncles and aunts, nephews and nieces, cousins, and so on. Human beings evolved from highly social primates in a complex social matrix, and inherited highly refined and tender social sentiments and sympathies. With the acquisition of the power of speech and some capacity for abstraction, our ancestors began to codify the kinds of behavior concordant and discordant with their inherited communal-emotional bonds. They dubbed the former good and the latter evil. Ethics, thus, came into being.

As human gens began to merge to form tribes, and social organization and relationships grew more varied and complex, the circle of morally enfranchised persons expanded apace and ethical prescriptions and precepts grew more varied and complex in response to and as a reflection of the newer, more varied and

---

[55] Aldo Leopold, *A Sand County Almanac* (New York: Oxford University Press, 1949).

complex social structures. Capping off his description of this process, Darwin writes,

> As man advances in civilization, and small tribes are united into larger communities, the simplest reason would tell each individual that he ought to extend his social instincts and sympathies to all the members of the same nation, though personally unknown to him. This point being once reached there is only an artificial barrier to prevent his sympathies extending to the men of all nations and races. If, indeed, such men are separated from him by great differences of appearance and habits, experience unfortunately shews us how long it is, before we look at them as our fellow-creatures.[56]

Ontogeny recapitulates phylogeny in our social lives and moral institutions as well as in our phenotypic growth and development. The primitive family, clan, and tribal communities, which Darwin imagines to have gradually evolved, did not simply disappear upon their merger into larger and looser social wholes. They remained intact and became, rather, encircled by these larger communal spheres. Not only do we still retain the more ancient bonds, we feel them to be stronger than those of more recently evolved associations. Correspondingly, we feel the mores of the more venerable and intimate communities to be more binding.

Darwin himself anticipated the recent layering of the human-animal community orbit, and quite appropriately so, since it was he who first suggested that we conceptually reorganize contemporary animals as members of a wider community or kinship group, i.e., as beings descended along parallel evolutionary lines from common ancestors. "Sympathy beyond the confines of man," he writes,

> that is humanity to the lower animals, seems to be one of the latest moral acquisitions. . . . This virtue, one of the noblest with which man is endowed, seems to arise incidentally from our sympathies becoming more tender and widely diffused until they are extended to all sentient beings.[57]

Half a century later, Charles Elton added the most recent addition to the nested social circles to which we now regard ourselves to belong. Elton suggested that we conceive of ecological relationships as uniting plants, animals, soil, airs, waters, and so on into "biotic communities."[58] Leopold simply plugged Elton's

---

[56] Charles Darwin, *The Descent of Man and Selection in Relation to Sex* (New York: J. A. Hill and Company, 1904), p. 124.
[57] Darwin, *Descent*, p. 124.
[58] Charles Elton, *Animal Ecology* (New York: Macmillan, 1927).

community concept in ecology into Darwin's analysis of the origin and evolution of ethics, and articulated a land or environmental ethic.[59]

We have before us then the bare bones of a *univocal* ethical theory embedded in a coherent world view that provides, nevertheless, for a *multiplicity* of hierarchically ordered and variously "textured" moral relationships (and thus duties, responsibilities, and so on) each corresponding to and supporting our multiple, varied, and hierarchically ordered social relationships. If we accept it, we can then discard the competing and inconsistent metaphysics of morals—Kant's, Bentham's, and the lot—that make up the theoretical menagerie of moral pluralism and, in the last analysis, that only serve to obfuscate the actual basis of our multiple moral sensibilities, the interplay between them, and the lexical principle of their delicate arrangement.[60]

Borrowing an image that seems to have been original with Richard Sylvan and Val Plumwood, I suggest that we graphically represent the expansion of our moral sensibilities from narrower to wider circles, not as Peter Singer would have us represent it, like the expansion of the circumference of a balloon, but like the annular growth rings of a tree.[61] In such a figure the inner rings remain visible and present and the outer are added on, each more remote from the center, from the moral heartwood.[62]

The Hume-Darwin-Leopold line of social, humane, and environmental ethics—univocal in its world view and moral philosophy, but multiple in its moral domains—has not been widely endorsed or even critically debated as an alternative to moral pluralism. Because this community-centered complex is rooted in a theory of moral sentiments, it was confused with emotivism, after the ascendency of logical positivism, then associated with "rank relativism" (as Stone calls it), and finally went completely out of fashion in philosophy. Ethics grounded in a theory of moral sentiments emerged as a part of the romantic revolt against the apotheosis of reason in the Enlightenment and has recently survived primarily in the biological literature, having been established there by no less great a figure than Darwin. It has reemerged in contemporary sociobiology—which by itself makes such a theory a philosophical pariah.

I emphatically do not think that such a moral philosophy as I have here outlined—although it involves a multiplicity of overlapping and competing community entanglements—is pluralistic. It is not—because it involves one metaphysics of morals: one concept of the nature of morality (as rooted in moral

---

[59] See Aldo Leopold, *Sand County Almanac;* and Callicott, "The Conceptual Foundations of the Land Ethic."

[60] I have more fully expounded this theory in *In Defense of the Land Ethic.*

[61] See Peter Singer, *The Expanding Circle: Ethics and Sociobiology* (New York: Farrar, Straus, and Giroux, 1982).

[62] See Richard and Val Routley, "Human Chauvinism and Environmental Ethics."

sentiments), one concept of human nature (that we are social animals voyaging with fellow creatures in the odyssey of evolution), one moral psychology (that we respond in subtly shaded ways to the fellow members of our multiple, diverse, tiered communities and to those communities *per se*). Certainly, it does not suggest that we follow Kant here, Bentham and Mill there, Singer and/or Regan yonder, and Leopold at the frontier. It posits a single coherent strand of moral thought: David Hume and Adam Smith set out its elements in the eighteenth century, Charles Darwin grounded them in an evolutionary account of human nature in the nineteenth, and Leopold (making moral hay of Elton's ecological paradigm) provided its outermost "accretion" in the twentieth.

[32]

No Holism Without Pluralism

Gary E. Varner*

> In his recent essay on moral pluralism in environmental ethics, J. Baird Callicott exaggerates the advantages of monism, ignoring the environmentally unsound implications of Leopold's holism. In addition, he fails to see that Leopold's view requires the same kind of intellectual schitzophrenia for which he criticizes the version of moral pluralism advocated by Christopher D. Stone in *Earth and Other Ethics*. If it is plausible to say that holistic entities like ecosystems are directly morally considerable—and that is a very big *if*—it must be for a very different reason than is usually given for saying that individual human beings are directly morally considerable.

J. Baird Callicott's essay on moral pluralism provides a useful overview of the growing interest in pluralism among environmental ethicists and a challenging statement of certain philosophical problems facing the advocates of pluralism.[1] However, by ignoring a problem I have raised for his theory, Callicott presents a distorted picture of the advantages of clinging to his own version of moral monism, and by focussing on the very multilayered pluralism of Christopher D. Stone in *Earth and Other Ethics*, Callicott presents a distorted picture of the reasons for embracing pluralism in environmental ethics.

In the final section of the essay, Callicott describes his view as "a *univocal* ethical theory" which is "multiple in its moral domains"

> that provides, nevertheless, for a *multiplicity* of hierarchically ordered and variously 'textured' moral relationships . . . each corresponding to and supporting our multiple, varied, and hierarchically ordered social relationships. (p. 123)

According to Callicott, the Leopold land ethic is just the last in a series of "accretions" by which the nested social relationships in which all human beings live have come to be reflected in a series of ethics which acknowledge increasingly wider spheres of obligation.

---

*Department of Philosophy and Humanities, Texas A & M University, College Station, TX 77843-4237. Varner holds a joint appointment in Texas A & M's Center for Biotechnology Policy and Ethics. He has previously published papers on environmental ethics, animal rights, the philosophy of environmental law, and agricultural research policy. He is currently writing a book on environmental ethics and animal rights, tentatively titled *In Nature's Interests: Interests, Animal Rights, and Environmental Ethics*.

[1] J. Baird Callicott, "The Case against Moral Pluralism," *Environmental Ethics* 12 (1990): 99–124. Page references in the text are to this essay.

For present purposes, the key issue is Callicott's claim that these spheres are "hierarchically ordered." In defending the land ethic against the charge that it is misanthropic, Callicott has repeatedly claimed that our obligations to family members and fellow human beings trump our obligations to nonhuman animals and ecosystems. In "The Search for an Environmental Ethic," Callicott said that the land ethic

> creates additional, less urgent obligations to additional, less closely related beings. Hence, our obligations to family and friends—and to human rights and human welfare generally—come first; they are not challenged or undermined by an ecocentric environmental ethic.[2]

In "The Conceptual Foundations of the Land Ethic" he said that "duties correlative to the inner social circles to which we belong eclipse those correlative to the rings farther from the heartwood when conflicts arise,"[3] and in "Animal Liberation and Environmental Ethics: Back Together Again" he said that "We are still subject to all the other more particular individually oriented duties to the members of our various more circumscribed and intimate communities. And since they are closer to home, they come first."[4]

In repeating this now familiar claim in his most recent essay, Callicott ignores a problem which I raised in my review of Stone's book,[5] a problem which, unless and until Callicott answers it, utterly trivializes the land ethic. Suppose that an environmentalist enamored with the Leopold land ethic is considering how to vote on a national referendum to preserve the spotted owl by restricting logging in Northwest forests. According to Callicott, he or she would be required to vote, not according to the land ethic, but according to whatever ethic governs closer ties to a human family and/or the larger human community. Therefore, if a relative is one of 10,000 loggers who will lose jobs if the referendum passes, the environmentalist is obligated to vote against it. Even if none of the loggers is a family member, the voter is more closely related to any of them than any spotted owl, and is still obligated to vote against the referendum.

In fairness to Callicott, I must note that he also has claimed that the hierarchy holds only "as a general rule."[6] He has insisted that "the outer orbits of our various moral spheres exert a gravitational tug on the inner ones."[7] Although "in

---

[2] J. Baird Callicott, "The Search for an Environmental Ethic," in Tom Regan, ed., *Matters of Life and Death*, 2d ed. (New York: Random House, 1986), p. 420.

[3] J. Baird Callicott, "The Conceptual Foundations of the Land Ethic," in Callicott, ed., *Companion to A Sand County Almanac* (Madison: University of Wisconsin Press, 1987), p. 208.

[4] J. Baird Callicott, "Animal Liberation and Environmental Ethics: Back Together Again," in Callicott, ed., *In Defense of the Land Ethic* (Albany: State University of New York Press, 1989), p. 58.

[5] Gary E. Varner, review of Christopher D. Stone, *Earth and Other Ethics*, *Environmental Ethics* 9 (1987): 264.

[6] Callicott, "Search," p. 208; Callicott, "Back Together Again," p. 58.

[7] Callicott, "Back Together Again," p. 58; Callicott, "Search," secs. 3–4.

principle" it may be possible "to assign priorities and relative weights and thus to resolve such conflicts in a systematic way,"[8] he has yet to supply even an outline of how these conflicts could be resolved without appealing to some consideration other than communal relatedness. If, as Callicott claims in his most recent essay, community is the sole criterion of moral considerability, it certainly is hard to see how anything but the relative closeness of two communal relationships could be used to decide which one has priority.

Callicott's simplistic hierarchical ordering rule robs the land ethic of any practical force, for it makes it appear that wherever human interests are at stake—and they almost always are—the land ethic is preempted and one is required to apply a good old-fashioned anthropocentric ethic of the kind which Callicott supposes leads to environmental havoc.[9]

In light of Callicott's critique of Stone, it is interesting to note, further, how closely our hypothetical environmental monist's thinking resembles Callicott's farcical account of Stone's moral pluralist senator, whom Callicott characterizes as "blithely" abandoning one ethical theory for another "over lunch" (p. 115). In accordance with Callicott's monism, an environmental holist by lunch would have to become an individualist anthropocentrist in the voting booth. Callicott's monism thus requires the same kind of intellectual gymnastics which he criticizes Stone's theory for requiring.

For the foregoing reasons, Callicott exaggerates the advantages of clinging to his own version of moral monism. In addition, by focussing his critique of pluralism on Stone, he also distorts the reasons for embracing pluralism in environmental ethics. To see clearly how it is that he does this, we need to be clear about what moral pluralism is. By a pluralist ethical theory I mean one which acknowledges distinct, theoretically incommensurable bases for direct moral consideration. Because this definition will not be familiar to readers of Stone's book and Callicott's essay, let me explain some of what I take to be its merits.

First, Stone himself sometimes wavers between a robust theoretical pluralism and a pragmatic pluralism. Sometimes he stresses the utility of attacking different kinds of ethical quandries separately, as if pluralism were a pragmatic strategy for theory construction in ethics rather than a characteristic of completed ethical theories.[10] Most of the time, however, he means the latter, which is what Callicott apparently intended to discuss. My definition makes it clear that we are discussing theoretical rather than pragmatic pluralism.

---

[8] Callicott, "Back Together Again," p. 59.

[9] This supposition has been repeated so often, by Callicott and others, that it is almost gospel, but Bryan Norton has convincingly called it into question in *Why Preserve Natural Variety?* (Princeton: Princeton University Press, 1987).

[10] Christopher D. Stone, *Earth and Other Ethics: The Case for Moral Pluralism* (New York: Harper and Row, 1987), pp. 251–52.

Second, Stone sometimes writes as if imperfect decidability were part of the definition of moral pluralism. For instance, he says that "determinateness," by which he means "the ambition . . . [of] yield[ing] for each quandary one right answer," is "a sort of corollary" to the monism of the dominant ethical theories.[11] Although it is an implication of pluralism as I have defined it that the theory will remain undecided in some possible situations, the inclusion of indecidability in the definition of pluralism prejudices philosophers against it.

At one point, Callicott comes close to identifying environmental pluralism with the kind of very multilayered, and therefore more often undecided system which Stone seems to embrace in his book, when he characterizes moral pluralism as "invit[ing]" one to adopt a different ethical theory to guide one's actions in almost every different facet of one's life (pp. 104 and 115). A major virtue of my definition is that it attenuates the tendency to identify pluralism with such a multilayered system. I think this tendency threatens to scare many philosophers away from pluralism before they have given it a fair hearing, and I think that once the tendency is abandoned, philosophers are more likely to see and take seriously what I take to be the real incentive for embracing moral pluralism in environmental ethics.

What, then, is this incentive? Callicott writes as if it were primarily either intellectual laziness or philosophical charlatanism. He characterizes Stone's pluralism as "happy-go-lucky," "an easy and appealing alternative" to monism, and he calls Stone's argument for pluralism "seductive," suggesting that it is just as suspect as he takes deconstructive postmodernism to be (pp. 116, 104, 102, and 118–20). These may be two incentives for embracing pluralism, and given Stone's very multilayered theory and his sometimes pragmatic characterization of pluralism, I can see how one could get the impression that pluralism is a substitute for a fully worked-out monism. Nevertheless, there are other reasons for siding with pluralism, at least one of which expresses a serious philosophical challenge which I think ultimately cannot be met by the monists, and certainly not by the kind of monism which Callicott advocates.

As Callicott understands, the bowhead whale example in *Earth and Other Ethics* is intended to dramatize Stone's suspicion that the broad range of entities which some environmental philosophers want to say are morally considerable cannot be claimed to be directly morally considerable on any single ground. Stone may turn out to be wrong. It may be that many of the entities in his example (e.g., corporations and states) are not morally considerable at all, and it may turn out, as Callicott argues, that the ones which are can all be said to be directly morally considerable on the same ground. However, if Stone is right, if at least some of the entities on his list both are directly morally considerable and cannot be considerable on commensurable grounds, then pluralism as I have

---

[11] Ibid., p. 116.

defined it is required, and to insist on giving a monist account of what are distinct and incommensurable moral realms is not parsimony but dogmatism.

Is Stone right? I am convinced that he is. Although I cannot go into the details of my argument here, the following brief sketch illustrates what is wrong with the land ethic as Callicott interprets it, and how this failing suggests that holism requires pluralism.

Leopold said that "a land ethic . . . implies respect for [the] fellow members [of one's biotic community], and also respect for the community as such."[12] Callicott's theory will generate a truly holistic ethic of the kind described by Leopold only if his "Humean-Smithian moral psychology" can generate concern for one's biotic community as such—as opposed to concern for the other members of one's biotic community—when combined with modern ecological science. However, *pace* Callicott, sympathetic concern for communities as such has no historical antecedent in David Hume or Adam Smith, and because modern ecological theory provides no account of what is and is not good for an ecosystem, it would appear to be impossible to be concerned about one's biotic community as such.[13] Although Callicott criticizes Holmes Rolston for trying to base a Leopoldian-style holism on a conative view of ecosystems (pp. 108–09), doing so makes no sense, for only by taking some such tack can one hope to make Callicott's Humean-Smithian moral psychology generate a holistic environmental ethic. Only if I can know what is and is not good for another entity can I be concerned about its welfare.

It is because an ecosystem has no welfare of its own, in the sense that each individual member of an ecosystem has a welfare of its own, that a holistic environmental ethic must be pluralistic. If it is plausible to say that ecosystems (or biotic communities as such) are directly morally considerable—and that is a very big *if*—it must be for a very different reason than is usually given for saying that individual human beings are directly morally considerable (and, perhaps, higher animals or all individual living organisms).[14]

Succinctly: no holism without pluralism.

---

[12] Aldo Leopold, *A Sand County Almanac* (London: Oxford University Press, 1949), p. 204.

[13] Detailed arguments to these conclusions are contained in Gary E. Varner, "A Critique of Environmental Holism," in preparation.

[14] In my recent paper, "Biological Functions and Biological Interests," *Southern Journal of Philosophy* 27 (1990): 251–70, I argue that the biological functions of a living organism's component subsystems provide a nonarbitrary criterion of what is and is not in its interests. Nevertheless, as Harley Cahen has shown in "Against the Moral Considerability of Ecosystems," *Environmental Ethics* 10 (1988): 195–216, ecosystems are not goal directed in the way that organisms are. An alternative basis for ecosystem moral considerability is defended by Eugene C. Hargrove in *Foundations of Environmental Ethics* (Englewood Cliffs: Prentice Hall, 1989). Hargrove uses G. E. Moore's thought experiment in *Principia Ethica* to show that the existence of beautiful objects is a moral good, independently of anyone's perception of them, and he uses an analysis of landscape painting to explain how some believe that naturally evolving ecosystems are always beautiful. He thus establishes direct moral considerability for ecosystems (or at least naturally evolving ones) that is wholly independent of whatever (presumably very different) considerations he would use to ground the moral considerability of persons.

[33]

Moral Pluralism and the Environment

ANDREW BRENNAN

Philosophy Department
University of Western Australia
Nedlands
Perth
WA 6009 Australia

ABSTRACT: Cost-benefit analysis makes the assumption that everything from consumer goods to endangered species may in principle be given a value by which its worth can be compared with that of anything else, even though the actual measurement of such value may be difficult in practice. The assumption is shown to fail, even in simple cases, and the analysis to be incapable of taking into account the transformative value of new experiences. Several kinds of value are identified, by no means all commensurable with one another – a situation with which both economics and contemporary ethical theory must come to terms. A radical moral pluralism is recommended as in no way incompatible with the requirements of rationality, which allows that the business of living decently involves many kinds of principles and various sorts of responsibilities. In environmental ethics, pluralism offers the hope of reconciling various rival theories, even if none of them is universally applicable.

KEYWORDS: Cost-benefit analysis, pluralism, preferences, rationality, transformative values.

1. WHAT IS WRONG WITH COST-BENEFIT ANALYSIS?

It is normal for writers to make extravagant claims for their disciplines. Perhaps there is no discipline at present of which this is more true than economics. In environmental discussions, the economist often takes the part of the sensible, rational being, the person who wants to be objective, and base judgments on solid fact. Yet the appeal to economic rationality is highly dangerous – some would say immoral.

Let us start by thinking about a case where economic rationality may make some sense. Suppose we want to reduce the carbon and sulphur pollution associated with a range of industries. Is there any sensible way we can allocate a level of taxation on these pollutants which will be fair? Let us further suppose that part of our idea of what is fair involves being able to justify the proposed level of tax to the industries which produce the pollution.

We can approach this problem by considering first of all just how much damage is caused by the pollution and how to put a monetary value on it. Thus for the pollutants mentioned, we can look at effects on health, in terms of treatment costs for patients with respiratory disease directly due to the pollution. We can consider the cost of repairing damaged buildings, the losses suffered by forestry and the loss of agricultural production. All of these costs can be put in monetary terms (more or less), and give us a measure of the costs associated with air pollution. Put another way, the same figures indicate the scale of benefits that could be achieved by restriction of such pollution.

Now it is not a simple matter to fix the appropriate level of sulphur and carbon taxes in the light of the above information. But we can make sense of the idea that quantifiable costs are associated with pollution and quantifiable benefits are to be gained from controlling it. However, economists typically want to count in other effects of air pollution apart from the ones just mentioned. Consider, for example, the loss of pleasure due to impaired viewing conditions. If air pollution is bad enough in an area with a tourist industry, then there may well be loss of tourist revenue to count in with the other losses. But what of the people who already live or work in the area? Would they not also count in the economic equation? Even if their health is not directly affected, are they not suffering other losses of amenity?

In trying to count in the losses of this last sort, economists usually resort to a technique known as 'contingent valuation' or 'shadow pricing'. But this technique is not without problems. How, for example, can we put a price on reduction of visibility in everyday life? Some American economists from the University of Wyoming tried to do this. They showed a number of people photographs of their surroundings in which air quality was better and worse. They then asked how much the subjects would be willing to pay in addition to their normal electricity bills to preserve a particular level of visibility rather than the next lower one.[1]

This attempt to elicit willingness-to-pay is regularly used by economists. In another study, economists managed to put values on grizzly bears and bighorn sheep by quizzing hunters about how much they were willing to pay to maintain sufficient stocks of these animals.[2] But, as Mark Sagoff points out in a critique of such studies, they are often undermined by the large number of people who refuse co-operation. Thus, instead of agreeing some level of payment to ensure better visibility, many of those questioned simply refused to play the economists' game. They would either refuse to make bids, or they would lodge protest bids which were in excess of any sums they could actually afford to pay. In cases where subjects are asked about levels of compensation for some loss or disbenefit, protest bids – as Sagoff points out – sometimes include a demand for infinite compensation.

The existence of protest bids is uncomfortable for the economist. But the fact that some people show discomfort about the whole exercise of contingent valuation is itself interesting. For what the exercise is supposed to reveal is something about preferences. We are all consumers of various goods and services, and in making consumer choices we reveal, so it is thought, our preferences. It seems logical, then, to try to find out what our preferences about visibility are, just as it is worthwhile to find out what our preferences are on hospital treatment, the colour of toothpaste or whatever. The moral that Sagoff draws, however, is rather different. For he argues that the existence of protest bids in the contingent valuation experiments reveals that the issues involved are not simply ones about preferences.

2. ENVIRONMENTAL POLICY AND ITS FOUNDATIONS

Sagoff's general strategy is to argue that matters of public policy involve values as well as preferences. He has a general objection to economic analysis on the grounds that it pretends there is no difference between matters of preference and matters of value. However, I want to consider whether economic analysis can even get to grips with preferences themselves. It turns out, if I am correct, that the economist's claim to rationality is a feature of a wider view about rationality which is quite false. But the falsity of that wider view has important consequences for ethics as well as economics.[3]

It is widely recognized in human communities that not all preferences are of equal weight. If my preference is to make money by stealing, or by murdering those who are wealthy, then the laws of every society will be against me. Passing laws against murder and stealing is one way a community can protect itself against citizens who might otherwise develop unworthy preferences. From the point of view of the law-abiding individual, illegal acts do not find a place on that person's preference-map.

Economists have long recognized that the preferences of individuals are hard to map. For example, as Mark Sagoff has pointed out, a person may rationally bribe a judge on one occasion (in order to save their driving licence) yet later help to vote the judge out of office (because the same person disapproves of corruption in the law). One way of explaining this behaviour is to suppose that in the role of *consumer* I operate with quite different preferences from those I have in the role of *citizen*. In taking this approach, Sagoff develops arguments originally put forward by other writers (See Tullock 1967, Sen 1977, and Sagoff 1988). If Sagoff and these other writers are correct, then there is no single order in which all my preferences can be placed, for there is no single role which embraces all my roles. It would then follow that there is no single preference-map which can be ascribed to me. Yet it is a feature of standard neo-classical economic theory that all of us do possess a single set of ordered preferences.

I would be inclined to go further than Sagoff and argue that even within a single social role, I lack an ordered preference set. Consider, for example, our behaviour concerning books or music. Do I prefer Schumann's *Études Symphoniques* to Chopin's *Préludes*? I have no answer to give to this question, not because I am indifferent between the Schumann and the Chopin but because each has its own strengths. The Chopin is more pianistic, while Schumann's work makes, so to speak, an orchestra out of the piano. Sometimes I would rather listen to the Schumann, and other times to the Chopin. But such an observation shows nothing about the possibility of placing the two works in a single map of ordered preferences. The difficulty here is noteworthy, for we might have expected two standard works from the romantic piano repertoire to be commensurable in many ways.

One way of explaining this difficulty about preferences is by resort to the idea of *value*. In the case of music, we might say that different pieces have different mixes of values. The aesthetics of music will thus not be simple. Even in comparing two romantic composers of the piano, like Chopin and Schumann, we will encounter complexes of value which make rankings difficult, if not impossible. Now, a simple story about preferences, let us suppose, is that we try to accord the higher preference to the item of greater value. According to this story, the acquisition of musical taste is in part the process of learning to prefer music of greater value to music of lesser value. But if values are themselves mixed, and fail to condense into a single order, there is no single set of values for our preferences to latch on to.[4]

3. KINDS OF VALUE

If we lay the question of mixed values aside for a moment, we can try to make sense of ordering our preferences in a different way. If we can distinguish a number of kinds of value, then perhaps we can try to ground our preferences in the different values that things have. When we try to defend our *considered* preferences, we can do so perhaps by pointing to the different sources for these preferences in the things valued.

Theorists of environmental policy and ethics usually distinguish three kinds of values. First, there is a major distinction between *intrinsic* and *instrumental* value. Something is of intrinsic value if it has value in its own right, or for its own sake. Education, for example, is often held up as an example of something of intrinsic value, as is the study of music, literature, the sciences and the visual arts. By contrast, something is of instrumental value when its existence is necessary for the preservation or realisation of some other value. A good violin is of instrumental value because without it we could not appreciate certain kinds of fine music. A forest is of instrumental value if it yields timber for building and paper making. Notice that the forest may also be of value in its own right: the categories of instrumental and intrinsic value are not exclusive.

Bryan Norton has pointed out that there is a class of instrumentally valuable goods which stand in a peculiar relation to our preferences (see Norton 1987). For they are things which do not simply satisfy our considered felt preferences. Rather, they provide an occasion for examining and sometimes revising these preferences themselves. In this way, they contrast with what might be called 'demand values'. The latter are characteristic of things which either have a use for us, or which may conceivably yield a use for us in future. Norton contrasts demand values of this sort with non-demand values. See the following table:

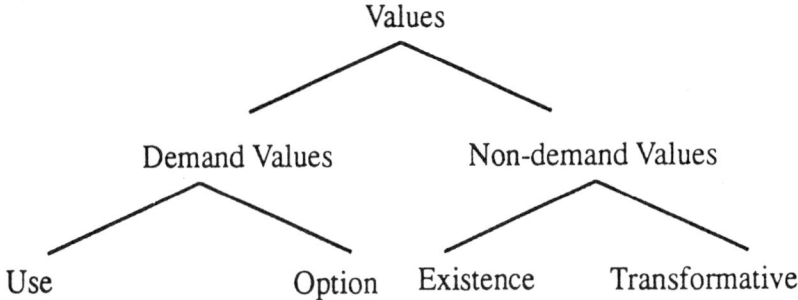

The items we buy at the market or in the shops are typical examples of what have a use value. To argue that rainforest species should be preserved on the grounds that they may one day yield useful medicines is to claim that such species have an option value for us.

But some people would like rainforest species to be preserved independent of any future industrial or pharmaceutical value we might find for them. What they are recognizing, then, is value in the existence of such wild places. Now to recognize such existence values, we do not need to argue that rainforests have value in their own right. Rather, it may be that the existence of rainforest species is instrumentally valuable, in that without them other things of value would be lost.[5]

Finally, the category of transformative value represents a further instrumental type of value that something may possess. Norton gives the example of a teenager who is forced, by circumstances beyond her control, to go to a classical music concert. Until then, we are to suppose that she was keen on popular music and had no time for other musical forms. Surprisingly, the classical concert appeals to her; she subsequently develops an interest in Mozart and starts collecting classical music. Attending the concert has transformed her considered preferences. We can thus classify the three forms of value in general as follows:

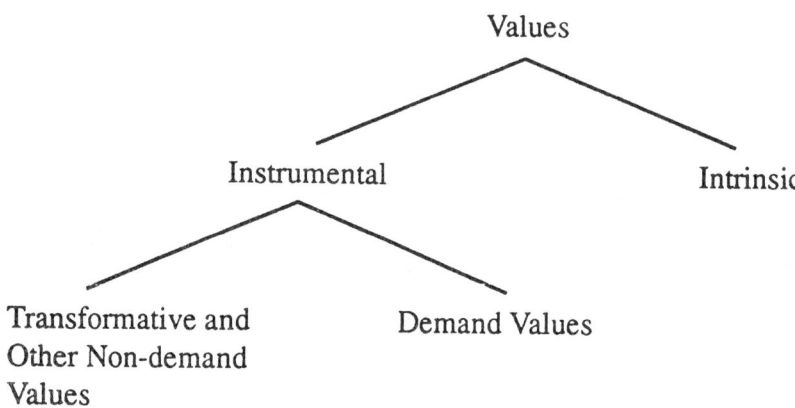

Now a very difficult problem is posed by the existence of transformative value, at least as far as the economist is concerned. For if we try to put a price on transformative value, we shall fail. The impact of transformative values on people's preferences is completely unpredictable, and the degree of impact (when it occurs) will vary massively from person to person. However good cost-benefit analysis is when dealing with demand values, it is bound to leave out transformative value. If we now agree with Norton that natural things and systems themselves possess transformative values, then it will follow immediately that they cannot be priced by the standard techniques of economics.

Norton's results are very much what we might expect. It would be foolish to think that an economic technique which has important application in the field of pollution control can be extended to all environmental policy areas. Yet this is precisely what is attempted by those economists who aim to include all values in their calculations. They are pursuing a phantom; for however hard they try, there will be values which forever elude them.

These thoughts about value also provide the answer to an earlier problem. Recall the suggestion that we might try to make our preferences line up, so to speak, with the order of values in the things around us. If there is no single order of values, then we cannot look for help here in constructing our preferences. We saw already that musical compositions may display mixes of different values, as do novels, paintings, scientific treatises and natural objects. These various values, we can now see, break down further according to whether they are intrinsic, instrumental or transformative. But since some of these forms of value will forever elude quantification, it would be folly to try to set up a single order of values, or of corresponding preferences. Much though economics may like to present itself as a rational discipline, commitment to a single order of values and of preferences would be quite the opposite of rationality.

4. PLURALISM DEFENDED

The suggestions I have just made may seem to give grounds for pessimism. How can we approach issues of public policy in medical ethics, allocation of scarce resources and environmental protection in the absence of shared, objective standards by means of which to judge between different values and different sets of preferences? Unfortunately, it will do no good to turn for help to ethics rather than economics. For ethical theory, as understood by the majority of contemporary moral philosophers, parallels economics in its attempt to reduce complexities about value to simple principles and single measures. Even people who are sympathetic to what I have argued so far may prefer to stick with the methods of cost-benefit analysis, and try to extend these to all forms of environmental impact assessment. The alternative looks like muddling through in an impressionistic, irrational way.

It is important not to underestimate the power of the conception of rationality that is built into the fear just mentioned. That conception is central to economics, and to standard moral philosophy.[6] Bernard Williams has described it in the following terms:

> [there is] an assumption about rationality, to the effect that two considerations cannot be rationally weighed against each other unless there is a common consideration in terms of which they can be compared. This assumption is at once very powerful and utterly baseless ... The drive toward a *rationalistic conception of rationality* comes ... from social features of the modern world ... (Williams, 1985, 16-17)

Williams goes on to explain that the social features to which he is referring impose on *personal* deliberations a model drawn from a particular understanding of public rationality. According to that model, every decision must be based on grounds which can be laid out by appeal to certain general principles which are comparable with each other.

I think it is clear both that the phenomenon exists and that it exerts a peculiarly compelling force on us. If we want to regard ourselves as operating in *principled* ways, we expect that our behaviour falls under certain principles, even if we are not terribly good at articulating them clearly. Likewise, if there are regularities in natural processes and systems, we hope that our sciences at least approximate to some of these, even if we can never be entirely sure that we have hit upon the ultimately *true* theory. It is a short step from these thoughts to the idea that the sciences are ultimately concerned with one kind of object or event in terms of which everything else can be explained. This is what has motivated atomism in its various guises over the years. It is also one of the motives in the positivists' search for a unified science. Likewise, economic theory – as we have seen – orders values according to one monetary weighting and orders our preferences in a corresponding way. Finally, moral philosophy often purports to deal with just one kind of state of affairs in terms of which the good or the right is defined. For example, in utilitarianism all morally relevant states will be compared in terms of overall pleasure (or in terms of some other single measure of utility).

Once these ideas are out in the open, it is easy to see why Williams regards the shared notions as baseless. Few modern scientists would want to make the objectivity of science depend on there being a single set of objects (or events), and a single theory of them, which explains everything.[7] And once we are free from the lures of economic theory we recognize that economic considerations are different from, and not commensurate with, moral or aesthetic ones. So the attempt to reduce all science to a common coin, like the attempt to reduce all values to monetary ones, looks doomed.

But we could agree to what has just been said while still defending monism in morality itself. We could argue that there will be just one set of principles concerning just one form of value that provides ultimate government for our actions. But why should we want to do so? It cannot be to defend notions of objectivity and impartiality. For, once we give up the monistic model of rationality, we recognize that there is nothing whimsical, or unreasonable, in deciding on one issue to be swayed by economic considerations, while on another to follow aesthetic ones. If we use statistical mechanics for studying the behaviour of gases, this by no means determines the theory to be used in studying the ecology of a salt marsh. Nor is there anything irrational about preferring Schumann one day and Chopin the next. For, as we have seen, within the aesthetic sphere there are mixes of values and mixes of preferences. If we have really shaken off the rationalistic picture of rationality, then it is not a lapse of rationality that there is no higher set of comprehensive considerations under which all aesthetic considerations fall, let alone both aesthetic and economic ones. Still less would we want to reduce the aesthetic to the economic.

If we can be objective and rational in adjudicating the competing claims of aesthetics and economics, then we can be equally objective and rational, within the moral enterprise itself, when faced with competing claims. That we come to moral decisions does not mean, then, that we must somehow have reduced all the competing claims to a common measure, or have seen them as falling under some single hierarchy of principles. *Moral pluralism* – to give a name to the position opposed to monism – allows that the business of living decently involves many kinds of principles and various sorts of responsibilities. It recognizes that our feelings and responses to situations are drawn from many sources and cannot be simplified without distortion. Moreover, it maintains that the absence of any clear principles we can articulate for a given case is no evidence of the absence of moral significance. Once we have succeeded in resisting the lure of the rationalistic conception of rationality, it is hard to see why moral monism should commend itself over a pluralist perspective.

It remains true that a pluralist perspective will not be easy to use. If many different sets of values are in play when environmental issues are being discussed, the role of the policy-maker becomes much more complicated. But life is complicated, and we will not make progress in tackling the grave difficulties we face unless we learn to avoid shallow thinking and simple solutions. Although Sagoff has expressed the hope that democratic communities are well placed for establishing procedures by which we can start to give weight to the many complexities of environmental problems, it remains to be seen whether his optimism is justified.

5. ENVIRONMENTAL ETHICS

I conclude with a brief survey of the main underlying theories of environmental ethics and with a suggestion about how we should think about the challenge they pose to conventional ethical theory. There are two main approaches: ethics that are human-based (anthropocentric) and those that are non-anthropocentric. The human-based ethics focus on human beings not only as the actors in morality, but also as the proper subjects of morality too. Now some human beings value natural things and processes in their own right. An anthropocentric ethic can take account of this, since it regards human interests as the only ethically significant ones. Likewise, an anthropocentric ethic can make sense of preserving buildings, paintings or other artefacts for the benefit of those humans who enjoy or study them.

But over the last twenty years or so, environmental ethics has been largely occupied with exploring an alternative, non-anthropocentric approach to morality. According to this approach, things apart from human agents might be proper subjects of moral concern. Those theories that are *biocentric* claim that at least some other living things are possessors of value in their own right. But other accounts of moral value suggest that the possession of life is not itself morally significant. For these last theorists, there can be ethical (and aesthetic) value in a lake, a landscape or a mountain range, even though none of these things is itself alive.

The range of non-anthropocentric theories has posed something of a problem for moral philosophy as it is usually conceived. Contrasted with the conventional treatment of human beings as both moral agents and moral subjects, the new theories have forced reflection on the moral, legal and aesthetic standing of many kinds of non-human beings. These include artefacts (such as paintings, buildings and corporations), living and non-living natural objects, and also systems and processes:

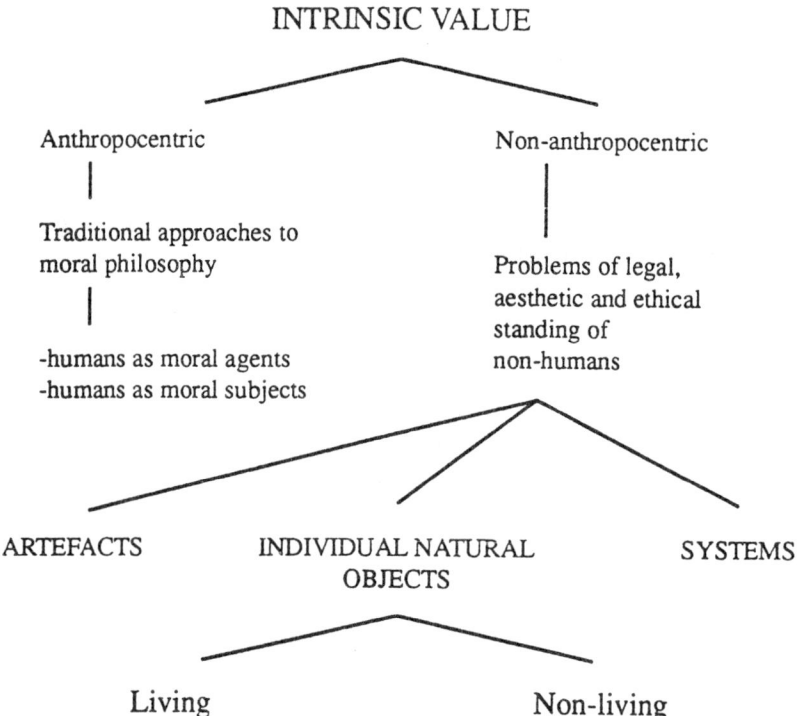

We can map the various kinds of non-anthropocentric value theory by noting that some of them are very much individually-based, while others are holistic in the sense that they attribute intrinsic value to aggregates of individuals. Under the impact of writers like the Scot John Muir and the American preservationist, Aldo Leopold, several more recent theorists have argued that whole communities, ecosystems and even the land itself are complexes which deserve respectful treatment (see Muir, 1988, Leopold, 1949, Rolston, 1988). Of course, the distinction between an individual and an aggregate is not clear-cut, but it is possible to make a rough and ready distinction between holistic and individualistic theories, thus:

NON-ANTHROPOCENTRIC VALUE

	Holistic	Individualistic
Biocentric	*Deep Ecology* *Gaia*	*Animal Rights* *Respect for Trees*
Non-biocentric	*Systemic Value*	*Art works* *Rocky crests*

These positions are by no means exclusive. For example, Holmes Rolston has argued that there is moral value in individual lives, in species, and in ecosystems. Other theorists, Paul Taylor for one, have adopted a resolutely individualistic, life-centred ethic, while those who follow Arne Naess's thoughts about deep ecology generally support the view that living things are knots in a larger web of value (see Taylor, 1984 and Naess 1973).

What is striking about all the non-anthropocentric value positions is that they challenge centuries of orthodox philosophical theory. The fact that things other than human beings have a place in our moral thinking in their own right is something on which all the major theories of morality – whether rights-based, utilitarian, or virtue-based – have been largely silent.[8] Environmental ethics thus differs from other areas of applied philosophy, in that it does not call for the extension of existing value categories and moral analysis. Rather, it challenges the standard categories and analyses themselves. Environmental ethics does not call for expanding the circle of beings recognized as having feelings, for example, or capable of feeling pain. In this way it contrasts with the position of those who argue for the extension of moral consideration to non-human animals (see Callicott, 1980). Rather, some environmental thinkers have posed the question of why feelings should be so important to morality. Forests, lakes and rocky crests have no feelings, nor have they, as far as we know, any other interests we can take into account. But that fact alone does not mean that they have to be left out of the moral reckoning.

Many workers in the field would disagree strongly with what has just been suggested. For some of them, the heart of our environmental concerns lies precisely in a kind of extended individual benevolence. Writers like Stephen Clark defend vegetarianism and the elimination of animal experimentation on the grounds that present human practices do wrong to other animals (Clark, 1977). It is perhaps natural to extend this concern, as Clark does himself. If animals can be wronged, then perhaps so can plants, forests, lakes and rocky crests. Tom Regan, to take another notable example, holds that non-human animals have what he calls 'inherent value' through being subjects of a life. Such value does not come in degrees: if rabbits have it at all, they have it equally with us. For Regan, our duties towards other animals, and their rights, are founded on this inherent value. For the environment, Regan articulates an ethic of general non-intervention in nature, based on the inherent value of all natural objects. These examples could be multiplied.

Without in any way wishing to challenge the moral sincerity of the writers just mentioned, I would like to propose that theirs is not a route worth following. I have given detailed reasons elsewhere for thinking that species, natural systems and landscapes cannot plausibly be regarded as having interests of any sort, nor do they have modes of flourishing, nor can we make sense of the claim that they have directions of development.[9] I will not repeat these arguments now but instead I would like to suggest a partial explanation for why so many writers have taken the route, however ill-advised it is to do so. They have, I conjecture, simply followed tradition. The recent tradition of moral philosophy has regarded the interests of the parties concerned as central to their role in the moral enterprise. For the utilitarian, the morally relevant parties in any situation are those whose pleasures or pains will be affected by what is done in that situation. A standard prescription for the agent is that he or she should act so as to ensure that pains are minimized or pleasures maximized. For the contract theorist, the moral community consists of a group of beings with interests, and principles of morality and justice are devised so that these interests are respected without special regard to status or special position. Rights theorists who are not contractarians normally have in mind a moral community of items who can be represented in their own right, and thus as possessors of interests.[10]

In the face of this near-unanimity on the features of the moral community, we are faced with a real problem if we try to introduce considerations on behalf of items that lack interests in the sense of having no primary goods, no projects, no directions of development, no possibilities of pleasure, pain or other utility and no other features in terms of which they can be represented in their own right in moral debate. How could such items possibly have a moral claim on us? In desperation, heroic efforts are sometimes made to try to establish that trees, rocks, landscapes, and even natural systems have interests. A great deal of nonsense has been written in the attempt to characterise ecosystems as having possibilities of flourishing, directions of development or ways of being more or less healthy. Moreover, the need to introduce considerations about preserving trees, species and landscapes into debates on public policy has seemingly added urgency to the underlying philosophical project. Even though the works of philosophers have little direct bearing on public policy, it is not helpful to our reflections on these matters to find no room for valued environmental features in our ethical deliberations.

6. PROSPECTS FOR A 'NEW ETHIC'

There has been a tendency in recent years for concerned people to suggest that what we need is a new ethic to moderate our dealings with nature. If my diagnosis of the challenge posed by environmental ethics is correct, we can begin to see why this is so. The conventional model for moral deliberations has been built on the understanding that the moral community is constituted of a number of individuals each with their own interests. Extending that community from humans first to other animals and ultimately to ecosystems has seemed to be an attractive option. But even early in the process there were symptoms that something was amiss. The claim by some animal liberationists that the world would be a better place if there were no predators and no parasites is entirely in keeping with the utilitarian arguments in favour of changing our modes of treating animals. Yet such a claim is anathema to many environmental philosophers. Over the last fifteen years, the animal rights and environmental philosophy camps have showed ever wider divisions, despite some attempts to heal the rift. The rift itself signifies a tension in the way the extension of individual benevolence has been perceived. Finally, attempts like Regan's to suggest that we should adopt an ethic of non-intervention in nature seem to constitute a *reductio* of the whole project of extending the circle of benevolence (Regan, 1981). Such non-intervention has never been an option for humans, any more than it has been an option for other living species.

In the face of the challenge to find a new ethic, what can moral philosophy do? And, more urgently, what can the philosopher say to those who are concerned to frame environmentally sensitive policy and regulations? If we follow the monistic tradition and the rationalistic conception of rationality, we may perhaps look to interests or some other single feature to fund the value of trees, rocks, ecosystems, bacteria, dogs and people. Such a feature would be one in terms of which natural things and systems could gain entry to policy debates and be recognized as having standing in their own right. I now think there is probably no such feature.[11] But even if such a feature could be found, it is hard to believe that we could develop any plausible ranking of the relative value of all the things which possess it. Alternatively we can give up the rationalistic conception of rationality and the associated monistic point of view. We can start to think in terms of a plurality of values, and an associated plurality of principles. More radically, we will recognize that not all matters of morality can even be thought of in terms of principles, rules, contracts and the rest of the apparatus of conventional moral theory.[12]

To make clear just what is at stake here, notice that there are several kinds of pluralism. Liberal democracies are supposed to admit that there are competing conceptions of the good, and an associated plurality of values, held by their citizens. What might be called *moral liberalism* would be the ethical equivalent of this political position. It would accept that different people might bring different ethical perspectives to bear on an individual case (noting that in some cases the different perspectives would agree on what counts as the *right* course of action). This is *not* the kind of pluralism that I am advocating here. By contrast, in the present paper, moral pluralism is meant as a philosophical, not a moral, thesis. In its philosophical sense, we can still distinguish a number of forms for pluralism to take. I will look at two different forms, noting that the first form itself permits of several varieties.

The first kind of moral pluralism recognizes the possibility that *different considerations apply in different cases*. This somewhat ambiguous notion of pluralism has been at the centre of attention in recent discussions of moral pluralism in environmental ethics.[13] It is surprising that any controversy at all has been aroused by this idea. Take three simple cases. Consider, first, the proper response to an injured animal. If it is badly injured and in obvious pain, the most humane course might be to kill it speedily. By contrast, there are circumstances in which it is decent to preserve the life of a human being who is suffering extreme pain. Third, trees, as far as we know, are not sentient beings at all, yet it may be proper to take steps to preserve a tree or trees by actions which cause pain to a sentient being (for example, when a vandal is forcibly restrained). What the first form of pluralism suggests is that no one set of considerations provides the rationale for these three cases, let alone for the multitude of others that face us daily. Yet what is involved in treating different cases differently? Someone who holds only a single, structured theory of ethics with a single standard of right and wrong might well count considerations involving human interests and welfare differently from those affecting other animals and plants. Although the considerations are not the same, this kind of pluralist would hold that – at a suitably abstract level – the principles are always the same.

As pointed out already, this version of pluralism has several varieties. Perhaps a more interesting variant is associated with the thought that principles drawn from one kind of understanding of ethics may apply in one case, while those drawn from a separate understanding apply in another. For example, maybe we could approach issues of public policy from a utilitarian point of view, yet be committed to some quite different point of view in our dealings with friends and relatives. Even if I think about public policy in terms of the balance of benefits or happiness over disbenefits, when fulfilling a promise to a friend, it may never occur to me to consider whether greater good overall might be achieved by breaking the promise and doing something else instead. Notice two points that arise from this example. First, the discrepancy between utilitarianism on public policy and some quite different approach to private morality does not indicate inconsistency on the part of the thinker or moral agent. Second, even if I do think about public policy in ways different from the way I think about obligations to my friends, it is not clear that the conventional distinctions of standard moral theories are best for describing such a difference.

More generally, it seems extremely simplistic to try to characterise our thinking about issues of public policy in terms of any one theory of ethics. For we are caught up in a number of very different considerations as soon as we try to give detailed attention to whether, for example, a leisure development should take place on farmland recently set aside from arable use. In such cases, we typically think of impacts on local employment, the many different interests and needs of those living in nearby towns and villages, effects on local ecosystems and biological communities – both physical and chemical – as well as aesthetic issues. Then there are notions we have about the sense of place shared by inhabitants of an area, and to what extent any proposed development will fit in with a continuing narrative history that can be given for that place. To reduce all these considerations to ones of utility, duty or beauty may be, in its own way, just as simplistic as the attempt to reduce all values to monetary ones.

These ideas give rise, finally, to a second development of the concept of moral pluralism. According to this, any particular situation in which we face a decision, or have to act, is complex. Its complexity could be described in terms of the idea of mixed values introduced earlier. Alternatively, it can be claimed that there is no single activity of valuing involved in assessing any situation. Pluralism of the second sort maintains that there is no single theoretical lens which provides a privileged set of concepts, principles and structure in terms of which a situation is to be viewed. Furthermore, complexity is a feature of the situations that arise in public debate and also in the most private of moral deliberations. This second kind of pluralism seems to me more interesting than the first sort, although related to it. As opposed to the claim that different cases call for different treatment, this new form recognizes that one and the same case can properly be viewed in many different ways.

There is a way into the thesis that the ethical enterprise is pluralistic in this second way which draws on the first, less controversial form of pluralism. Suppose, for the sake of argument, that the simplistic philosophical models have some merit. I think about public policy as a utilitarian, let us imagine, but I think about duties to my friends in terms of their moral claims on me as Kantian ends-in-themselves. By contrast, I cultivate a caring and responsible attitude to my local environment by way of a conception of worthwhile human living in nature which does not reduce either to utilitarianism or Kantianism. I do not, for a moment, intend that any of these descriptions does more than caricature the moral situation of an agent. But if we accept the caricature for the moment notice how the switch of perspective from one case to the next can reveal something we may overlook when we concentrate on a single case viewed through a single lens. This is that the various perspectives brought to bear on the several cases can be brought to bear – to some degree – on the individual case as well. If each case can be viewed from more than one perspective, then the business of being morally engaged with the world around us involves a multiplicity of perspectives and a value complexity which is ignored in the standard, reductionist accounts found in textbooks. That a certain perspective tends to dominate in a particular kind of case does not mean that the others are inapplicable to it. Moral pluralism as a philosophically interesting thesis is the claim that valuing things is pluralistic in just this way.[14]

It is the second form of moral pluralism which is not only interesting, but which urgently requires exploration. By adopting the pluralist stance, we not only start to do justice to the complexity of real situations, but we also can start to look for ways by which environmental ethics can be linked up with other modes of valuing and ways of responding to our surroundings. Utilitarianism and its rivals need not be abandoned, but can be considered as partial accounts of the moral life. There is scope, for example, for developing notions such as attention, humility and selflessness in our dealings with nature as part of the story of what makes a worthwhile human life. These notions should not be thought of as *the truth* about morality – any more than utilitarianism is. Rather, they provide greater depth in characterising our situation.[15] Abandoning reductive monism about values and valuing makes even more sense once the force of moral pluralism in this latest form is recognized.

If we accept moral pluralism as a philosophical position, the project of environmental ethics can be seen in a new light. The challenge of non-anthropocentric ethics to the western, human-centred tradition need not be described as an attempt to supplant one set of principles (ones regarding human welfare, or human virtues or whatever) with some new overarching set that embrace not only *human* concerns but also the interests, whatever they are, of other natural things. Instead, exploring non-anthropocentric ethics is to be seen as adding further sophistication to our moral discourse and helping us understand a further dimension to our lived experience. Seen in this way, environmental ethics is less a competitor for a certain moral position, but an investigation of a more sophisticated turn that moral philosophy has taken. Embarking upon it is a partial recognition of the complexity of our moral situation. Note, once more, that the complexity in question is intrinsic to the business of being moral. Moral pluralism is a philosophical, not a moral, thesis.

To say this may seem not to help the policy-maker, for it involves admitting that we face a challenge to which philosophy has not so far found a solution. None the less, it provides at least some negative advice. We can caution against the use of reductionist methods, and the trap of thinking that policy decisions can always be reduced to some common, comprehensive weighting. Environmental ethics can provide an antidote to theories which encourage the idea that our moral situation is a simple one. So it can be an antidote to a tradition of systematic, but simplistic, theorizing. When we turn our attention to the challenge within philosophy itself, it will hardly be surprising if we find that many of our philosophical concepts do not quite fit us for meeting the challenge of moral pluralism. But this is certainly not the first time that efforts have to be made to see just which of our concepts are adequate to map newly encountered terrain. Indeed, it remains to be seen to what extent a developed ethics of the environment will fit in with the complex of other theories, intuitions and feelings we have about value. But this exploration is not something for rationality to avoid, but one which an appropriate sense of imagination and discovery should commend to us.

NOTES

Versions of this paper have been read in Singapore, Stirling, Lancaster and London, and I am grateful to members of all those audiences whose comments have been important in establishing the final shape of the paper.

[1] *See* Rowe, D'Arge and Brookshire, 1980. This study is described and commented on in chapter 4 of Sagoff, 1988.

[2] For references to this and other studies, *see* Pearce, Markandya and Barbier, 1989, chapter 3.

[3] That economists do lay claim to rationality is shown by statements like the following: 'By trying to value environmental services we are forced into a rational decision-making frame of mind. Quite simply, we are forced to think about the gains and losses, the benefits and costs of what we do. If nothing else, economic valuation has made a great advance in that respect'. (Pearce, Markandya and Barbier, 1989, 81).

[4] See the recent discussion in Raz and Griffin, 1991. A sensitive treatment of value pluralism is found in Nussbaum, 1986. Clearly, in a brief attack on common assumptions about preferences, I am unable to give detailed attention to all the possible kinds of orderings for preferences. The difficulties suggested here arise well before the stage at which it would be necessary to investigate different orders for individual preferences.

[5] There is a dispute about whether existence values are to be considered as a kind of demand value (given that people clearly indicate preferences regarding the existence of things remote from them and which do not impinge in a direct way on their lives). I am not intending to take sides on this issue in the present paper.

[6] It is also to be found elsewhere, I would argue, particularly in the positivists' conception of a *unified science*, and particularly in Rudolf Carnap's early work on scientific objectivity. Mandelbaum points to the influence of the economist Emmanuel Herrmann on Carnap's thinking (Mandelbaum, 1971).

[7] At the height of his Vienna Circle involvement, Carnap put forward a view of science which related its unity to just the conception of objectivity which I think is implausible. Michael Friedman has pointed out that Carnap's view of how concepts are to be discriminated from one another is deeply linked with the idea that all concepts are part of a single interconnected system. Such a single organization can only be possible if, in Carnap's own words, 'there is only one object domain and each scientific statement is about the objects in this domain'. *See* Friedman, 1987.

[8] An important exception to this claim is the work of Iris Murdoch, which, although focusing on the virtues of attention and care, does not rule out the importance of what she calls the 'sheer alien otherness' of non-humans (Murdoch, 1970). I also ignore dissenting voices within the western tradition itself, such as Heidegger's.

[9] *See* Brennan, 1986, and further discussion in Brennan, 1988.

[10] One standard statement of the link between interests, rights and the capacity to be represented is given in Feinberg, 1974.

[11] Although if there is such a feature it would have to be something like the lack of function characteristic of natural objects – see Brennan, 1984, and Brennan, 1988, chapter 13, paragraphs 13.2-13.3.

[12] Feminist and other critiques of conventional moral philosophy have made this point repeatedly. Although the status of feminism as an epistemological and metaphysical position is unclear, I am happy to be aligned with feminists in their objection to how the project of morality has been followed in post-renaissance western philosophy. For a useful overview of ecological feminism, *see* Warren, 1987.

[13] This is the conception of pluralism which seems to be attacked in Callicott, 1990. Callicott's principal target is Stone, 1988 (and Stone, 1987).

[14] Christopher Stone uses an analogy with the multiplicity of maps which can be produced for a single territory. See Stone, 1987, chapter 5.

[15] In an interesting, but so far unpublished essay, Tom Birch has tried to develop an account of meaningful attention which would fund our valuation not only of other living things, but also of rocks and lakes. Many of the remarks in Murdoch's *Sovereignty of Good* seem to merit exploration in an environmental context.

REFERENCES

Brennan, Andrew 1986 "Ecological Theory and Value in Nature", *Philosophical Inquiry* **8**: 66-95.

Brennan, Andrew 1988 *Thinking About Nature*. London, Routledge.

Brennan, Andrew 1984 "The Moral Standing of Natural Objects", *Environmental Ethics* **6**: 35-56.

Callicott, J. Baird 1980 "Animal Liberation: A Triangular Affair", *Environmental Ethics* **2**: 311-38.

Callicott, J. Baird 1990 "The Case Against Moral Pluralism", *Environmental Ethics* **12**.

Clark, Stephen 1977 *The Moral Status of Animals*. Oxford, Clarendon Press.

Feinberg, Joel 1974 "The Rights of Animals and Unborn Generations", in *Philosophy and Environmental Crisis*, edited by W.T. Blackstone, pp.43-68. Athens, University of Georgia Press.

Friedman, M. 1987 "Carnap's *Aufbau* Reconsidered", *Nous* **21**: 521-45.

Leopold, Aldo 1949 *A Sand County Almanac*. Oxford, Oxford University Press.

Mandelbaum, Maurice 1971 *History, Man and Reason*. Baltimore, Johns Hopkins Press.

Muir, John 1988 *My First Summer in the Sierra*. Edinburgh, Canongate Publishing.

Murdoch, Iris 1970 *The Sovereignty of Good*. London, Routledge and Kègan Paul.

Naess, Arne 1973 "The Shallow and the Deep, Long-Range Ecology Movement: A Summary", *Inquiry* **16**: 95-100.

Norton, Bryan 1987 *Why Preserve Natural Variety?* Princeton University Press.

Nussbaum, Martha 1986 *The Fragility of Goodness*. Cambridge University Press.
Pearce, D., Markandya, A., and Barbier, E. 1989 *Blueprint for a Green Economy*. London, Earthscan.
Raz, J. and Griffin, J. 1991 "Mixing Values", *Proceedings of the Aristotelian Society* supplementary volume 65: 83-118.
Regan, Tom 1981 "The Nature and Possibility of an Environmental Ethic", *Environmental Ethics* 3: 19-34.
Rolston III, Holmes 1988 *Environmental Ethics*. Philadelphia, Temple University Press.
Rowe, Robert D., D'Arge, R.C. and Brookshire, D. 1980 "An Experiment on the Economic Value of Visibility", *Journal of Environmental Economics and Management* 7: 1-19.
Sagoff, Mark 1988 *The Economy of the Earth*. Cambridge University Press.
Sen, Amartya K. 1977 "Rational Fools: A Critique of the Behavioural Foundations of Economic Theory", *Philosophy and Public Affairs* 6: 317-44.
Stone, Christopher D. 1987 *Earth and Other Ethics*. New York, Harper and Row.
Stone, Christopher D. 1988 "Moral Pluralism and the Course of Environmental Ethics", *Environmental Ethics* 10: 139-54.
Taylor, Paul 1984 *Respect for Nature*. Princeton University Press.
Tullock, Gordon 1967 *Toward a Mathematics of Politics*. Ann Arbor, University of Michigan Press.
Warren, Karen 1987 "Feminism and Ecology: Making Connections", *Environmental Ethics* 9: 3-21.
Williams, Bernard 1985 *Ethics and the Limits of Philosophy*. London, Fontana.

Name Index

Allen, T.F.H. 302
Allison, A.C. 93
Andrews, Peter 194
Aquinas, St Thomas 542
Arbor, J.L. 120, 121, 122, 129
Aristotle 65, 68, 145, 367, 377, 542, 545
Attfield, Robin 128, 142, 144, 390, 391, 529
Attig, T. xxii
Augustine, St 542
Austin, Mary 103
Ayala, Francisco J. 98

Bahro, Rudolf 248, 256
Barney, Gerald 167
Barrett, S.W. 437
Bennett, Jonathan 151
Bentham, Jeremy 391, 421, 533, 542, 551, 552
Bentkover, Judith 467
Berry, Wendell 250, 342
Biehl, Janet 343
Birch, Thomas H. 219–38, 436
Bookchin, Murray xvii, 322, 343
Boorse, Christopher 127
Botkin, D.B. 436, 438
Bradford, George 343
Brennan, Andrew xviii, 33–54, 142, 223, 234, 304–5, 403, 407, 533, 534, 535, 559–77
Browne, Sir Thomas 95
Brundtland, G.H. 438, 439, 495, 496, 497
Buber, Martin 193, 196–9 *passim*, 200, 202, 203, 207, 208, 216
Burger, Justice 467
Burke, Edmund 347, 501

Cahen, Harley 137–58
Callicott, J. Baird xviii, xix, xxii, 36, 37, 44, 126, 142, 207–8, 271, 306, 309, 342, 394, 431–43, 445, 446, 447, 448, 449, 450, 451, 452, 453–4, 527–52, 553–7 *passim*, 569
Campbell, John H. 99
Camus, Albert 232
Carson, Rachel xv
Casebeer, R.L. 441
Cebik, L.B. 405
Cheney, Jim xx, 223, 299–317, 330, 335, 346, 347, 351–3 *passim*, 544–5, 546, 547, 548
Clark, Stephen xxi, xxiii, 50, 52, 395, 569, 570
Clemens, J.L. 301
Cockburn, A. 439
Cohen, Michael 245
Collard, Andree 342
Collier, John 190–1
Columbus, Christopher 432, 449, 454
Conaway, J. 440
Constanza, R. 438
Cordell, H.K. 436, 438, 448
Cousteau, Jean 477
Cracraft, J. 80
Crick, Frances 96

Daley, Mary 359
Darwin, Charles 44–5, 67, 79, 87, 98, 108, 376, 436, 453, 527, 549, 550, 551, 552
Dasmann, R.F. 439
Davidow, Ellen Messer 355, 356
da Vinci, Leonardo 43
Davis, Don 346
Davis, Mark 312
Dawkins, Richard 90, 93, 155
Day, G.M. 437
DeAngelis 302
d'Eaubonne, Françoise 319
de Beauvoir, Simone 283
DeBonis, J. 440
de Groot, W. xvi
Deleuze, Gilles 394
Dennett, D. 40
Derrida, Jacques 546
Descartes, René 436, 545, 546
Devall, Bill 250, 341, 345, 437
Diamond, Irene 304
Dickinson, G.L. 376–7, 387
Dillard, Annie 112–13
Disney, Walt 396
Dobson, Andrew xx
Dobyns, H.F. 437
Dobzhansky, Theodosious 98
Dodson-Gray, Elizabeth 351
Doubiago, Sharon 343
Duchen, Claire 289
DuPont 371

Edwards, Steven 457–61 *passim*, 463–5 *passim*
Ehrenfeld, David 422, 423, 439
Ehrlich, A.H. 77, 422, 423
Ehrlich, P.R. 77, 341, 422
Einstein, Albert 85, 547
Eisner, Thomas 77
Eldredge, N. 80
Eliot, T.S. 219
Elliot, Robert xxii, 400–2 *passim*, 404, 405, 412–15 *passim*, 417, 422, 423, 426, 427, 428
Elton, Charles 550, 552
Emerson, Ralph W. 366–7, 379, 385, 432, 433
Enç, Berent 40
Engels, Freidrich 355
Euclid 519

Faber, M. 504
Falwell, Jerry 542
Feinberg, Joel xvi, 34, 47, 77, 82, 117–23 *passim*, 125, 127, 128, 141
Fernald, M.L. 79
Ferré, Frederick 545, 546
Flader, Susan L. 431
Foltz, B. xvii
Fox, Warwick xix, 195, 346, 347, 348, 354, 358, 359
Frankena, William 6
Fraser, G.R. 82
Freidman, Marilyn 309
French, Marilyn 359
Frye, Marilyn 237, 330–1

Galbraith, J.K. 250
Gandhi, Mahatma xxiv, 239, 244
Gauss 301
Gearhart, Sally M. 287, 288, 289
Gewirth, Alan 35
Gharton, Per 349
Ghiselin, M.T. 80
Gilligan, Carol 309
Glover, Jonathan 488–9
Godfrey-Smith, William 530
Gomez-Pompa, A. 439
Goodin, R.E. xx
Goodman, N. 341
Goodpaster, Kenneth 33, 117, 123, 129, 130, 138, 141–2, 145, 220
Gordon, J.C. 438
Gorz, André 263
Gould, Stephen J. 92, 96, 97
Gray, Elizabeth D. 305, 308
Griffin, Susan 305, 308, 348, 358

Guha, Ramachandra 239–51, 253–8 *passim*, 260, 272, 434–5, 438
Gunn, Alistair S. 411–30

Hampshire, S. 77
Haraway, Donna 311–13 *passim*
Hardin, Garrett 205, 206
Hare, Nathan 337
Hargrove, Eugene C. xvii, 529, 533, 535, 538, 540, 548
Harper, John L. 156–7
Harrison, Lea 342
Hays, Samuel 246
Hecht, S. 439
Hefferman, James 140
Hegel, Georg W.F. 42
Heidegger, Martin xvii, 350
Heisenberg, Werner K. 547
Heizer, R.F. 437
Heraclitus 219
Hirschman, A.O. 459
Hughes, J.D. 438
Hull, D.L. xxii, 80
Hume, David 63, 64, 104, 396, 527, 549, 552, 557
Huxley, Aldous 241, 395
Huxley, Thomas H. 87

Iltis, Hugh H. 194
Inden, Ronald 245–6

Jaggar, Alson M. 300, 309
James, William 214, 366, 382, 399
Janzen, Daniel 243, 244
Jeffers, Robinson 366
Johns, David M. 253
Jones, Arthur W. 92

Kant, Immanuel 9, 43, 515, 533, 535, 542, 543, 548, 551, 552
Kantor, Jay 144
Kasperson, R.E. 482
Katz, Eric 142, 150, 399–409
Kaus, A. 439
Kennedy, Margrit 342
Khomeini, Ayatollah Ruholla 542
King, Ynestra 299, 300, 305, 343, 355
Kipling, Rudyard 395, 396
Knight, Frank 459–60
Kokopeli, Bruce 293
Kroeber, A.L. 437
Kvaloy, Sigmund 168

Lakey, George 293

Lecky 390
Lem, Stanislaw 45
Lemons, J. 437
Leopold, Aldo xv, xvi, xviii, xix, 49, 101, 137, 140, 173, 241, 255, 431–5 *passim*, 448, 450, 452, 502, 503, 505, 514, 527, 530, 531, 533, 535, 542, 549, 550, 551, 552, 553, 554, 557, 568
Levine, Michael xvii
Levi-Strauss, Claude 356
Lewis, H.T. 437
Lindeman, C. 302
Lippman, Walter 247
List, Peter C. xvii, xxiv
Lloyd, Genevieve 277, 284, 288
Locke, John xxiii, 356
Lopez, Barry 104
Losonsky, Michael 402
Lotka, J. 301
Loucks, Orie L. 194
Lovejoy, Thomas 416
Lovelock, J.E. 394–5
Lugones, Maria 326, 331
Luke, Tim 343

Maathai, Wangari 342
McDowell, J. 63
McIntosh, Robert 156
McKibben, Bill xv
Mackie, J. 63, 64
Magurran, A.E. 447
Manstetten, R. 504
Marcuse, Herbert 263
Marshall, Robert 431, 452, 453
Martin, C. 437, 438
Martin, P.S. 454
Marx, Karl 264
Maser, Christopher 403–4, 407
Masserman, J.H. 453
Mathews, Freya xix, 183–91
Matthews, G.B. 33
May, Robert M. 152
Mayr, E. 80
Meadows, Donella H. xv
Meeker, Joseph 168, 194
Meine, C. 452
Merchant, C. 277, 437
Merigliano, L. 436
Merola, Giovanni 348
Metzger, Deena 342
Mill, John S. 87, 347, 366, 370, 375, 376, 379, 388, 542, 552
Miller, Becca 347
Miller, G. Tyler 168, 177–8

Mish'alani, James K. 144
Mollison, W. 341
Moore, Gordon E. 56, 59–61 *passim*, 382
Mowat, F. 453
Mozart, Wolfgang A. 208
Muir, John xviii, 61, 102–3, 108, 240, 241, 245, 431, 432, 433, 453, 568
Murdoch, Iris 53, 395
Murie, M. 417, 431
Murphy, Patrick 356
Myers, N. 77, 440

Nabhan, G.P. 439, 447, 448
Naess, Arne xix, 55, 161–82, 193, 195, 211–18, 240, 341, 345, 569
Nagel, Ernest 122, 146, 147, 148, 150
Nash, Roderick 247, 432, 433, 437
Newton, Isaac 85, 545
Nietzsche, Friedrich 377
Norton, Bryan xviii, 143, 495–509, 563, 564
Noss, R.F. 438, 440

Odum, E.P. 302
Oelschlaeger, M. 436
Ollman, Bertell 264
O'Neill, John 55–73
O'Neill, R.V. 301, 302, 303
Orenstein, Gloria F. 304
Otto, Rudolf 193, 196, 198, 199, 200, 202, 203, 207, 216

Paley, William 422
Parsons, D.J. 436
Partridge, E. xviii
Paske, Gerald H. 25–31
Passell, P. 439
Passmore, John xvii, 34, 49, 390
Perry, Ralph B. 120
Peters, C.M. 439–40
Pielou, E.C. 447
Pierce, C. xxii
Pietila, Hilkka 348
Pinchot, Gifford xviii, 240, 433, 434, 533
Planck, Max 547
Plant, Judith 303
Plato 542, 545
Plumwood, Val xix, 275–98, 530, 551
Ponting, Clive xxiv
Posey 439
Posner, Richard A. 460
Proops, J. 504
Pyne, S.J. 437

Randall 433

Raup, D.M. 84
Raven, Peter H. 78, 79
Rawls, John 77, 130
Reagan, Ronald 546
Reed, Peter 193–209, 211–14 *passim*, 215–16, 217
Reed, P.C. 436, 438, 448
Regan, Tom xvi, xxi, xxii, 82, 120, 133, 219, 224–7 *passim*, 235, 238, 341, 527, 531, 532, 552, 570, 571
Reich, Wilhelm 263
Rescher, N. 82
Rich, Adrienne 355
Ricklefs, Robert 152
Rodman, John xviii, xix, 48, 140, 141, 144, 151, 168, 200, 224, 226
Rolston III, Holmes xviii, xix, xxii, 77–85, 87–115, 140, 200, 304, 365–88, 445–52, 453–4 *passim*, 527, 529, 531, 536, 537, 557, 568
Rorty, Richard 470, 546
Roshi, Robert A. 244
Ross, W.D. 59
Rousseau, Jean-Jacques xvii, xxiii, 277
Routley, Richard xvi
Routley, Val *see* Plumwood
Ruether, Rosemary R. 292, 305, 355, 359
Russell, Bertrand 546
Russow, Lilly-Marlene 128

Sagoff, Mark xxiv, 48, 457–76, 560, 561, 562, 567
Salamone, Connie 342
Sale, Kirkpatrick 239, 360–1
Salleh, Ariel 341–62
Sandburg, Carl 388
Santayana, George 241
Sapontzis, Steve F. 90
Savile, A. 42–3
Scherer, D. xxii
Schweitzer, Albert 530, 542
Scott, Peter 420
Seattle, Chief 37–9 *passim*
Sen, Amartya 561
Sepkoski, J.J. jr 84
Sessions, George 165, 250
Shakespeare, William 388
Shaw, A.B. 80
Shaw, A.J. 79
Shiva, Vandana 300
Shrader-Frechette, Kristin 477–93
Sidgwick, Henry 130, 131
Simon, Herbert A. 100–1
Simpson, G.G. 80

Singer, Peter xxi, xxii, 81, 82, 140, 141, 143, 144, 341, 390, 391, 531, 542, 551, 552
Sjoo, Monica 342
Skutch, Alexander F. 92
Smith, Adam 527, 541, 552, 557
Smith, Frederick E. 366
Smith, Michael F. xx, 389–98
Smith, Robert L. 419
Sober, Elliott 121, 156
Socrates 178, 236, 388
Sophocles 236
Soulé, M.E. 438
Snyder, Gary 165, 241, 446
Spinoza, Benedict xxiv, 187, 214, 394
Spretnak, Charlene 355
Squires, Radcliffe 366
Starhawk 342
Stevens, W.K. 439
Stevenson, C.L. 57
Stewart, R.D. 486
Stone, Christopher D. 47, 342, 511–26, 527, 528, 530, 531, 532, 533, 534, 535, 538–44 *passim*, 545, 547, 549, 553, 554–7 *passim*
Sylvan, Richard 530, 551

Tansley 302
Taylor, Harriet 283
Taylor, Paul W. xvi, xviii, 3–24, 25–31 *passim*, 122, 123, 144, 147, 148, 150, 224, 225, 226, 389, 391–3 *passim*, 400, 401, 406, 407, 529, 533; 535–7 *passim*, 542, 548, 569
Tennyson, Alfred 90
Thales 547
Thomas, Lewis 199
Thompson, T.S. 438
Thoreau, Henry D. 241, 341, 432
Tribe, L.H. 48
Tullock, Gordon 561
Tzu, Lao 244, 245

Ulanowicz, R. 438

Valentine, J.W. 97
van der Post, Laurens 34
Van De Veer, D. xxii
Varner, Gary E. 117–36, 534, 553–7
von Droste, B. 435
von Wright, G.H. 65, 66

Waide, J.B. 302
Waismann, F. 33
Warnock, G.J. 33
Warren, Karen J. 299–317, 319–40, 347

Warren, Mary Anne 532, 553, 534
Watkins, T.H. 440
Watts, Alan 394
Wenz, Peter 400, 401, 533, 534, 535, 537, 538, 540
Westman, W.E. 438
White, Lynn xvii
White, R.J. 441
Whittaker, R.H. 84, 301
Williams, Bernard 228, 229, 232, 565
Williams, George C. 87, 93, 94, 98, 149, 151, 154, 155
Wilson, Edward O. 92, 194, 530
Wittbecker, Alan 346, 351, 352, 354, 356, 357, 358, 359
Witthoft, J. 437

Wollstonecraft, Mary 279, 281, 283
Wordsworth, William 395
Worster, Donald 56–7, 152
Wright, A. 434
Wright, H.E. 453
Wright, Larry 41, 124, 126, 127, 147–9 *passim*, 150, 155
Wuerther, G. 440
Wynne-Edwards, V.C. 154, 155

Young, Iris 309

Zahnizer, Howard 431
Zapffe, Peter W. 193, 194, 206, 211, 217, 218
Zimmerman, Michael xxii, 346–50 *passim*

GE 42 .E845 1995

THE ETHICS OF THE

ENVIRONMENT

DATE DUE